Lecture Notes in Applied and Computational Mechanics

Volume 94

This series aims to report new developments in applied and computational mechanics—quickly, informally and at a high level. This includes the fields of fluid, solid and structural mechanics, dynamics and control, and related disciplines. The applied methods can be of analytical, numerical and computational nature. The series scope includes monographs, professional books, selected contributions from specialized conferences or workshops, edited volumes, as well as outstanding advanced textbooks.

Indexed by EI-Compendex, SCOPUS, Zentralblatt Math, Ulrich's, Current Mathematical Publications, Mathematical Reviews and MetaPress.

More information about this series at http://www.springer.com/series/4623

Takanori Hino · Frederick Stern ·
Lars Larsson · Michel Visonneau ·
Nobuyuki Hirata · Jin Kim
Editors

Numerical Ship Hydrodynamics

An Assessment of the Tokyo 2015 Workshop

Springer

Editors
Takanori Hino
Faculty of Engineering
Yokohama National University
Yokohama, Japan

Frederick Stern
IIHR – Hydroscience & Engineering
University of Iowa
Iowa City, IA, USA

Lars Larsson
Mechanics and Maritime Sciences
Chalmers University of Technology
Gothenburg, Sweden

Michel Visonneau
Fluid Mechanics Laboratory
Ecole Centrale de Nantes
Nantes, France

Nobuyuki Hirata
National Maritime Research Institute
Mitaka, Tokyo, Japan

Jin Kim
Korea Research Institute of Ships and Oc
Daejeon, Korea (Republic of)

ISSN 1613-7736 ISSN 1860-0816 (electronic)
Lecture Notes in Applied and Computational Mechanics
ISBN 978-3-030-47571-0 ISBN 978-3-030-47572-7 (eBook)
https://doi.org/10.1007/978-3-030-47572-7

This Springer imprint is published by the registered company Springer Nature Switzerland AG
The registered company address is: Gewerbestrasse 11, 6330 Cham, Switzerland

Preface

Since 1980, workshops on CFD in Ship Hydrodynamics have been held regularly. The main purpose of these workshops is to assess the state of the art in CFD for hydrodynamic applications. Active researchers in the field worldwide are invited to provide computed results for a number of well-specified test cases, and the organizers collect and present the results such that comparisons between different methods can be made easily. Detailed information about each method is also reported via a questionnaire provided by the organizers. All results are discussed at a meeting, and a final assessment of the workshop is made by the organizers.

The Tokyo 2015 Workshop attracted 36 groups from all over the world, and different types of computations were carried out for three hulls. It was the largest of the workshops in the series so far. All computed results were compiled in a volume, called Proceedings II, and distributed at the meeting, which was held in Tokyo in December 2015. The volume also includes short papers describing the computational methods and the results in more detail.

In the present book, in-depth evaluations of all computed results are presented. For some of the test cases, additional computations by the organizers are presented on topics of particular interest found at the meeting. All experimental data are reported, as well as a comprehensive set of new data is obtained after the workshop. The book has been written by the organizers and their co-workers. Supplementary materials are available for free at https://www.springer.com/gp/book/9783030475710 the book constitutes the final documentation of the Tokyo 2015 Workshop and gives a state-of-the-art assessment of the CFD capabilities within the area of Ship Hydrodynamics.

Yokohama, Japan — Takanori Hino
Mitaka, Japan — Nobuyuki Hirata
Iowa City, USA — Frederick Stern
Gothenburg, Sweden — Lars Larsson
Nantes, France — Michel Visonneau
Daejeon, South Korea — Jin Kim

February 2020

Contents

Introduction, Conclusions and Recommendations 1
Takanori Hino, Nobuyuki Hirata, Frederick Stern, Lars Larsson, Michel Visonneau, and Jin Kim

Experimental Data for JBC Resistance, Sinkage, Trim, Self-Propulsion Factors, Longitudinal Wave Cut and Detailed Flow with and without an Energy Saving Circular Duct 23
Nobuyuki Hirata, Hiroshi Kobayashi, Takanori Hino, Yasuyuki Toda, Moustafa Abdel-Maksoud, and Frederick Stern

Experimental Data for KCS Resistance, Sinkage, Trim, and Self-propulsion 53
Jin Kim

Experimental Data for KCS Added Resistance and ONRT Free Running Course Keeping/Speed Loss in Head and Oblique Waves 61
Yugo Sanada, Claus Simonsen, Janne Otzen, Hamid Sadat-Hosseini, Yasuyuki Toda, and Frederick Stern

Evaluation of Resistance, Sinkage, Trim and Wave Pattern Predictions for JBC 139
Lars Larsson

Analysis of the Local Flow around JBC 159
Michel Visonneau

Evaluation of Self-propulsion and Energy Saving Device Performance Predictions for JBC 279
Takanori Hino

Evaluation of Resistance, Sinkage, Trim and Self-propulsion Predictions for KCS 311
Jin Kim

Assessment of CFD for KCS Added Resistance and for ONRT Course Keeping/Speed Loss in Regular Head and Oblique Waves 333
Frederick Stern, Hamid Sadat-Hosseini, Timur Dogan, Matteo Diez, Dong Hwan Kim, Sungtek Park, and Yugo Sanada

Contributors

Moustafa Abdel-Maksoud Hamburg University of Technology, Hamburg, Germany

Matteo Diez CNR-INM, National Research Council-Institute of Marine Engineering, Rome, Italy

Timur Dogan University of Iowa and IIHR-Hydroscience & Engineering, Iowa City, IA, USA

Takanori Hino Yokohama National University, Yokohama, Japan

Nobuyuki Hirata National Maritime Research Institute, Mitaka, Japan

Dong Hwan Kim Korea Research Institute of Ocean Science & Technology, Daejeon, South Korea

Jin Kim Korea Research Institute of Ships and Ocean Engineering (KRISO), Daejeon, South Korea

Hiroshi Kobayashi National Maritime Research Institute, Mitaka, Japan

Lars Larsson Chalmers University of Technology, Gothenburg, Sweden

Janne Otzen FORCE Technology, Lyngby, Denmark

Sungtek Park University of Iowa and IIHR-Hydroscience & Engineering, Iowa City, IA, USA

Hamid Sadat-Hosseini University of North Texas, Denton, TX, USA

Yugo Sanada University of Iowa and IIHR-Hydroscience & Engineering, Iowa City, IA, USA

Claus Simonsen FORCE Technology, Lyngby, Denmark

Frederick Stern University of Iowa and Iowa Institute of Hydraulic Research (IIHR), Iowa City, IA, USA

Yasuyuki Toda Osaka University, Suita, Japan

Michel Visonneau CNRS/Centrale Nantes, Nantes, France

Introduction, Conclusions and Recommendations

Takanori Hino, Nobuyuki Hirata, Frederick Stern, Lars Larsson, Michel Visonneau, and Jin Kim

Abstract The Tokyo 2015 Workshop on CFD in Hydrodynamics was the seventh in a series started in 1980. The purpose of the Workshops is to regularly assess the state of the art in Numerical Hydrodynamics and to provide guidelines for further developments in the area. The 2015 Workshop offered 16 test cases for three ship hulls. A total of 36 participating groups of CFD specialists submitted their computed results during the fall of 2015. The results were compiled by the organizers and discussed at a meeting in Tokyo in December 2015. In this chapter the background and development of the Workshops since the start are presented. The three hulls used in the 2015 Workshop are introduced and the computations requested from the participants are specified. Based on a questionnaire sent to all participants the details of their CFD methods are listed, and finally the general conclusions from each chapter and recommendations for future Workshops are presented. The detailed results of the computations are discussed in subsequent Chapters.

T. Hino (✉)
Yokohama National University, Yokohama, Japan
e-mail: hino-takanori-nf@ynu.ac.jp

N. Hirata
National Maritime Research Institute, Mitaka, Japan

F. Stern
University of Iowa and Iowa Institute of Hydraulic Research (IIHR), Iowa City, IA, USA

L. Larsson
Chalmers University of Technology, Gothenburg, Sweden

M. Visonneau
CNRS/Centrale Nantes, Nantes, France

J. Kim
Korea Research Institute of Ships & Ocean Engineering (KRISO), Daejeon, South Korea

T. Hino et al. (eds.), *Numerical Ship Hydrodynamics*, Lecture Notes in Applied and Computational Mechanics 94, https://doi.org/10.1007/978-3-030-47572-7_1

1 Background

The history of CFD Workshops dates to 1980. That year, the first Workshop was held in Gothenburg, Sweden (Larsson 1981). The objective of the workshop was the assessment of up-to-date numerical methods for ship hydrodynamics to aid code development and guide industry.

Several test cases were set for viscous flows around ship hulls and the participants were requested to submit their numerical results. Through the comparison with experimental results both for integral variables such as resistance or self-propulsion factors and local flow quantities such as wake or pressure distributions, the state-of-the-art of ship CFD methods was evaluated. The objective and the style of the workshop remained the same in the following workshops, which were held approximately every five years between Gothenburg and Tokyo, i.e. at Gothenburg in 1990 (Larsson et al. 1991), (Larsson et al. 2002, 2003) and (Larsson et al. 2014) and at Tokyo in 1994 (Kodama et al. 1994) and (Hino 2005).

In the 35 years from the first workshop, computer technology has progressed exponentially, and available computing resources have increased several orders of magnitude. Numerical flow analysis methods have also evolved from boundary-layer-theory-based methods to Navier-Stokes methods and from resistance prediction to unsteady complex flow simulations with waves, propellers and ship motions.

In 2015, again at Tokyo, the seventh CFD workshop in the series was organized. The purpose of the workshop was the same as at the preceding workshops and it was to assess state-of-the-art of the current CFD codes for ship hydrodynamics. In the present workshop, three ship hulls, two of which are new, were selected and a total of 16 test cases were specified by the organizers. One of the new ship hulls is the Japan Bulk Carrier (JBC) which is a cape-size bulk carrier equipped with a stern duct as an energy saving device. The other new hull is the ONR Tumblehome model 5613 (ONRT) which is a preliminary design of a modern surface combatant. The third hull is the KRISO Container Ship (KCS) which is a 3,600 TEU container ship and has been used in the preceding workshops. Like in all previous workshops, participants were asked to provide computed results, information about the method used and a paper summarizing the computations. 36 groups submitted their computed results for one or more cases.

The present chapter gives a background to the Workshop, specifications of the hulls and test cases followed by the summary of the numerical methods the participants used. General conclusions and recommendations for future work are also included in this chapter. Chapter 2 is a report of the experimental data for JBC, including resistance and self-propulsion tests, local flow measurement using stereo particle image velocimetry (SPIV) with and without a stern duct and wave profiles. Chapter 3 is a summary of the resistance and self-propulsion tests of the KCS. Chapter 4 is devoted to the experimental data for added resistance in waves of the KCS and the ONRT hulls. Chapters 5–9 are the evaluation reports of the submitted data by the organizers. Chapters 5 and 7 are the evaluations of the resistance and self-propulsion simulations, respectively, for the JBC hull. Chapter 6 evaluates the local flow predictions for the JBC. Chapter 8 is the evaluations of resistance and self-propulsion simulations for the KCS. Chapter 9 is the evaluations of added resistance and course keeping simulations for the KCS and the ONRT. It should be noted that these reports are based solely on results submitted for evaluation at the Workshop. No results from submitted papers or from computations carried out afterwards are included except the additional analysis of the SPIV measurement in Chapter 2 (Figs. 17, 19-21, 35–38) and the vortex flow analysis in Chap. 6 (Figs. 50–53, 70–79).

Similarly with the previous Gothenburg 2010 Workshop, this book replaces Proceedings, Part 1 of the Workshop. The detailed results and papers by all participants are presented in the Proceedings, Part 2, distributed at the Workshop. These Proceedings can be downloaded from the web site SpringerExtras (for address, see the book cover) together with much other supplementary information, such as the spreadsheet with detailed information about all methods, discussions at the Workshop and some photos.

2 Hulls

The following three hulls were used in the Workshop:

(1) The JBC, Japan Bulk Carrier
(2) The KCS, KRISO Container Ship
(3) The ONRT, ONR Tumblehome model 5613.

The JBC (Japan Bulk Carrier) is a cape-size bulk carrier equipped with a stern duct as an energy saving device. National Maritime Research Institute (NMRI), Yokohama National University and the Ship Building Research Centre of Japan (SRC) were jointly involved in the design of a ship hull, a duct and a rudder. The hull design and measurements were conducted with the support of ClassNK as part of the ClassNK joint R&D for Industry Program (Hino et al. 2016). Towing tank experiments were carried out at NMRI (Hino et al. 2016) and Osaka University (Jufuku et al. 2015), including resistance tests, self-propulsion tests and SPIV measurements of stern flow fields. Additionally, LDV/SPIV measurement data in the wind tunnel are available from Technical University of Hamburg (Shevchuk et al. 2020).

The KCS (KRISO Container Ship) was conceived to provide data for both explication of flow physics and CFD validation for a modern container ship with a bulbous bow, and towing tank tests were carried out at Korea Research Institute of Ships and Ocean Engineering (KRISO) in Daejeon (Van et al. 1998, Kim et al. 2001). Self-propulsion tests were carried out at the Ship Research Institute (now NMRI) in Tokyo and are reported in the proceedings of Tokyo 2015 Workshop (Hino 2005). Later, resistance tests were also reported by NMRI (See Larsson et al. 2014). Data for pitch, heave, and added resistance are also available from Force/DMI measurements (Simonsen et al. 2008).

The ONRT (ONR Tumblehome model 5613) is a preliminary design of a modern surface combatant, which is publicly accessible for fundamental research. The 1/49

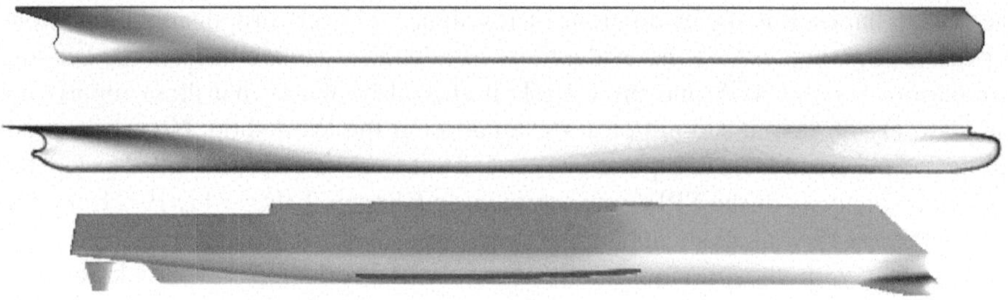

Fig. 1 The three ships used in the workshop (top: JBC; middle: KCS; bottom: ONRT)

scaled ship model is appended with skeg and bilge keels. The model has a wave piercing hull design with 10° tumblehome sides and a transom stern. The model also has rudders, shafts and propellers with propeller shaft brackets, but the superstructure is not attached. Free-running tests including course keeping, zig-zag and turning in calm water and in regular waves were performed at IIHR Hydraulics Wave Basin Facility (Sanada et al. 2013, 2019).

Side views of the three hulls are seen in Fig. 1 and the main particulars are given in Table 1. No full-scale ships exist for all three hulls.

The Cartesian coordinate system adopted in the workshop has its origin at the forward perpendicular and still water level, x is backwards, y to starboard and z vertically upwards, as shown in Fig. 2.

3 Test Cases

Several types of computations were requested, namely:

(1) Resistance, sinkage and trim either at zero sinkage and trim (denoted FX_0) or dynamic sinkage and trim ($FR_{z\theta}$)
(2) Local flow at fixed condition, either at FX_0 or $FR_{z\theta}$. In some cases FX_0 is simulated by pre-ballasting the hull to obtain zero sinkage and trim in the free condition at the correct Froude number (FR_0).
(3) Self-propulsion at FX_0, $FR_{z\theta}$ or FR_0
(4) Captive tow in waves either at $FR_{z\theta}$ or with free surge, heave, roll and pitch ($FR_{xz\Phi\theta}$)
(5) Free running with all degrees of freedom (FR_{all}).

All test cases for the three hulls are listed in Table 2. The measured data are provided by the organizations listed. A total of 16 cases are specified and the participants were free to select which cases to compute.

Table 1 Main particulars of the three ships

Main particulars (Full scale)		JBC	KCS	ONRT
Length between perpendiculars	L_{PP} (m)	280.0	230.0	154.0 (L_{WL})
Maximum beam of waterline	B (m)	45.0	32.2	18.78
Draft	T (m)	16.5	10.8	5.494
Displacement volume	∇ (m^3)	178369.9	52030.0	8507 ton
Wetted area of a bare hull	S_W (m^2)	19556.1	9424.0	
Wetted surface area of a rudder	S_R (m^2)	164.3	115.0	
Wetted surface area of ESD	S_{ESD} (m^2)	77.8	–	–
Block coefficient (C_B)	$\nabla/(L_{PP} \cdot B \cdot T)$	0.8580	0.6505	0.535 (L_{WL})
Propeller center, long. location(from FP)	x/L_{PP}	0.9846	0.9825	0.9267 (L_{WL})
Propeller center, vert. location (below WL)	$-z/L_{PP}$	0.0404214	0.02913	0.03565 (L_{WL})
Propeller center, lateral location (from CL)	y/L_{PP}	0	0	±0.02661 (L_{WL})
LCB (%L_{PP}), fwd+	–	2.5475	−1.48	79.52 m from FP
Vertical center of gravity(from keel)	KG (m)	13.29	14.35	7.63
Metacentric height	GM (m)	5.30	0.60	2.07
Moment of inertia	K_{xx}/B	0.40	0.40	0.444
Moment of inertia	K_{yy}/L_{PP}, K_{zz}/L_{PP}	0.25	0.25	0.25 (L_{WL})
Speed	U (knots)	14.5	24.0	15.1
Froude number	Fr	0.142	0.26	0.20

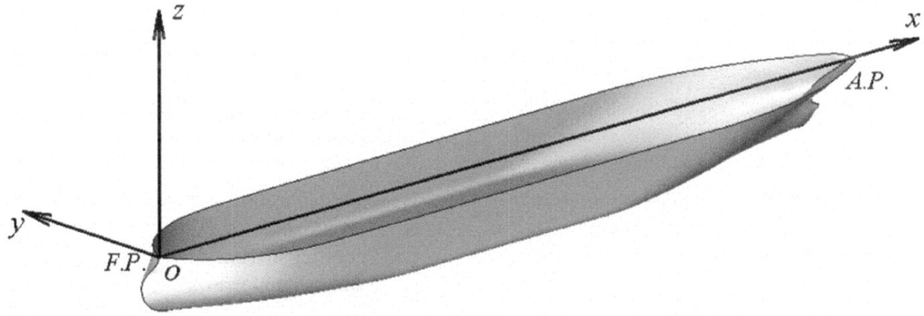

Fig. 2 Cartesian coordinate system

Table 2 Test cases

Case	Hull	Condition	Attitude	Validation variables	Data provider
1.1a	JBC w/o ESD	Towed in calm water	$FR_{z\theta}$	Resistance, sinkage and trim	NMRI
1.2a	JBC with ESD	Towed in calm water	$FR_{z\theta}$	Resistance, sinkage and trim	NMRI
1.3a	JBC w/o ESD	Towed in calm water	FRzθ	Mean velocities, turbulence and wave patterns	NMRI
1.3b	JBC w/o ESD	Double model in wind tunnel	FX0	Mean velocities	TUHH
1.4a	JBC with ESD	Towed in calm water	FRzθ	Mean velocities	NMRI
1.5a	JBC w/o ESD	Self-propelled in calm water	FRzθ	Thrust, torque, RPM or force balance and resistance	NMRI
1.6a	JBC with ESD	Self-propelled in calm water	FRzθ	Thrust, torque, RPM or force balance and resistance	NMRI
1.7a	JBC w/o ESD	Self-propelled in calm water	FRzθ	Mean velocities	NMRI
1.8a	JBC with ESD	Self-propelled in calm water	FRzθ	Mean velocities	NMRI
2.1	KCS	Towed in calm water	$FR_{z\theta}$	Resistance, sinkage and trim	KRISO/NMRI
2.5	KCS	Self-propelled in calm water	FR_0	Thrust, torque and propulsive coefficients	NMRI
2.7	KCS	Self-propelled in calm water	FR_0	Mean velocities	NMRI
2.10	KCS	Captive towed in head waves	$FR_{z\theta}$	Resistance, heave and pitch	FORCE
2.11	KCS	Captive towed in beam, follow and oblique waves	FRxzϕθ	Resistance, surge, heave, roll and pitch	IIHR
3.9	ONRT	Free running in calm water	FR_{all}	Thrust, torque, propulsive coefficients, sinkage and trim	IIHR
3.12	ONRT	Free running in head waves	FR_{all}	Thrust, torque, rpm, motions and trajectory	IIHR

(continued)

Table 2 (continued)

Case	Hull	Condition	Attitude	Validation variables	Data provider
3.13	ONRT	Free running in beam, follow and oblique waves	FR_{all}	Thrust, torque, rpm, motions and trajectory	IIHR

4 Participants and Methods

The workshop participants together with the main features of their methods are listed in Table 3. In the first column the acronym of the participating group is given. This is used in combination with the code name of column three to identify each submission. The cases computed are given in column two. In the remaining columns the features of each method are given.

Types of the codes used and their numbers are in-house codes (12), open-source codes (11) and commercial codes (13). In the previous Workshop, they were in-house (14), open-source (3) and commercial (16). The number of open-source code users has increased considerably.

For turbulence models, the majority of methods use two-equation models, k-ω SST or k-ε. There are also some one-equation models, either Spalart-Allmaras or Menter. The anisotropic models are either of the algebraic stress or Reynolds stress type. Note that there are also some LES/DES methods.

Most of the participants use no-slip wall boundary conditions, but there are also several methods, particularly for open-source and commercial codes, with wall functions, both with and without pressure gradient corrections. The Volume of Fluid (VOF) technique is the most popular one for the free-surface modeling, but there are also several level set (LS) methods. There is only one entry with surface fitting approach. The propeller is represented either as an actual rotating propeller or through a body force approximation.

Most simulations were performed using finite volume discretization but there are a few methods which adopt the finite difference or finite element method. For spatial accuracy, 2nd order accurate schemes were used in most codes and limited studies used 3rd or 4th order schemes. Time accuracy is either 1st or 2nd order. The most common grids were unstructured. Multi-block structured or overlapping structured grids were also used to some extent. Most methods are pressure based but there are also several solving the equations directly or with an artificial compressibility.

A complete specification of each method and application is given in the supplementary material on SpringerExtras based on the replies to a questionnaire answered by all participating groups.

Table 3 Workshop participants and methods

Organization	Cases submitted	Code	Turbulence model	Wall model	Free surface	Propeller	Grid type	Spatial disc.	Spatial accuracy	Time accuracy	Vel-press coupling
ABS	1.1a-8a 3.9 3.12 3.13	OpenFOAM	SST	W	VOF	A	OS, U	FV	2	2	PR
Chalmers/FLOWTECH	1.1a-8a	SHIPFLOW5.2	EASM	N		BL	OS	FV	M		AC
CNR-INSEAN	1.3a-8a	Xnavis	SA	N	LS	BX,A	OS	FV	2	2	AC
CSSRC	1.5a 1.6a 3.9 3.12	FLUENT	RNG-KE	W	VOF	A	MS + MU	FV	2	2	PR
CTO	2.1 2.5	STAR-CCM+	2E		VOF		U	FV	2	1	PR
Damen	2.1	FINE/Marine	SST	N	VOF		U	FV	2	2	PR
ECN/CNRS	1.1a-8a 1.3b 3.9 3.12 3.13	ISIS-CFD	SST, EASM	W	VOF	BX, A	U	FV	2	2	PR
FOI	1.3b	OpenFOAM	LES	N, W	N/A		U	FV	2	2	PR
FORCE	2.1	StarCCM+	k-ε/k-ω SST	N	VOF		MU	FV	2	2	PR
HHI	1.1a 1.2a 1.5a 1.6a	HiFoam	Realizable-KE	N	VOF	A	U	FV	2	1	PR
HHI	1.1a 1.2a 1.5a 1.6a 2.1 2.5 3.9 3.12	STAR-CCM+	RS	W	VOF	A	U	FV	2	1	PR
HSVA	1.1a-8a 2.10	FreSCo+	2E	W, WO	VOF	BV, BX	U	FV	2	2	PR

(continued)

Table 3 (continued)

Organization	Cases submitted	Code	Turbulence model	Wall model	Free surface	Propeller	Grid type	Spatial disc.	Spatial accuracy	Time accuracy	Vel-press coupling
IIHR	2.10 2.11 3.9 3.12 3.13	CFDShip-Iowa	Hybrid	N	LS	BX	OS	FD	4	2	PR
IIHR	1.5a-8a 3.9 3.12 3.13	REX	Hybrid	N	LS	A	OS	FD/FV	4	2	PR
KRISO	1.1a-8a 2.1-7 2.10	WAVIS	Realizable-KE,EARSM		LS	BS	MS	FV	3	1	PR
MARIC	1.3a-6a	ISIS-CFD	SST	N	VOF	BX	U	FV	2	2	PR
MARIN	2.5	PARNASSOS	Menters 1E model with Dacles-Mariani correction	N	Fitting	BS	MS	FD	2		D
MARIN	1.1a-8a 2.10	ReFRESCO	SST	N	VOF	A	U	FV	2	1	PR
MHI	1.1a-4a	Fluent v14.5	RNG-KE, SST	N	VOF		U	FV	2		PR
MIJAC	1.1a 1.2a 1.5a 1.6a	OpenFOAM v2.3.0	SST	N	N/A	BX	U	FV	2	N/A	PR
NMRI	1.1a-8a	NAGISA	EASM	N, WO	LS	BX	OS	FV	3	1	A
NUMECA	2.1 2.10	ISIS-CFD	SST	WO	VOF		U	FV	2	2	PR
PNU	1.3a 1.4a 1.7a 1.8a 2.1 2.5	FLUENTv15	2E	W	VOF	A	U	FV	2	2	PR

(continued)

Table 3 (continued)

Organization	Cases submitted	Code	Turbulence model	Wall model	Free surface	Propeller	Grid type	Spatial disc.	Spatial accuracy	Time accuracy	Vel-press coupling
SHIME-CFD	1.1a 1.2a 1.5a 1.6a	OpenFOAM	SST	N	N/A		MU	FV	2	2	PR
SJTU	1.1a 1.2a 1.5a 1.6a 2.1 3.9	naoe-FOAM-SJTU	2E	N	VOF	BX, A	MU	FV	M	2	PR
SRC	1.3b	FrontFlow/blue	LES	N	VOF		U	FE	2	2	C
Southampton	1.3a 1.4a 1.7a 1.8a	OpenFOAM	SST	N	N/A	BP	MS	FV	2	N/A	PR
Southampton	1.3a 1.4a 1.7a	Star CCM+	SST	N	N/A	BP	MS	FV	2	N/A	PR
UDE	2.1	ISIS-CFD	SST	N	VOF		U	FV	2	2	PR
UDE	2.1	OpenFOAM	2E	N, W	VOF		U	FV	2	2	PR
ISMT/UDE	1.5a 1.6a	STAR-CCM+	SST	N	VOF	A	U	FV	2	2	PR
UM	2.7	OF23x	SST	WO	VOF	A	U	FV	2	1	PR
LeMoS/Uni Rostock	1.3a 1.7a	OpenFOAM	SST/LES	N	N/A	A	S	FV	2	2	PR
UNIZAG-FSB	2.1	navalFoam	SST	N	VOF		U	FV	M	1	PR
UNIZAG-FSB	2.10 2.11	swenseFoam	SST	N	LS		U	FV	M	2, 1	PR
YNU	1.1a-8a	SURFv7	EASM	N	N/A	BX	OS	FV	2	1, 2	AC

A–Actual propeller, BL–Body force propeller (Lifting line), BP–Body Force Propeller (Prescribed), BS–Body force propeller (lifting surface), BV–Body force propeller (Vortex lattice), BX–Body force propeller (Other), AC–Artificial compressibility; C–Compressibility with low Mach number assumption, D–Direct method; FD–Finite difference; FV–Finite volume; MS–Multiblock structured; MU–Multiblock unstructured; N–No slip; OS–Overlapping structured; PR–Pressure correction; S–Single block structured; U–Unstructured; W/WO–Wall functions with/without pressure gradient correction

5 Conclusions

The analysis of the experimental and computational data from the Tokyo 2015 Workshop on CFD in Hydrodynamics are given in the following chapters. The summary of the main conclusions of each chapter is shown below. The detailed discussions are included in each chapter. Some chapters provide the extensive conclusions as well.

Chapter 2 Experimental Data for JBC Resistance, Sinkage, Trim, Self-Propulsion Factors, Longitudinal Wave Cut and Detailed Flow with and without an Energy Saving Circular Duct
In this Chapter, the measured data for Japan Bulk Carrier (JBC) with an energy saving circular duct are presented. The measurement is conducted in three facilities, i.e., the towing tanks of National Maritime Research Institute (NMRI) and Osaka University (OU) and the wind tunnel of Technical University of Hamburg (TUHH).

- Resistance, sinkage, trim, self-propulsion factors with and without the duct are acquired by the tank tests at NMRI and OU. Wave field data is measured at the NMRI towing tank.
- The detailed flows fields in seven stations in a stern region are measured by using SPIV in the tanks of NMRI and OU. The data are acquired in towed and self-propelled conditions without and with the duct. In addition, LDV/SPIV measurement is conducted at the TUHH wind tunnel.
- For the SPIV measurement at NMRI, the error estimates are carried out using the uniform flow test results and the statistical convergence analysis of the actual measurement. It turned out that uncertainty of the velocity measurement is approximately 2–3% of the uniform flow and the uncertainty of TKE measurement is approximately 1% of the square of the uniform flow magnitude. Uncertainty of v velocity is largest and this may be attributed to the problem with the reflection mirror setting.
- For the mean velocity components, the data of three facilities are generally in good accordance in spite of some problems such as the laser reflection in TUHH measurement or the reflection mirror problem in NMRI measurement.
- On the other hand, it is found that there is a large difference in the measured TKE levels between NMRI and OU/TUHH. Examination of the statistical convergence of NMRI measurement shows that the difference between NMRI and other facilities does not seem to come from the statistical convergence, though apparently the more frames are needed for the accurate estimation of turbulent quantities. The effect of Reynolds number difference is also investigated for the local TKE distributions along the horizontal lines at the shaft height near the propeller plane. The distributions in the bare-hull towing condition are compared in the various scaling. It appears that the TKE distributions look more wake-like rather than boundary-layer-like from the comparisons with the flat plate data. However, it is rather difficult to specify the exact reason for differences of the extremely higher TKE of NMRI data. Further investigations, such as additional measurements in

different facilities and/or different measuring systems, are desirable for obtaining the reasonable distributions of TKE.

- Finally, the recommendations to the future workshops at present are as follows: For the resistance and self-propulsion, sinkage and trim and the wave profiles, the measured data from NMRI can be considered to be appropriate. For the local flow data, the OU data seems to have no serious deficiencies and thus can be recommended as the reliable measured data of mean velocities. Further works will be needed to establish reliable turbulence data of this JBC case.

Chapter 3 Experimental Data for KCS Resistance, Sinkage, Trim, and Self-propulsion

- Additional experimental data for KCS calm water resistance obtained after the workshop G2010 are reported. All KCS calm water models are listed at Table 1 in Chap. 4. In the present chapter, all experimental data are obtained with 7.3 m model from the KRISO and NMRI towing tanks. These experimental data have been widely used as a standard benchmarking case for CFD validation after G2010. However, turbulence quantities are still missing in the set of experimental data and we also feel inconvenienced by the lack of local flow measurement for self-propulsion including a rudder. It is desirable to have more data to make up the full set of experimental data, especially for the standard size KCS model ship.

Chapter 4 Experimental Data for KCS Added Resistance and ONRT Free Running Course Keeping/Speed Loss in Head and Oblique Waves

- Evaluation is performed of the data used for T2015 test cases for the KCS captive added resistance σ_{AR} and ONRT free running course keeping/speed loss in head and oblique waves.
- For KCS calm water resistance, the individual facility N-order level testing uncertainty is $U_{Xi} = 1\%D$ and the multiple facility standard deviation based on three institutes using 2 model sizes ($L = 7.3$ and 6.1 m) is $SD = 0.74\%D$, such that the individual facility MxN-order level testing uncertainty is $U_{Di} = 1.75\%D$. U_{Xi} was not reported for sinkage and trim; however, the $SD = 4$ and $7\%D$, respectively. D corresponds to either the individual or mean facility, as implied by the context of the discussion.
- For KCS head waves, the analysis was based on three institutes with model sizes $L = 6.1$, 3.2 and 2.7 m. For the 6.1 m model, FORCE provided $U_{Xi} = 8$, 4 and 4% D, respectively, for σ_{AR} and first harmonic heave z_1/ζ_1 and pitch $\theta_1/\zeta_1 k$ amplitudes, whereas for the 2.7 m model, FORCE/IIHR provided $U_{Xi} = 7/18$, 8/2 and 9/5% D, respectively. In consideration of the differences in model sizes and rigid vs. surge free mounts the agreement between the facilities is reasonable: for the primary variables, $SD = 3$, 20 and 10% D for z_1/ζ_1 and $\theta_1/\zeta_1 k$ amplitudes and σ_{AR}, respectively; and for secondary variable $SD = 66\%$ D for first harmonic resistance C_{T1}.

- For KCS oblique waves, the data is only available from IIHR for tests in 2015 and 2016. The agreement is good, i.e., SD values are comparable to the corresponding values for head waves: for primary variables, SD = 6 and 63%D for z_1/ζ_1 and $\theta_1/\zeta_1 k$ amplitudes and σ_{AR}, respectively; and for secondary variable SD = 13%D for C_{T1}.
- For ONRT self-propulsion and head and oblique waves, the data is only available from IIHR and only preliminary uncertainty analysis is available.
- The evaluation showed reasonably reliable data for both KCS and ONRT. However, clearly it is desirable to have data from more facilities including uncertainty analysis for assessment of facility biases, which will provide more robust data and uncertainty analysis for CFD validation.

Chapter 5 Evaluation of Resistance, Sinkage, Trim and Wave Pattern Predictions for JBC

The Workshop submissions for JBC contained four cases: the hull with or without an ESD both in towing and self-propulsion. Altogether 88 predictions were provided. In this chapter, resistance, sinkage and trim, as well as wave pattern results are evaluated. Conclusions are as follows:

- Like at the previous workshops the mean signed comparison error for resistance is very small, around 1%, which is also the accuracy reported for the experimental data. However, more interesting is the mean absolute comparison error, since this represents the average deviation from the data. This is around 2% for the towed cases and 3% for self-propulsion. The scatter is represented by the standard deviation and is about 2% and 4% respectively for the two conditions. In the 2005 and 2010 Workshops the standard deviations were 6% and 1%. There was thus a significant reduction in scatter between 2005 and 2010, but no further reduction in 2015.
- A resistance reduction around 1% is predicted with the ESD fitted. This is confirmed by the measurements.
- For the comparison error to be smaller than 4% the grids have to contain more than 10 M cells. In 2010 the corresponding limit was 3 M cells. Thus, more grid cells were required for the same accuracy in 2015.
- There is hardly any influence of the grid type on the predicted resistance. If any, there is a slight advantage of the structured grids, compared to hex-dominated and completely unstructured grids.
- Like in 2010 the two-equation turbulence models predict better results than the more advanced models. The k-ε and k-ω models have mean signed comparison errors less than 1% and mean absolute errors just below 2%. EASM has gained in popularity and its mean absolute error is now 2%.
- Resistance predictions using wall-functions are more accurate than those with wall resolved flows. Both the absolute error and the standard deviation are about twice as large with wall resolution.
- Grid convergence is demonstrated for 64% of the unstructured grids and 77% of the structured grids. The achieved order of accuracy in the grid refinement studies

is in many cases far from the theoretical accuracy. Most methods predict a too high order of accuracy, 2.3 times the theoretical one, on the average. For those with a too low order of accuracy the mean value is 0.7 times the nominal one.

- As stated above, the mean absolute comparison error in all resistance predictions is 2%, which is considerably smaller than the mean validation uncertainty, 5%, composed of the mean numerical uncertainty of 4.9% and the stated experimental accuracy of 1%. The methods may thus be considered validated in a mean sense.
- The comparison errors for sinkage were very large in 2010, but are now reduced considerably. The mean signed error, absolute error and standard deviation are now -1.9%, 4.7% and 5.6% respectively.
- For trim the corresponding errors are also reduced considerably and are now 0.6%, 3.1% and 3.5%, respectively.
- Very accurate predictions are reported for the wave profile along the hull, while less correspondence with measurements is demonstrated for two wave cuts. There is a consistent phase shift in all results, which may indicate an error in the data. A small over prediction of the bow wave height is also noted in all results.

Chapter 6 Analysis of the Local Flow around JBC
In this chapter, we have reviewed computations of the flow around the Japanese Bulk Carrier (JBC) provided by the contributors to the Tokyo T2015 workshop from the standpoint of the local flow analysis. Five flow configurations were studied: (i) the naked hull without propeller nor duct (case 1–3a), (ii) the double body model (case 1–3b), (iii) the hull with a duct but without propeller (case 1–4), (iv) the hull with a propeller and without duct (case 1–7) and finally, (v) the hull with a ducted propeller (case 1–8). This chapter aimed at understanding the physics of the flow from a local point of view with the help of the flow measurements performed by NMRI at three stations S2, S4 and S7. But several additional objectives were pursued during this analysis. The first one, in the continuity of the G2010 Gothenburg workshop concerned the verification and validation of stern flows over the JBC hull. This hull is characterised by a relatively high block coefficient, which means that the flow is more complex in the stern region. An intense bilge vortex was detected and some questions arose about the steadiness of the flow in the stern region. It was therefore decided to focus the study on the bilge vortex core and try to compare the various contributions in this specific region. The second objective was to understand the mechanisms which might improve the propulsive efficiency when a ship is equipped with a ducted propeller. To separate physical effects, several configurations with and without duct combined with the presence or absence of a propeller were studied with the help of the measurements made by NMRI at stations S2, S4 and S7. This gave us the opportunity (i) to compare the various turbulence closures in presence of a duct or a propeller, (ii) to assess the reliability of the various propeller representations implemented by the contributors, and (iii) to check if the observed influence of the duct on the local flow is accurately captured by the computations. The main conclusions that stem from all these studies are summarized as follows

- For the case 1–3a, since the typical grids which are used by the contributors (5-10 M points) are fine enough, the influence of the grid discretisation is moderate for RANSE, which means that the numerical error appears to be under control. It was feasible therefore to focus the study on the influence of the turbulence closures and see if it was possible to draw general conclusions form all the contributions. As noticed during the Gothenburg 2010 workshop for the analysis of the local flow around the KVLCC2, the major influence comes from the turbulence closure. One noticed that the linear isotropic closures significantly under-predict the longitudinal vorticity while full RSM closures slightly over-predict it. Non-linear anisotropic closures (EARSM) seemed to offer a good compromise from the standpoint of the local flow although they slightly underpredict the vorticity at the key station S2. Contributions employing hybrid RANS-LES were also taken into account. The results appeared promising although IDDES seemed to over-predict the vorticity again. No spectacular advantage was noticed compared to the best RANS models at the measurement stations.
- This study was completed by a local core vortex analysis which provided a first interesting attempt to carry out a more local analysis of the time-averaged bilge vortex. Globally, the agreement between most of the computations and the rather coarse experiments was satisfactory in terms of global trends. The longitudinal vorticity was somewhat underestimated and the longitudinal velocity distribution was fairly reproduced. These experiments pointed out very large differences on the turbulence kinetic energy and this was the most striking (and unexpected) result of this comparison. Only the hybrid RANS/LES closures were able to reproduce this characteristic, thanks to the contribution of the resolved turbulence kinetic energy. Whether the NMRI tke measurements are reliable or not to provide an accurate turbulence kinetic energy at this location is still a matter of intense debate which is partly addressed in Chap. 2 of this book. Further much more detailed measurements will be necessary to draw safer conclusions.
- In the case 1–3b, the turbulence closure study led to very similar conclusions; i.e. that the major influence on the computed distribution of isowakes in the stern region comes from the turbulence closure. Linear isotropic closures under-predict the longitudinal vorticity at station S2 while full RSM closures tend to overpredict it at the same station. But what was remarkable and quite unique was that two LES computations were presented and especially, one from SRC on an extremely fine grid (4 billion cells) which was in good agreement with measurements and predicted the level of turbulence kinetic energy measured by NMRI in the core of the bilge vortex (at least three times higher than what was modeled by EASM). Unfortunately, it was not possible to exploit further the results of the (almost) wall-resolved LES simulation, due to the unavailability of this flow database.
- For the case 1–4, it was again noticed that the turbulence models have a critical role to play once the grid is fine enough. The main modifications of the flow created by the presence of the duct were correctly reproduced by linear isotropic models which provided the correct qualitative trend. A better quantitative agreement was reached by anisotropic closures although the inner part of the vortex was not

perfectly reproduced as illustrated by the occurrence of a reversed flow region which was over-estimated by both RSTM and EARSM turbulence closures.

- For the test case 1–7, it was observed that the computations based on the simulation of the actual propeller were the only ones able to represent all the characteristics of the flow in the wake of propeller, specifically at station S7. Various codes using different meshes and different turbulence models yielded globally comparable results in terms of isowake distribution when the real rotating propeller is taken into account in the simulation. However, alternate coupling strategies based on simplified propeller models appeared to produce a local flow which was quite satisfactory for some of them, even if they were not able to capture all the characteristics of the wake flow. Hybrid RANS/LES model used by IIHR/Rex yielded very good results but not necessarily way superior for the averaged flow field to what is obtained with usual $k-\omega$-like models to justify the very high cost associated with this time-accurate approach. However, this hybrid RANS/LES model was the only one able to capture most of the details of the flow revealed by the phase-averaged measurements.
- For the test case 1–8, the computations confirmed the fact that when a duct is mounted before the propeller, the flow in the wake of the propeller is more homogeneous. This influence was correctly captured by most of the contributors, whatever the turbulence model or propeller representation used. As expected, the isowake distribution behind the ducted propeller was better represented when the actual rotating propeller was accounted for. However, some simplified propeller models worked reasonably well and were able to indicate the right trend concerning the influence of the duct. RANS models provided a reasonably accurate prediction of the time-averaged flow behind the propeller. For phase-averaged quantities, some URANS-based computations were in fair agreement with the experiments while some others did not exhibit any unsteadiness in the wake of the propeller, probably due to the combined influence of a locally too coarse grid and a too diffusive turbulence model. From that standpoint, the hybrid RANS/LES model appeared again more consistent with the physics of the flow.

Chapter 7 Evaluation of Self-propulsion and Energy Saving Device Performance Predictions for JBC

Conclusions from the analysis of submitted data for JBC self-propulsion cases are summarized as follows:

- Grid uncertainty U_G in general is about 4–5% S for K_T and K_Q.
- Averages of absolute comparison errors of K_T and K_Q are 3–4% D for a bare hull case and 2–3% D with ESD case. ESD case is slightly smaller than a bare hull case. Actual propeller models give slightly better results than body force models though the difference is smaller in case of a bare hull.
- Averages of absolute comparison errors of self-propulsion factors, $1-t$ and $1-w$ are 7–10% D in a bare hull case, which are considerably larger than the errors of thrust and torque. This is due to the increased number of quantities involved to

estimate the factors. Errors become smaller with the presence of a duct, maybe due to the smoothening of stern flow fields by a duct.

- Average absolute errors of model scale delivered power (*DP*) is in the range of 5–6%, about the same level of other quantities. However, the estimated *DP* reduction rates due to a duct scatter from 0.88 to 1.0, whereas the experimental value is 0.94. The current accuracy of CFD estimation of ESD performance seems to be reasonably well in terms of the magnitude, although this may not be sufficient for design/performance prediction of an energy saving duct which aims the reduction rate of a few percent.

Chapter 8 Evaluation of Resistance, Sinkage, Trim and Self-propulsion Predictions for KCS

- For the predictions of resistance, sinkage and trim, the number of grid points are varied from 1.1 M to 6.6 M. The submitted results do not show large dependency of grid points. It means that CFD users who have efficient grid making technique can have an acceptable simulation results for low block coefficient hull form such as KCS.
- Most participants use two equation turbulence model and only few participants use an advanced turbulence model. Based on these submissions, there is no visible improvement in accuracy of the resistance prediction for more advanced turbulence model than the two equation models.
- The comparison error for resistance prediction at the design speed ($Fr = 0.26$) is -0.2% of the data and the standard deviation is 1.5% of the data value. The mean comparison error for all 6 speeds is 0.43% and the mean standard deviation is 2.48%. The increased error and standard deviation are caused by the results for low Fr simulation submissions. In the meanwhile, the absolute error $|E|$ is also analyzed. The mean absolute error $|E|$ is 2%, which is slightly larger than 1.64% the results of Gothenburg 2010.
- The comparison errors and standard deviation of the sinkage and trim are much larger than for the resistance. The errors come from the results in a speed range below $Fr = 0.2$. Two possible reasons can be considered. This range generates very short waves and so it requires better grid distribution to resolve the waves. Other problems are likely to be due to the difficulties of measuring two quantities at low speed. The data might have larger uncertainties than for high speed.
- Self-propulsion submissions are slightly improved than the results of Gothenburg 2010 Workshop. The mean comparison errors of K_T and K_Q are 0.5% and $-$3.5% respectively (-0.6% and -4.6% for Gothenburg 2010 workshop). The standard deviations for of K_T and K_Q are 2.7% and 2.4% respectively, which are reduced the results of 7.2% and 6.1% respectively for Gothenburg 2010 workshop. However, K_Q is still over predicted than predictions of the resistance. Self-propulsion parameters are slightly better predicted by body force models (FD fixed) but local flow characteristics are comparatively well predicted by actual propeller rotating simulation (rps fixed).

- For the future workshops, new experimental data for KCS or other container hull would be recommended to simulate not only captive but free running 6DoF calm water self-propulsion condition.

Chapter 9 Assessment of CFD for KCS Added Resistance and for ONRT Course Keeping/Speed Loss in Regular Head and Oblique Waves

- CFD is assessed for added resistance for KCS (captive test cases 2.10 and 2.11) and course keeping/speed loss for ONRT (free running test cases 3.9/3.12/3.13) in head and oblique waves. The number of submissions were 10, 2, and 8 for test cases 2.10, 2.11, and 3.9/3.12/3.13, respectively.
- The assessment approach uses both solution and N-version validation. The former considers whether the absolute error $|E_i| = |D-S_i|$ is less, equal or greater than the validation uncertainty, which is the root sum square of the numerical and experimental uncertainties, i.e., $|E_i| \leq U_{V_i} = \sqrt{U_{SN_i}^2 + U_D^2}$. The latter considers whether the absolute error is less, equal or greater than the state-of-the-art SoA_i uncertainty, i.e., $|E_i| \leq U_{SoA_i} = \sqrt{U_{V_i}^2 + P_{|E_i|}^2}$ where $P_{|E_i|} = k\sigma_{|E|}$ is the uncertainty due to the scatter in the solution absolute error. Errors and uncertainties are normalized using both the data value D and its dynamic range DR. The analysis uses $k = 2$.
- The captive resistance C_T and free running self-propulsion propeller revolutions RPS $|\overline{E}|$ <2% D with U_D less than but comparable $P_{|Ei|}$ < 3% D such that only 3/1 solutions were validated but 8/4 codes/solutions were N-version validated for C_T/RPS.
- The head waves captive and free running heave and pitch $|\overline{E}|$ is less than 8% DR with U_D less than 5% DR and $P_{|Ei|}$ less than 13% DR such that about 5 for captive and 2 for free running solutions were validated and about 7 for captive and 5 for free running codes/solutions were N-version validated. The errors for added resistance and speed loss were less than 13% DR with U_D less than 7% DR and $P_{|Ei|}$ = 25 and 13% DR such that about 4 for captive and 1 for free running solutions were validated and 7 for captive 4 for free running codes/solutions were N-version validated. For captive head waves, the errors and scatter are smaller than those for potential flow.
- The oblique waves captive and free running motion errors $\overline{|E|}$ are less than 10% DR except for roll with U_D large 23% DR for captive pitch and other wise small <2% DR. The errors for added resistance and speed loss were <8% DR with U_D < 9% DR. The largest errors were for roll, which had errors for captive of 12% DR and for free running of 20% DR.
- None of the KCS and ONRT test cases were able to achieve a programmatic requirement of 5% D and 5% DR for calm water and waves, respectively. Note that not all the experimental uncertainties are able to meet this requirement either such that both experiments and CFD need to reduce their uncertainties and in addition CFD its errors. Nonetheless, in view the comparable CFD capability for

ONRT free running vs. KCS captive conditions, the prognosis for CFD capability is excellent.
- Future CFD assessment should extend the current test cases for consideration of added power in regular and irregular waves. Verification should be required, and more limited/most important validation variables should be confirmed and used for the V&V and SoA assessment.

6 Future Workshops

The Tokyo 2015 Workshop was the seventh in a series started in 1980. Since 1990 the workshops have been held essentially every 5 years. This is a period long enough to allow significant progress to take place, but short enough to enable an evaluation of the new developments without undue delay. Therefore, the 5-year period should be maintained in principle. Following the selection of the organizer and venue of the next workshop by the previous Steering Committee, the new Steering Committee for the CFD Workshops in Ship Hydrodynamics was formed in April 2016. Its missions are to define the test cases and to coordinate campaigns to obtain more data and to support the local Organizing Committee in the organization of each workshop. The members of the Steering Committee are the organizer of the next workshop (presently Dr. Serge Toxopeus, MARIN), area representatives of the USA (presently Prof. Frederick Stern, IIHR), Europe (presently Dr. Michel Visonneau, ECN/CNRS) and Asia (presently Prof. Takanori Hino, YNU, Dr. Jin Kim, KRISO and Prof. Decheng Wan, SJTU). The present Steering Committee is chaired by Prof. Stern.

The date and venue of the next workshop are tentatively set to Fall of 2021 at the Maritime Research Institute Netherland (MARIN), Wageningen, Netherland. Test cases are now under development, but the current plan is as follows. Some test cases of the previous workshop will remain such as JBC or other hull for turbulence modeling assessment, but there will be new cases of breaking bow waves for air-water interface flow modeling assessment, free running added power in regular and irregular head and oblique waves for motions and hull-propeller-rudder interaction modeling assessment and full-scale simulations for scale effects and "real-world" vs. "laboratory" modeling assessment.

7 Closing Remarks

The workshops clearly document the progression of prediction capability for viscous ship hydrodynamics from integral and differential boundary layer to current CFD methods with test case complexity progressing from double body resistance and local flow to captive calm water resistance, sinkage and trim, self-propulsion and local flow to captive added resistance and ship motions in head and oblique waves to free running self-propulsion and speed loss in head and oblique waves. The number of participants/submissions has increased as has the use, i.e., number of research-institute vs. commercial and open-source codes submissions. The assessment methods have also progressed from consideration of errors only to quantitative verification and validation.

The original workshop objectives were to provide guidance for code developers which were mostly research institutes that were also members of ITTC in support of institute prediction capability and to assess the SoA similarly as ITTC committee reports did for experimental ship hydrodynamics. However, with increased submissions using commercial and open source codes the assessment should include guidelines here again with similarity to the more recent ITTC focus on its Quality Manual; however, such an effort would require resources beyond those currently available without financial support, which is also needed to support the procurement of reliable benchmark validation data and to support the host costs. The organization of the CFD workshops has also progressed from several interested persons to the present CFD Workshop Organizing Committee with representatives from US, Europe and Asia. The documentation of the CFD Workshops progressed from institute reports to proceedings to journal articles to most recently books. The planning of the CFD Workshop extends over the period between workshops which has been 5 or 6 years.

The advent of ever more complex test cases and increased number of submissions (especially using commercial and open-source codes rather than research-institute codes) puts greater burden on the Organizing Committee for procurement of benchmark data, preparation of submission instructions, data reduction for CFD assessment, more emphasis on SoA assessment and the need for guidelines, all of which affect the motivation of the Organizing Committee members.

In view of the progress of both CFD and the CFD Workshops, the following recommendations are made:

1. Experiments including uncertainty analysis should be conducted by multiple facilities for each test case to establish reliable benchmark validation data, including facility biases and scale effects.
2. Fewer and only challenging test cases should be used for the assessment.
3. Fewer and only primary validation variables should be used for the assessment.
4. All test cases should uniformly use the established verification and validation methods.
5. Improved submission instructions are needed that enable automated data reduction that the organizers can more easily use for the assessments.
6. If financial support can be obtained, the Organizing Committee should be formalized; ideally with ability for certification of CFD performance capability.
7. The Organizing Committee should liaison with and leverage the resources of the ITTC but should continue as voluntary and retain its independence so as not to be influenced by commercial interests.

Acknowledgements The workshop was organized by the International Steering Committee of six members, i.e. the authors of this book and also by the local organizers at National Maritime Research

Institute, Japan: Dr. Hiroshi Kobayashi, Dr. Yusuke Tahara, Dr. Kunihide Ohashi and Dr. Naoyuki Onodera who significantly contributed to the success of the workshop. Finally, the great efforts by all workshop participants in the preparation and delivery of all computed results are gratefully acknowledged.

References

Hino, T. (Ed.) (2005). Proceedings of CFD workshop Tokyo 2005, NMRI report.

Hino, T., Hirata, N., Ohashi, K., Toda, Y., Zhu, T., Makino, K., Takai, M., Nishigaki, M., Kimura, K., Anda, M., & Shingo, S. (2016). Hull form design and flow measurements of a bulk carrier with an energy-saving device for CFD validations. In: Proc. 13th International Symposium on Practical Design of Ships and Other Floating Structures (PRADS' 2016), Copenhagen, Denmark.

Jufuku, N., Hori, M., Itou, S., Toda, Y., & Hinatsu, M. (2015). SPIV stern flow measurement around operating propeller for with and without duct condition of japan bulk carrier (in Japanese). *Conference Proceedings of the Japan Society of Naval Architects and Ocean Engineers, 21,* 309–312.

Kim, W. J., Van, D. H., & Kim, D. H. (2001). Measurement of flows around modern commercial ship models. *Experiment in Fluids, 31,* 567–578.

Kodama, Y., Takeshi, H., Hinatsu, M., Hino, T., Uto, S., Hirata, N., et al. (1994). *Proceedings, CFD Workshop.* Tokyo, Japan: Ship Research Institute.

Larsson, L. (Ed.). (1981). *SSPA-ITTC workshop on ship boundary layers.* Gothenburg, Sweden: SSPA Report 90.

Larsson, L., Patel, V. C., & Dyne, G. (Eds.). (1991). *SSPA-CTH-IIHR workshop on viscous flow.* Gothenburg, Sweden: Flowtech Research Report 2, Flowtech Int. AB.

Larsson, L., Stern, F., & Bertram, V. (Eds.). (2002). *Gothenburg 2000-A workshop on numerical hydrodynamics.* Gothenburg, Sweden: Department of Naval Architecture and Ocean Engineering, Chalmers University of Technolog.

Larsson, L., Stern, F., & Bertram, V. (2003). Benchmarking of computational fluid dynamics for ship flow: the Gothenburg 2000 Workshop. *Journal Ship Research, 47,* 63–81.

Larsson, L., Stern, F., & Visonneau, M. (Eds.) (2014). *Numerical ship hydrodynamics—An assessment of the Gothenburg 2010 Workshop.* Springer. https://doi.org/10.1007/978-94-007-7189-5

Sanada, Y., Tanimoto, K., Takagi, K., Toda, Y., & Stern, F. (2013). Trajectories and local flow field measurements around ONR tumblehome in maneuvering motion. *Ocean Engineering, 72*(2013), 45–65.

Sanada, Y., Elshiekh, H., Toda, Y., & Stern, F. (2019). Effects of waves on course keeping and manoeuvring for surface combatant ONR tumblehome. *Journal Marin Science Technology, 24*(3), 948–967.

Shevchuk, I., Sahab, A., Stern, F., & Abdel-Maksoud, M. (2020). Experimental and numerical studies of the flow around the JBC hull form at straight ahead condition and 8° drift angle. In Proc. the 33rd Symposium on Naval Hydrodynamics. Osaka, Japan.

Simonsen, C., Otzen, J., & Stern, F. (2008). EFD and CFD for KCS heaving and pitching in regular head waves. In Proc. 27th Symp. Naval Hydrodynamics. Seoul, Korea.

Van, S. H., Kim, W. J., Yim, G. T., Kim, D. H., & Lee, C. J. (1998). Experimental investigation of the flow characteristics around practical hull forms. In: *Proceedings 3rd Osaka Colloquium on Advanced CFD Applications to Ship Flow and Hull Form Design.* Osaka, Japan.

Experimental Data for JBC Resistance, Sinkage, Trim, Self-Propulsion Factors, Longitudinal Wave Cut and Detailed Flow with and without an Energy Saving Circular Duct

Nobuyuki Hirata, Hiroshi Kobayashi, Takanori Hino, Yasuyuki Toda, Moustafa Abdel-Maksoud, and Frederick Stern

Abstract In this Chapter, measured data of resistance, sinkage, trim, self-propulsion factors, longitudinal wave cut and detailed flow are summarized for Japan Bulk Carrier (JBC) with and without an energy saving circular duct. JBC is a newly designed ship for CFD validation of a ship with an energy saving device (ESD). Resistance and self-propulsion tests are conducted in towing tanks of National Maritime Research Institute (NMRI) and Osaka University (OU) using model ships of different sizes. Detailed local flow data are acquired using SPIV (Stereo Particle Image Velocimetry) for several cross sections at towing tanks of NMRI and OU. The local flow data are also measured by SPIV in a wind tunnel at Hamburg University of Technology (TUHH).

1 Introduction

New test cases associated with hydrodynamics of flows around a ship equipped with an energy saving device (ESD) are adopted in T2015 Workshop. A ship hull called Japan Bulk Carrier (JBC) which is a Cape-size bulk carrier was designed together with a circular duct in front of a propeller as an ESD.

N. Hirata · H. Kobayashi
National Maritime Research Institute, Mitaka, Japan

T. Hino (✉)
Yokohama National University, Yokohama, Japan
e-mail: hino-takanori-nf@ynu.ac.jp

Y. Toda
Osaka University, Suita, Japan

M. Abdel-Maksoud
Hamburg University of Technology, Hamburg, Germany

F. Stern
University of Iowa and Iowa Institute of Hydraulic Research (IIHR), Iowa, IA, USA

T. Hino et al. (eds.), *Numerical Ship Hydrodynamics*, Lecture Notes in Applied and Computational Mechanics 94, https://doi.org/10.1007/978-3-030-47572-7_2

Validation data for the test cases were obtained by model tests in the multiple facilities including National Maritime Research Institute (NMRI), Osaka University (OU) and Hamburg University of Technology (TUHH). Resistance and self-propulsion tests were conducted at the towing tanks of NMRI and OU. In addition, local flow measurements are carried out at two towing tanks on NMRI and OU, and also at the wind tunnel of TUHH.

In this chapter, description of the ship hull, the propeller and the energy saving duct is given first. It is followed by the summary of towing tank tests. The next section is devoted to the local velocity measurements. Accuracy estimations are given for NMRI measurement and credibility of SPIV (Stereo Particle Image Velocimetry) data of three facilities is discussed. The next section gives the wave height measurement results at NMRI towing tank. Conclusions of the chapter are summarised in the last section.

2 JBC (Japan Bulk Carrier)

2.1 *Ship Hull*

Japan Bulk Carrier (JBC) is a Cape-size bulk carrier designed for the validation of CFD analysis of a ship with an energy saving device (ESD). A ship type of a bulk carrier is selected since it is one of the major cargo vessels in international shipping and since a blunt ship hull is considered to be appropriate for the examination of an effect of ESDs. A circular duct placed ahead of a propeller is adopted as an ESD. The ship hull and the duct have been newly designed in the collaborative research project in Japan which were organized by universities, research organizations and shipyards in Japan under the sponsorship of ClassNK (Hino et al. 2016).

The principal particulars of a ship are determined following the representative values of current vessels as shown in Table 1. The design speed is set to 14.5 knots which corresponds to $F_n(L_{PP}) = U_0/\sqrt{gL_{PP}} = 0.142$. Based on the comparative study (Hino et al. 2016) of resistance and wake distributions at a propeller plane by numerical simulations at the design Froude number $F_n = 0.142$ and the model scale Reynolds number $R_n(L_{PP}) = 7.245 \times 10^6$, the final hull form is designed as shown in Fig. 1.

2.2 *Propeller*

A propeller used is a conventional five-bladed MAU type propeller. The particulars of the model propeller for a model ship of $L_{PP} = 7.000$ m are shown in Table 2.

Table 1 Principal particulars of JBC in actual ship scale

Length between perpendiculars	L_{PP} (m)	280.0
Length of waterline	L_{WL} (m)	285.0
Maximum beam of waterline	B_{WL} (m)	45.0
Depth	D (m)	25.0
Draft	T (m)	16.5
Displacement volume	∇ (m3)	178369.9
Wetted surface area w/o ESD	$S_{\_w/oESD}$ (m2)	19556.1
Wetted surface area with ESD	$S_{\_withESD}$ (m2)	19633.9
Block coefficient (C_B)	$\nabla/(L_{PP}B_{WL}T)$	0.8580
Midship section coefficient (C_M)		0.9981
LCB ($\%L_{PP}$), fwd+		2.5475
Vertical Center of Gravity (from keel)*1	KG (m)	13.29
Metacentric height*1	GM (m)	5.3
Moment of Inertia*1	K_{xx}/B_{WL}	0.40
Moment of Inertia*1	K_{yy}/B_{WL}, K_{zz}/B_{WL}	0.25
Propeller center, longitudinal location (from FP)	x/L_{PP}	0.9864
Propeller center, vertical location (below waterline)	$-z/L_{PP}$	0.0404214
Propeller rotation direction (view from stern)		Clockwise
Design speed	(knots)	14.5

*1: These values are determined after the workshop. Please note that these values were not available in computations for the workshop

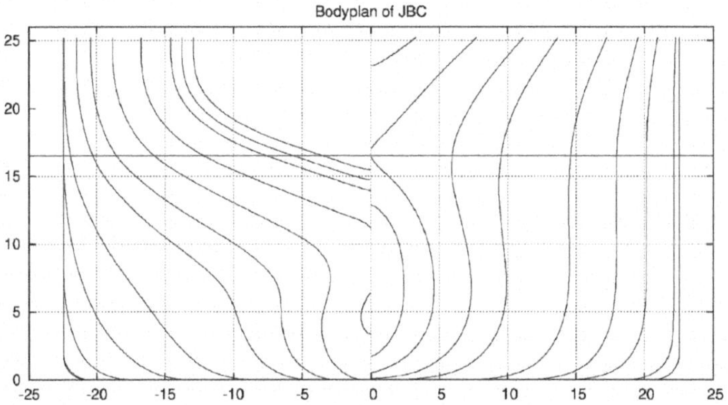

Fig. 1 Body plan of JBC

Table 2 Principal particulars of a model propeller

Diameter (mm)	203.0
Boss ratio	0.1800
Pitch ratio	0.7500
Expanded area ratio	0.5000
Max blade width ratio	0.2262
Blade thickness ratio	0.050
Angle of rake	5.00 deg
Number of blades	5
Blade section	MAU
Direction of rotation	Clockwise

2.3 Energy Saving Device

A circular duct installed ahead of a propeller is selected as an energy saving device, considering that a computational setup is easy due to its simple geometry. A duct is designed based on a parametric study using CFD simulations aiming to better propulsive efficiency. A series of computations are made for various configurations of ducts with diameters at trailing edge ranging from $0.70D_P$ to $1.10D_P$ where D_P is a diameter of a propeller, opening angles ranging from 5 deg. to 20 deg. and several vertical positions at model scale Reynolds number $R_n(L_{PP}) = 7.245 \times 10^6$. Final selection is made between two candidates based on self-propulsion test as shown below. Principal particulars of the designed duct are listed in Table 3. The duct is attached to the stern-tube using a vertical strut as shown in Fig. 2 in such a way that the leading edge of the duct is located at S.S. 1/4.

The geometry data of the hull, the propeller and the duct can be found at the Workshop web site (https://t2015.nmri.go.jp/).

Table 3 Principal particulars of a duct

Section	NACA4420
Diameter at trailing edge	$0.55D_P$*1
Chord length	$0.3D_P$*1
Opening angle	20 deg.

*1 D_P: propeller diameter

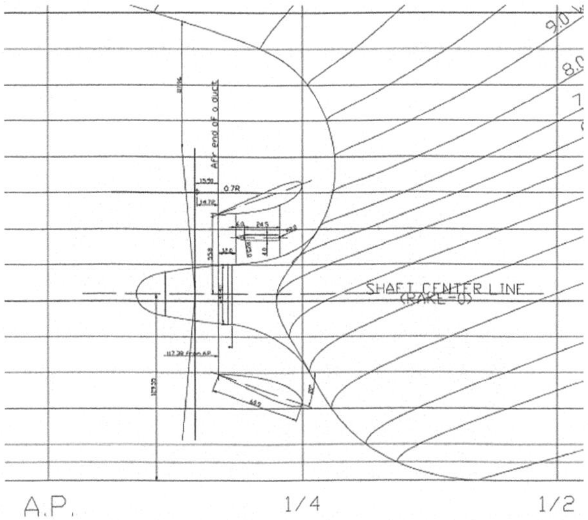

Fig. 2 Energy saving duct layout

2.4 Models

Three different models are used by National Maritime Research Institute, Japan (NMRI), Osaka University (OU) and Hamburg University of Technology (TUHH). Table 4 shows the principal particulars of each model used in three facilities. Two models of NMRI and OU are used in their tank tests (resistance and self-propulsion) and SPIV flow measurement. A model by TUHH is used to measure flow field by LDV and PIV in its wind tunnel. Note that the TUHH model is a double model shape. A rudder is not installed in all measurements to avoid interference with measuring

Table 4 Principal particulars of NMRI, OU and TUHH models

		NMRI	OU	TUHH
Length between perpendiculars	L_{PP}(m)	7.000	3.200	3.513
Length of waterline	L_{WL} (m)	7.125	3.257	3.576
Maximum beam of waterline	B_{WL} (m)	1.1250	0.5143	0.5646
Depth	D (m)	0.6250	0.2857	–
Draft	d (m)	0.4125	0.1886	0.2070
Wetted surface area w/o ESD	$S_{\_\mathrm{w/oESD}}$ (m^2)	12.223	2.554	3.078
Wetted surface area with ESD	$S_{\_\mathrm{withESD}}$ (m^2)	12.271	2.564	3/091
Block coefficient	C_{B}	0.8580		
Midship section coefficient	C_{M}	0.9981		
LCB (%L_{pp}), fwd+		2.55		

instruments. Figures 3 and 4 are photographs of the NMRI model and the OU model, respectively. Figure 5 shows the TUHH model in the wind tunnel.

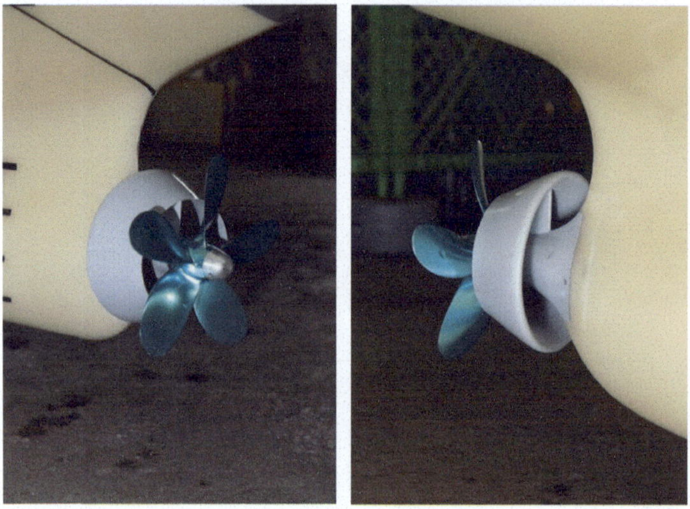

Fig. 3 NMRI model

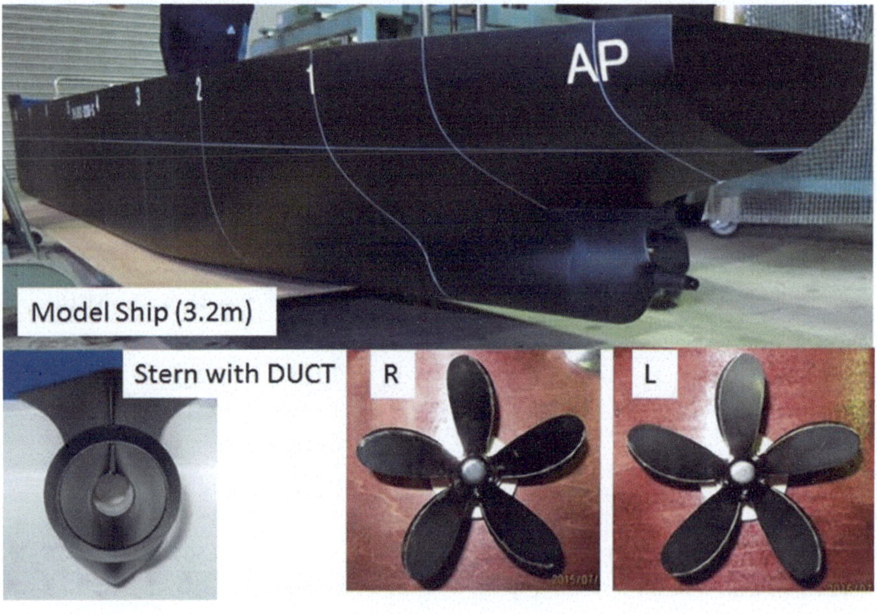

Fig. 4 OU model

Fig. 5 TUHH model

Model propellers used in self-propulsion tests and SPIV measurements at NMRI and OU are the MAU propeller shown in Table 3 with the diameters being 0.203 m for NMRI and 0.0928 m for OU, respectively. Propeller open water characteristics of the NMRI model propeller are shown in Fig. 6.

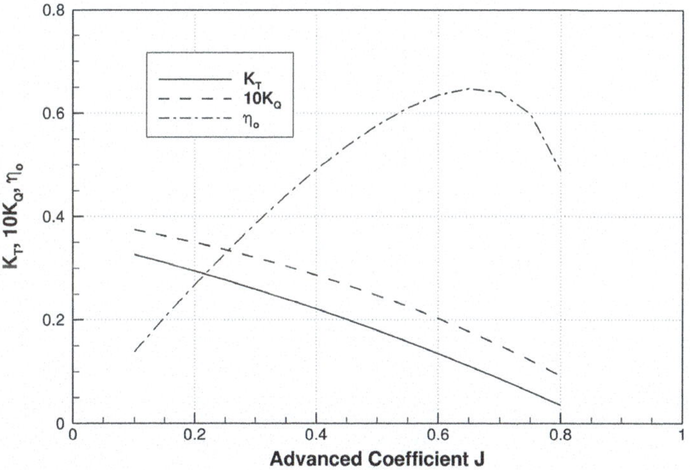

Fig. 6 Propeller open water characteristics of the NMRI model propeller

3 Resistance and Self-propulsion Tests

Resistance and self-propulsion tests are carried out in towing tanks of NMRI (Hino et al. 2016) and OU (Jufuku et al. 2015). The dimension (length, width and water depth) of tanks are 400 × 18 × 8 m for NMRI and 100 × 8 × 4.35 m for OU. The tests are conducted in the conditions with and without a duct while a rudder is not installed throughout the measurements.

The results at the design Froude number $F_n = 0.142$ are shown in Table 5 for both NMRI and OU. The form factors $1 + k$ are obtained using ITTC 1957 correlation line. Note that resistance coefficients of OU are based on the same wetted area (the value with the duct) for both configurations. The self-propulsion tests in both facilities are carried out at the ship point, where the roughness allowance ΔC_F is set to 0.12×10^{-3}.

NMRI data is obtained at the Reynolds number $R_n = 7.569 \times 10^6$ for both with and without the duct. However, test cases 1.1a and 1.2a of T2015 Workshop adopted $R_n = 7.46 \times 10^6$, the same as the SPIV measurement. This makes it possible to use the same computations between cases for integral values and for local flow data. On the other hand, the C_T, K_T, K_Q or propeller revolution values given as the validation data at the Workshop are slightly different (approximately 0.3% in case of C_T) from the values with specified Reynolds number. Nevertheless, this difference is within the nominal uncertainty of the resistance tests (1% D).

Table 5 Results of resistance and self-propulsion tests at design speed

	NMRI		OU	
ESD	Without Duct	With Duct	Without Duct	With Duct
V_m(m/s)	1.179	1.179	0.795	0.795
$F_n(L_{pp})$	0.142	0.142	0.142	0.142
$R_n(L_{pp})$	7.569×10^6	7.569×10^6	2.17×10^6	2.17×10^6
C_T	4.29×10^{-3}	4.26×10^{-3}	5.27×10^{-3}	5.28×10^{-3}
$1 + k$	1.314	1.305	1.26	1.26
C_W	1.50×10^{-4}	1.52×10^{-4}	2.447×10^{-4}	2.547×10^{-4}
sinkage/L_{PP}(%)	-8.57×10^{-2}	-8.46×10^{-2}	-7.65×10^{-2}	-7.45×10^{-2}
trim/L_{PP}(%)	-1.80×10^{-1}	-1.82×10^{-1}	-1.75×10^{-1}	-1.75×10^{-1}
n(1/s)	7.810	7.516	11.8	11.8
K_T	0.217	0.233	0.217	0.230
$10K_Q$	0.279	0.295	–	–
$1 - t$	0.804	0.810	0.748	0.751
$1 - w_T$	0.552	0.481	0.400	0.346
η_R	1.015	1.009	–	–

Sinkage is positive upward and trim is positive bow up

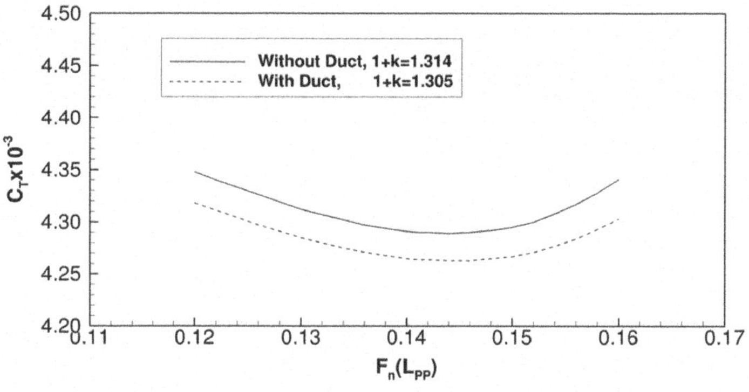

Fig. 7 Total resistance coefficient (C_T) of NMRI model

Figures 7 through 10 show all the measured results of resistance tests at NMRI. Figure 7 shows total resistance coefficient $\left(C_T = \frac{R_T}{0.5\rho U_0^2 S}\right)$ curves of the cases with and without the duct. C_T with the duct is smaller than that without the duct at every Froude number, while wave making resistance coefficient $\left(C_W = \frac{R_W}{0.5\rho U_0^2 S}\right)$ curves shown in Fig. 8 are not so different between with and without the duct. Trim and sinkage are shown in Figs. 9 and 10, respectively. Trim τ and sinkage σ are defined as $\tau = (d_a - d_f)/L_{pp}$ and $\sigma = -(d_a + d_f)/2L_{pp}$, where d_f and d_a are dipping at FP and AP, respectively. Differences of trim and sinkage between with and without duct are very small.

Figures 11 through 15 are the results of self-propulsion tests at NMRI in which a range of Froude numbers from 0.12 to 0.16 are covered. The thrust deduction coefficient $1 - t$ in Fig. 11 and relative rotative efficiency η_R in Fig. 13 show small differences between with and without the duct. $1 - t$ is larger with the duct than without the duct, while η_R shows the opposite trend. $1 - w_T$ with w_T being wake

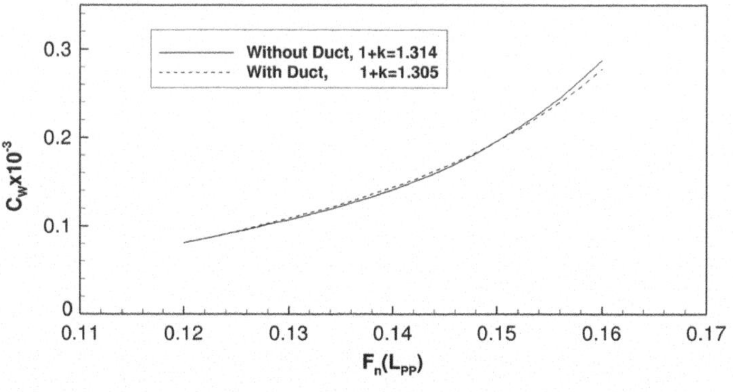

Fig. 8 Wave making resistance coefficient (C_W) of NMRI model

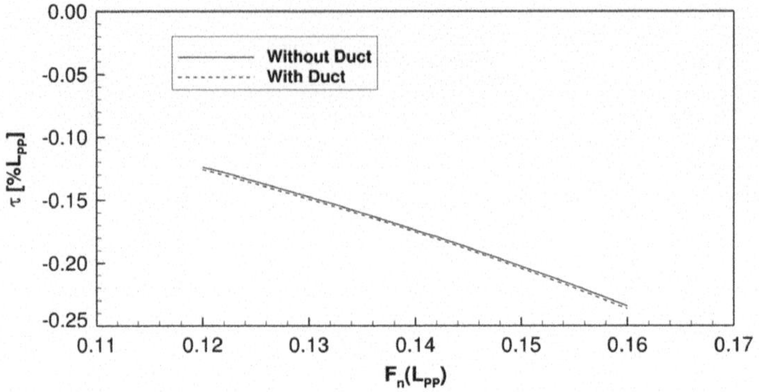

Fig. 9 Trim of NMRI model

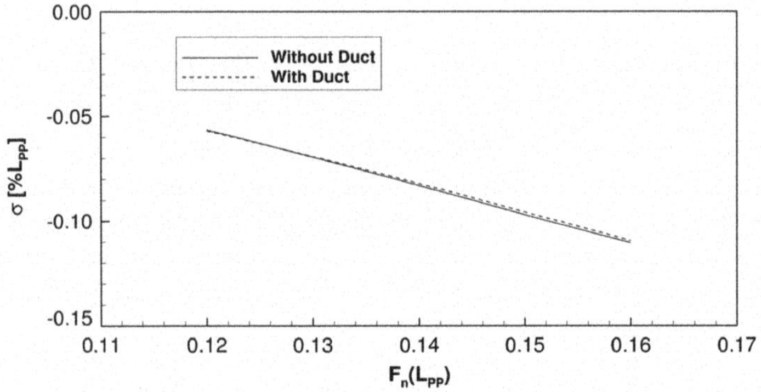

Fig. 10 Sinkage of NMRI model

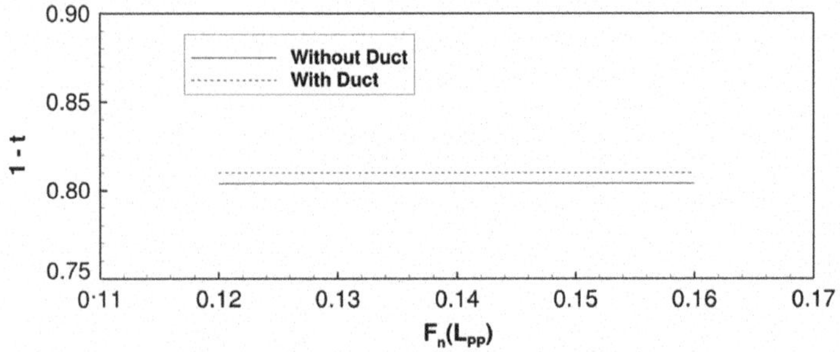

Fig. 11 Thrust deduction coefficient $(1 - t)$ of NMRI model

fraction is largely improved by the presence of the duct as shown in Fig. 12. Both K_T in Fig. 14 and K_Q in Fig. 15 increase with the duct.

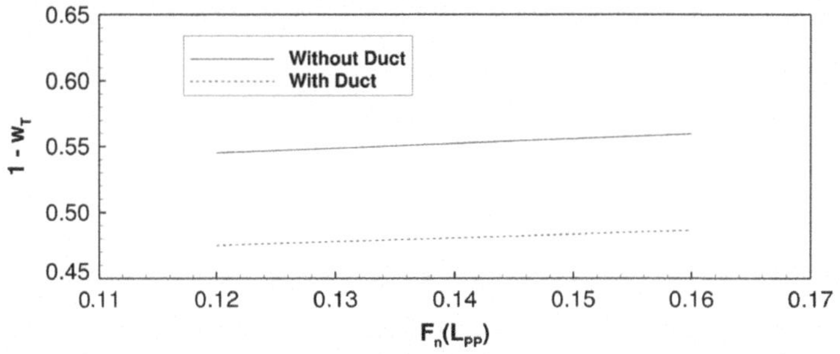

Fig. 12 Thrust wake coefficient $(1 - w_T)$ of NMRI model

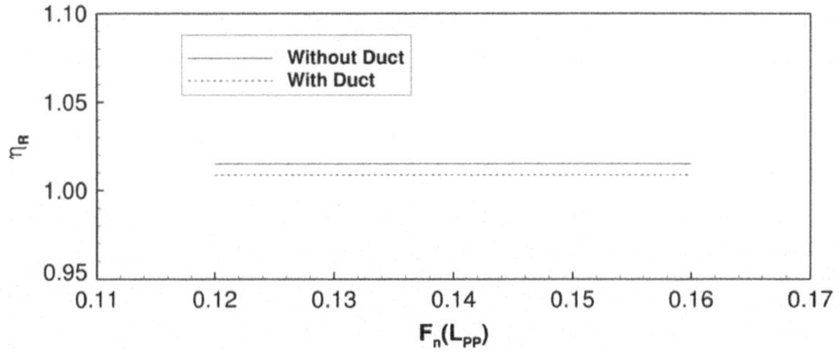

Fig. 13 Relative rotative efficiency (η_R) of NMRI model

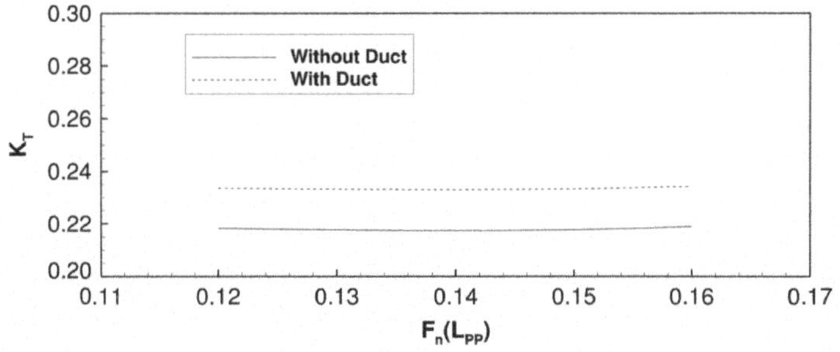

Fig. 14 Propeller thrust coefficient (K_T) of NMRI model

4 Detailed Flow Measurements

4.1 Overview

For the validation of computed results, not only the integrated values such as C_T, the attitude of the hull and self-propulsion factors but also detailed flow field data are desired. To this end, Stereo Particle Image Velocimetry (SPIV) measurements

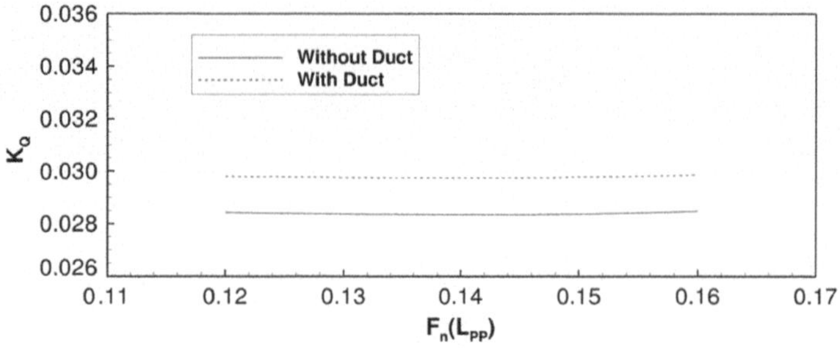

Fig. 15 Propeller torque coefficient (K_Q) of NMRI model

are carried out at the towing tanks of NMRI (Hino et al. 2016) and OU (Jufuku et al. 2015). Also, LDV/PIV measurement is conducted at the TUHH wind tunnel (Shevchuk et al. 2020).

The models used in NMRI, OU and TUHH measurements are specified in Table 4. Measuring sections are common in three facilities. Seven cross sections with constant x/L_{PP} ahead and behind a duct ranging from S.S. 1/2 to AP are adopted as shown in Fig. 16 for NMRI measurements. Section 1 (S1) is at S.S. 1/2 ($x/L_{PP} = 0.950$), Sect. 2 (S2) at S.S. 3/8 ($x/L_{PP} = 0.9625$), Sect. 3 (S3) is at the duct mid-chord ($x/L_{PP} = 0.9788$), Sect. 4 (S4) is between the duct and the propeller ($x/L_{PP} = 0.9843$), Sect. 5 (S5) is at the propeller plane ($x/L_{PP} = 0.9864$), Sect. 6 (S6) is behind the propeller boss ($x/L_{PP} = 0.9923$) and Sect. 7 is at AP ($x/L_{PP} = 1.0000$).

The measurement in NMRI is carried in its middle towing tank (length × width × depth = 150 × 7.5 × 3.5 m) using 7.0 m model, while that in OU is done in its towing tank (length × width × depth = 100 × 8 × 4.35 m) using 3.2 m model. Froude number is $F_n = 0.142$ in both NMRI and OU measurements. Reynolds numbers are 7.46×10^6 and 2.17×10^6 in NMRI and OU, respectively.

LDV/SPIV measurement is conducted in the TUHH wind tunnel. The TUHH low-speed wind tunnel is 40 m long with the test section (length × width × height = 5.5 × 3 × 2 m). The double model of length 3.513 (m) is used and the blockage coefficient is 0.04. The wind velocity $U_0 = 11.8$ m/s which corresponds to Reynolds number of 2.74×10^6 for LDV measurement and $U = 10.0$ m/s and Reynolds number of 2.42×10^6 for SPIV measurement.

4.2 Accuracy Estimation of SPIV Measurement at NMRI

The uniform flow measurements without a ship model using SPIV system are conducted at NMRI for clarification of the accuracy. Figure 17 shows the measured data of uniform flow at $U_0 = 1.0$ m/s which consists of three-components of velocity u, v and w and turbulent kinetic energy TKE. The accuracy of the averaged velocities can be estimated to be 2–3% of U_0 for u, 3–4% of U_0 for v and 1% of U_0 for w.

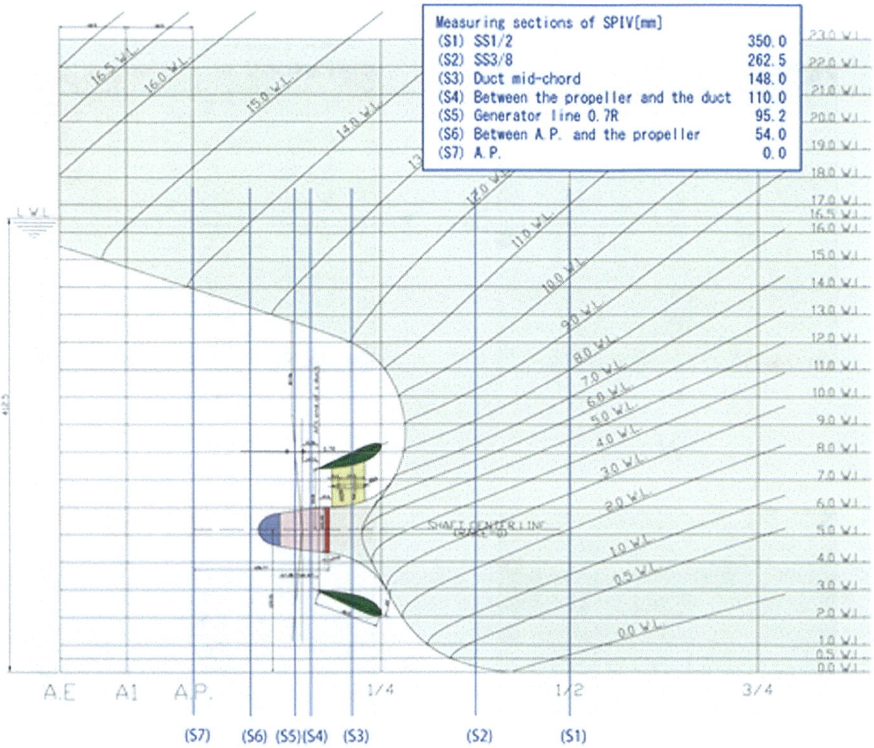

Fig. 16 Measuring sections of velocity fields

The reason why the accuracy of v is somewhat large is that the reflection mirror in the PIV system is not adjusted correctly. For the accuracy of TKE, 0.3–0.5% of U_0^2 can be estimated. Note that all the data is acquired using 250 images.

To assess the accuracy of NMRI measurement further, the statistical convergence of PIV data is examined. Measurement at the S4 section without a propeller and without a duct is taken as an example. Figure 18 is the contours of the axial velocity u and TKE k, respectively, measured at NMRI. Figure 19 is the cumulative moving errors of velocity components and TKE with respect to the number of images. Figure 20 is the histogram of three velocity components. Figure 21 shows the statistical convergence errors of the mean velocity components and TKE. The sample point is at $y/L_{PP} = -0.009814$ and $z/L_{PP} = -0.03926$ also shown as a red circle in Fig. 18. The velocity components and the TKE values in Fig. 19 do not converge completely since the number of images is limited to about 1000, though the data variations are small. The histograms in Fig. 20 show that the fluctuations of velocity components are large. This may be also due to the small number of images. The statistical convergence of the mean velocity components in Fig. 21 is defined using the sample standard deviation s_x as

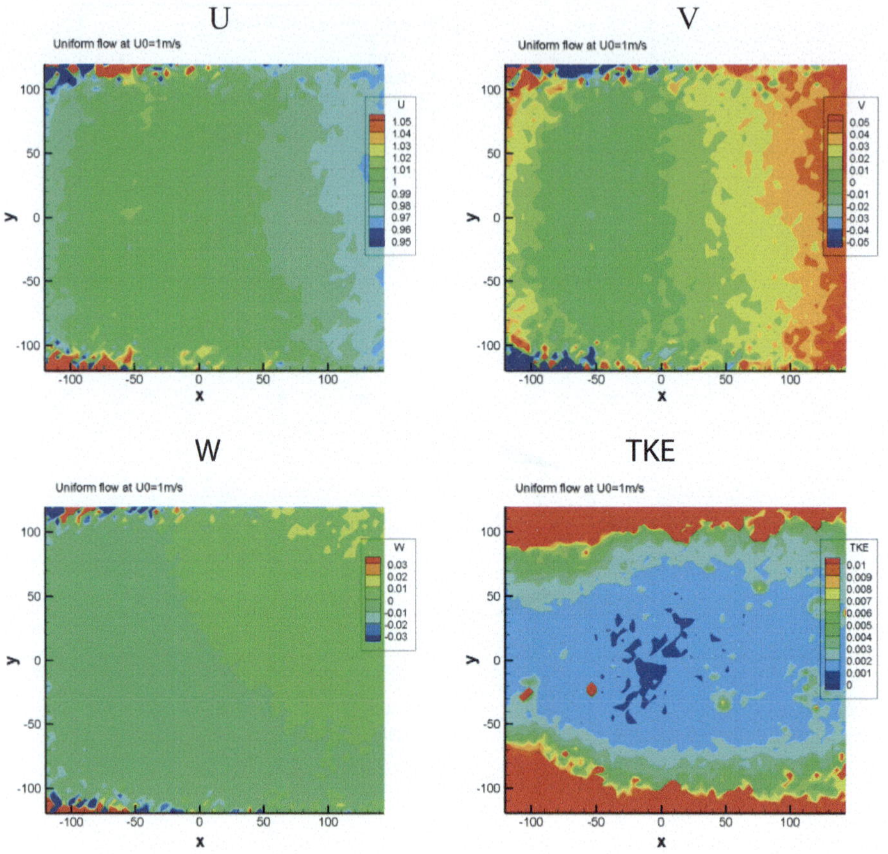

Fig. 17 SPIV measurement of uniform flow ($U_0 = 1.0$ m/s) at NMRI

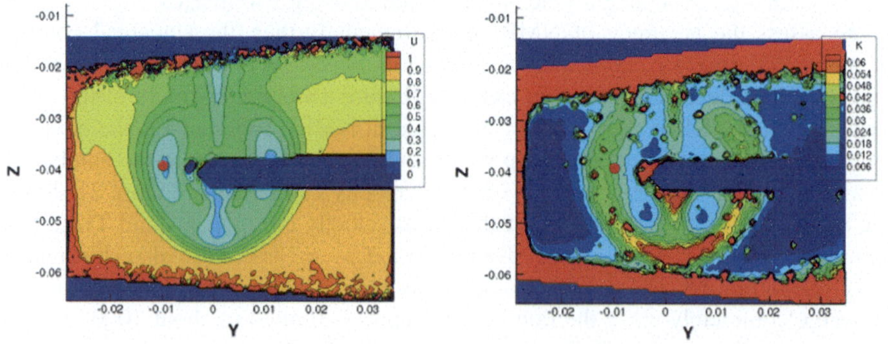

Fig. 18 Contours of the axial velocity u (left) and TKE k (right) in S4 ($x/L_{PP} = 0.9843$) station of NMRI measurement

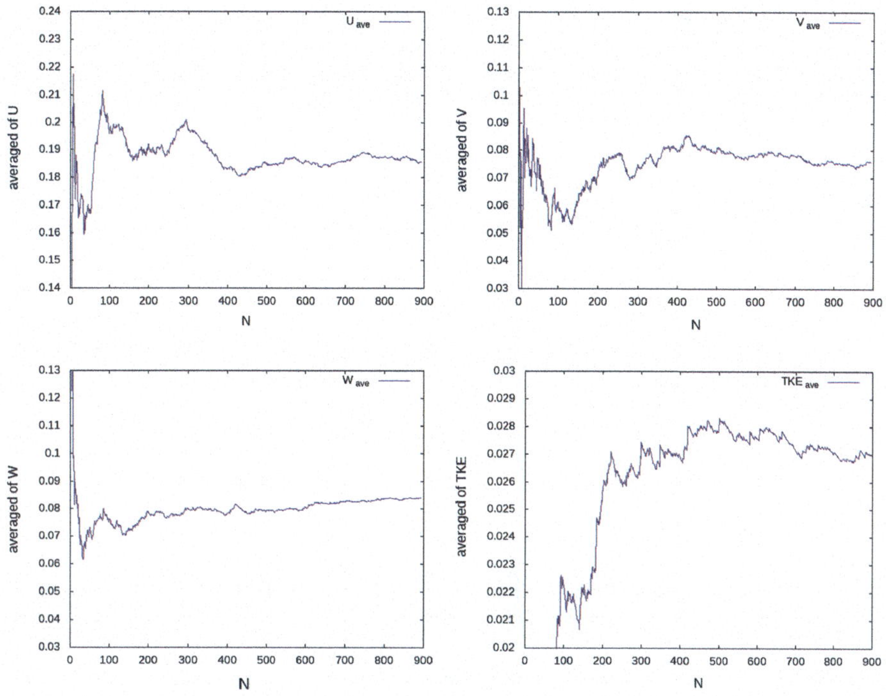

Fig. 19 Cumulative moving averages of velocity components and TKE in NMRI measurement

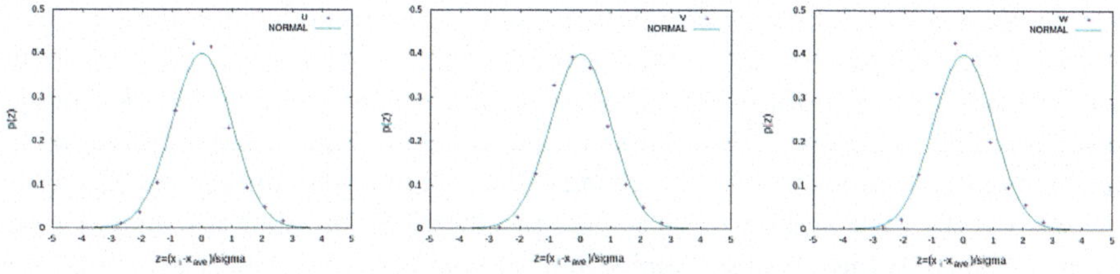

Fig. 20 Histogram of velocity components in NMRI measurement

$$E_{SC}(\%) = \frac{t_{\text{n};\alpha/2}}{\sqrt{N}} \frac{s_x}{x_{\text{ref}}} \times 100$$

where N is the sample size and $t_{\text{n};\alpha/2}$ is the Student-t variable of $n = N-1$ degrees of freedom for a $100(1-\alpha)$ percent level of confidence and x_{ref} is the reference velocity equal to the uniform flow magnitude, U_0 (Yoon et al. 2015). For the convergence of TKE, the χ^2 -distribution is assumed and the upper and lower limits of the confidence interval is defined based on the sample variance s_x^2 as

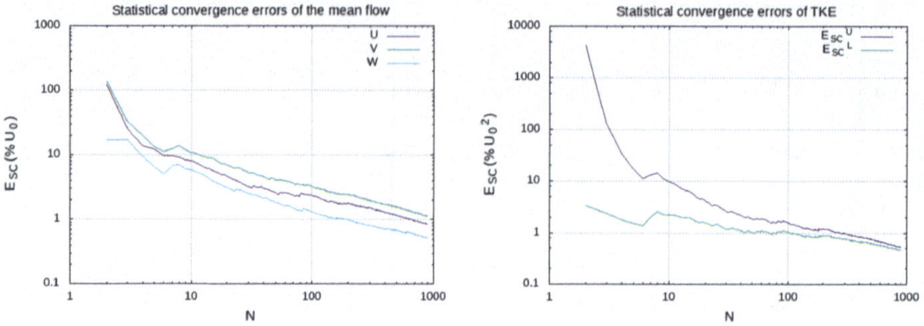

Fig. 21 Statistical convergence errors of mean velocity components (left) and TKE (right) at in NMRI measurement at and $(y/L_{pp}, z/L_{pp}) = (-0.009814, -0.03926)$

$$E_{SC}^{U}\ (\%) = \left(\frac{n}{\chi^2_{n;1-\alpha/2}} - \right)\frac{s_x^2}{s_{ref}^2} \times 100$$

$$E_{SC}^{L}\ (\%) = \left(1 - \frac{n}{\chi^2_{n;\alpha/2}}\right)\frac{s_x^2}{s_{ref}^2} \times 100$$

for a $100(1-\alpha)$ percent level of confidence. The $\chi^2_{\mathrm{n};1-\alpha/2}$ and $\chi^2_{\mathrm{n};\alpha/2}$ are the χ^2 variables and s^2_{ref} is the reference velocity-squared value for which the square of the uniform velocity magnitude U_0^2 is taken (Yoon et al. 2015). Note that 95% level of confidence ($\alpha = 0.05$) is set for all variables. When the sample size N is approaching to 1000, the statistical convergence of the mean velocity component is around 1% of U_0 and that of TKE is below 1% of U_0^2. Figure 22 depicts the contours of the statistical convergence errors E_{SC} for the mean velocity components and E_{SC}^U for TKE in the S4 section. The convergence error of the mean velocity u, v and w are proportional to their standard deviations which correspond to the square-root of the normal Reynolds stresses, $\sqrt{u'u'}$, $\sqrt{v'v'}$ and $\sqrt{w'w'}$. The patterns, therefore, are similar to that of TKE in Fig. 18. The error levels of u, v and w are estimated as 1.2, 1.4 and 0.6% of U_0 for the average and 1.8, 2.0 and 0.9% of U_0 for the maximum, respectively. The error of TKE is 1.0% of U_0^2 for the average and 1.6% of U_0^2 for the maximum. The errors of v are slightly larger than those of u, which is also observed in Fig. 22. Though the reason is not clear, it may be related to the mirror alignment problem.

Total uncertainty of PIV measurement is estimated using the analysis above. The bias errors E_b are estimated from the uniform flow test and the precision errors E_p from the statistical convergence errors and the uncertainty is defined as $\sqrt{E_b^2 + E_p^2}$. Results are shown in Table 6 and the average uncertainty of the mean velocity is 1–3% of U_0 and that of TKE is 1.0% of U_0^2. The maximum uncertainty is slightly larger than the average.

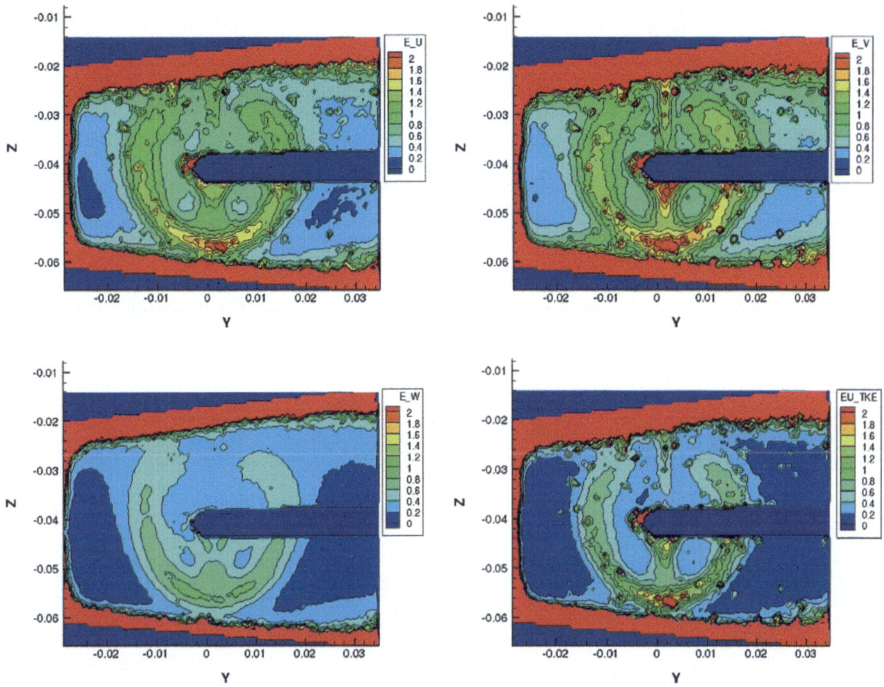

Fig. 22 Contours of the statistical convergence errors mean velocity u (top left), v (top right), w (bottom left) and TKE k (bottom right) in S4 ($x/L_{pp} = 0.9843$) station of NMRI measurement

Table 6 Uncertainty of PIV measurement

	Bias error	Precision error (average)	Precision error (max.)	Uncertainty (average)	Uncertainty (max.)
u ($\%U_0$)	2.0	1.2	1.8	2.3	2.7
v ($\%U_0$)	3.0	1.4	2.0	3.3	3.6
w ($\%U_0$)	1.0	0.6	0.9	1.2	1.3
TKE ($\%U_0^2$)	0.3	1.0	1.6	1.0	1.6

4.3 Credibility of SPIV Measurement at NMRI, OU and TUHH

In order evaluate the credibility of measurement, the measured data of three facilities are compared. The test case 1.3 in the workshop, i.e. the towed condition without a propeller and without an ESD is chosen since this configuration is simplest.

In the S4 section ($x/L_{PP} = 0.9843$), the contours of the axial velocity u, the horizontal velocity v and the vertical velocity w are shown in Figs. 23, 24 and 25,

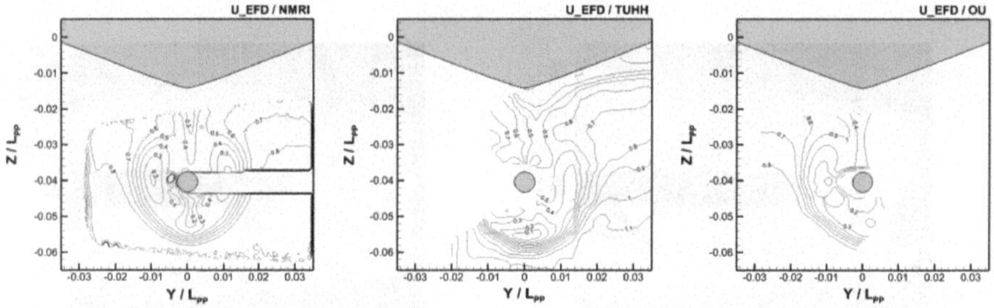

Fig. 23 Contours of measured axial velocity (u) at S4 section. Left: NMRI, middle: TUHH and right: OU

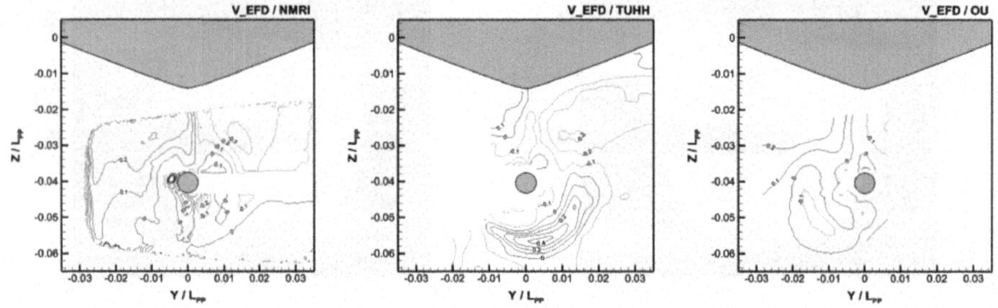

Fig. 24 Contours of measured horizontal velocity (v) at S4 section. Left: NMRI, middle: TUHH and right: OU

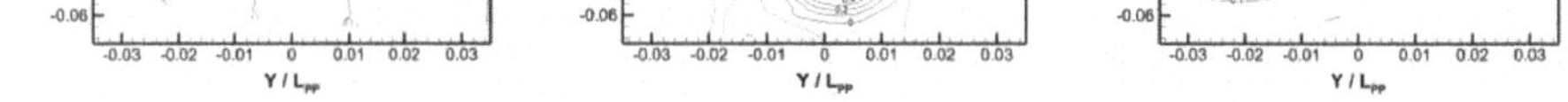

Fig. 25 Contours of measured vertical velocity (w) at S4 section. Left: NMRI, middle: TUHH and right: OU

respectively. Comparisons of the crossflow vectors are shown in Fig. 26 and contours of the x-vorticity ω_x and TKE k are shown in Figs. 27 and 28, respectively.

All the velocity contours are similar to each other and the so-called hook shapes can be observed. While NMRI data is almost symmetric in y direction, the measured areas of TUHH and OU are too narrow to exhibit symmetry.

Lateral locations of the longitudinal vorticies of OU and TUHH are almost same. That of NMRI, on the other hand, is a little closer to a symmetry plane. This is due

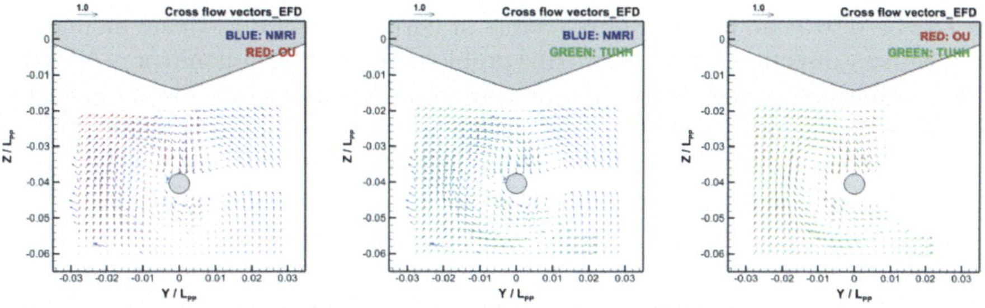

Fig. 26 Comparisons of cross flow vectors (v, w) at S4 section. Left: NMRI/OU, middle: NMRI/TUHH and right: OU/TUHH

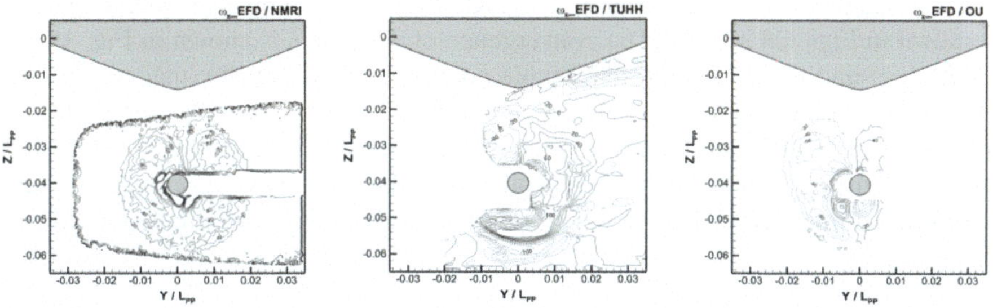

Fig. 27 Contours of measured vorticity (ω_x) at S4 section. Left: NMRI, middle: TUHH and right: OU

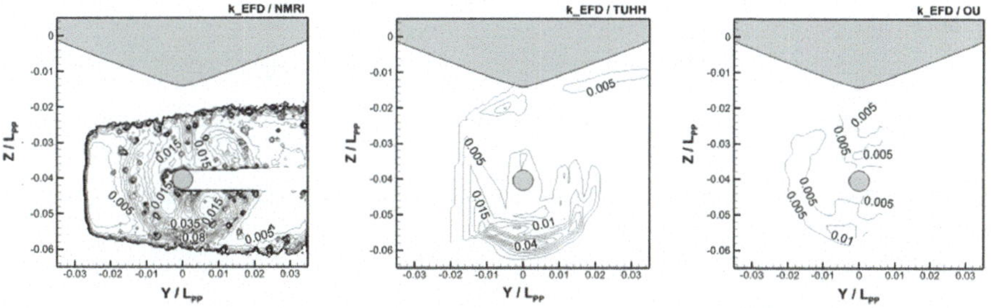

Fig. 28 Contours of measured turbulent kinetic energy (k) at S4 section. Left: NMRI, middle: TUHH and right: OU

to the difference of Reynolds numbers. Reynolds numbers of OU and TUHH are almost the same, 2.17×10^6 and 2.42×10^6, respectivley, and that of NMRI is larger, 7.46×10^6. It is known that the boundary layer is thicker in low Reynolds number flows.

In the narrow region near the center of measuring area ($-0.01 < y/ L_{PP} < 0.01$, $-0.05 < z/ L_{PP} <$ -0.03), axial and horizontal velocities u and v agree well between NMRI and OU. However, the discrepancies between NMRI and OU become large in

$|y/L_{PP}| > 0.015$ and the crossflow vectors of NMRI data in Fig. 26 are inclined to the positive y direction. This is due to the problem of the reflection mirror in NMRI's SPIV system as described above. The uniform flow measurement in Fig. 17 indicates v is overestimated by about 5% of uniform velocity $U_0 = 1.00$ m/s near both side edges of the measuring area. In TUHH case, all flow components under the propeller shaft show asymmetric distributions which is due to the laser reflection effects. The contours of x-vorticity ω_x in Fig. 27 are similar between NMRI and OU. TUHH data is different from the other two, particularly under the propeller shaft, which is again attributed to the laser reflection. The distributions of TKE in Fig. 28 show that the peak value of NMRI data is much larger than that of OU, although the patterns are similar. Figures 29 through 34 are the same plots in the S7 section as Figs. 23 through 28 in the S4 section. General trends are the same in case of the S4 section.

The large differences are found in the TKE distributions between NMRI and OU as shown in Figs. 28 and 34. The convergence of TKE data is shown in Fig. 19 for NMRI case and it exhibits the reasonable convergence with 1000 images. TKE of OU data is acquired using 3000 images and it is believed the convergence is similar or better than NMRI data. Therefore, the difference of TKE values between NMRI and OU seems to come from the reason other than the statistical convergence though more images may be needed for the accurate measurement of turbulent quantities.

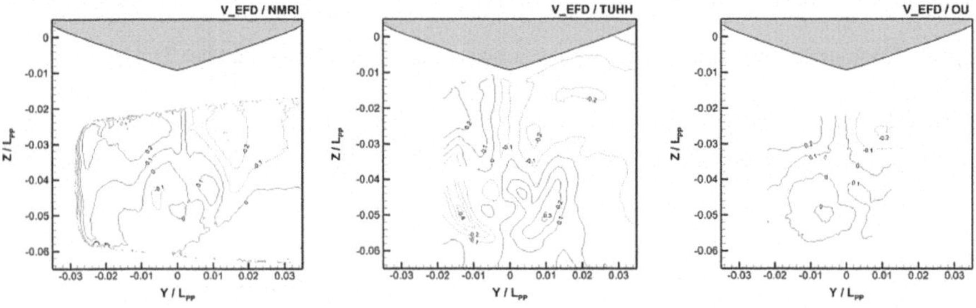

Fig. 29 Contours of measured axial velocity (u) at S7 section. Left: NMRI, middle: TUHH and right: OU

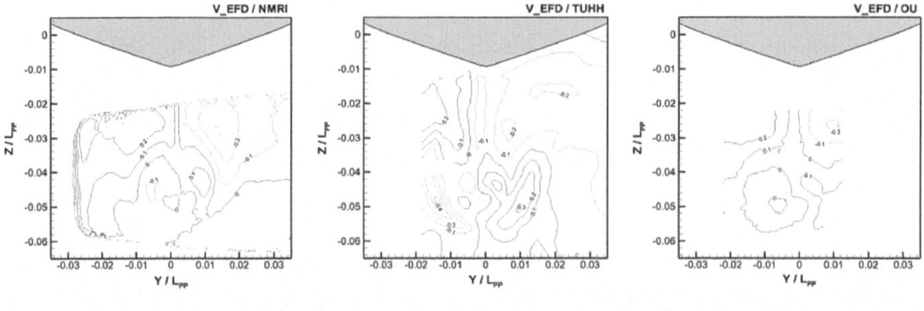

Fig. 30 Contours of measured horizontal velocity (v) at S7 section. Left: NMRI, middle: TUHH and right: OU

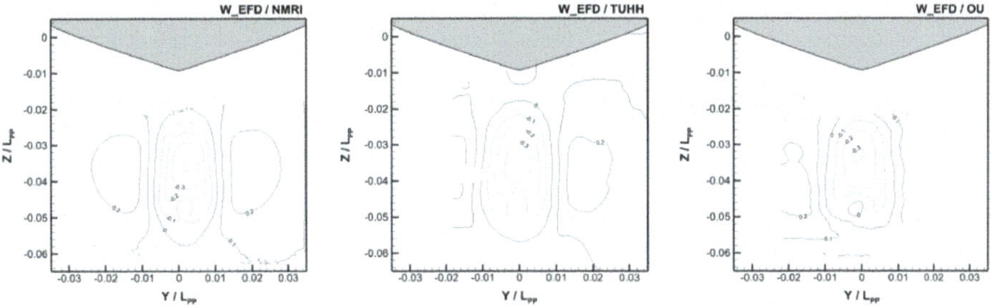

Fig. 31 Contours of measured vertical velocity (w) at S7 section. Left: NMRI, middle: TUHH and right: OU

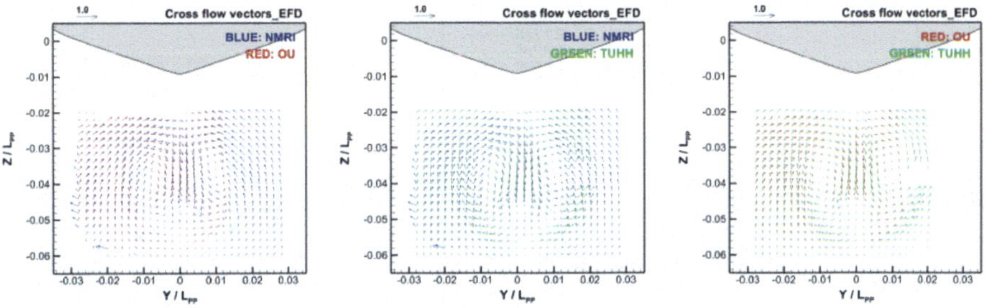

Fig. 32 Comparisons of cross flow vectors (v, w) at S7 section. Left: NMRI/OU, middle: NMRI/TUHH and right: OU/TUHH

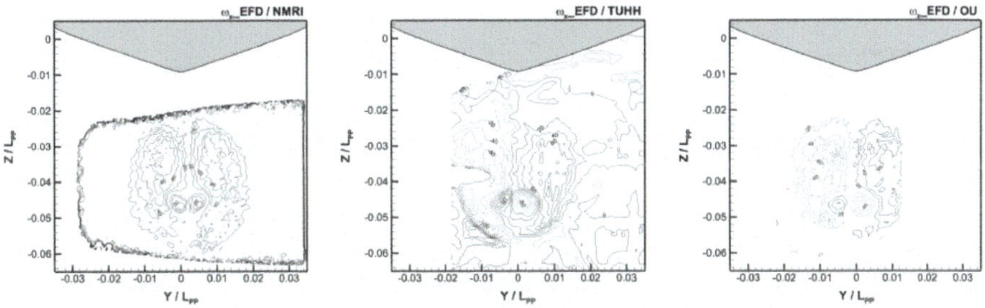

Fig. 33 Contours of measured vorticity (ω_x) at S7 section. Left: NMRI, middle: TUHH and right: OU

In order to assess the Reynolds number effect, three TKE data are compared by the scaled variables. TKE distributions along the horizontal line at the shaft height shown in Fig. 35 are extracted from NMRI, OU and TUHH measurement. The distributions are plotted as $\sqrt{\mathrm{TKE}}/u_\tau \sim y^+$ in left of Fig. 35. Also plotted is the same scaled data from the propeller plane of KVLCC2. The data is taken from the wind tunnel measurement by a hot-wire (Lee et al. 2003) where Reynolds number is $R_n = 4.6 \times 10^6$. The wall distance y is measured from the shaft edge and the friction

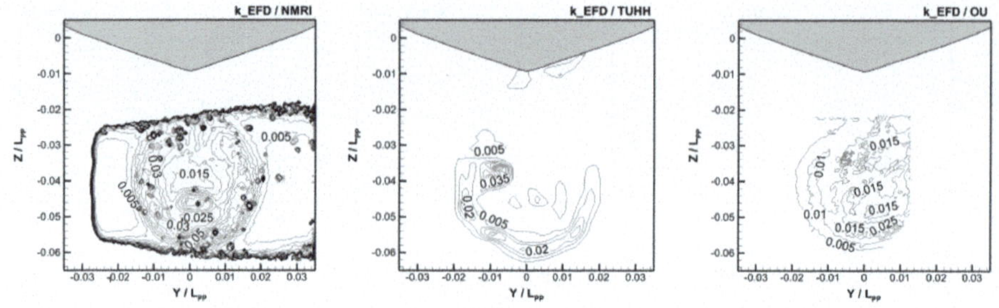

Fig. 34 Contours of measured turbulent kinetic energy (k) at S7 section. Left: NMRI, middle: TUHH and right: OU

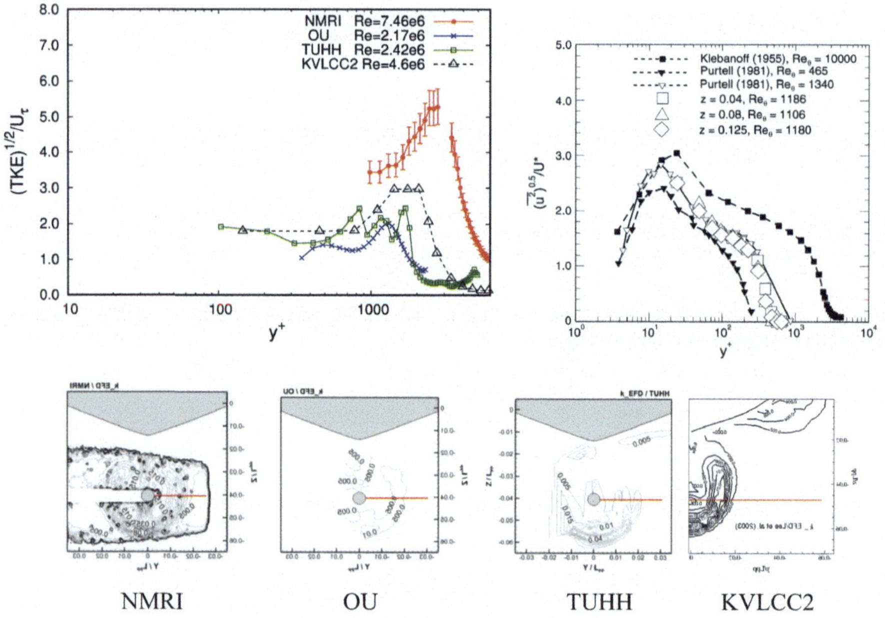

Fig. 35 Left: Distributions of scaled TKE along a horizontal line at the shaft height in S4 section of JBC and the propeller plane of KVLCC2. Right: Distributions of scaled streamwise normal Reynolds stress $\overline{u'u'}$ of a flat plate boundary layer (Longo et al. 1998). Bottom: Data-extracted lines

velocity u_τ is estimated from the local c_f of a turbulent boundary layer of a flat plate using the formula of Prandtl- Schlichting as below

$$c_f = (2.0\ \log_{10}(Re_x) - 0.65)^{-2.3}, \quad u_\tau/U = \sqrt{c_f/2}$$

where Re_x is set equal to the Reynolds number of each case. Actual data of u_τ/U_0 in each case are 0.0367, 0.0405, 0.0401 and 0.0381 for NMRI, OU, TUHH and KVLCC2, respectively. The plots show the similarity in the distribution patterns of all cases. Note that the irregular data of NMRI measurement are omitted. The peak locations and peak values of OU and TUHH are almost the same since the Reynolds numbers of two cases are similar. The peak value of NMRI is larger than those of OU and TUHH. KVLCC2 data with the moderate Reynolds number is in-between OU/TUHH and NMRI for the peak location and value. Right of Fig. 35 is the scaled streamwise normal Reynolds stress $\overline{u'u'}$ in the boundary layer (Longo et al. 1998). The distributions show the different distribution pattern than the JBC case, though the general tendency of the turbulence increase with Reynolds number seems to be the same as in the JBC case.

Figure 36 shows the maximum TKE values of $k^+(= k/u_\tau^2)$ in the profiles of the present horizontal lines. Again, the peak values of TKE increase with the Reynolds number. The NMRI data has the highest peak value but it is rather difficult to tell if this follows the trend of the other data or not since the date points is quite few.

Figure 37 shows the wake-scale distributions of u velocity. The wake half width b is determined by manual fitting of the velocity distributions. Due to the presence of the vortex core, the velocity distributions have the minimum away from the center. The TKE distributions in the wake-scale are shown in left of Fig. 38. Right of Fig. 38 is the similar plot for the wake of a flat plate (Longo et al. 1998). The distributions of the JBC case look closer to the wake profiles rather than the boundary layer profiles in Fig. 35.

In summary, it is rather difficult to assess the quantitative Reynolds number effect from the data currently available. Although the tendencies seem to be reasonable, the extremely higher peak of NMRI data compared with other cases cannot be fully justified. Further investigations, such as additional measurements in different facilities and/or different measuring systems, are desirable for obtaining the proper distributions of TKE.

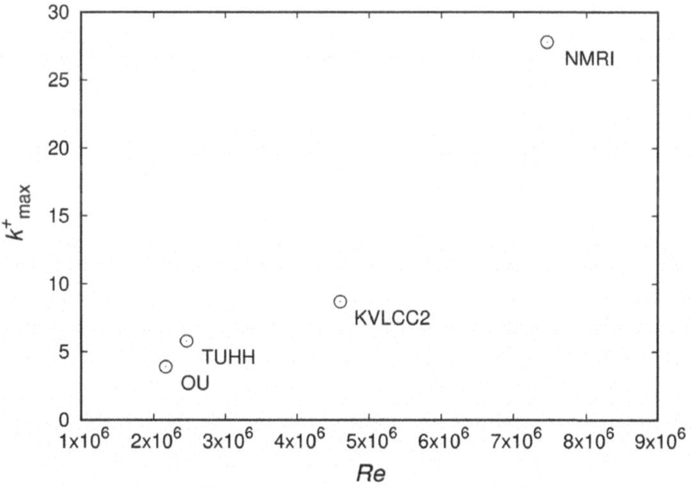

Fig. 36 Relation of R_n and the maximum TKE k^+

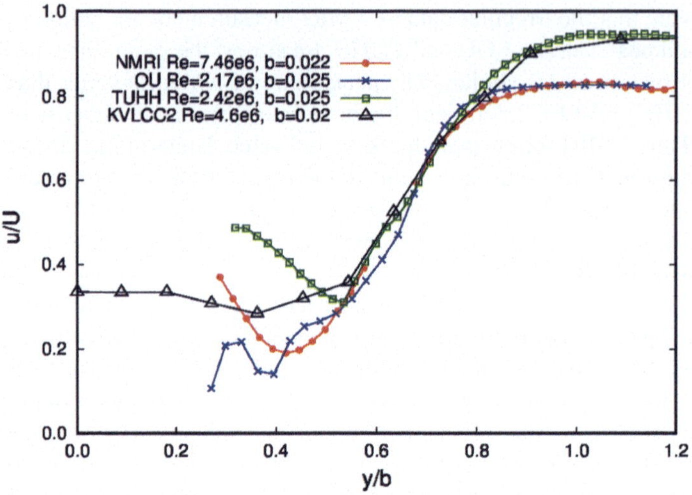

Fig. 37 Wake-scale distribution of u velocity along a horizontal line at the shaft height in S4 section shown in Fig. 35

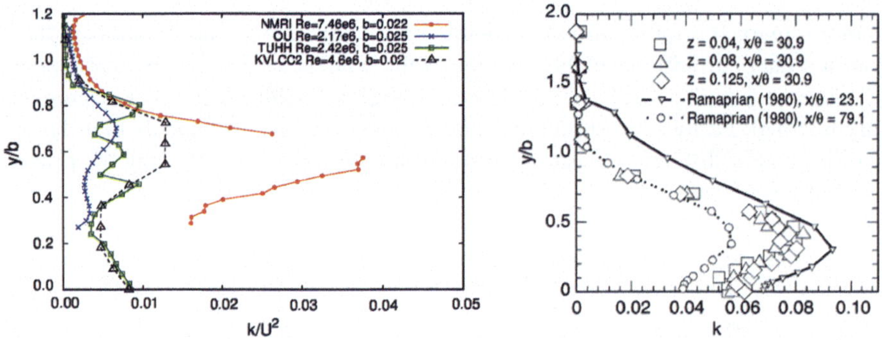

Fig. 38 Left: Wake-scale distribution of TKE along a horizontal line at the shaft height in S4 section shown in Fig. 35. Right: Distributions of turbulent kinetic energy in the wake of a flat plate (Longo et al. 1998)

On the other hand. for the mean velocity components, the data of three facilities are generally in good accordance. However, TUHH measurement has the problem of laser reflection under the propeller shaft and NMRI data has some errors (about 3–4% of the uniform axial velocity) in v near the side edges of measuring area though the errors can be negligible in the center. In total, the OU data seems to have no serious deficiencies and thus this can be considered as the reliable measured data.

Table 7 Test cases of JBC local flow fields

Test case	Conditions
Cases1.3a and 1.3b	Towed in calm water condition w/o rudder w/o propeller and w/o ESD
Case1.4a	Towed in calm water condition w/o rudder w/o propeller and with ESD
Case1.7a	Self-propelled at the ship point in calm water w/o rudder, with propeller and w/o ESD
Case1.8a	Self-propelled at the ship point in calm water w/o rudder, with propeller and with ESD

4.4 *Local Flow Data of NMRI, OU and TUHH Used in Test Cases*

In the Workshop, the test cases for local flow fields are set up as Table 7 based on the measurement described above. SPIV data of the measuring sections S2, S4 and S7 shown in Fig. 16 are picked up for all the test cases. Cases 1.3 and 1.4 are towed condition without and with ESD, respectively. Similarly, Cases 1.7 and 1.8 are the self-propelled condition without and with ESD. Note that the NMRI data were used for all the cases with the identification "a" and the part of the TUHH data of LDV measurement was used in Test Case 1.3b. The OU data and SPIV data of TUHH were not used in the Workshop. All the data both used and not used in the Workshop are listed in Table 8. In Cases 1.7 and 1.8 with a rotating propeller, 'averaged' indicates the mean flow data and 'prop000' to 'prop048' indicate the phase averaged data with the blade angles of 0 to 48 degrees, respectively. Actual figures of the data used in the Workshop data are shown in Chap. 6 together with the submitted numerical results.

5 Wave Height Measurement

Wave height distributions around JBC advancing at the design speed $F_n = 0.142$ are measured in the large towing tank of NMRI. The ship is towed in a trim-free condition without a propeller and a duct. The wave profile on the hull is acquired from the photographs of the hull surface on which the ordinate and the abscissa are marked. The longitudinal wave cuts are measured using the capacitance type wave gauge at $y/L_{PP} = 0.1043$ and 0.1900 where y is the lateral distance from the center line of the hull. Figure 39 shows the wave profile on the hull. Figures 40 and 41 show the longitudinal wave cuts at $y/L_{PP} = 0.1043$ and 0.1900, respectively.

Table 8 List of local veocity data

Testcase	Section	Flow variables	NMRI	TUHH/LDV	TUHH/SPIV	OU
Case1.3	S2	u, v, w	O			*
		ω_x , k	O			*
	S4	u, v, w	O	O	*	*
		ω_x , k	O		*	*
	S7	u, v, w	O	O	*	*
		ω_x , k	O		*	*
Case1.4	S2	u, v, w	O			*
		ω_x , k	O			*
	S4	u, v, w	O			*
		ω_x , k	O			*
	S7	u, v, w	O			*
		ω_x , k	O			*
Case1.7 Averaged	S2	u, v, w	O			*
Case1.7 prop000	S4	u, v, w	O			*
Case1.7 prop024	S4	u, v, w	O			*
Case1.7 prop048	S4	u, v, w	O			*
Case1.7 Averaged	S4	u, v, w	O			*
Case1.7 prop000	S7	u, v, w	O			*
Case1.7 prop024	S7	u, v, w	O			*
Case1.7 prop048	S7	u, v, w	O			*
Case1.7 Averaged	S7	u, v, w	O			*
Case1.8 Averaged	S2	u, v, w	O			*
Case1.8 prop000	S4	u, v, w	O			*
Case1.8 prop024	S4	u, v, w	O			*
Case1.8 prop048	S4	u, v, w	O			*
Case1.8 Averaged	S4	u, v, w	O			*
Case1.8 prop000	S7	u, v, w	O			*
Case1.8 prop024	S7	u, v, w	O			*
Case1.8 prop048	S7	u, v, w	O			*
Case1.8 Averaged	S7	u, v, w	O			*

O: data used in the Workshop *: data not used in the Workshop

6 Conclusions

In this Chapter, the measured data for Japan Bulk Carrier (JBC) with an energy saving circular duct are presented. The measurement is conducted in three facilities, i.e., the towing tanks of National Maritime Research Institute (NMRI) and Osaka University (OU) and the wind tunnel of Hamburg University of Technology (TUHH).

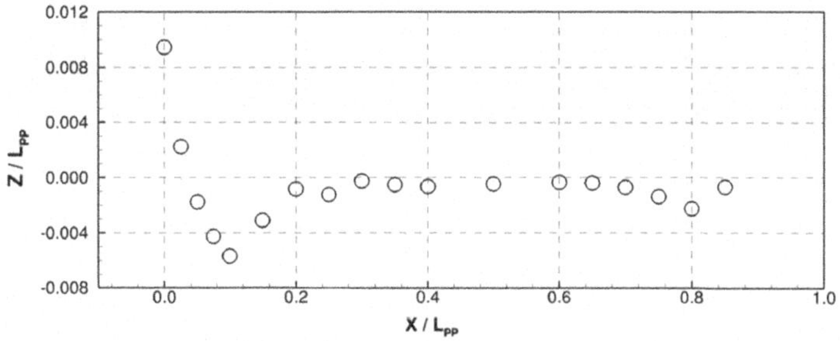

Fig. 39 Wave profile on the hull

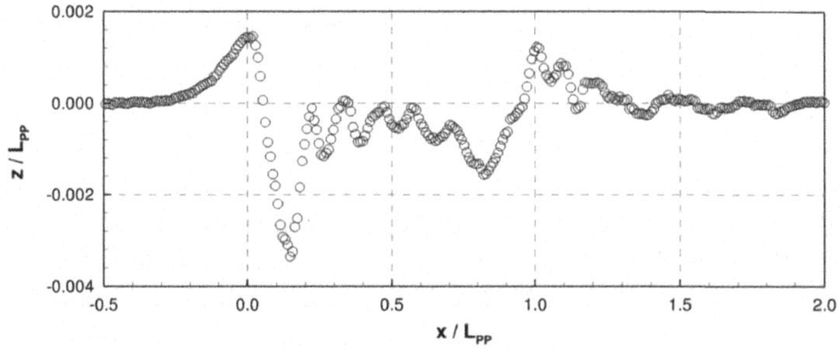

Fig. 40 Longitudinal wave cut at $y/L_{PP} = 0.1043$

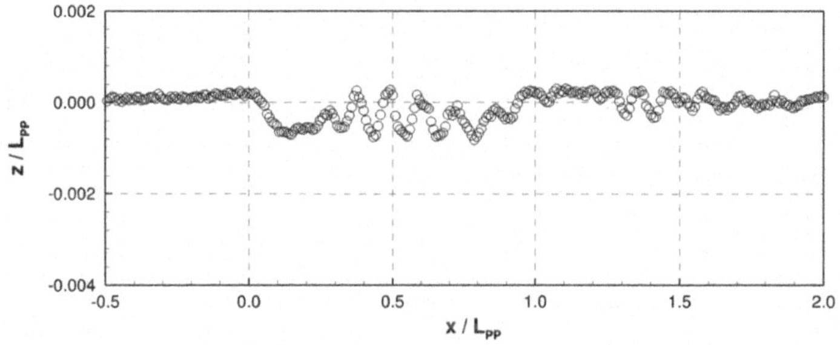

Fig. 41 Longitudinal wave cut at $y/L_{PP} = 0.1900$

Resistance, sinkage, trim, self-propulsion factors with and without the duct are acquired by the tank tests at NMRI and OU. Wave field data is measured at the NMRI towing tank.

The detailed flows fields in seven stations in a stern region are measured by using SPIV in the tanks of NMRI and OU. The data are acquired in towed and self-propelled conditions without and with the duct. In addition, LDV/SPIV measurement is conducted at the TUHH wind tunnel.

For the SPIV measurement at NMRI, the error estimates are carried out using the uniform flow test results and the statistical convergence analysis of the actual measurement. It turned out that uncertainty of the velocity measurement is approximately 2–3% of the uniform flow and the uncertainty of TKE measurement is approximately 1% of the square of the uniform flow magnitude. Uncertainty of v velocity is largest and this may be attributed to the problem with the reflection mirror setting.

For the mean velocity components, the data of three facilities are generally in good accordance in spite of some problems such as the laser reflection in TUHH measurement or the reflection mirror problem in NMRI measurement.

On the other hand, it is found that there is a large difference in the measured TKE levels between NMRI and OU/TUHH. Examination of the statistical convergence of NMRI measurement shows that the difference between NMRI and other facilities does not seem to come from the statistical convergence, though apparently the more frames are needed for the accurate estimation of turbulent quantities. The effect of Reynolds number difference is also investigated for the local TKE distributions along the horizontal lines at the shaft height near the propeller plane. The distributions in the bare-hull towing condition are compared in the various scaling. It appears that the TKE distributions look more wake-like rather than boundary-layer-like from the comparisons with the flat plate data. However, it is rather difficult to specify the exact reason for differences of the extremely higher TKE of NMRI data. Further investigations, such as additional measurements in different facilities and/or different measuring systems, are desirable for obtaining the reasonable distributions of TKE.

Finally, the recommendations to the future workshops at present are as follows:

For the resistance and self-propulsion, sinkage and trim and the wave profiles, the measured data from NMRI can be considered to be appropriate. For the local flow data, the OU data seems to have no serious deficiencies and thus can be recommended as the reliable measured data of mean velocities. Further works will be needed to establish reliable turbulence data of this JBC case.

References

Hino, T., Hirata, N., Ohashi, K., Toda, Y., Zhu, T., Makino, K., et al. (2016). Hull form design and flow measurements of a bulk carrier with an energy-saving device for CFD validations. In *Proc. 13th International Symposium on Practical Design of Ships and Other Floating Structures*. Denmark: Copenhagen.

Jufuku, N., Hori, M., Itou, S., Toda, Y., & Hinatsu, M. (2015). SPIV stern flow measurement around operating propeller for with and without duct condition of Japan bulk carrier (in Japanese). In *Conference Proceedings of the Japan Society of Naval Architects and Ocean Engineers* (Vol. 21, pp. 309 – 312).

Shevchuk, I., Sahab, A., Stern, F., & Abdel-Maksoud, M. (2020). Experimental and numerical studies of the flow around the JBC hull form at straight ahead condition and 8° drift angle. In *Proc. the 33rd Symposium on Naval Hydrodynamics*. Japan: Osaka.

Yoon, H., Longo, J., Toda, Y., & Stern., F. (2015). Benchmark CFD validation data for surface combatant 5415 in PMM maneuvers – Part II: Phase-averaged stereoscopic PIV flow field measurements. *Ocean Engineering, 109*, 735–750.

Longo, L., Huang, H. P., & Stern, F. (1998). Solid/free-surface juncture boundary layer and wake. *Experiments in Fluids, 25*, 283 – 297.

Lee, S. -J., Kim, H. -R., Kim, W. -J., & Van, S. –H. (2003). Wind tunnel tests on flow characteristics of the KRISO 3,600 TEU containership and 300 K VLCC double-deck ship models. *Journal of Ship Research, 47*(1), 24 – 38.

Experimental Data for KCS Resistance, Sinkage, Trim, and Self-propulsion

Jin Kim

Abstract This chapter is dedicated to the experimental results of resistance, sinkage, trim, and self-propulsion tests of the KCS hull form. The KCS was conceived to provide data for both explication of flow physics and CFD validation for a modern container ship with a bulbous bow. The Korea Research Institute of Ships and Ocean engineering (KRISO) performed towing tank experiments to obtain resistance, mean flow data and free surface waves. Self-propulsion tests were carried out at the National Maritime Research Institute (NMRI) and later, additional resistance tests were also reported in Gothenburg 2010 Workshop on CFD in ship hydrodynamics. The main information of the KCS hull and its propeller is introduced in Sect. 1.1 and then the results of resistance and hull attitudes measurement are followed by self-propulsion results and some local flow measurements.

1 KCS (KRISO Container Ship)

1.1 Ship Model

The ship hull used in Tokyo 2015 Workshop is a 3,600 TEU container carrier called KRISO Container Ship (KCS) (Van et al. 1998) which has been designed for the validation of CFD predictions. This hull form was selected also in the Gothenburg 2010 Workshop on CFD in ship hydrodynamics (Larsson et al. 2010; Larsson et al. 2013). KCS has a low block coefficient and the design speed is 24kn (Kim et al. 2001). The model ship was made with a scale ratio of 1/31.6 (i.e. Lpp = 7.2786 m). A conventional five-bladed propeller with the NACA66 section is adopted. The geometries of the hull and the propeller are shown in the Figs. 1 and 2, respectively. The principal particulars of the ship and the propeller are listed in Tables 1 and 2.

J. Kim (✉)
Korea Research Institute of Ships and Ocean Engineering (KRISO), Daejeon, South Korea
e-mail: jkim@kriso.re.kr

T. Hino et al. (eds.), *Numerical Ship Hydrodynamics*, Lecture Notes in Applied and Computational Mechanics 94, https://doi.org/10.1007/978-3-030-47572-7_3

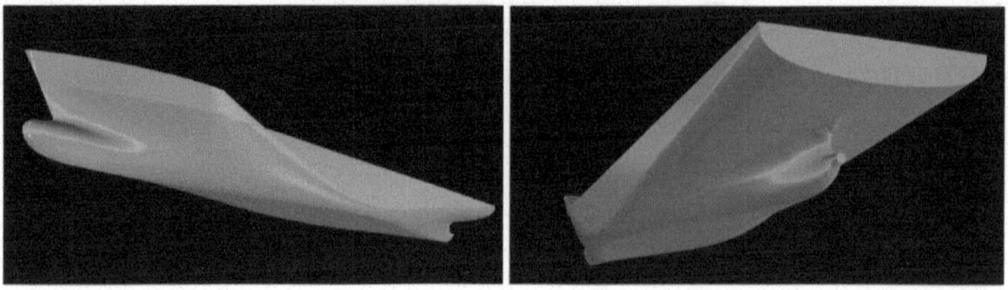

Fig. 1 Geometry of the KCS (bow and stern view)

Fig. 2 Geometry of the KP505 (KCS propeller)

Table 1 Principal particulars of KCS

	Lpp [m]	Lwl [m]	B [m]	D [m]	T [m]	CB
Full scale	230.0	232.5	32.2	19.0	10.8	0.6505
Model	7.2786	7.3577	1.0190	0.6013	0.3418	

Table 2 Principal particulars of the model propeller

Dp[m]	Hub ratio	Pitch type	EAR	No. of blades	Blade section
0.25	0.18	FPP	0.50	5	NACA66

1.2 Test for Resistance, Sinkage, and Trim

The global force and the hull attitudes were measured in the towing tank of the KRISO (Van et al. 1998). This test corresponds to case 2.1. The rudder was fitted, and the measurement was carried out for the 6 speeds shown in Table 3. The obtained values of C_T, sinkage, and trim angle is listed in Table 4 where *Sf* and *Sa* mean sinkage at FP and AP, respectively, and σ means the mean sinkage. Figure 3 shows

Table 3 Test conditions for case 2.1

No.	1	2	3	4	5	6
Speeds [m/s]	0.915	1.281	1.647	1.922	2.196	2.379
Fr	0.108	0.152	0.195	0.227	0.260	0.282
Re	5.23×10^6	7.33×10^6	9.42×10^6	1.10×10^7	1.26×10^7	1.36×10^7

Table 4 The measured values of C_T, sinkage, and Trim angle

Fr	$C_T \times 10^3$	$Sf \times 10^2$ [m]	$Sa \times 10^2$ [m]	$\sigma \times 10^2$ [m]	τ °
0.108	3.796	−0.195	0.015	−0.090	−0.017
0.152	3.641	−0.613	0.063	−0.275	−0.053
0.195	3.475	−1.213	0.015	−0.599	−0.097
0.227	3.467	−1.750	−0.138	−0.944	−0.127
0.260	3.711	−2.465	−0.322	−1.394	−0.169
0.282	4.501	−2.710	−0.695	−1.702	−0.159

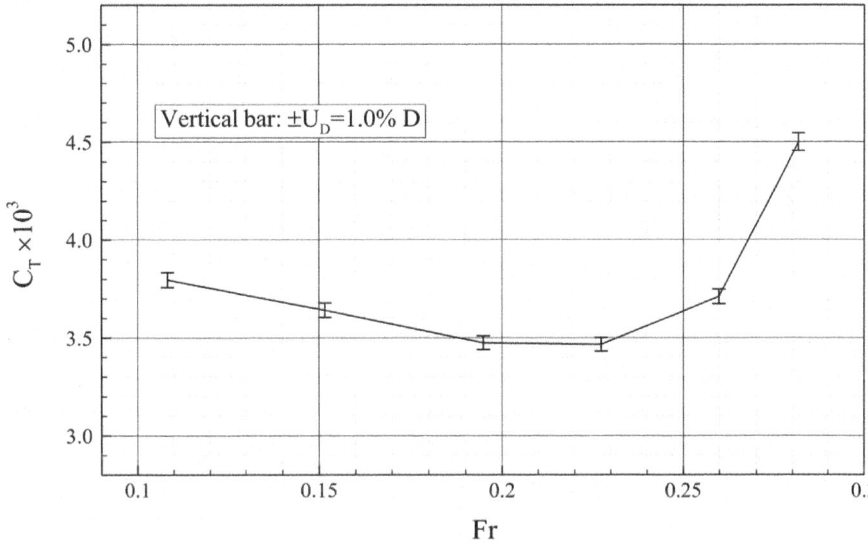

Fig. 3 The measured total resistance coefficients

the measured total resistance coefficients. The data uncertainty, U_D, of C_T is approximately 1.0%D. Figures 4 and 5 show the measured sinkage and trim angle results. The data uncertainty of the sinkage and trim was not reported at Van et al. (1998). We assumed 1.0%D, but these values are reasonable only in the high Fn region. The data uncertainty in the low Fn region might be much higher than 1.0%D. In Chap. 8, all participants show large comparison errors in the low Fn region.

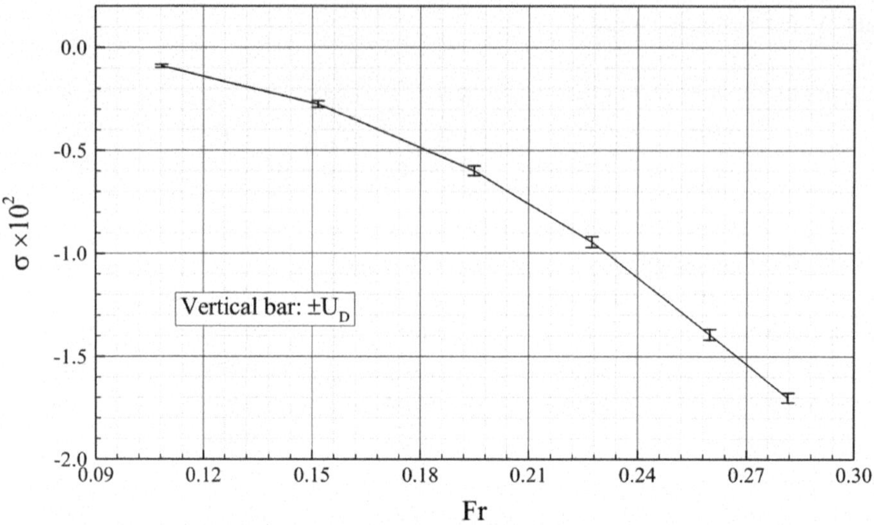

Fig. 4 The measured mean sinkage

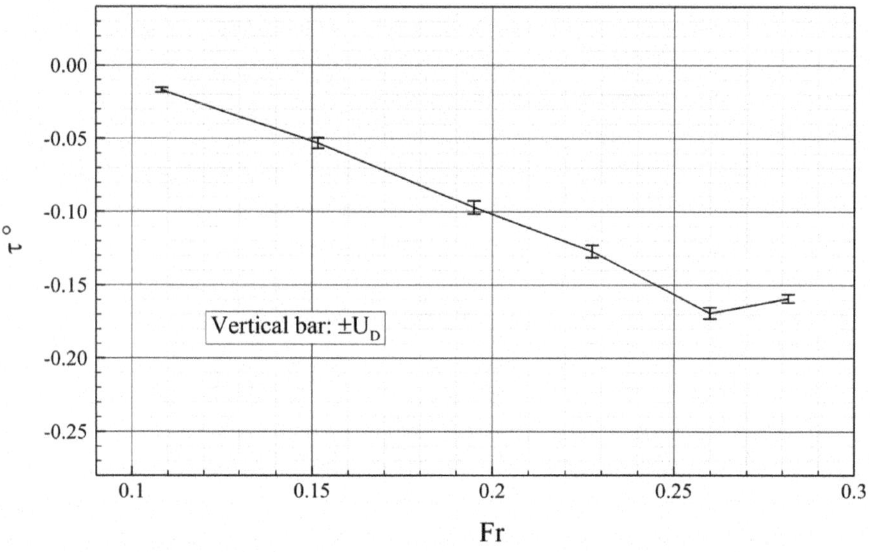

Fig. 5 The measured trim angle

1.3 Test for Self-propulsion

The self-propulsion test was carried out at the NMRI (Hino 2005). This test corresponds to case 2.3 and case 2.7. The hull is free to heave and pitch but ballast weights are arranged to give zero speed attitude at the design speed and no rudder is fitted. The towing speed is set corresponding to Fr is 0.26. The self-propulsion test is conducted at the ship point, which means that one should adjust the rate of rotation of the propeller to get a force equilibrium taking into account the additional towing force (Skin Friction Correction, SFC). The revolution rate of the propeller model is set to 9.5 rps at the ship point and the SFC is 30.25 N at that time. The three-dimensional velocity field is measured around the stern by a spherical type eight-hole Pitot tube (Fujisawa et al. 2000). For the surface pressure measurements on the stern 144 pressure taps are employed (Tsukada et al. 2000). The measured self-propulsion factors are summarized in Table 5. The measurement of the local velocity field is carried out downstream of the propeller at x/Lpp = 0.991 and the corresponding results are presented in Fig. 6. The axial velocity contours are characterized by two regions inside and outside of the propeller disk. A moon crescent-like region of high velocity up to U = 1.1 caused by the rotating propeller is shown inside. A more detailed value on the transversal evolution of the three components of the velocity is also measured at x/Lpp = 0.991, z/Lpp = −0.03 as shown in Fig. 7. Measurements of the surface pressure distribution for the validation of CFD results are conducted on the port side of the hull (ref. Fig. 8).

Table 5 The measured values obtained from the self-propulsion test

$C_T \times 10^3$	K_T	K_Q	n(rps) (for given SFC)	R_T-T(N) (for given n)
3.966	0.170	0.0288	9.5	30.25

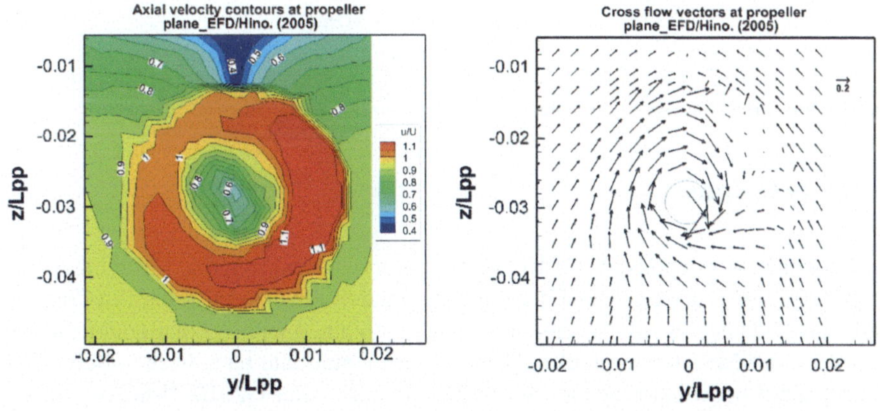

Fig. 6 Measured axial velocity contours and cross flow vectors at x/Lpp = 0.9911 (KCS)

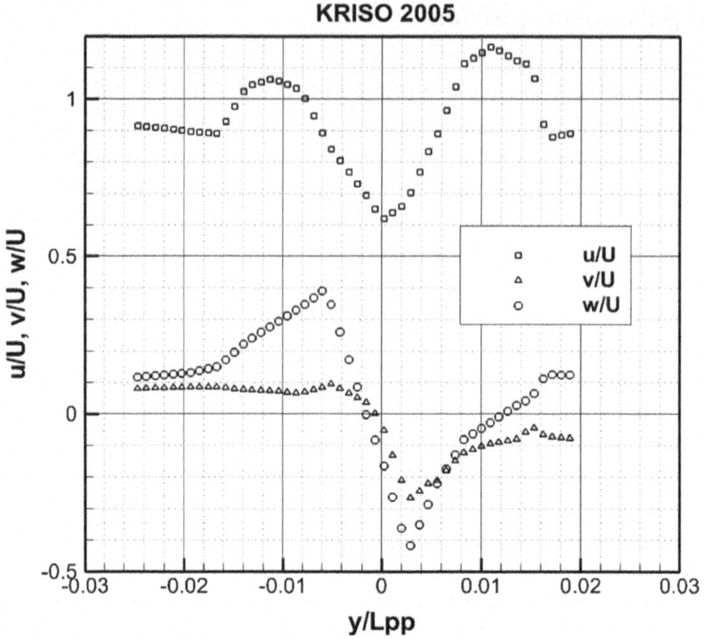

Fig. 7 Velocity profiles downstream of the propeller

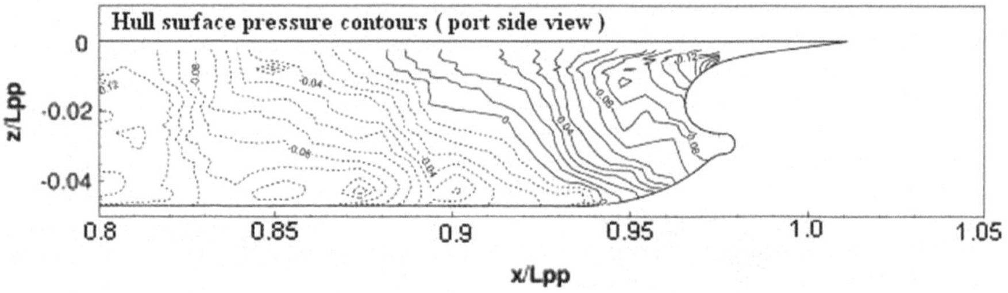

Fig. 8 Hull surface pressure coefficient contours with working propeller

2 Conclusions

Additional experimental data for KCS calm water resistance obtained after the workshop G2010 are reported. All KCS calm water models are listed at Table 1 in Chap. 4. In the present chapter, all experimental data are obtained with 7.3 m model from the KRISO and NMRI towing tanks. These experimental data have been widely used as a standard benchmarking case for CFD validation after G2010. However, turbulence quantities are still missing in the set of experimental data and we also feel inconvenienced by the lack of local flow measurement for self-propulsion including a rudder. It is desirable to have more data to make up the full set of experimental data, especially for the standard size KCS model ship.

References

Fujisawa, J., Ukon, Y., Kume, K., & Takeshi, H. (2000). Local velocity field measurements around the KCS model (SRI M. S. No. 631) in the SRI 400 m towing tank. NMRI report 2000.

Hino, T. (ed.) (2005). *Proceedings of CFD Workshop Tokyo 2005*. NMRI report 2005.

Kim, W. J., Van, S. H., & Kim, D. H. (2001). Measurement of flows around modern commercial ship models. *Experiment in Fluids, 31*, 567–578.

Larsson, L., Stern, F., & Visonneau, M. (2010). Proceedings of Gothenburg 2010 a workshop on numerical ship hydrodynamics. Gothenburg, Sweden.

Larsson, L., Stern, F., & Visonneau, M. (ed.) (2013). Numerical ship hydrodynamic –An assessment of the Gothenburg 2010 Workshop. Springer.

Tsukada, Y., Hori, T., Ukon, Y., Kume, K., & Takeshi, H. (2000). Surface pressure measurements on the KCS model (SRI M. S. No. 631) in the SRI 400 m towing tank. NMRI report 2000.

Van, S.H., Kim, W.J., Yim, G.T., Kim, D.H., & Lee, C.J. (1998). Experimental investigation of the flow characteristics around practical hull forms. In: *Proceedings* 3rd *Osaka Colloquium on Advanced CFD Applications to Ship Flow and Hull Form Design*. Osaka, Japan.

Experimental Data for KCS Added Resistance and ONRT Free Running Course Keeping/Speed Loss in Head and Oblique Waves

Yugo Sanada, Claus Simonsen, Janne Otzen, Hamid Sadat-Hosseini, Yasuyuki Toda, and Frederick Stern

Abstract Evaluation is performed of the data used for T2015 test cases for the KCS captive added resistance σ_{AR} and ONRT free running course keeping/speed loss in head and oblique waves. For KCS calm water resistance, the individual facility N-order level testing uncertainty is $U_{Xi} = 1\%D$ and the multiple facility standard deviation based on three institutes using 2 model sizes (L = 7.3 and 6.1 m) is SD = 0.74%D, such that the individual facility MxN-order level testing uncertainty is $U_{Di} = 1.75\%D$. U_{Xi} was not reported for sinkage and trim; however, the SD = 4 and 7%D, respectively. For KCS head waves, the analysis was based on three institutes with model sizes L = 6.1, 3.2 and 2.7 m. For the 6.1 m model, FORCE provided $U_{Xi} = 8$, 4 and 4%D, respectively, for σ_{AR} and first harmonic heave z_1/ζ_1 and pitch $\theta_1/\zeta_1 k$ amplitudes, whereas for the 2.7 m model, FORCE/IIHR provided U_{Xi} = 7/18, 8/2 and 9/5%D, respectively. In consideration of the differences in model sizes and rigid vs. surge free mounts the agreement between the facilities is reasonable: for the primary variables, SD = 3, 20 and 10%D for z_1/ζ_1 and $\theta_1/\zeta_1 k$ amplitudes and σ_{AR}, respectively; and for secondary variable SD = 66%D for first harmonic resistance C_{T1}. For KCS oblique waves, the data is only available from IIHR for tests in 2015 and 2016. The agreement is good, i.e., SD values are comparable to the corresponding values for head waves: for primary variables, SD = 6 and 63%D for z_1/ζ_1 and $\theta_1/\zeta_1 k$ amplitudes and σ_{AR}, respectively; and for secondary variable SD = 13%D for C_{T1}. For ONRT self-propulsion and head and oblique waves, the data is only available from IIHR and only preliminary uncertainty analysis is available. The evaluation showed reasonably reliable data for both KCS and ONRT. However,

---

Y. Sanada · F. Stern (✉)
University of Iowa and IIHR-Hydroscience & Engineering, Iowa City, IA, USA
e-mail: frederick-stern@uiowa.edu

C. Simonsen · J. Otzen
FORCE Technology, Lyngby, Denmark

H. Sadat-Hosseini
University of North Texas, Denton, TX, USA

Y. Toda
Osaka University, Suita, Japan

T. Hino et al. (eds.), *Numerical Ship Hydrodynamics*, Lecture Notes in Applied and Computational Mechanics 94, https://doi.org/10.1007/978-3-030-47572-7_4

clearly it is desirable to have data from more facilities including uncertainty analysis for assessment of facility biases, which will provide more robust data and uncertainty analysis for CFD validation.

Nomenclature

A	Desired wave amplitude (m)
b_R	Bias error
B	Breadth (m)
B_{WL}	Maximum beam at the waterline (m)
C_B	Block coefficient
C_F	Frictional resistance coefficient
$C_F^{15°}$	Nominal frictional resistance coefficient
C_R	Residual resistance coefficient
C_T	Total resistance coefficient
$C_T^{15°}$	Nominal total resistance coefficient (Converted from measured temperature to nominal temperature 15 °C)
C_{Ti}	Ith harmonic of total resistance coefficient
D	Depth (m)
DR	Dynamic range
f_{nz}	Natural heave frequency (Hz)
$f_{n\theta}$	Natural pitch frequency (Hz)
$f_{n\phi}$	Natural roll frequency (Hz)
FORCE	Technology, Denmark
Fr	Froude number
GM	Metacentric Height (m)
H	Wave height (m)
H_S	Significant wave height (m)
ITTC	International Towing Tank Conference
k	Wave number
$1 + k$	Form factor
KG	Vertical center of gravity (m)
k_{xx}	Radius of gyration in roll (m)
k_{yy}	Radius of gyration in pitch (m)
KRISO	Korea Research Institute of Ships and Ocean Engineering
KCS	KRISO Container Ship
L	Ship length (m)
LCB	Longitudinal center of buoyancy
L_{PP}	Length between perpendiculars (m)
L_{WL}	Length at the waterline (m)
n	Propeller revolutions (rps)
NMRI	National Maritime Research Institute
NRIFE	National Research Institute of Fisheries Engineering

ONRT	ONR Tumblehome
OU	Osaka University
Re	Reynolds number
S_R	Random error
S	Wetted surface area (m^2)
SD	Standard deviation
S_R	Wetted suface area of rudder (m^2)
S_W	Wetted surface area of hull (m^2)
t	Time (s)
T	Draft (m)
T_e	Wave encounter period (s)
T_P	Peak wave period (s)
u	Surge velocity (m/s)
u_i	I-th harmonic surge velocity amplitude (m/s)
u_R	Total uncertainty
U	Ship speed (m/s)
U_{Di}	Individual facility experimental MxN uncertainty
U_{Xi}	Individual facility experimental uncertainty
$U_{95,R}$	95% confidence experimental uncertainty
v	Sway velocity (m/s)
v_i	I-th harmonic sway velocity amplitude (m/s)
V	Ship speed (m/s)
x	Surge (m)
X	X force (N)
Z_i	I-th harmonic heave amplitude (m)
δ	Rudder angle (deg)
ΔFP	Draft changes at FP (m)
ΔAP	Draft changes at AP (m)
ζ	Wave elevation (m)
ζ_i	I-th harmonic wave amplitude
θ	Pitch angle (rad)
θ_i	I-th harmonic pitch amplitude (rad)
λ	Wave length (m)
ρ	Density of water (kg/m^3)
σ_{AR}	Added resistance
σ/L	Sinkage
τ	Trim(deg)
ϕ	Roll angle (rad)
ϕ_i	I-th harmonic roll amplitude (rad)
χ,μ	Wave encounter angle (deg)
ψ	Yaw angle (rad|deg)
ψ_i	I-th harmonic yaw amplitude (rad)
∇	Displacement (m^3)

1 Introduction

The experimental data used for T2015 test cases for KCS added resistance in regular head (2.10) and oblique (2.11) waves and ONRT free running course keeping/speed loss in regular head (3.12) and oblique (3.13) waves is evaluated, including in the former case calm water resistance, sinkage and trim and in the latter case calm water self-propulsion (3.9). The Froude numbers Fr for KCS and ONRT are 0.26 and 0.2, respectively. H/λ = 1/60 and 1/50 for the KCS and ONRT test cases, respectively. The evaluation compares data and uncertainty analysis from multiple facilities and model sizes when available. Only FORCE and IIHR provided experimental uncertainty estimates following ITTC (2008) and ASME (2013), respectively.

The individual facility data value is Xi and when available its individual facility N-order level testing uncertainty estimate is U_{Xi}. The multiple facility MxN order level testing average value $\overline{X} = \frac{1}{M}\sum_{i=1}^{M} X_i$ and precision limit are $P_{Xi} = 2\sigma_{Xi}$, where $\sigma_{X_i}^2 = \frac{1}{M-1}\sum_{i=1}^{M}\left(X_i - \overline{X}\right)^2$ is the standard deviation SD such that the individual facility MxN-order level testing uncertainty estimate is $U_{Di} = \sqrt{U_{Xi}^2 + P_{Xi}^2}$. The discussions use both Xi and $\overline{X}$, which are often simply referred to as D and similarly for U_{Xi} and U_{Di} as U_D for convenience. Note that herein SD was calculated using the population definition with factor 1/M vs/the sample definition with factor 1/(M-1) such that the values are under conservative by a factor of 3/2.

The validation variables for calm water KCS are resistance, sinkage and trim and for ONRT self-propulsion are propeller revolutions, sinkage and trim. Resistance and propeller revolutions are considered primary and motions secondary variables. For KCS head waves, added/0th harmonic resistance, and 1st harmonic heave and pitch are considered primary and 1st harmonic resistance as secondary variables; with additionally 1st harmonic roll as primary variable for oblique waves. Similarly, for ONRT free running in head and oblique waves primary variables are speed loss (0th harmonic axial velocity) and first harmonic motions which additionally include yaw are the primary variables and other 0th and 2nd harmonic variables as secondary.

2 KCS Calm Water

Table 1 lists the KCS full and model scale dimensions, mass properties, natural frequencies and form factors. The full-scale and NMRI/KRISO 7.3 m model data is from Zou and Larsson (2014). The FORCE 4.3 m model data is from Simonsen et al. (2013). The FORCE 6.1 m model data and uncertainty values are from Simonsen et al. (2014) and Otzen (2015a). The FORCE 2.7 m model data and uncertainty values are from Otzen (2015b). The IIHR 2.7 m model data is from both 2015 (Sadat-Hosseini

Table 1 Principle particulars for full scale, KRISO, NMRI, FORCE, and IIHR KCS models

		Full Scale	NMRI	KRISO	FORCE[1]	FORCE[2]	FORCE[3]	IIHR	OU
Length between perpendiculars	L_{PP} (m)	230	7.2786	7.2786	4.367	6.070 ± 0.001	2.700 ± 0.001	2.700 ± 0.001	3.200 ± 0.001
Length of waterline	L_{WL} (m)	232.482	7.3571	7.3571	4.414	6.135 ± 0.001	2.729 ± 0.001	2.729 ± 0.001	3.235 ± 0.001
Maximum beam of waterline	B_{WL} (m)	32.2	1.019	1.019	0.611	0.850 ± 0.001	0.378 ± 0.001	0.378 ± 0.001	0.4480 ± 0.0001
Depth	D (m)	19	0.601	0.601	–	–	–	0.223	0.2643 ± 0.0001
Draft	T (m)	10.8	0.3418	0.3418	0.205	0.2850 ± 0.0006	0.1268 ± 0.0006	0.1268 ± 0.0006	0.1503 ± 0.0001
Displacement volume	$\nabla/(m^3)$	52061.738	1.649	1.649	0.356	0.9571 ± 0.002	0.0842 ± 0.0005	0.0842 ± 0.0005	0.140
Wetted surface area w/o rudder	S_W (m^2)	9500.728	9.4379	9.4379	3.436	6.6177 ± 0.02	1.309 ± 0.008	1.309 ± 0.008	1.839
Wetted surface area of rudder	S_R (m^2)	115	0.0741	0.0741	–	0.08	0.016	0.016	0.022
Block coefficient (CB)	$\nabla/(L_{PP}B_{WL}T)$	0.6505	0.6505	0.6505	0.651	0.6505	0.643	0.643	0.651
Midship section coefficient (CM)		0.9849	0.9849	0.9849	0.985	0.9849	0.983	0.983	0.9849

(continued)

Table 1 (continued)

		Full Scale	NMRI	KRISO	FORCE[1]	FORCE[2]	FORCE[3]	IIHR	OU
LCB (%LPP), fwd+		−1.48	−1.48	−1.48	−1.48	−1.48	−1.48	−1.48	−1.48
Vertical Center of Gravity (from keel)	KG (m)	14.324				0.378		0.168 ± 0.02	
Metacentric height	GM (m)	0.623				0.016		0.0067 ± 0.0001	
Moment of Inertia	k_{xx}/B	0.4	0.4	0.4	0.4	0.4	0.4	0.39 ± 0.02	
Moment of Inertia	k_{yy}/L_{PP}	0.25	0.25	0.25	0.25	0.252	0.25	0.25 ± 0.01	0.25
Natural Frequency of Heave (Estimated)	f_{nz} (Hz)	0.121	0.121	0.681	0.880	0.746	1.119	1.119	
Natural Frequency of Roll (Estimated)	$f_{n\phi}$ (Hz)	0.031	0.031	0.168	0.217	0.185	0.276	0.283	
Natural Frequency of Pitch (Estimated)	$f_{n\theta}$ (Hz)	0.118	3.739	0.665	0.859	0.722	1.092	1.092	
Form factor	$1 + k$		1.082	1.083		1.087		1.088	1.283

et al. 2015a) and 2016 and uncertainty values are from Stocker (2016). The OU 3.2 m model data is from Toda (2018).

Adjustment of ship mass properties such as metacentric height (GM), radius of gyration in roll (k_{xx}) and pitch (k_{yy}) for the IIHR model was done as follows. (1) Ballasting weights were added to adjust the water line. The weights were distributed in a way that there was no pitch and roll. (2) Inclining tests were conducted in water. One of the ballasting weights that was placed around the ship center of gravity in the 1st step was selected as the moving weight and was moved transversally to induce a roll angle. If the calculated GM based on the roll angle is not within ±2% of the desired GM, the distribution of ballasting weights in the vertical direction would be changed and the test would be repeated. (3) The ship was removed from the water and the swing tests were conducted. k_{xx} and k_{yy} can be obtained by measuring the swing period. In the swing tests for k_{xx}, the k_{xx} was adjusted to the desired value by moving weights away from the centerline symmetrically. For the k_{yy} adjustment, the weights were moved away from the center of gravity in the longitudinal direction. The swing tests were repeated until the obtained k_{xx} and k_{yy} are within ±2% of the desired value. (4) Free roll decay experiments were conducted to evaluate the natural roll period in water.

IIHR made improvements to the experimental setup between 2015 and 2016. The improvements were using a reference wave gauge mounted to the surge free mount vs. sub-carriage and use of a laser tracking system and accelerometers for the resistance and x data reduction. The OU 3.2 m model data is from Sanada et al. (2018a). In summary, calm water resistance C_T, sinkage σ/L_{PP} and trim τ experimental data are available from five facilities (KRISO, NMRI, FORCE, IIHR and OU) using five different model sizes (7.3, 4.4, 6.1, 2.7 and 3.2 m) and different types of mounts.

The resistance, sinkage and trim data-reduction equations are:

$$C_T^{15^\circ} = C_R + C_F^{15^\circ}(1+k) \tag{1}$$

$$C_T = \frac{X}{1/2\rho S V^2} \tag{2}$$

$$C_F = .075/(\log_{10}\mathrm{Re} - 2)^2 \tag{3}$$

$$C_R = C_T - C_F(1+k) \tag{4}$$

$$\frac{C_T}{C_F} = m\frac{Fr^4}{C_F} + (1+k) \tag{5}$$

$$\frac{\sigma}{L_{PP}} = (\Delta FP + \Delta AP)/L_{PP} \tag{6}$$

$$\tau = (\Delta FP - \Delta AP)/L_{PP} \tag{7}$$

where ρ is the water density, S is the wetted surface area excluding wetted surface area of rudder ($S = S_W$), V is the ship velocity, L_{PP} is the ship length, ΔFP and ΔAP are the draft changes at the fore and aft perpendiculars, k is the form factor, X is the resistance in calm water, m is the slope of the line based on Prohaska method, $C_T^{15°}$ corrects the resistance coefficient for the same temperature at 15 °C, C_F is the frictional resistance coefficient, and C_R partially removes scale/Re effects. Form factor k was evaluated using Eq. (5), i.e., Prohaska method.

Tables 2 and 3 and Fig. 1 provide $C_T^{15°}$ without and with form factor. Figure 1 shows $C_T^{15°}$ increases from large to small model sizes and values are often higher with form factor. All the available data is tabulated for all Fr. Most data is for Fr = 0.26. The standard deviation SD is presented using both %D and %DR where D is the average data value and DR is the dynamic range and both with and without the 2.7 and 3.2 m model data. The scatter in the results is much larger including the 2.7 and 3.2 m models, but reasonably small for the larger models both without and with form factor, as shown in Fig. 1. For Fr = 0.26, the scatter in the results for models excluding the 2.7 m model is < 7% and 3%D without and with form factor, respectively. Including the 2.7 m model increases the scatter by factors of about 2 and 5 for without and with form factor, respectively. The differences without vs. with form factor are small if all models are included but large if the smaller models are excluded. Tables 4 and 5 and Fig. 2 report residual resistance C_R without and with form factor and they show similar results to those for $C_T^{15°}$, but somewhat larger %D and %DR as $C_R < C_T^{15°}$. In this case, the SD differences without vs. with the form factor are large both including and excluding the smaller models. Thus, the SD results without and with form factor do not show a consistent trend.

Experimental uncertainty U_{Xi} estimates for calm water tests are provided from only two institutes. For C_T, FORCE provided $U_{Xi} = 1$ and 4%D for 6.1 and 2.7 m models, respectively. IIHR provided $U_{Xi} = 2.5$%D for the 2.7 m model. For CFD validation studies in Chap. 9 for test case 2.10, the SD for C_R with form factor and without the 2.7 m model was normalized by D for FORCE and multiplied by 2 and was used as the MxN-order level testing precision limit P_{Xi} for C_T [Chap. 9, Eq. (29)], i.e., $4.7 \times (0.602/3.863) \times 2 = 1.46$%D.

Tables 6 and 7 and Fig. 3 show the sinkage and trim provided from different institutes. Figure 3 shows magnitudes of both sinkage and trim increase for increasing Fr. The trends for the sinkage and trim scatter are like those for $C_T^{15°}$ and C_R i.e. including the 2.7 and 3.2 m models increases the scatter. For Fr = 0.26, the scatter excluding the 2.7 and 3.2 m model is 4 and 7%D for sinkage and trim, respectively, i.e. larger than that for $C_T^{15°}$ with form factor.

Only IIHR provided U_{Xi} estimates of 1 and 66%D for sinkage and trim, respectively. For CFD validation studies in Chap. 9 for test case 2.10, SD = 4 and 7%D for the larger models was used for sinkage and trim, respectively, normalized by D for FORCE, i.e., for sinkage $3.8 \times (-0.2014/-0.2074) \times 2 = 7.38$%D and for trim $7 \times (-0.168/-0.165) \times 2 = 14.25$%D were used as the as the MxN-order level testing precision limit [Chap. 9, Eq. (29)]. U_{Xi} was not considered for CFD validation since it was not available for the large models. Note that FORCE and IIHR both used added

Table 2 Comparison of total resistance coefficient calculated without using Prohaska's method, $C_T^{15^\circ}$, for KCS in calm water

Institute	NMRI	KRISO	FORCE[1]	FORCE[2]	IIHR (Aug 2016)	OU (Oct 2016)	All Facilities		Without 2.7 m and OU model	
Fr	$C_T^{15^\circ} \times 10^3$						Ave (D)	SD%D {SD%DR}	Ave (D)	SD%D {SD%DR}
0.0867	3.841						**3.841**	**0.0**	**3.841**	
0.1084	3.740	3.729		3.916			**3.795**	**2.3**	**3.795**	**2.3**
0.1300	3.657						**3.657**	**0.0**	**3.657**	
0.1517	3.580	3.578		3.767			**3.641**	**2.4**	**3.641**	**2.4**
0.1734	3.503			3.678			**3.591**	**2.4**	**3.591**	**2.4**
0.1950	3.438	3.417		3.605			**3.487**	**2.4**	**3.487**	**2.4**
0.2059	3.415						**3.415**	**0.0**	**3.415**	
0.2167	3.404			3.593			**3.498**	**2.7**	**3.498**	**2.7**
0.2276	3.409	3.413					**3.411**	**0.0**	**3.411**	**0.0**
0.2384	3.440			3.639			**3.540**	**2.8**	**3.540**	**2.8**
0.2492	3.505						**3.505**	**0.0**	**3.505**	
0.2601	3.650	3.659	4.31	3.860	5.203	4.637	**4.220**	**13.4**	**3.870**	**6.9**
0.2709	3.981						**3.981**	**0.0**	**3.981**	
0.2817	4.459	4.447		4.561			**4.489**	**1.1**	**4.489**	**1.1**
Ave (Total)	**3.645**	**3.707**		**3.827**			3.719	**2.1.0{14.6*}**	3.726	**2.6{17.6 +}**

*DR = 0.539
+DR = 0.539

Table 3 Comparison of total resistance coefficient calculated using Prohaska's method, $C_T^{15°}$, for KCS in calm water

Institute	NMRI	KRISO	FORCE[2]	IIHR (Aug 2016)	OU (Oct 2016)	All Facilities		Without 2.7 m and OU model	
Fr	$C_T^{15°} \times 10^3$					Ave (D)	SD%D {SD%DR}	Ave (D)	SD%D {SD%DR}
0.0867	3.827					**3.840**	**0.4**	**3.840**	**0.4**
0.1084	3.727	3.721	3.919			**3.789**	**2.4**	**3.789**	**2.4**
0.1300	3.645	3.648				**3.647**	**0.0**	**3.647**	**0.0**
0.1517	3.568	3.573	3.770			**3.637**	**2.6**	**3.637**	**2.6**
0.1734	3.492	3.487	3.681			**3.554**	**2.5**	**3.554**	**2.5**
0.1950	3.428	3.412	3.608			**3.482**	**2.6**	**3.482**	**2.6**
0.2059	3.404	3.392				**3.398**	**0.2**	**3.398**	**0.2**
0.2167	3.393	3.389	3.596			**3.460**	**2.8**	**3.460**	**2.8**
0.2276	3.399	3.406				**3.402**	**0.1**	**3.402**	**0.1**
0.2384	3.430	3.441	3.642			**3.504**	**2.8**	**3.504**	**2.8**
0.2492	3.495	3.514				**3.504**	**0.3**	**3.504**	**0.3**
0.2601	3.640	3.652	3.863	4.888	4.676	**4.144**	**12.8**	**3.718**	**2.8**
0.2709	3.972	3.948				**3.960**	**0.3**	**3.960**	**0.3**
0.2817	4.449	4.443	4.564			**4.485**	**1.2**	**4.485**	**1.2**
Ave (Total)	**3.634**	**3.634**	**3.830**	**4.888**		**3.7**	**2.2{15.1*}**	**3.670**	**1.6{10.7 +}**

*DR = 0.544
+DR = 0.544

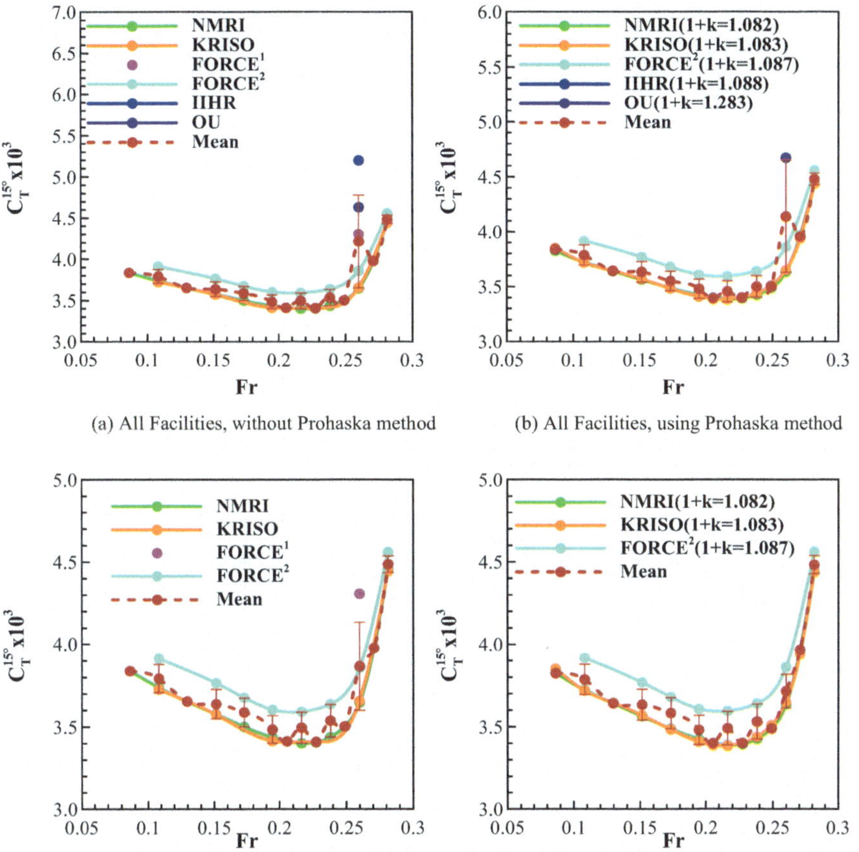

(a) All Facilities, without Prohaska method (b) All Facilities, using Prohaska method

(c) Excluding 2.7m model, without Prohaska method (d) Excluding 2.7m model, using Prohaska method

Fig. 1 Comparison of $C_T^{15°} \times 10^3$ calculated with and without Prohaska's method, at NMRI, KRISO, FORCE, IIHR, OU and mean of the facilities. Standard deviation between facilities is included with the mean line

resistance mounts designed for large motions and not conventional calm water test setups. Further FORCE used mount with fixed surge, while IIHR used free surge.

The IIHR and OU surge free mounts for calm water resistance, sinkage and trim measurements were designed primarily for ship motions studies, and possibly small model size are reasons for its large scale-effect/facility bias. IIHR and OU sinkage and trim were measured by a 2DOF mount (sinkage and trim free) installed at the hull center of gravity. The displacement measurement at the bow (FP) and stern (AP) performed in the ordinary calm water resistance test is more accurate in measuring small displacements than this method. IIHR and OU should re-take the measurements using rigid mounts in their towing tanks to provide better calm water reference data. All FORCE calm water measurements are also performed with added resistance mounts designed for large motions, but with fixed surge and not with conventional

Table 4 Comparison of residuary resistance coefficient calculated without using Prohaska's method, C_R, for KCS in calm water

Institute	NMRI	KRISO	FORCE[1]	FORCE[2]	IIHR (Aug 2016)	OU (Oct 2016)	All Facilities		Without 2.7 m and OU model	
Fr	$C_R \times 10^3$						Ave (D)	SD%D {SD%DR}	Ave (D)	SD%D {SD%DR}
0.0867	0.403						**0.403**	**0.0**	**0.403**	
0.1084	0.440	0.427		0.448			**0.438**	**1.9**	**0.438**	**1.9**
0.1300	0.464						**0.464**	**0.0**	**0.464**	
0.1517	0.473	0.472		0.505			**0.483**	**3.2**	**0.483**	**3.2**
0.1734	0.469			0.495			**0.482**	**2.7**	**0.482**	**2.7**
0.1950	0.465	0.444		0.485			**0.465**	**3.6**	**0.465**	**3.6**
0.2059	0.470						**0.470**	**0.0**	**0.470**	
0.2167	0.484			0.532			**0.508**	**4.7**	**0.508**	**4.7**
0.2276	0.514	0.516					**0.515**	**0.2**	**0.515**	
0.2384	0.568			0.629			**0.599**	**5.1**	**0.599**	**5.1**
0.2492	0.654						**0.654**	**0.0**	**0.654**	
0.2601	0.820	0.828	1.07	0.897	1.177	1.111	**0.984**	**14.4**	**0.905**	**11.3**
0.2709	1.170						**1.170**	**0.0**	**1.170**	
0.2817	1.666	1.655		1.635			**1.652**	**0.8**	**1.652**	**0.8**
Ave (Total)	**0.647**	**0.724**		**0.703**			**0.663**	**2.6{2.8*}**	**0.658**	**4.2{4.4 +}**

*DR = 0.625
+DR = 0.625

Table 5 Comparison of residuary resistance coefficient calculated using Prohaska's method, C_R, for KCS in calm water

Institute	NMRI	KRISO	FORCE[2]	IIHR (Aug 2016)	OU (Oct 2016)	All Facilities		Without 2.7 m and OU model	
Fr	$C_R \times 10^3$					Ave (D)	SD%D {SD%DR}	Ave (D)	SD%D {SD%DR}
0.0867	0.107					**0.118**	**9.9**	**0.118**	**9.9**
0.1084	0.156	0.148	0.150			**0.151**	**2.4**	**0.151**	**2.4**
0.1300	0.190	0.190				**0.190**	**0.1**	**0.190**	**0.1**
0.1517	0.207	0.209	0.224			**0.213**	**3.6**	**0.213**	**3.6**
0.1734	0.209	0.201	0.220			**0.210**	**3.7**	**0.210**	**3.7**
0.1950	0.211	0.192	0.216			**0.206**	**5.1**	**0.206**	**5.1**
0.2059	0.217	0.202				**0.210**	**3.6**	**0.210**	**3.6**
0.2167	0.234	0.227	0.268			**0.243**	**7.3**	**0.243**	**7.3**
0.2276	0.266	0.270				**0.268**	**0.8**	**0.268**	**0.8**
0.2384	0.322	0.330	0.369			**0.340**	**6.1**	**0.340**	**6.1**
0.2492	0.410	0.427				**0.418**	**2.0**	**0.418**	**2.0**
0.2601	0.577	0.587	0.642	0.862	0.152	**0.564**	**40.8**	**0.602**	**4.7**
0.2709	0.930	0.903				**0.917**	**1.5**	**0.917**	**1.5**
0.2817	1.428	1.419	1.383			**1.410**	**1.4**	**1.410**	**1.4**
Ave (Total)	**0.390**	**0.388**	**0.434**			**0.390**	**6.3{3.8*}**	**0.393**	**3.7{2.3 +}**

*DR = 0.646
+DR = 0.646

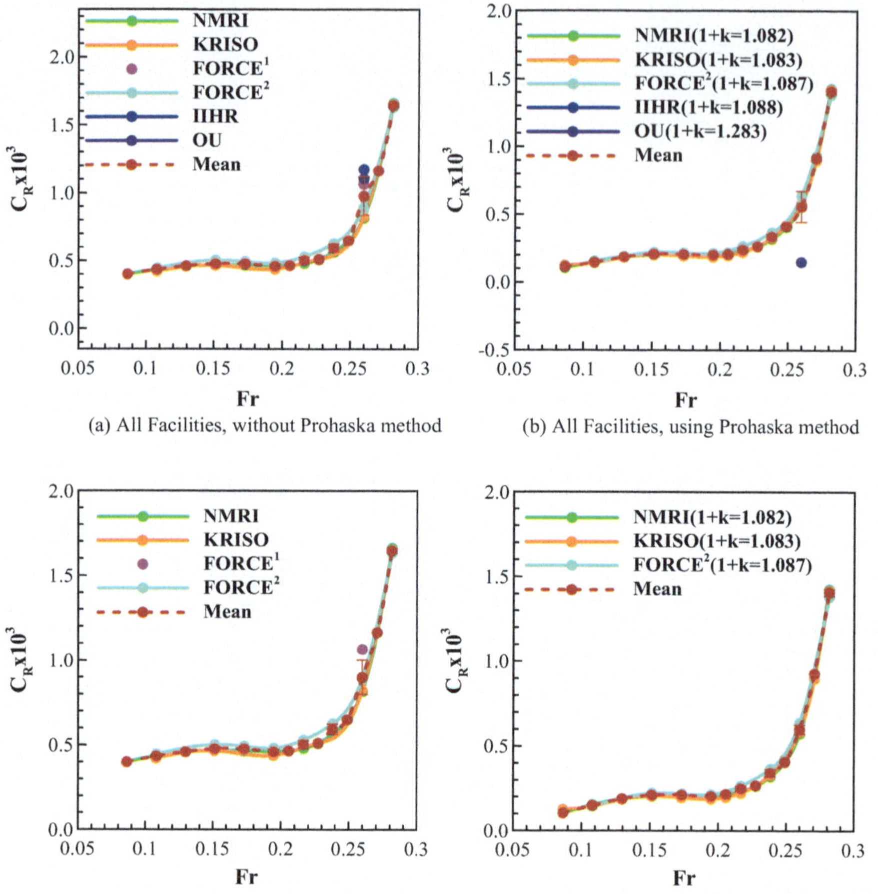

(a) All Facilities, without Prohaska method (b) All Facilities, using Prohaska method

(c) Excluding 2.7m model, without Prohaska method (d) Excluding 2.7m model, using Prohaska method

Fig. 2 Comparison of C_R calculated with and without Prohaska's method, at NMRI, FORCE, IIHR, OU and mean of the facilities. Standard deviation between facilities is included with the mean line

calm water test setup. Ideally comparison of calm water resistance across facilities should be made based on conventional resistance test setups, but that was not possible, since data was taken in connection with added resistance test campaigns.

For KCS calm water resistance based on three institutes using 2 model sizes with $L \geq 6$ m, the individual facility N-order level testing uncertainty $U_{Xi} = 1\%D$ (only reported one facility) and the MxN-order level testing standard deviation SD = 0.74%D, such that the best estimate multiple facility uncertainty is $U_{Di} = [U_{Xi}^2 + (2SD)^2]^{1/2} = 1.75\%D$. U_{Xi} was not reported for sinkage and trim; however, SD = 4 and 7%D such that $U_{Di} = 2SD = 7.38$ and 14.15%D, respectively.

Table 6 Comparison of sinkage for KCS in calm water

Institute	NMRI	KRISO	FORCE[1]	FORCE[2]	IIHR (Aug 2016)	OU (Oct 2016)	All Facilities		Without 2.7 m and OU model	
Fr	$\sigma/L_{PP} \times 10^2$						Ave (D)	SD%D {SD%DR}	Ave (D)	SD%D {SD%DR}
0.0867	−0.0216						**−0.0105**	**105.7**	**−0.0105**	**105.7**
0.1084	−0.0316	−0.0124		−0.0382			**−0.0274**	**40.0**	**−0.0274**	**40.0**
0.1300	−0.0447	−0.0219					**−0.0333**	**34.2**	**−0.0333**	**34.2**
0.1517	−0.0612	−0.0378		−0.0646			**−0.0545**	**21.8**	**−0.0545**	**21.8**
0.1734	−0.0807	−0.0608		−0.0902			**−0.0772**	**15.9**	**−0.0772**	**15.9**
0.1950	−0.1030	−0.0823		−0.1089			**−0.0980**	**11.7**	**−0.0980**	**11.7**
0.2059	−0.1154	−0.1026					**−0.1090**	**5.9**	**−0.1090**	**5.9**
0.2167	−0.1288	−0.1185		−0.1315			**−0.1263**	**4.4**	**−0.1263**	**4.4**
0.2276	−0.1434	−0.1297					**−0.1365**	**5.0**	**−0.1365**	**5.0**
0.2384	−0.1594	−0.1529		−0.1736			**−0.1619**	**5.4**	**−0.1619**	**5.4**
0.2492	−0.1771	−0.1716					**−0.1743**	**1.6**	**−0.1743**	**1.6**
0.2601	−0.1967	−0.1915	−0.2100	−0.2074	−0.1704	−0.1747	**−0.1918**	**7.8**	**−0.2014**	**3.8**
0.2709	−0.2184	−0.2138					**−0.2161**	**1.1**	**−0.2161**	**1.1**
0.2817	−0.2423	−0.2339		−0.2572			**−0.2445**	**3.9**	**−0.2445**	**3.9**
Ave (Total)	**−0.1231**	**−0.1092**		**−0.1340**			**−0.1187**	**18.9{19.2*}**	**−0.1193**	**18.6{19.0 + }**

*DR = 0.117
+DR = 0.117

Table 7 Comparison of trim for KCS in calm water

Institute	NMRI	KRISO	FORCE[1]	FORCE[2]	IIHR (Aug 2016)	OU (Oct 2016)	All Facilities		Without 2.7 m and OU model	
Fr	τ [deg]						Ave (D)	SD%D {SD%DR}	Ave (D)	SD%D {SD%DR}
0.0867	−0.011						**−0.009**	**28.8**	**−0.009**	**28.8**
0.1084	−0.019	−0.017		−0.029			**−0.022**	**25.0**	**−0.022**	**25.0**
0.1300	−0.031	−0.040					**−0.035**	**13.0**	**−0.035**	**13.0**
0.1517	−0.045	−0.053		−0.056			**−0.051**	**9.3**	**−0.051**	**9.3**
0.1734	−0.062	−0.077		−0.082			**−0.073**	**11.7**	**−0.073**	**11.7**
0.1950	−0.082	−0.097		−0.103			**−0.094**	**9.2**	**−0.094**	**9.2**
0.2059	−0.095	−0.105					**−0.100**	**5.2**	**−0.100**	**5.2**
0.2167	−0.108	−0.113		−0.128			**−0.116**	**7.3**	**−0.116**	**7.3**
0.2276	−0.122	−0.127					**−0.124**	**2.1**	**−0.124**	**2.1**
0.2384	−0.135	−0.144		−0.168			**−0.149**	**9.3**	**−0.149**	**9.3**
0.2492	−0.146	−0.158					**−0.152**	**4.0**	**−0.152**	**4.0**
0.2601	−0.152	−0.169	−0.185	−0.165	−0.022	0.006	**−0.114**	**66.7**	**−0.168**	**7.0**
0.2709	−0.150	−0.171					**−0.160**	**6.8**	**−0.160**	**6.8**
0.2817	−0.132	−0.159		−0.221			**−0.171**	**21.9**	**−0.171**	**21.9**
Ave (Total)	**−0.092**	**−0.102**		**−0.119**			**−0.098**	**15.7{19.0*}**	**−0.102**	**11.5{14.4 +}**

*DR = 0.081
+DR = 0.081

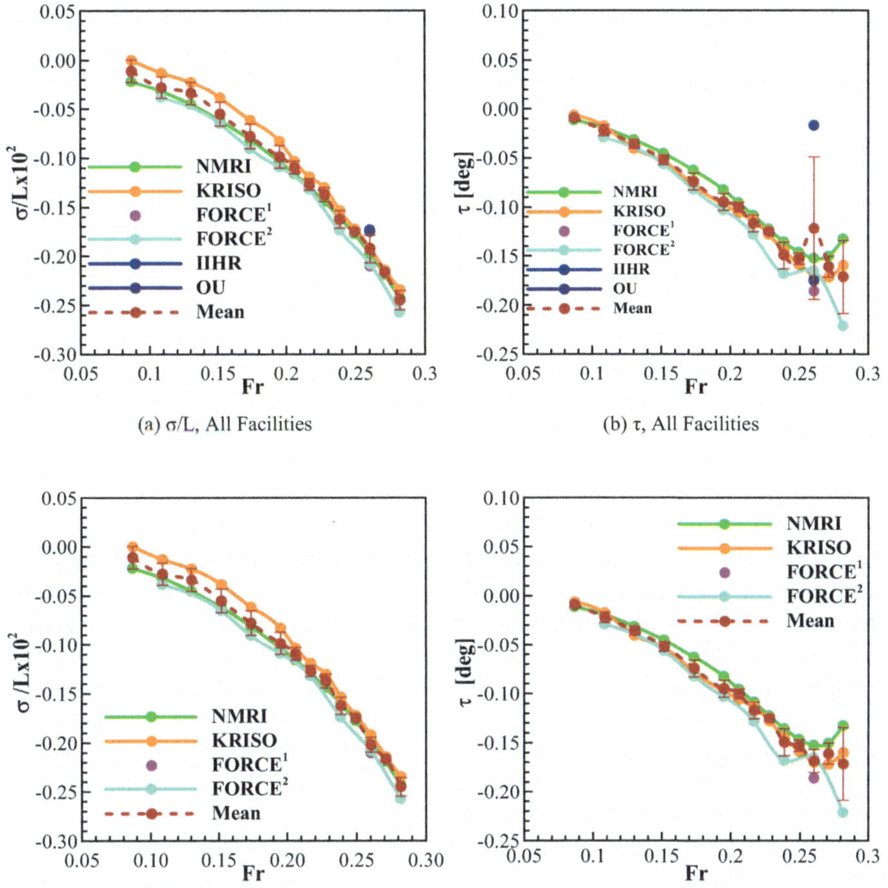

(a) σ/L, All Facilities

(b) τ, All Facilities

(c) σ/L, Excluding 2.7m model

(d) τ, Excluding 2.7m model

Fig. 3 Comparison of σ/L and τ, at NMRI, KRISO, FORCE, IIHR, OU and mean of the facilities. Standard deviation between facilities is included with the mean line

3 KCS Head Waves

Regular head wave experimental data are available from three facilities using three model sizes (6.1, 2.7 and 3.2 m). The FORCE 6.1 m model data and uncertainty values are from Simonsen et al. (2014) and Otzen (2015a). The FORCE 2.7 m model data and uncertainty values are from Otzen (2015b). The IIHR 2.7 m model data is from both 2015 (Sadat-Hosseini et al. 2015a) and 2016 and uncertainty values are from Stocker (2016). IIHR made improvements to the experimental setup between 2015 and 2016, as discussed previously. Unfortunately, the FORCE 4.3 m model data from Simonsen et al. (2013) is not available. Data is also recently available from Osaka University (Toda 2018), although not used at T2015 useful for assessing the scale effects/facility biases.

Table 8 provides SD analysis for calm water resistance, sinkage, trim and added resistance variables in head waves for Fr = 0.26. Here again the SD trends are scattered and generally larger than for the larger models only, especially for C_R with the Prohaska method and trim.

Figures 4, 5, 6, 7 and 8 compare time histories from FORCE 6.1 m, IIHR 2.7 m, OU 3.2 m models for wavelengths of $\lambda/L = 0.65, 0.85$, 1.15, 1.37, and 1.95 for wave elevation (ζ/A), resistance (C_T), surge (x/ζ_1), heave (z/ζ_1), and pitch ($\theta/\zeta_1 k$). Motions are non-dimensional by either wave amplitude (ζ_1) or wave slope ($\zeta_1 k$). Unlike FORCE and OU, IIHR used a roll free condition since roll measurement was important for other headings for test case 2.11. Results show differences between FORCE, IIHR and OU data for C_T, heave, and pitch for short wave length conditions. Note that FORCE experienced structural resonance in the rigid mount close to the heave and pitch resonance condition. This is most clearly seen in C_T for λ/L equal to 0.85 and 1.15 in Figs. 5 and 6.

Figure 9 compares AR and 0th, 1st and 2nd harmonic amplitudes for ζ_1/A, C_T, z/ζ_1, $\theta/\zeta_1 k$; and Fig. 10 compares their 1st and 2nd harmonic phases. For the IIHR data the natural surge damping effects of the surge free mount is removed by a least square method (Stocker 2016) and the harmonic analysis window size for all λ/L in head waves is 5Te. FORCE, IIHR and OU data agree well for AR, C_{T0}, z_0/ζ_1 and 1st harmonic amplitudes, except C_{T1} (OU is outlier). The $\theta_0/\zeta_1 k$ and 2nd harmonic amplitudes for C_T (2nd harmonic amplitudes for heave and pitch are small) and C_{T1} 1st and all 2nd harmonic phases show differences between the three measurements. All three measurements show similar trends vs. wavelength for all 0th harmonic amplitudes and heave and pitch 1st harmonic phases.

Tables 9, 10, 11, 12 and 13 provide SD analysis for added resistance $\sigma_{AR} = \frac{\overline{X} - X_{calm}}{\rho g A^2 B^2 / L}$ ($\overline{X}$ is the mean resistance in waves, $A = \zeta_1$ is the wave amplitude, and B is the ship beam) and 1st and 2nd harmonic amplitudes for ζ/A, C_T, z/ζ_1, and $\theta/\zeta_1 k$, respectively. Table 14, 15 and 16 provide SD analysis for 0th harmonic amplitudes for C_T, z/ζ_1 and $\theta/\zeta_1 k$, respectively; and Tables 17, 18, 19 and 20 provide SD analysis for 1st and 2nd harmonic phases for ζ/A, C_T, z/ζ_1 and $\theta/\zeta_1 k$, respectively. The scatter for the 1st harmonic amplitudes is mostly small, i.e., 4, 66, 3 and 20%D, respectively, for ζ/A, C_T, z/ζ_1, $\theta/\zeta_1 k$. The scatter for 1st harmonic amplitudes are often larger for short waves conditions. The scatter for 0th harmonic for C_T (SD = 5%D) and z/ζ_1 (SD = 16%D) are relatively small. The scatter for the σ_{AR} (SD = 10%D), 0th harmonic $\theta/\zeta_1 k$ (SD = 163%D), 1st harmonic phases, and 2nd harmonic amplitudes and phases are mostly large.

FORCE provided U_{Xi} for σ_{AR}, heave and pitch, i.e., 8, 4 and 4%D, respectively, for the 6.1 m model. FORCE/IIHR provided U_{Xi} for AR, heave and pitch, i.e., 7/18, 8/2 and 9/5%D, respectively, for the 2.7 m model. For CFD validation studies in Chap. 9 test case 2.10, SD was not considered, and the U_{Xi} values provided by FORCE for the 6.1 m model were used, including estimates for the uncertainty due to statistical convergence.

Table 8 Comparison of total resistance coefficient, residuary resistance coefficient, sinkage and trim in calm water and added resistance variables in head waves (Fr = 0.26)

Institute	L_{PP} [m]	Without Prohaska		With Prohaska		$\sigma/L_{PP} \times 10^2$	τ [deg]	Added resistance in head waves[a]				
		$C_T^{15°} \times 10^3$	$C_R \times 10^3$	$C_T^{15°} \times 10^3$	$C_R \times 10^3$			σ_{AR}	ζ_1/A	C_{T1}	z_1/ζ_1	$\theta_1/\zeta_1 k$
FORCE	6.07	3.860	0.897	3.863	0.642	−0.2074	−0.165	4.850	1.010	9.790	0.613	0.598
IIHR	2.7	5.203	1.177	4.888	0.862	−0.1704	−0.022	5.180	1.070	9.300	0.610	0.606
OU	3.2	4.637	1.111	4.676	0.152	−0.1747	0.006	5.580	0.983	0.500	0.595	0.536
Ave. (D)		**4.567**	**1.062**	**4.476**	**0.552**	**−0.1842**	**−0.060**	**5.200**	**1.020**	**6.530**	**0.606**	**0.580**
SD%D		**12.1**	**11.2**	**9.9**	**53.8**	**9.0**	**124.3**	**10.2**	**4.3**	**66.4**	**2.7**	**20.3**
DR%D		**14.7**	**13.2**	**11.5**	**64.3**	**10.1**	**131.9**	**14.0**	**8.5**	**142.3**	**3.0**	**12.1**

[a] Average of all 5 different λ/L

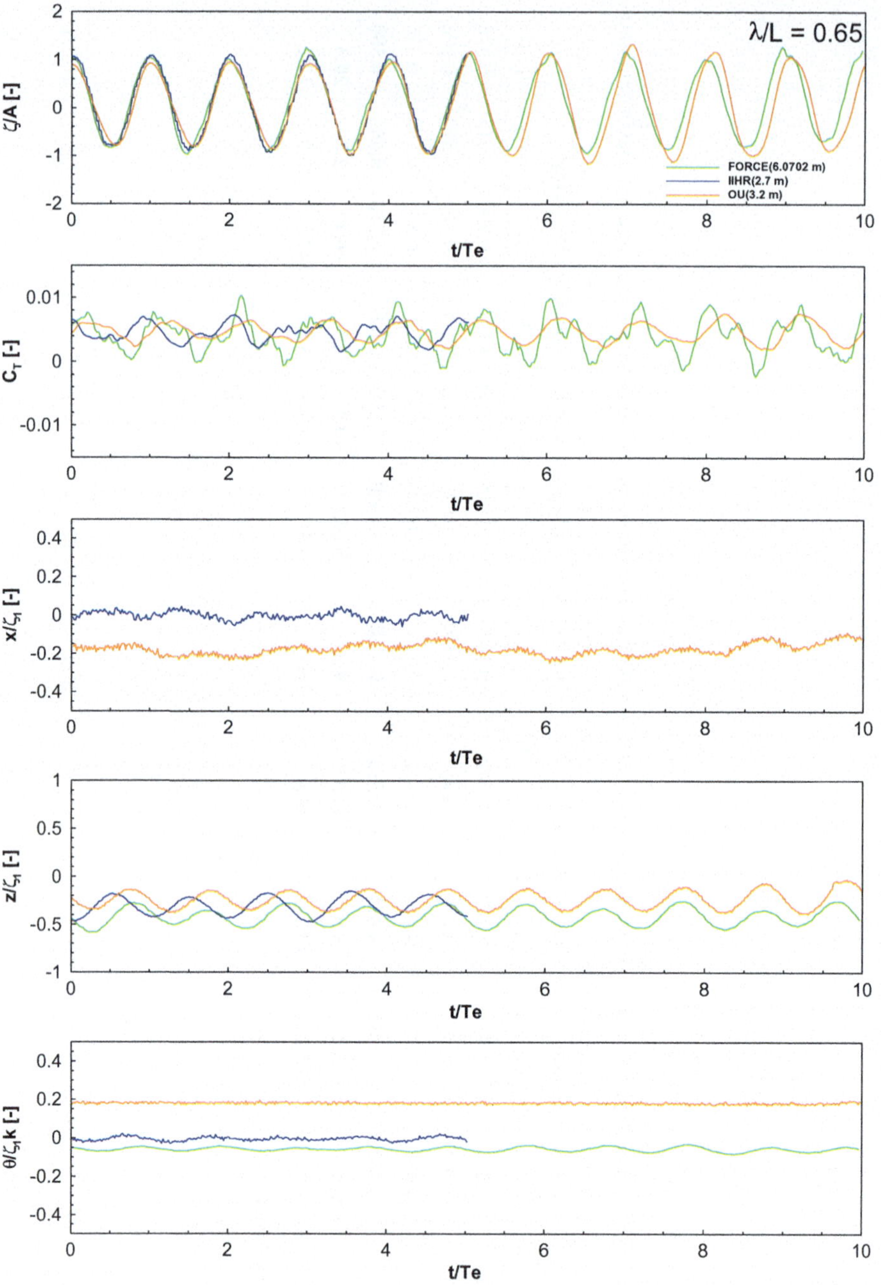

Fig. 4 Time series for force and motions of KCS in regular head waves at λ/L = 0.65 for FORCE[2] IIHR and OU

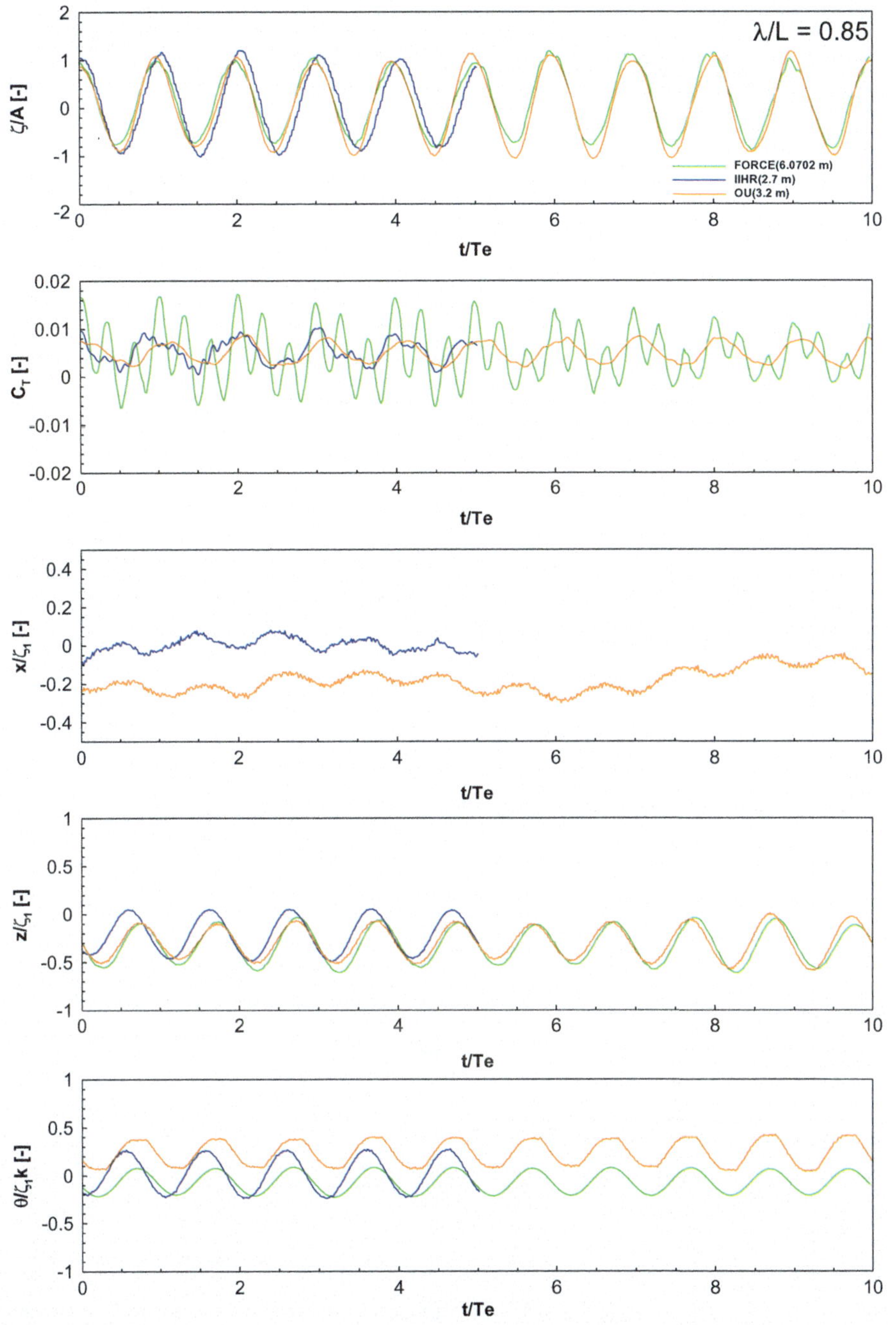

Fig. 5 Time series for force and motions of KCS in regular head waves at λ/L = 0.85 for FORCE[2], IIHR and OU

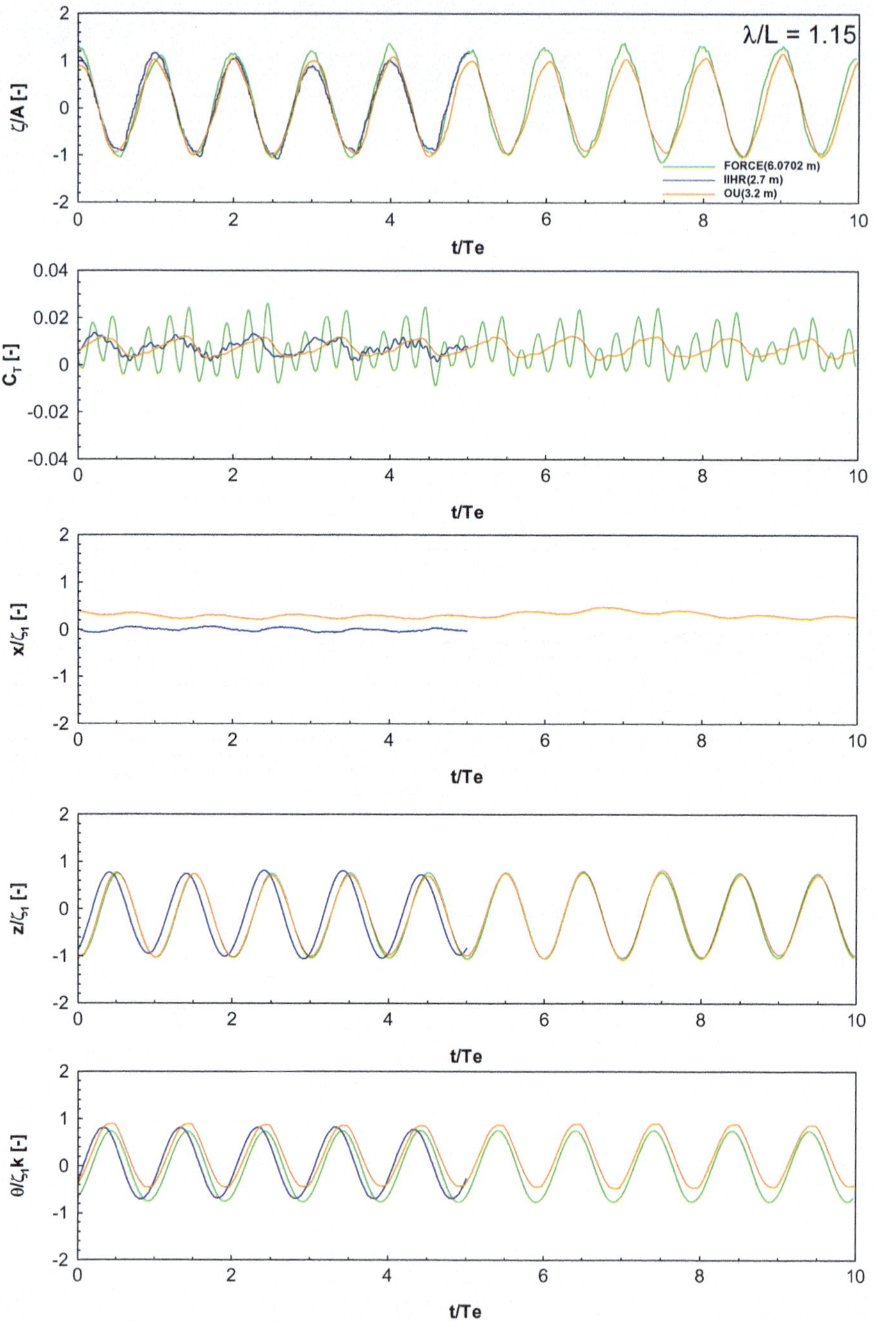

Fig. 6 Time series for force and motions of KCS in regular head waves at $\lambda/L = 1.15$ for FORCE[2],IIHR and OU

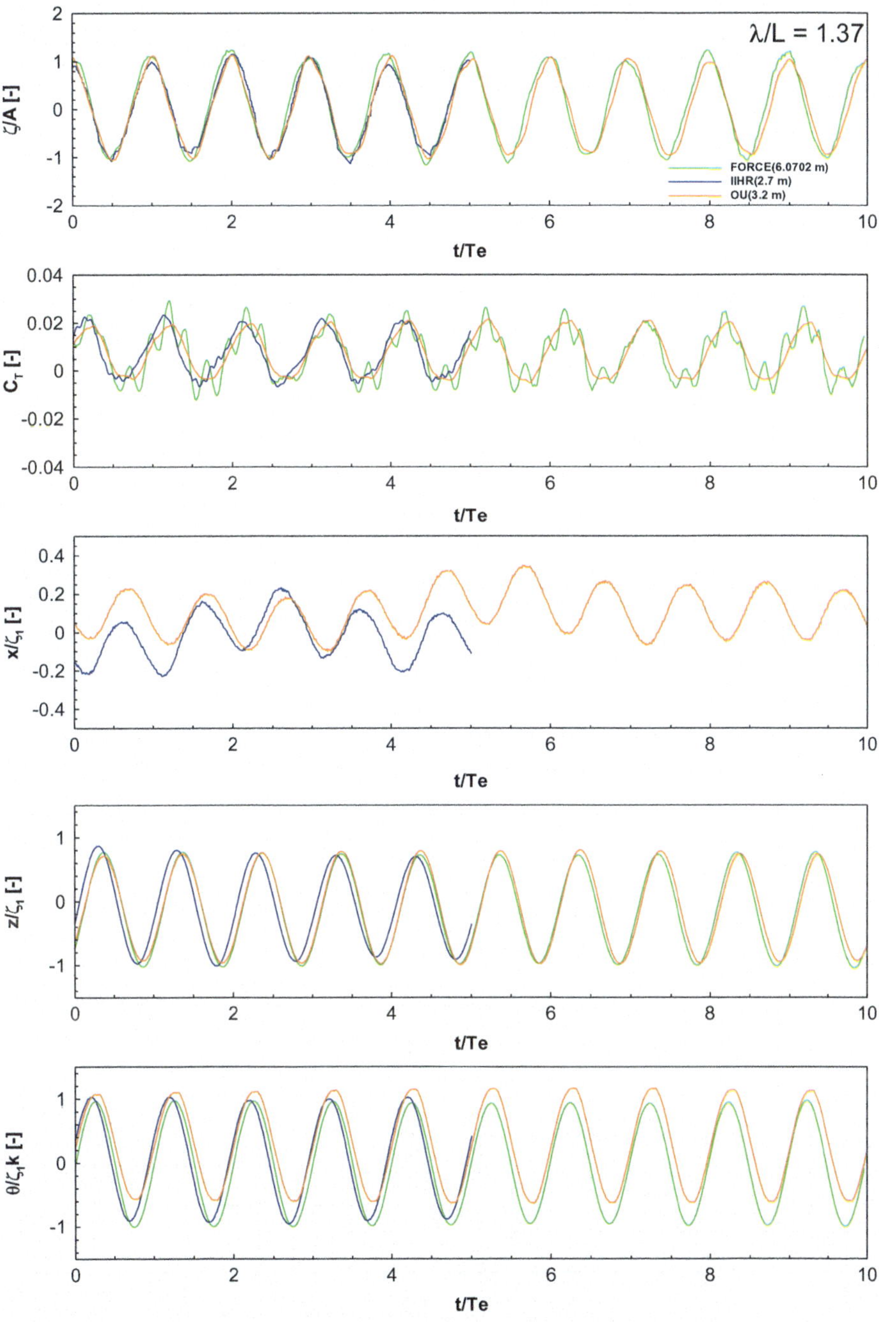

Fig. 7 Time series for force and motions of KCS in regular head waves at λ/L = 1.37 for FORCE[2],IIHR and OU

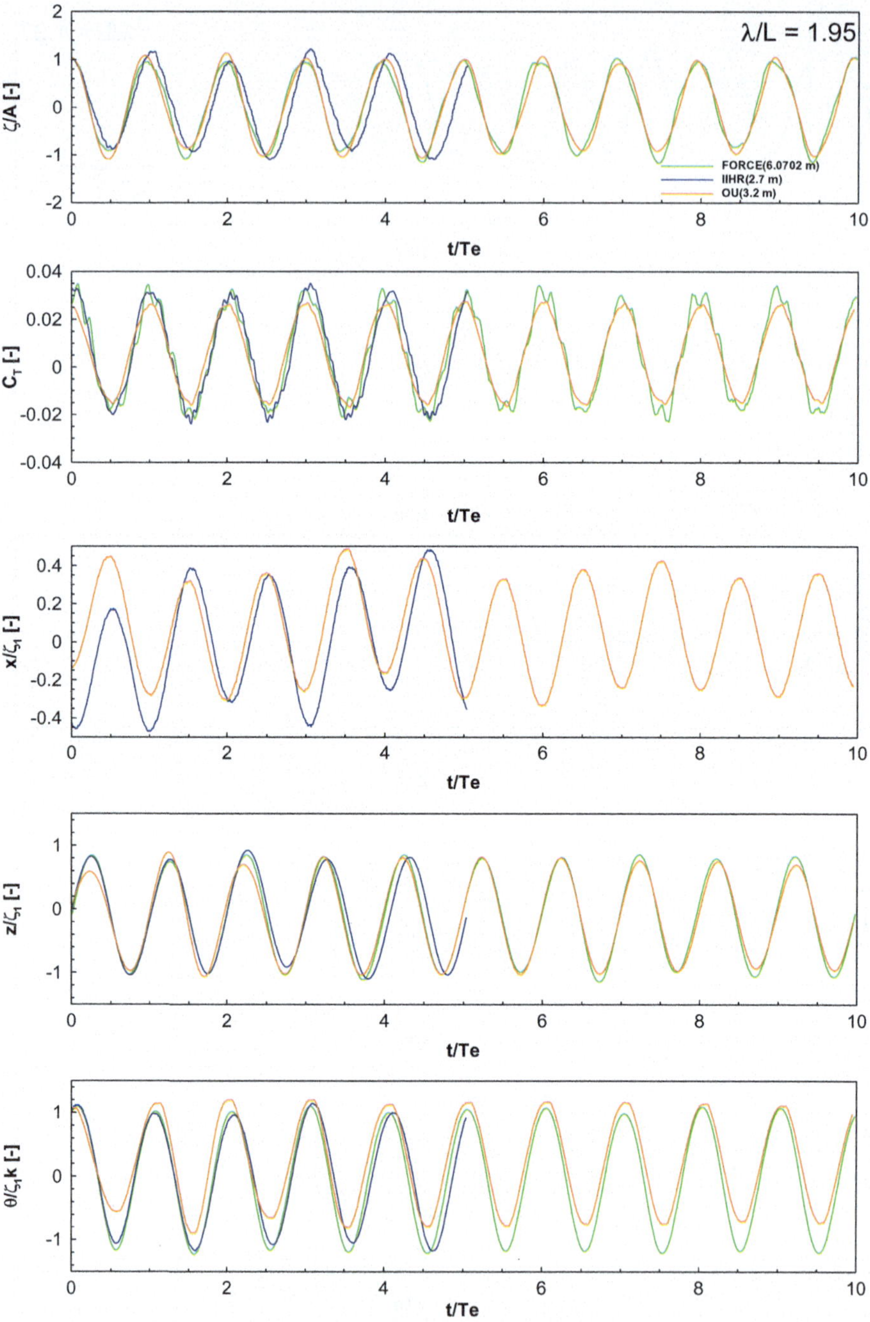

Fig. 8 Time series for force and motions of KCS in regular head waves at $\lambda/L = 1.95$ for FORCE[2] IIHR and OU

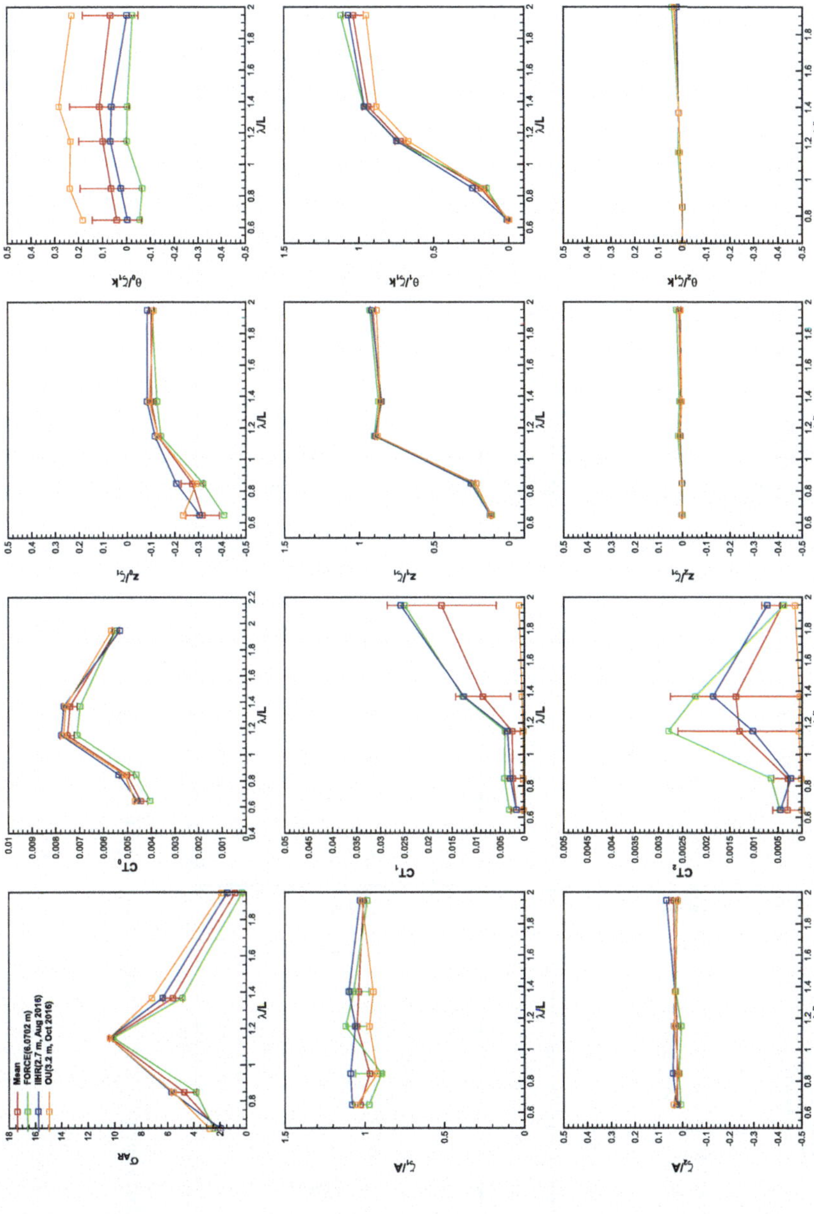

Fig. 9 Harmonic amplitudes of resistance and motions for KCS in regular head waves

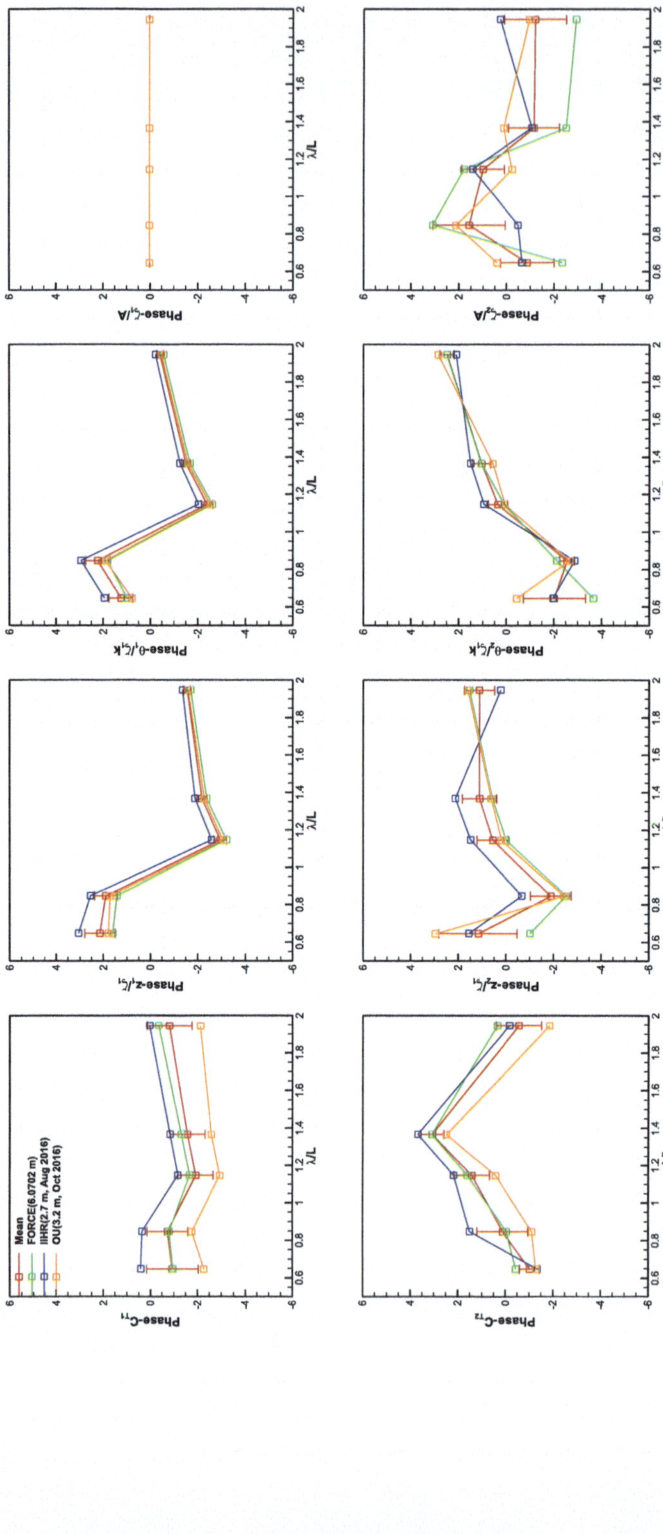

Fig. 10 Harmonic phases of resistance and motions for KCS in regular head waves

Table 9 SD analysis on added resistance for KCS in head waves

σAR						
λ/L	FORCE (6.0702 m)	IIHR (2.7 m, Aug 2016)	OU (3.2 m, Oct 2016)	AVERAGE (D)	SD	SD%D
0.65	2.06E + 00	1.96E + 00	2.80E + 00	2.27E + 00	3.75E-01	16.5
0.85	4.17E + 00	5.72E + 00	5.59E + 00	5.16E + 00	7.00E-01	13.6
1.15	9.92E + 00	1.04E + 01	1.04E + 01	1.03E + 01	2.34E-01	2.3
1.37	6.22E + 00	6.37E + 00	7.21E + 00	6.60E + 00	4.34E-01	6.6
1.95	1.87E + 00	1.43E + 00	1.89E + 00	1.73E + 00	2.13E-01	12.3
Average (Total)	**4.85E + 00**	**5.18E + 00**	**5.58E + 00**	**5.20E + 00**	**3.91E-01**	**10.24**

Table 10 SD analysis on 1st and 2nd harmonic amplitude of ζ/A for KCS in head waves

ζ_1/A

λ/L	FORCE (6.0702 m)	IIHR (2.7 m, Aug 2016)	OU (3.2 m, Oct 2016)	AVERAGE (D)	SD	SD%D
0.65	9.74E-01	1.08E + 00	1.05E + 00	1.03E + 00	4.35E-02	5.0
0.85	8.95E-01	1.09E + 00	9.26E-01	9.72E-01	8.71E-02	10.2
1.15	1.12E + 00	1.06E + 00	9.74E-01	1.05E + 00	5.97E-02	2.8
1.37	1.07E + 00	1.10E + 00	9.50E-01	1.04E + 00	6.60E-02	1.3
1.95	9.85E-01	1.03E + 00	1.02E + 00	1.01E + 00	1.90E-02	2.2
Average (Total)	**1.01E + 00**	**1.07E + 00**	**9.83E-01**	**1.02E + 00**	**5.51E-02**	**4.31**

ζ_2/A

λ/L	FORCE (6.0702 m)	IIHR (2.7 m, Aug 2016)	OU (3.2 m, Oct 2016)	AVERAGE (D)	SD	SD%D
0.65	7.18E-03	2.60E-02	3.82E-02	2.38E-02	1.27E-02	53.6
0.85	1.48E-02	4.01E-02	1.72E-02	2.40E-02	1.14E-02	47.4
1.15	4.55E-03	3.58E-02	3.41E-02	2.48E-02	1.43E-02	57.8
1.37	2.89E-02	3.32E-02	3.32E-02	3.17E-02	2.01E-03	6.3
1.95	2.67E-02	6.74E-02	2.28E-02	3.89E-02	2.02E-02	51.8
Average (Total)	**1.64E-02**	**4.05E-02**	**2.91E-02**	**2.87E-02**	**1.21E-02**	**43.40**

Table 11 SD analysis on 1st and 2nd harmonic amplitudes of C_T for KCS in head waves

C_{T1}						
λ/L	FORCE (6.0702 m)	IIHR (2.7 m, Aug 2016)	OU (3.2 m, Oct 2016)	AVERAGE (D)	SD	SD%D
0.65	3.11E-03	1.65E-03	2.36E-04	1.67E-03	1.17E-03	70.5
0.85	4.11E-03	2.91E-03	2.76E-04	2.43E-03	1.60E-03	65.9
1.15	4.00E-03	3.49E-03	2.79E-04	2.59E-03	1.65E-03	63.6
1.37	1.28E-02	1.26E-02	5.63E-04	8.66E-03	5.72E-03	66.1
1.95	2.50E-02	2.58E-02	1.15E-03	1.73E-02	1.14E-02	66.1
Average (Total)	**9.79E-03**	**9.30E-03**	**5.00E-04**	**6.53E-03**	**4.32E-03**	**66.43**
C_{T2}						
λ/L	FORCE (6.0702 m)	IIHR (2.7 m, Aug 2016)	OU (3.2 m, Oct 2016)	AVERAGE (D)	SD	SD%D
0.65	4.63E-04	4.51E-04	1.36E-05	3.09E-04	2.09E-04	67.6
0.85	6.45E-04	2.34E-04	2.43E-05	3.01E-04	2.58E-04	85.6
1.15	2.79E-03	1.02E-03	7.48E-05	1.30E-03	1.12E-03	86.8
1.37	2.23E-03	1.86E-03	5.79E-05	1.38E-03	9.49E-04	68.6
1.95	3.85E-04	7.25E-04	1.59E-04	4.23E-04	2.32E-04	55.0
Average (Total)	**1.30E-03**	**8.60E-04**	**6.60E-05**	**7.42E-04**	**5.54E-04**	**72.70**

Table 12 SD analysis on 1st and 2nd harmonic amplitudes of z/ζ_1 for KCS in head waves

Z_1/ζ_1						
λ/L	FORCE (6.0702 m)	IIHR (2.7 m, Aug 2016)	OU (3.2 m, Oct 2016)	AVERAGE (D)	SD	SD%D
0.65	1.16E-01	1.27E-01	1.23E-01	1.22E-01	4.31E-03	3.5
0.85	2.46E-01	2.55E-01	2.21E-01	2.40E-01	1.43E-02	6.0
1.15	9.01E-01	8.95E-01	8.82E-01	8.92E-01	8.01E-03	0.9
1.37	8.73E-01	8.53E-01	8.62E-01	8.63E-01	7.87E-03	0.9
1.95	9.29E-01	9.18E-01	8.87E-01	9.11E-01	1.80E-02	2.0
Average (Total)	**6.13E-01**	**6.10E-01**	**5.95E-01**	**6.06E-01**	**1.05E-02**	**2.65**
Z_2/ζ_1						
λ/L	FORCE (6.0702 m)	IIHR (2.7 m, Aug 2016)	OU (3.2 m, Oct 2016)	AVERAGE (D)	SD	SD%D
0.65	9.21E-04	3.98E-03	3.22E-03	2.71E-03	1.30E-03	48.0
0.85	2.68E-03	2.73E-03	4.96E-03	3.46E-03	1.06E-03	30.7
1.15	1.82E-02	1.05E-02	1.22E-02	1.36E-02	3.31E-03	24.3
1.37	1.32E-02	5.28E-03	5.01E-03	7.82E-03	3.79E-03	48.5
1.95	2.45E-02	9.86E-03	1.04E-02	1.49E-02	6.76E-03	45.3
Average (Total)	**1.19E-02**	**6.47E-03**	**7.16E-03**	**8.51E-03**	**3.24E-03**	**39.36**

Table 13 SD analysis on 1st and 2nd harmonic amplitudes of $\theta/\zeta_1 k$ for KCS in head waves

$\theta_1/\zeta_1 k$						
λ/L	FORCE (6.0702 m)	IIHR (2.7 m,Aug 2016)	OU (3.2 m, Oct 2016)	AVERAGE (D)	SD	SD%D
0.65	1.34E-02	1.12E-02	9.23E-04	8.52E-03	5.45E-03	63.9
0.85	1.46E-01	2.41E-01	1.71E-01	1.86E-01	4.06E-02	21.8
1.15	7.51E-01	7.49E-01	6.73E-01	7.24E-01	3.65E-02	5.0
1.37	9.66E-01	9.63E-01	8.83E-01	9.38E-01	3.84E-02	4.1
1.95	1.12E + 00	1.07E + 00	9.53E-01	1.04E + 00	6.81E-02	6.5
Average (Total)	**5.98E-01**	**6.06E-01**	**5.36E-01**	**5.80E-01**	**3.78E-02**	**20.28**
$\theta_2/\zeta_1 k$						
λ/L	FORCE (6.0702 m)	IIHR (2.7 m,Aug 2016)	OU (3.2 m, Oct 2016)	AVERAGE (D)	SD	SD%D
0.65	7.36E-04	9.40E-04	5.98E-04	7.58E-04	1.40E-04	18.5
0.85	1.26E-03	7.46E-04	1.96E-03	1.32E-03	4.97E-04	37.6
1.15	1.73E-02	1.11E-02	1.10E-02	1.31E-02	2.93E-03	22.3
1.37	1.56E-02	1.60E-02	1.47E-02	1.54E-02	5.23E-04	3.4
1.95	4.37E-02	2.37E-02	3.56E-02	3.43E-02	8.24E-03	24.0
Average (Total)	**1.57E-02**	**1.05E-02**	**1.28E-02**	**1.30E-02**	**2.47E-03**	**21.17**

Table 14 SD analysis on 0th harmonic amplitudes of C_T for KCS in head waves

C_{T0}						
λ/L	FORCE (6.0702 m)	IIHR (2,7 m, Aug 2016)	OU (3.2 m, Oct 2016)	AVERAGE (D)	SD	SD%D
0.65	4.05E-03	4.58E-03	4.70E-03	4.45E-03	2.81E-04	6.3
0.85	4.63E-03	5.37E-03	5.12E-03	5.04E-03	3.09E-04	6.1
1.15	7.10E-03	7.77E-03	7.68E-03	7.52E-03	2.97E-04	3.9
1.37	6.97E-03	7.66E-03	7.59E-03	7.41E-03	3.11E-04	4.2
1.95	5.50E-03	5.32E-03	5.68E-03	5.50E-03	1.47E-04	2.7
Average (Total)	**5.65E-03**	**6.14E-03**	**6.15E-03**	**5.98E-03**	**2.69E-04**	**4.66**

Table 15 SD analysis on 0th harmonic amplitudes of z/A for KCS in head waves

Z_0/ζ_1						
λ/L	FORCE (6.0702 m)	IIHR (2.7 m, Aug 2016)	OU (3.2 m, Oct 2016)	AVERAGE (D)	SD	SD%D
0.65	−4.10E-01	−3.05E-01	−2.33E-01	−3.16E-01	7.26E-02	16.6
0.85	−3.20E-01	−2.06E-01	−2.89E-01	−2.72E-01	4.82E-02	21.0
1.15	−1.41E-01	−1.16E-01	−1.29E-01	−1.28E-01	1.01E-02	9.6
1.37	−1.25E-01	−8.35E-02	−9.26E-02	−1.00E-01	1.80E-02	20.9
1.95	−1.05E-01	−8.47E-02	−1.10E-01	−9.99E-02	1.10E-02	10.1
Average (Total)	**−2.20E-01**	**−1.59E-01**	**−1.71E-01**	**−1.83E-01**	**3.20E-02**	**15.64**

Table 16 SD analysis on 0th harmonic amplitudes of θ/Ak for KCS in head waves

$\theta_0/\zeta_1 k$						
λ/L	FORCE (6.0702 m)	IIHR (2.7 m,Aug 2016)	OU (3.2 m, Oct 2016)	AVERAGE (D)	SD	SD%D
0.65	−5.50E-02	−3.21E-03	1.84E-01	4.20E-02	1.03E-01	244.6
0.85	−6.60E-02	2.44E-02	2.38E-01	6.56E-02	1.28E-01	194.5
1.15	−1.57E-03	6.70E-02	2.35E-01	1.00E-01	9.93E-02	99.2
1.37	−3.74E-03	6.22E-02	2.84E-01	1.14E-01	1.23E-01	107.8
1.95	−2.47E-02	−1.37E-03	2.30E-01	6.79E-02	1.15E-01	169.1
Average (Total)	**−3.02E-02**	**−2.98E-02**	**−2.34E-01**	**−7.80E-02**	**−1.14E-01**	**−163.06**

Table 17 SD analysis on 2nd harmonic phases of ζ/A for KCS in head waves

2nd Phase ζ_2/A						
λ/L	FORCE (6.0702 m)	IIHR (2.7 m, Aug 2016)	OU (3.2 m, Oct 2016)	AVERAGE (D)	SD	SD%D
0.65	−2.36E + 00	−6.76E-01	3.84E-01	−8.85E-01	1.13E + 00	127.8
0.85	3.07E + 00	−5.09E-01	2.09E + 00	1.55E + 00	1.51E + 00	97.5
1.15	1.78E + 00	1.41E + 00	−2.63E-01	9.74E-01	8.88E-01	91.1
1.37	−2.52E + 00	−1.08E + 00	9.65E-02	−1.17E + 00	1.07E + 00	91.5
1.95	−2.93E + 00	2.59E-01	−9.81E-01	−1.22E + 00	1.31E + 00	107.8
Average (Total)	**−5.94E-01**	**−1.20E-01**	**2.64E-01**	**−1.50E-01**	**1.18E + 00**	**103.15**

Table 18 SD analysis on 1st and 2nd harmonic phases of C_T for KCS in head waves

1st Phase C_{T1}						
λ/L	FORCE (6.0702 m)	IIHR (2.7 m, Aug 2016)	OU (3.2 m, Oct 2016)	AVERAGE (D)	SD	SD%D
0.65	−9.76E-01	3.99E-01	−2.30E + 00	−9.58E-01	1.10E + 00	114.9
0.85	−8.16E-01	3.38E-01	−1.79E + 00	−7.57E-01	8.71E-01	115.0
1.15	−1.69E + 00	−1.20E + 00	−2.96E + 00	−1.95E + 00	7.43E-01	38.1
1.37	−1.34E + 00	−8.65E-01	−2.61E + 00	−1.60E + 00	7.36E-01	45.9
1.95	−3.69E-01	1.27E-02	−2.15E + 00	−8.37E-01	9.44E-01	112.8
Average (Total)	**−1.04E + 00**	**−2.63E-01**	**−2.36E + 00**	**−1.22E + 00**	**8.79E-01**	**85.36**
2nd Phase C_{T2}						
λ/L	FORCE (6.0702 m)	IIHR (2.7 m, Aug 2016)	OU (3.2 m, Oct 2016)	AVERAGE (D)	SD	SD%D
0.65	−4.55E-01	−1.36E + 00	−1.31E + 00	−1.04E + 00	4.16E-01	39.9
0.85	−5.64E-02	1.48E + 00	−1.14E + 00	9.52E-02	1.07E + 00	1127.4
1.15	1.59E + 00	2.16E + 00	3.88E-01	1.38E + 00	7.37E-01	53.5
1.37	3.07E + 00	3.65E + 00	2.45E + 00	3.05E + 00	4.90E-01	16.0
1.95	3.41E-01	−1.79E-01	−1.88E + 00	−5.74E-01	9.50E-01	165.6
Average (Total)	**8.97E-01**	**1.15E + 00**	**−2.98E-01**	**5.82E-01**	**7.33E-01**	**280.50**

Table 19 SD analysis on 1st and 2nd harmonic phases of z/A for KCS in head waves

1st Phase Z_1/A						
λ/L	FORCE (6.0702 m)	IIHR (2.7 m, Aug 2016)	OU (3.2 m, Oct 2016)	AVERAGE (D)	SD	SD%D
0.65	1.59E + 00	3.04E + 00	1.76E + 00	2.13E + 00	6.47E-01	30.3
0.85	1.41E + 00	2.54E + 00	1.69E + 00	1.88E + 00	4.78E-01	25.4
1.15	−3.22E + 00	−2.60E + 00	−3.09E + 00	−2.97E + 00	2.68E-01	−9.0
1.37	−2.39E + 00	−1.91E + 00	−2.30E + 00	−2.20E + 00	2.11E-01	−9.6
1.95	−1.72E + 00	−1.39E + 00	−1.63E + 00	−1.58E + 00	1.43E-01	−9.0
Average (Total)	**−8.67E-01**	**−6.34E-02**	**−7.15E-01**	**−5.48E-01**	**3.49E-01**	**5.62**
2nd Phase Z_2/A						
λ/L	FORCE (6.0702 m)	IIHR (2.7 m, Aug 2016)	OU (3.2 m, Oct 2016)	AVERAGE (D)	SD	SD%D
0.65	−1.03E + 00	1.53E + 00	2.92E + 00	1.14E + 00	1.64E + 00	143.4
0.85	−2.55E + 00	−6.90E-01	−2.48E + 00	−1.91E + 00	8.62E-01	−45.2
1.15	−1.75E-02	1.46E + 00	1.73E-01	5.38E-01	6.56E-01	121.9
1.37	5.64E-01	2.10E + 00	6.06E-01	1.09E + 00	7.12E-01	65.5
1.95	1.52E + 00	2.18E-01	1.57E + 00	1.10E + 00	6.25E-01	56.7
Average (Total)	**−3.03E-01**	**9.22E-01**	**5.57E-01**	**3.92E-01**	**8.98E-01**	**68.45**

Table 20 SD analysis on 1st and 2nd harmonic phases of θ/Ak for KCS in head waves

1st Phase θ_1/Ak						
λ/L	FORCE (6.0702 m)	IIHR (2.7 m, Aug 2016)	OU (3.2 m, Oct 2016)	AVERAGE (D)	SD	SD%D
0.65	9.95E-01	1.93E + 00	7.55E-01	1.23E + 00	5.05E-01	41.2
0.85	1.80E + 00	2.93E + 00	1.88E + 00	2.21E + 00	5.15E-01	23.4
1.15	−2.62E + 00	−2.06E + 00	−2.54E + 00	−2.41E + 00	2.47E-01	10.3
1.37	−1.70E + 00	−1.28E + 00	−1.62E + 00	−1.53E + 00	1.84E-01	12.0
1.95	−5.90E-01	−2.58E-01	−5.07E-01	−4.51E-01	1.41E-01	31.2
Average (Total)	**−4.23E-01**	**2.52E-01**	**−4.06E-01**	**−1.92E-01**	**3.19E-01**	**23.61**
2nd Phase θ_2/Ak						
λ/L	FORCE (6.0702 m)	IIHR (2.7 m, Aug 2016)	OU (3.2 m, Oct 2016)	AVERAGE (D)	SD	SD%D
0.65	−3.68E + 00	−1.98E + 00	−4.66E-01	−2.04E + 00	1.31E + 00	64.2
0.85	−2.14E + 00	−2.90E + 00	−2.69E + 00	−2.58E + 00	3.19E-01	12.4
1.15	7.07E-02	9.24E-01	3.25E-02	3.43E-01	4.12E-01	120.2
1.37	1.03E + 00	1.48E + 00	5.48E-01	1.02E + 00	3.78E-01	37.2
1.95	2.47E + 00	2.08E + 00	2.82E + 00	2.46E + 00	3.04E-01	12.4
Average (Total)	**−4.49E-01**	**−8.02E-02**	**5.02E-02**	**−1.60E-01**	**5.45E-01**	**49.27**

In consideration differences in model sizes and rigid (FORCE) vs. surge free (IIHR and OU) mounts the agreement is reasonable: for primary variables, SD = 3, 20 and 10%D for heave and pitch amplitudes and added resistance, respectively; and for secondary variable SD = 66%D for C_{T1}.

4 KCS Oblique Waves λ/L = 1

Regular oblique wave experimental data are available from IIHR using one model size 2.7 m and from both 2015 (Sadat-Hosseini et al. 2015a) and 2016 and uncertainty values from Stocker (2016). Data is available for headings $\chi = 0, 45, 90, 135$ and 180 degrees with $0.5 \leq \lambda/L \leq 2$. The λ/L = 1 data was used for test case 2.11.

Figures 11, 12, 13, 14 and 15 compare the time histories from 2015 and 2016 for ζ/A, C_T, x/A, z/A, θ/Ak and ϕ/Ak for $\chi = 0, 45, 90, 135$ and 180 degrees and λ/L = 1. Figure 16 compares σ_{AR}, 1st and 2nd harmonic amplitude for ζ/A, 0th, 1st and 2nd harmonic amplitudes for C_T, x/A, z/A, ϕ/Ak and θ/Ak; and Fig. 17 compares 1st and 2nd harmonic phases for C_T, x/A, z/A, ϕ/Ak and θ/Ak. Harmonic analysis window size for χ = 0,45 degrees and χ = 90,135,180 degrees are 5Te and 3Te, respectively. Results show reasonable agreement between the time histories of the two sets of data.

Tables 21 and 22 provide SD analysis for σ_{AR} and 1st and 2nd harmonic amplitude for ζ/A. Tables 23, 24, 25, 26 and 27 provide SD analysis for 0th, 1st and 2nd harmonic amplitudes for C_T, x/A, z/A, ϕ/Ak and θ/Ak, respectively. Tables 28, 29, 30, 31, 32 and 33 provide SD analysis for 1st and 2nd harmonic phases for C_T, x/A, z/A, ϕ/Ak and θ/Ak, respectively. The scatter for the 1st harmonic amplitudes is relatively small 4, 13, 25, 6, 17, 6%D, respectively, for ζ/A, C_T, x/A, z/A, ϕ/Ak, θ/Ak. The scatter for 0th harmonic for C_T (SD = 1%D) and 1st harmonic phases for ζ/A (SD = 0.1%D) and z/A (SD = 4%D) are also small. The scatter for σ_{AR} (SD = 63%D) and other harmonic amplitudes and phases is mostly large.

The agreement between the two series of tests is reasonably good, i.e., SD values are comparable to the corresponding values for test case 2.10: for primary variables, SD = 6, 6 and 63%D for heave and pitch amplitudes and added resistance, respectively; and for secondary variable SD = 13%D for C_{T1}. T2015 test case 2.11 used the 2016 data.

5 KCS Oblique Waves All λ/L

Table 34 lists all IIHR KCS oblique wave data for all headings and wavelengths. Figures 18, 19, 20, 21 and 22 provide time histories and Figs. 23–24 harmonic amplitudes and phases. The harmonic analysis window size for oblique heading case are 2Te only for λ/L = 2.0 and $\mu = 135$ degree and 3Te to 5Te for other λ/L and χ. The figures show the trends are very different for different wave headings.

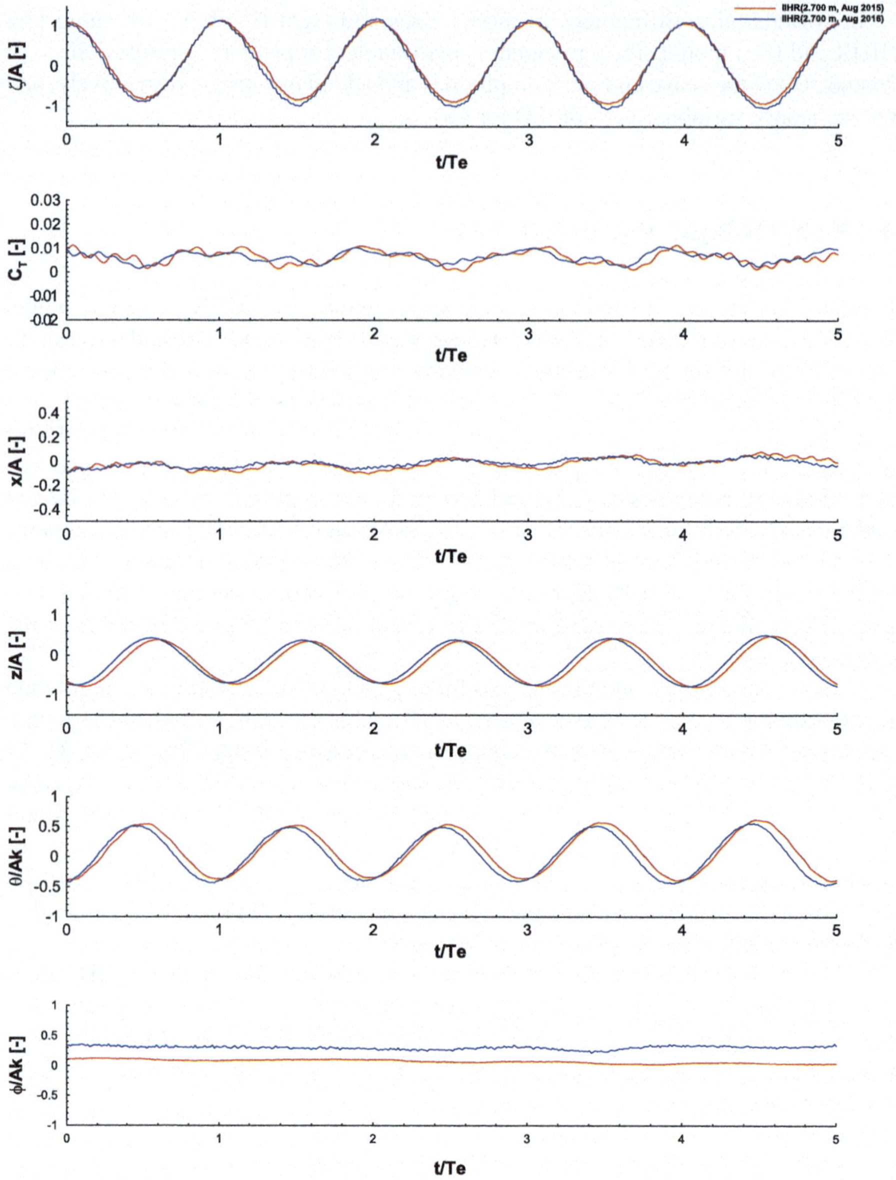

Fig. 11 Time series for force and motions of KCS in regular oblique waves at $\lambda/L = 1.00$, $\chi = 0°$ for IIHR Aug-15 and Aug-16

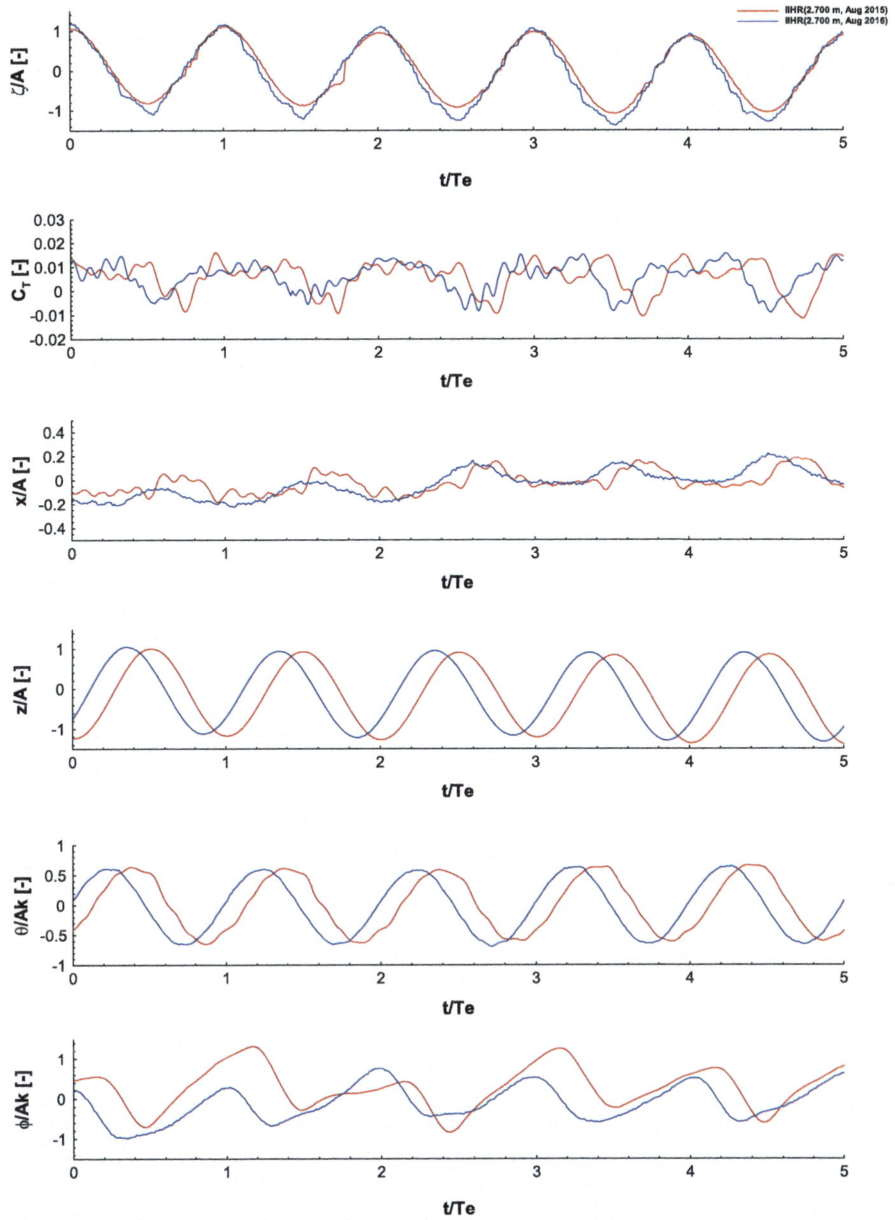

Fig. 12 Time series for force and motions of KCS in regular oblique waves at $\lambda/L = 1.00$, $\chi = 45°$ for IIHR Aug-15 and Aug-16

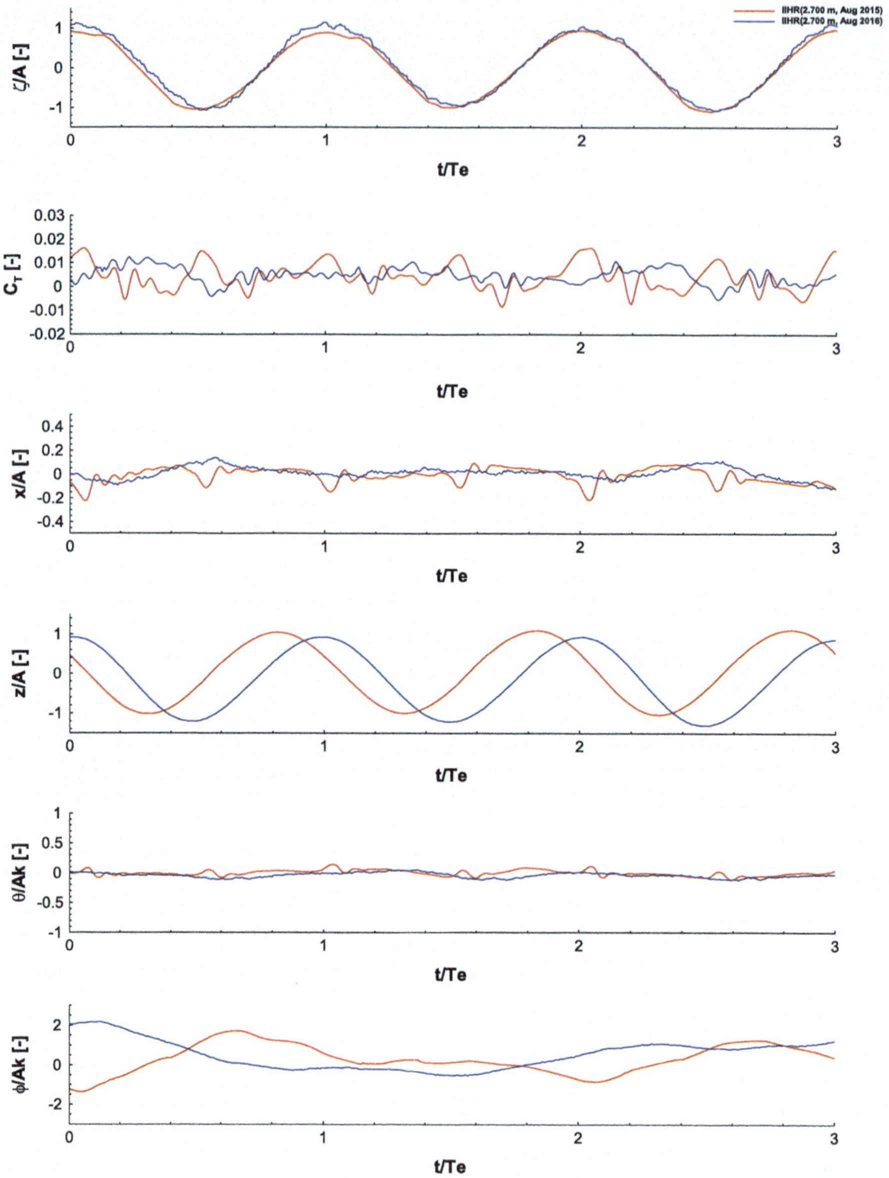

Fig. 13 Time series for force and motions of KCS in regular oblique waves at $\lambda/L = 1.00$, $\chi = 90°$ for IIHR Aug-15 and Aug-16

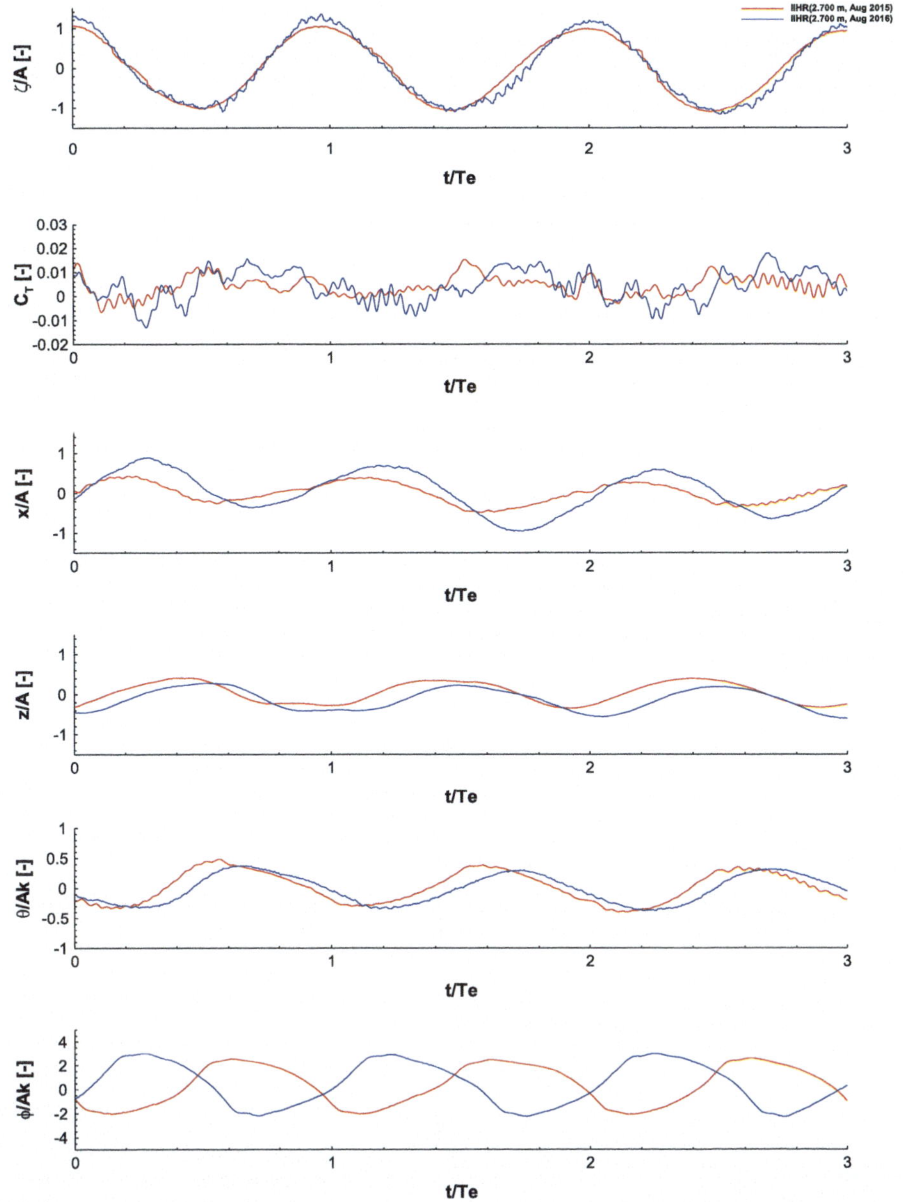

Fig. 14 Time series for force and motions of KCS in regular oblique waves at $\lambda/L = 1.00$, $\chi = 135°$ for IIHR Aug-15 and Aug-16

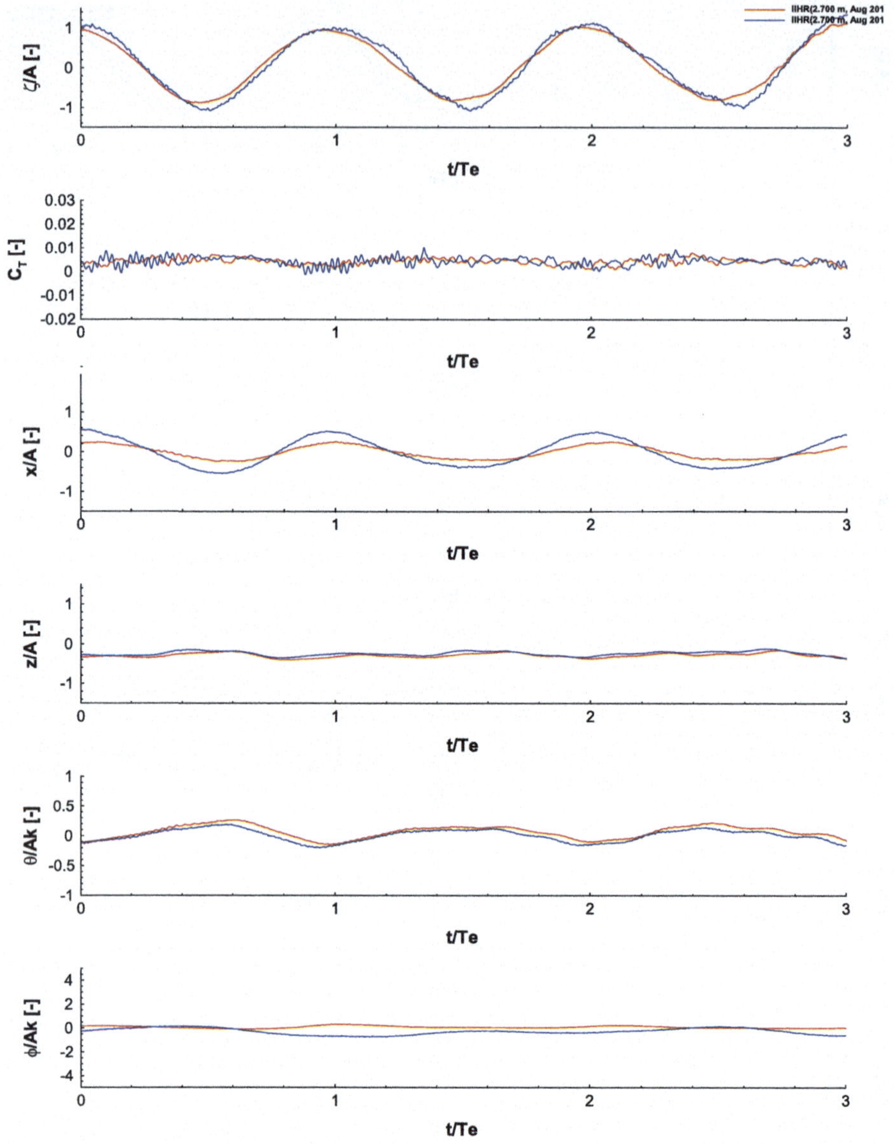

Fig. 15 Time series for force and motions of KCS in regular oblique waves at $\lambda/L = 1.00$, $\chi = 180°$ for IIHR Aug-15 and Aug-16

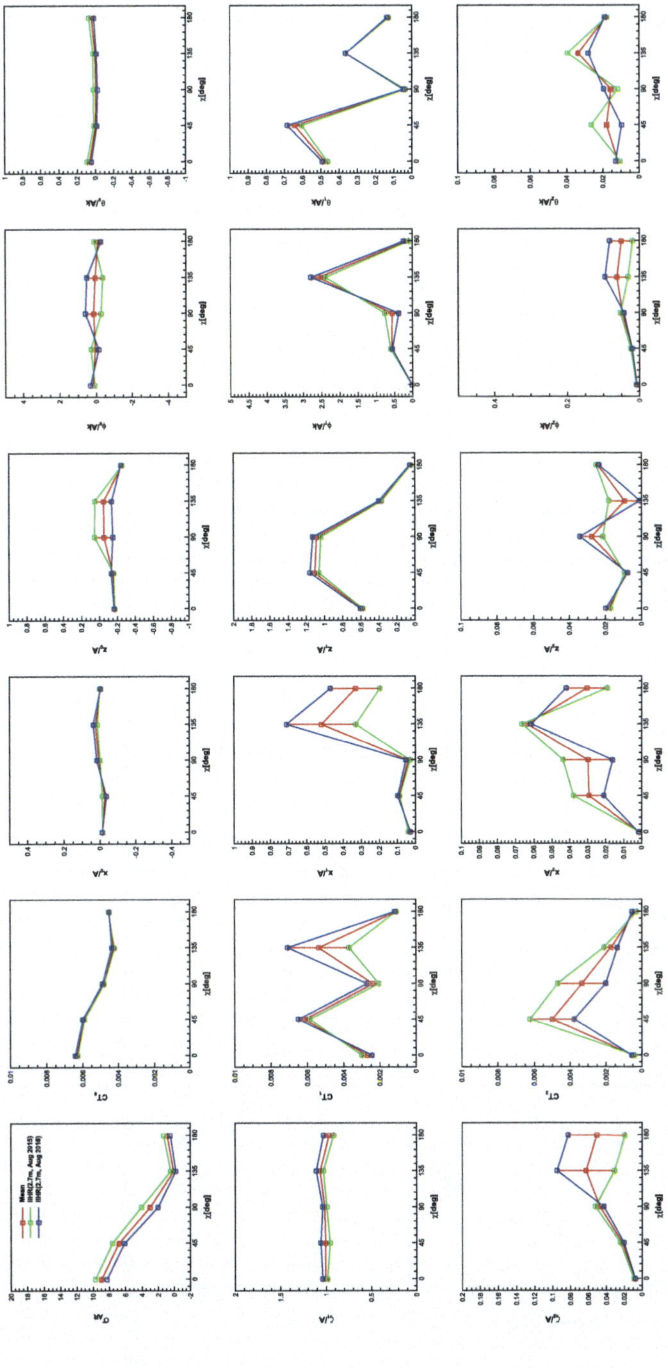

Fig. 16 Harmonic amplitudes of resistance and motions for KCS in regular oblique waves at Fr = 0.26

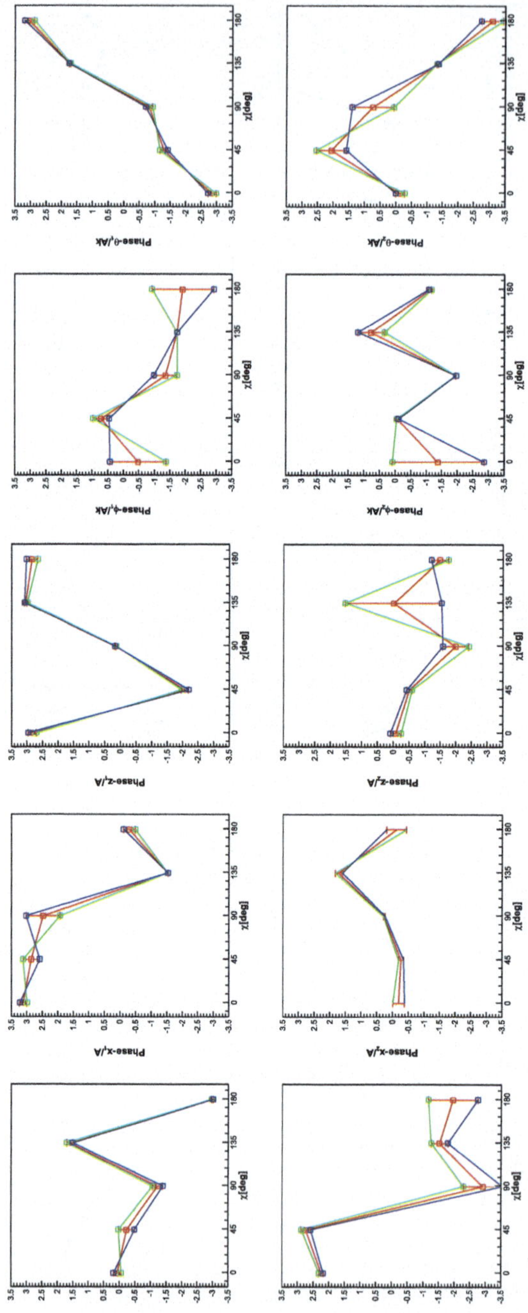

Fig. 17 Harmonic phases of resistance and motions for KCS in regular oblique waves at Fr = 0.26

Table 21 SD analysis on added resistance for KCS in oblique waves ($\lambda/L = 1$)

σ_{AR}

χ [deg]	λ/L	IIHR (2.7 m, Aug 2015)	IIHR (2.7 m, Aug 2016)	AVERAGE (D)	SD	SD%D
0	1.00	9.73	8.28	9.01	7.24E-01	8.0
45	1.00	7.58	6.12	6.85	7.30E-01	10.7
90	1.00	4.07	2.06	3.07	1.00E + 00	32.7
135	1.00	0.46	−0.17	0.14	3.16E-01	217.6
180	1.00	1.34	0.50	0.92	4.18E-01	45.3
0.00		**4.64**	**3.36**	**4.00**	**0.64**	**62.86**

Table 22 SD analysis on 1st and 2nd harmonic amplitudes of ζ/A for KCS in oblique waves ($\lambda/L = 1$)

ζ_1/A						
χ [deg]	λ/L	IIHR (2.7 m, Aug 2015)	IIHR (2.7 m, Aug 2016)	AVERAGE (D)	SD	SD%D
0	1.0	9.86E-01	1.04E + 00	1.01E + 00	2.91E-02	2.9
45	1.0	9.56E-01	1.06E + 00	1.01E + 00	5.12E-02	5.1
90	1.0	9.93E-01	1.04E + 00	1.02E + 00	2.52E-02	2.5
135	1.0	1.03E + 00	1.11E + 00	1.07E + 00	3.82E-02	3.6
180	1.0	9.18E-01	1.03E + 00	9.72E-01	5.43E-02	5.6
Average (Total)		**9.77E-01**	**1.06E + 00**	**1.02E + 00**	**3.96E-02**	**3.9**
ζ_2/A						
χ [deg]	λ/L	IIHR (2.7 m, Aug 2015)	IIHR (2.7 m, Aug 2016)	AVERAGE (D)	SD	SD%D
0	1.0	9.18E-03	7.56E-03	8.37E-03	8.07E-04	9.6
45	1.0	2.45E-02	2.05E-02	2.25E-02	1.99E-03	8.8
90	1.0	5.22E-02	4.30E-02	4.76E-02	4.63E-03	9.7
135	1.0	3.10E-02	9.50E-02	6.30E-02	3.20E-02	50.8
180	1.0	1.94E-02	8.26E-02	5.10E-02	3.16E-02	62.0
Average (Total)		**2.73E-02**	**4.97E-02**	**3.85E-02**	**1.42E-02**	**28.2**

Table 23 SD analysis on 0th, 1st and 2nd harmonic amplitudes of C_T for KCS in oblique waves ($\lambda/L = 1$)

C_{T0}						
χ [deg]	λ/L	IIHR (2.7 m, Aug 2015)	IIHR (2.7 m, Aug 2016)	AVERAGE (D)	SD	SD%D
0	1.0	6.29E-03	6.43E-03	6.36E-03	6.76E-05	1.1
45	1.0	5.91E-03	6.01E-03	5.96E-03	5.13E-05	0.9
90	1.0	4.81E-03	4.91E-03	4.86E-03	5.17E-05	1.1
135	1.0	4.22E-03	4.38E-03	4.30E-03	8.43E-05	2.0
180	1.0	4.56E-03	4.52E-03	4.54E-03	2.04E-05	0.5
Average (Total)		**5.16E-03**	**5.25E-03**	**5.20E-03**	**5.50E-05**	**1.1**
C_{T1}						
χ [deg]	λ/L	IIHR (2.7 m, Aug 2015)	IIHR (2.7 m, Aug 2016)	AVERAGE (D)	SD	SD%D
0	1.0	3.01E-03	2.45E-03	2.73E-03	2.79E-04	10.2
45	1.0	5.87E-03	6.50E-03	6.18E-03	3.18E-04	5.1
90	1.0	2.06E-03	2.73E-03	2.40E-03	3.35E-04	14.0
135	1.0	3.71E-03	7.09E-03	5.40E-03	1.69E-03	31.3
180	1.0	1.09E-03	1.18E-03	1.13E-03	4.66E-05	4.1
Average (Total)		**3.15E-03**	**3.99E-03**	**3.57E-03**	**5.33E-04**	**12.9**
C_{T2}						
χ [deg]	λ/L	IIHR (2.7 m, Aug 2015)	IIHR (2.7 m, Aug 2016)	AVERAGE (D)	SD	SD%D
0	1.0	4.25E-04	5.78E-04	5.02E-04	7.68E-05	15.3
45	1.0	6.23E-03	3.78E-03	5.01E-03	1.22E-03	24.5
90	1.0	4.69E-03	2.02E-03	3.36E-03	1.33E-03	39.7
135	1.0	2.12E-03	1.38E-03	1.75E-03	3.66E-04	20.9
180	1.0	3.05E-04	5.57E-04	4.31E-04	1.26E-04	29.2
Average (Total)		**2.75E-03**	**1.66E-03**	**2.21E-03**	**6.25E-04**	**25.9**

Table 24 SD analysis on 0th, 1st and 2nd harmonic amplitudes of x/A for KCS in oblique waves ($\lambda/L = 1$)

x_0/A						
χ [deg]	λ/L	IIHR (2.7 m, Aug 2015)	IIHR (2.7 m, Aug 2016)	AVERAGE (D)	SD	SD%D
0	1.0	−1.39E-02	−1.13E-02	−1.26E-02	1.34E-03	−10.6
45	1.0	−1.23E-02	−3.45E-02	−2.34E-02	1.11E-02	−47.3
90	1.0	−4.15E-04	1.87E-02	9.16E-03	9.58E-03	104.5
135	1.0	1.45E-02	3.85E-02	2.65E-02	1.20E-02	45.4
180	1.0	1.75E-03	−2.54E-03	−3.94E-04	2.15E-03	−544.5
Average (Total)		**−2.09E-03**	**1.80E-03**	**−1.43E-04**	**7.23E-03**	**−90.5**
x_1/A						
χ [deg]	λ/L	IIHR (2.7 m, Aug 2015)	IIHR (2.7 m, Aug 2016)	AVERAGE (D)	SD	SD%D
0	1.0	3.90E-02	2.95E-02	3.43E-02	4.78E-03	13.9
45	1.0	8.85E-02	9.89E-02	9.37E-02	5.19E-03	5.5
90	1.0	3.05E-02	5.33E-02	4.19E-02	1.14E-02	27.2
135	1.0	3.30E-01	7.12E-01	5.21E-01	1.91E-01	36.7
180	1.0	1.97E-01	4.71E-01	3.34E-01	1.37E-01	41.1
Average (Total)		**1.37E-01**	**2.73E-01**	**2.05E-01**	**7.00E-02**	**24.9**
x_2/A						
χ [deg]	λ/L	IIHR (2.7 m, Aug 2015)	IIHR (2.7 m, Aug 2016)	AVERAGE (D)	SD	SD%D
0	1.0	1.22E-03	1.89E-03	1.56E-03	3.36E-04	21.6
45	1.0	3.79E-02	2.12E-02	2.95E-02	8.38E-03	28.4
90	1.0	4.37E-02	1.63E-02	3.00E-02	1.37E-02	45.6
135	1.0	6.64E-02	6.14E-02	6.39E-02	2.48E-03	3.9
180	1.0	1.91E-02	4.20E-02	3.05E-02	1.14E-02	37.4
Average (Total)		**3.37E-02**	**2.85E-02**	**3.11E-02**	**7.26E-03**	**27.4**

Table 25 SD analysis on 0th, 1st and 2nd harmonic amplitudes of z/A for KCS in oblique waves ($\lambda/L = 1$)

z_0/A						
χ [deg]	λ/L	IIHR (2.7 m, Aug 2015)	IIHR (2.7 m, Aug 2016)	AVERAGE (D)	SD	SD%D
0	1.0	−1.70E-01	−1.61E-01	−1.66E-01	4.76E-03	−2.9
45	1.0	−1.58E-01	−1.30E-01	−1.44E-01	1.41E-02	−9.8
90	1.0	5.41E-02	−1.49E-01	−4.73E-02	1.01E-01	−214.3
135	1.0	5.08E-02	−1.34E-01	−4.17E-02	9.26E-02	−221.9
180	1.0	−2.53E-01	−2.37E-01	−2.45E-01	7.85E-03	−3.2
Average (Total)		**−9.52E-02**	**−1.62E-01**	**−1.29E-01**	**4.42E-02**	**−90.4**
z_1/A						
χ [deg]	λ/L	IIHR (2.7 m, Aug 2015)	IIHR (2.7 m, Aug 2016)	AVERAGE (D)	SD	SD%D
0	1.0	5.79E-01	6.08E-01	5.94E-01	1.45E-02	2.4
45	1.0	1.06E + 00	1.16E + 00	1.11E + 00	4.84E-02	4.4
90	1.0	1.04E + 00	1.13E + 00	1.09E + 00	4.52E-02	4.2
135	1.0	3.70E-01	4.04E-01	3.87E-01	1.68E-02	4.3
180	1.0	4.54E-02	5.95E-02	5.24E-02	7.04E-03	13.4
Average (Total)		**6.20E-01**	**6.73E-01**	**6.47E-01**	**2.64E-02**	**5.75**
z_2/A						
χ [deg]	λ/L	IIHR (2.7 m, Aug 2015)	IIHR (2.7 m, Aug 2016)	AVERAGE (D)	SD	SD%D
0	1.0	1.72E-02	1.98E-02	1.85E-02	1.31E-03	7.1
45	1.0	9.45E-03	7.85E-03	8.65E-03	8.01E-04	9.3
90	1.0	2.14E-02	3.42E-02	2.78E-02	6.40E-03	23.0
135	1.0	1.81E-02	1.03E-03	9.59E-03	8.56E-03	89.3
180	1.0	2.52E-02	2.36E-02	2.44E-02	7.94E-04	3.3
Average (Total)		**1.83E-02**	**1.73E-02**	**1.78E-02**	**3.57E-03**	**26.38**

Table 26 SD analysis on 0th, 1st and 2nd harmonic amplitudes of ϕ/Ak for KCS in oblique waves (λ/L = 1)

ϕ_0/Ak

χ [deg]	λ/L	IIHR (2.7 m, Aug 2015)	IIHR (2.7 m, Aug 2016)	AVERAGE (D)	SD	SD%D
0	1.0	7.84E-02	3.24E-01	2.01E-01	1.23E-01	61.0
45	1.0	2.83E-01	−1.32E-01	7.54E-02	2.07E-01	275.0
90	1.0	−2.79E-01	5.88E-01	1.55E-01	4.34E-01	280.0
135	1.0	−3.71E-01	5.13E-01	7.11E-02	4.42E-01	621.5
180	1.0	1.02E-01	−2.57E-01	−7.72E-02	1.80E-01	−232.7
Average (Total)		**−3.72E-02**	**2.07E-01**	**8.50E-02**	**2.77E-01**	**200.9**

ϕ_1/Ak

χ [deg]	λ/L	IIHR (2.7 m, Aug 2015)	IIHR (2.7 m, Aug 2016)	AVERAGE (D)	SD	SD%D
0	1.0	1.24E-02	1.19E-02	1.22E-02	2.49E-04	2.0
45	1.0	5.99E-01	5.58E-01	5.78E-01	2.07E-02	3.6
90	1.0	7.59E-01	3.78E-01	5.69E-01	1.90E-01	33.5
135	1.0	2.39E + 00	2.79E + 00	2.59E + 00	2.00E-01	7.7
180	1.0	1.06E-01	2.47E-01	1.77E-01	7.03E-02	39.8
Average (Total)		**7.73E-01**	**7.97E-01**	**7.85E-01**	**9.63E-02**	**17.3**

ϕ_2/Ak

χ [deg]	λ/L	IIHR (2.7 m, Aug 2015)	IIHR (2.7 m, Aug 2016)	AVERAGE (D)	SD	SD%D
0	1.0	2.74E-03	5.82E-03	4.28E-03	1.54E-03	36.0
45	1.0	1.89E-01	1.48E-01	1.68E-01	2.08E-02	12.3
90	1.0	1.47E-01	6.71E-02	1.07E-01	4.00E-02	37.4
135	1.0	8.21E-02	9.36E-02	8.79E-02	5.76E-03	6.6
180	1.0	3.49E-02	4.61E-02	4.05E-02	5.60E-03	13.8
Average (Total)		**9.12E-02**	**7.21E-02**	**8.16E-02**	**1.47E-02**	**21.2**

Table 27 SD analysis on 0th, 1st and 2nd harmonic amplitudes of θ/Ak for KCS in oblique waves (λ/L = 1)

θ_0/Ak						
χ [deg]	λ/L	IIHR (2.7 m, Aug 2015)	IIHR (2.7 m, Aug 2016)	AVERAGE (D)	SD	SD%D
0	1.0	9.12E-02	4.42E-02	6.77E-02	2.35E-02	34.7
45	1.0	1.81E-02	−1.35E-02	2.30E-03	1.58E-02	687.3
90	1.0	2.05E-02	−2.33E-02	−1.38E-03	2.19E-02	−1587.6
135	1.0	2.42E-02	−1.04E-02	6.88E-03	1.73E-02	251.1
180	1.0	6.84E-02	1.55E-02	4.19E-02	2.64E-02	63.1
Average (Total)		**4.45E-02**	**2.50E-03**	**2.35E-02**	**2.10E-02**	**−110.3**
θ_1/Ak						
χ [deg]	λ/L	IIHR (2.7 m, Aug 2015)	IIHR (2.7 m, Aug 2016)	AVERAGE (D)	SD	SD%D
0	1.0	4.64E-01	4.92E-01	4.78E-01	1.40E-02	2.9
45	1.0	6.05E-01	6.85E-01	6.45E-01	3.98E-02	6.2
90	1.0	3.53E-02	4.76E-02	4.14E-02	6.17E-03	14.9
135	1.0	3.66E-01	3.65E-01	3.65E-01	4.57E-04	0.1
180	1.0	1.28E-01	1.38E-01	1.33E-01	4.56E-03	3.4
Average (Total)		**3.20E-01**	**3.45E-01**	**3.32E-01**	**1.30E-02**	**5.5**
θ_2/Ak						
χ [deg]	λ/L	IIHR (2.7 m, Aug 2015)	IIHR (2.7 m, Aug 2016)	AVERAGE (D)	SD	SD%D
0	1.0	1.05E-02	1.30E-02	1.18E-02	1.27E-03	10.8
45	1.0	2.64E-02	9.74E-03	1.81E-02	8.32E-03	46.1
90	1.0	1.19E-02	1.96E-02	1.57E-02	3.88E-03	24.7
135	1.0	3.95E-02	2.81E-02	3.38E-02	5.68E-03	16.8
180	1.0	1.79E-02	1.91E-02	1.85E-02	5.96E-04	3.2
Average (Total)		**2.12E-02**	**1.79E-02**	**1.96E-02**	**3.95E-03**	**20.3**

Table 28 SD analysis on 1st and 2nd harmonic phases for C_T of KCS in oblique waves ($\lambda/L = 1$)

1st Phase C_{T1}						
χ [deg]	λ/L	IIHR (2.7 m, Aug 2015)	IIHR (2.7 m, Aug 2016)	AVERAGE (D)	SD	SD%D
0	1.0	−7.44E-02	1.65E-01	4.55E-02	1.20E-01	263.4
45	1.0	1.02E-02	−5.09E-01	−2.50E-01	2.60E-01	−104.1
90	1.0	−1.08E + 00	−1.41E + 00	−1.24E + 00	1.63E-01	−13.1
135	1.0	1.69E + 00	1.49E + 00	1.59E + 00	9.83E-02	6.2
180	1.0	−2.98E + 00	−3.03E + 00	−3.01E + 00	2.53E-02	−0.8
Average (Total)		**−4.88E-01**	**−6.58E-01**	**−5.73E-01**	**1.33E-01**	**30.3**
2nd Phase C_{T2}						
χ [deg]	λ/L	IIHR (2.7 m, Aug 2015)	IIHR (2.7 m, Aug 2016)	AVERAGE (D)	SD	SD%D
0	1.0	2.30E + 00	2.17E + 00	2.23E + 00	6.67E-02	3.0
45	1.0	2.86E + 00	2.57E + 00	2.72E + 00	1.44E-01	5.3
90	1.0	−2.31E + 00	−3.56E + 00	−2.94E + 00	6.22E-01	−21.2
135	1.0	−1.27E + 00	−1.81E + 00	−1.54E + 00	2.69E-01	−17.5
180	1.0	−1.17E + 00	−2.77E + 00	−1.97E + 00	8.02E-01	−40.7
Average (Total)		**8.32E-02**	**8.32E-02**	**−6.79E-01**	**−2.98E-01**	**3.81E-01**

Table 29 SD analysis on 1st and 2nd harmonic phases for x/A of KCS in oblique waves (λ/L = 1)

1st Phase x_1/A						
χ [deg]	λ/L	IIHR (2.7 m, Aug 2015)	IIHR (2.7 m, Aug 2016)	AVERAGE (D)	SD	SD%D
0	1.0	2.98E + 00	3.21E + 00	3.10E + 00	1.15E-01	3.7
45	1.0	3.11E + 00	2.57E + 00	2.84E + 00	2.68E-01	9.4
90	1.0	1.92E + 00	3.01E + 00	2.46E + 00	5.42E-01	22.0
135	1.0	−1.54E + 00	−1.55E + 00	−1.55E + 00	2.45E-04	0.0
180	1.0	−4.99E-01	−1.25E-01	−3.12E-01	1.87E-01	−60.1
Average (Total)		**1.19E + 00**	**1.19E + 00**	**1.42E + 00**	**1.31E + 00**	**2.23E-01**
2nd Phase x_2/A						
χ [deg]	λ/L	IIHR (2.7 m, Aug 2015)	IIHR (2.7 m, Aug 2016)	AVERAGE (D)	SD	SD%D
0	1.0	−2.52E-02	−3.97E-01	−2.11E-01	1.86E-01	−88.1
45	1.0	−2.03E-01	−3.59E-01	−2.81E-01	7.81E-02	−27.8
90	1.0	2.73E-01	2.18E-01	2.45E-01	2.77E-02	11.3
135	1.0	1.81E + 00	1.58E + 00	1.70E + 00	1.15E-01	6.8
180	1.0	−4.50E-01	1.73E-01	−1.38E-01	3.11E-01	−225.2
Average (Total)		**2.81E-01**	**2.43E-01**	**2.62E-01**	**1.44E-01**	**−64.6**

Table 30 SD analysis on 1st and 2nd harmonic phases for z/A of KCS in oblique waves (λ/L = 1)

1st Phase z_1/A						
χ [deg]	λ/L	IIHR (2.7 m, Aug 2015)	IIHR (2.7 m, Aug 2016)	AVERAGE (D)	SD	SD%D
0	1.0	2.69E + 00	2.95E + 00	2.82E + 00	1.32E-01	4.7
45	1.0	−1.98E + 00	−2.21E + 00	−2.09E + 00	1.16E-01	−5.5
90	1.0	1.29E-01	1.62E-01	1.46E-01	1.66E-02	11.4
135	1.0	2.98E + 00	3.07E + 00	3.03E + 00	4.35E-02	1.4
180	1.0	2.64E + 00	3.01E + 00	2.83E + 00	1.82E-01	6.5
Average (Total)		**1.29E + 00**	**1.40E + 00**	**1.34E + 00**	**9.81E-02**	**3.69**
2nd Phase z_2/A						
χ [deg]	λ/L	IIHR (2.7 m, Aug 2015)	IIHR (2.7 m, Aug 2016)	AVERAGE (D)	SD	SD%D
0	1.0	−2.58E-01	7.45E-02	−9.16E-02	1.66E-01	−181.3
45	1.0	−5.99E-01	−4.33E-01	−5.16E-01	8.30E-02	−16.1
90	1.0	−2.43E + 00	−1.60E + 00	−2.01E + 00	4.14E-01	−20.6
135	1.0	1.52E + 00	−1.56E + 00	−2.14E-02	1.54E + 00	−7222.9
180	1.0	−1.78E + 00	−1.23E + 00	−1.51E + 00	2.73E-01	−18.1
Average (Total)		**−7.08E-01**	**−9.51E-01**	**−8.29E-01**	**4.96E-01**	**−1491.79**

Table 31 SD analysis on 1st and 2nd harmonic phases for ϕ/Ak of KCS in oblique waves ($\lambda/L = 1$)

1st Phase ϕ_1/Ak						
χ [deg]	λ/L	IIHR (2.7 m, Aug 2015)	IIHR (2.7 m, Aug 2016)	AVERAGE (D)	SD	SD%D
0	1.0	−1.39E + 00	4.15E-01	−4.87E-01	9.02E-01	−185.1
45	1.0	9.56E-01	4.39E-01	6.98E-01	2.59E-01	37.1
90	1.0	−1.75E + 00	−9.94E-01	−1.37E + 00	3.76E-01	−27.4
135	1.0	−1.75E + 00	−1.76E + 00	−1.75E + 00	9.97E-04	−0.1
180	1.0	−9.29E-01	−2.94E + 00	−1.93E + 00	1.00E + 00	−52.0
Average (Total)		**−9.72E-01**	**−9.67E-01**	**−9.69E-01**	**5.08E-01**	**−45.5**
2nd Phase ϕ_2/Ak						
χ [deg]	λ/L	IIHR (2.7 m, Aug 2015)	IIHR (2.7 m, Aug 2016)	AVERAGE (D)	SD	SD%D
0	1.0	7.49E-02	−2.88E + 00	−1.40E + 00	1.48E + 00	−105.3
45	1.0	−5.58E-02	−1.20E-01	−8.80E-02	3.22E-02	−36.6
90	1.0	−1.97E + 00	−1.98E + 00	−1.97E + 00	3.91E-03	−0.2
135	1.0	3.31E-01	1.17E + 00	7.51E-01	4.20E-01	55.9
180	1.0	−1.18E + 00	−1.08E + 00	−1.13E + 00	5.10E-02	−4.5
Average (Total)		**−5.60E-01**	**−9.78E-01**	**−7.69E-01**	**3.97E-01**	**−18.2**

Table 32 SD analysis on 1st and 2nd harmonic phases for θ/Ak of KCS in oblique waves ($\lambda/L = 1$)

1st Phase θ_1/Ak

χ [deg]	λ/L	IIHR (2.7 m, Aug 2015)	IIHR (2.7 m, Aug 2016)	AVERAGE (D)	SD	SD%D
0	1.0	−3.00E + 00	−2.74E + 00	−2.87E + 00	1.31E-01	−4.6
45	1.0	−1.18E + 00	−1.44E + 00	−1.31E + 00	1.28E-01	−9.8
90	1.0	−9.41E-01	−7.17E-01	−8.29E-01	1.12E-01	−13.5
135	1.0	1.69E + 00	1.74E + 00	1.71E + 00	2.33E-02	1.4
180	1.0	2.87E + 00	3.18E + 00	3.03E + 00	1.55E-01	5.1
Average (Total)		**−1.13E-01**	**−1.13E-01**	**4.26E-03**	**−5.44E-02**	**1.10E-01**

2nd Phase θ_2/Ak

χ [deg]	λ/L	IIHR (2.7 m, Aug 2015)	IIHR (2.7 m, Aug 2016)	AVERAGE (D)	SD	SD%D
0	1.0	−2.94E-01	−1.20E-02	−1.53E-01	1.41E-01	−92.1
45	1.0	2.52E + 00	1.56E + 00	2.04E + 00	4.80E-01	23.5
90	1.0	5.53E-02	1.38E + 00	7.18E-01	6.62E-01	92.3
135	1.0	−1.34E + 00	−1.39E + 00	−1.36E + 00	2.45E-02	−1.8
180	1.0	−3.45E + 00	−2.76E + 00	v3.11E + 00	3.44E-01	−11.1
Average (Total)		**7.56E-01**	**−5.01E-01**	**−2.44E-01**	**−3.72E-01**	**3.30E-01**

Table 33 SD analysis on 1st and 2nd harmonic phases for ζ/A of KCS in oblique waves ($\lambda/L = 1$)

1st Phase ζ_1/A						
χ [deg]	λ/L	IIHR (2.7 m, Aug 2015)	IIHR (2.7 m, Aug 2016)	AVERAGE (D)	SD	SD%D
0	1.0	−6.19E-02	−3.89E-02	−5.04E-02	1.15E-02	−22.8
45	1.0	−9.49E-02	−2.14E-03	−4.85E-02	4.64E-02	−95.6
90	1.0	−7.74E-02	−1.26E-01	−1.02E-01	2.43E-02	−23.9
135	1.0	1.19E-01	3.00E-02	7.45E-02	4.45E-02	59.8
180	1.0	1.38E-01	1.28E-02	7.52E-02	6.24E-02	83.0
Average (Total)		**4.50E-03**	**−2.48E-02**	**−1.02E-02**	**3.78E-02**	**0.1**
2nd Phase ζ_2/A						
χ [deg]	λ/L	IIHR (2.7 m, Aug 2015)	IIHR (2.7 m, Aug 2016)	AVERAGE (D)	SD	SD%D
0	1.0	−2.42E + 00	7.47E-02	−1.17E + 00	1.25E + 00	−106.4
45	1.0	−5.40E-01	−6.97E-01	−6.19E-01	7.85E-02	−12.7
90	1.0	−2.85E + 00	2.22E + 00	−3.19E-01	2.53E + 00	−793.8
135	1.0	−2.19E + 00	3.57E-01	−9.16E-01	1.27E + 00	−138.9
180	1.0	1.69E + 00	1.85E + 00	1.77E + 00	7.98E-02	4.5
Average (Total)		**−1.26E + 00**	**7.61E-01**	**−2.51E-01**	**1.04E + 00**	**−209.5**

Table 34 Data lists of oblique wave cases

χ [°]	λ/L	15-Aug	16-Aug
45	0.50	0	
	1.00	0	0
	2.00	0	
90	0.25	0	
	0.30	0	
	0.40	0	
	0.50	0	
	1.00	0	0
	2.00	0	
135	0.25	0	
	0.30	0	
	0.40	0	
	0.50	0	
	1.00	0	0
	2.00	0	
180	0.50	0	
	1.00	0	0
	2.00	0	
	T2015		

Figure 25 compares the present and Fujii-Takahashi (1975) added resistance σ_{AR} vs. both wavelength and heading for reference. Fujii-Takahashi (1975) is the only available added resistance data for oblique headings but note that the data was taken at Fr = 0.25 using the S-175 container ship model which has different C_B and L/B from KCS. Table 35 lists the principal-particulars comparisons between the KCS and the S-175 model used in Fujii-Takahashi (1975). The IIHR trends vs. λ/L for $\chi = 0$ and 90° are consistent with expectation for the effects of f_e/f_n (for heave and pitch f_n), i.e., resonance for shorter waves for $\chi = 90°$ than $\chi = 0°$. Since $\chi = 180°$ is far from resonance it is not known why σ_{AR} is larger for short waves. The Fujii-Takahashi (1975) results for head waves are in reasonably close agreement with IIHR, but only roughly similar for the other headings. Figures 26 and 27 are like Figs. 23–24 but show the harmonic amplitudes and phases vs. heading. The data not was used at T2015 but was used for CFD assessment by Sadat-Hosseini et al. (2015a); and provided herein for use by others.

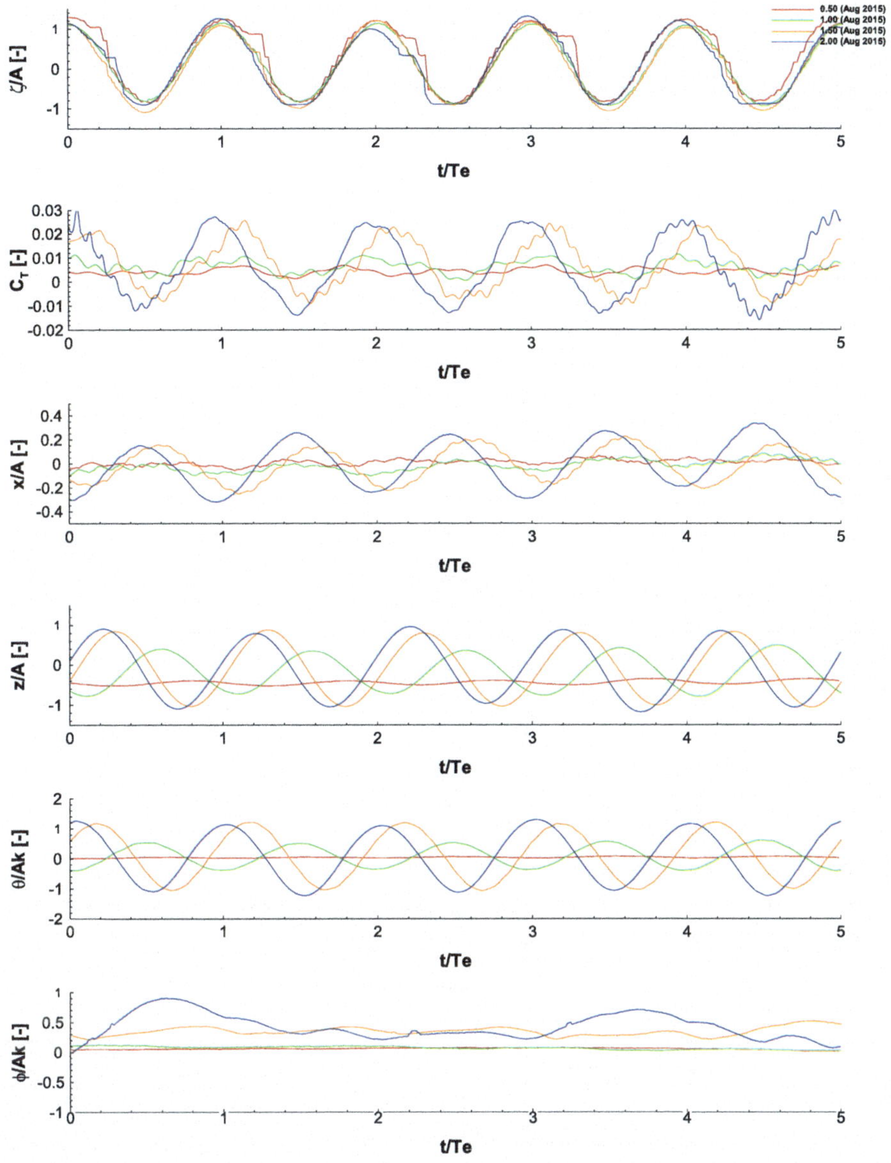

Fig. 18 Time series for force and motions of KCS in regular oblique waves at $\chi = 0°$ (IIHR Aug-15)

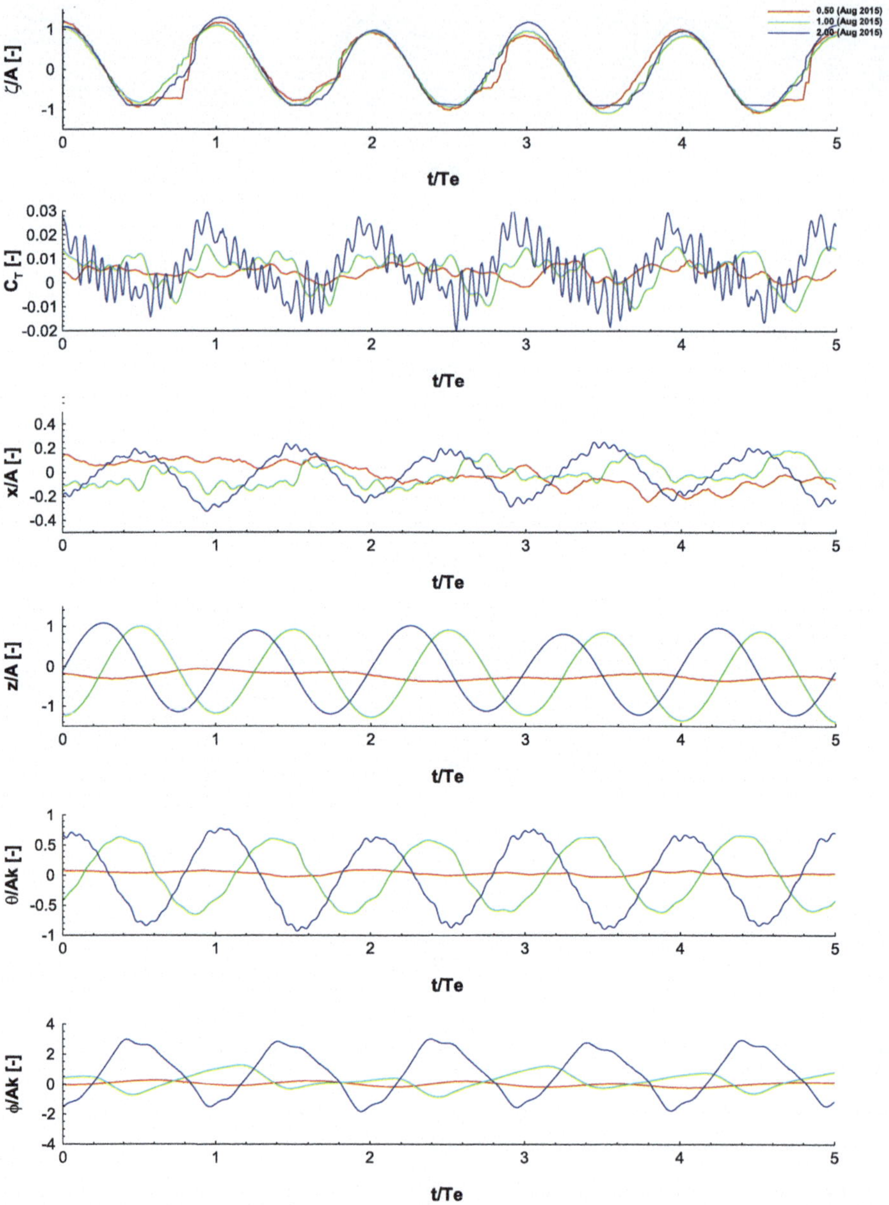

Fig. 19 Time series for force and motions of KCS in regular oblique waves at $\chi = 45°$ (IIHR Aug-15)

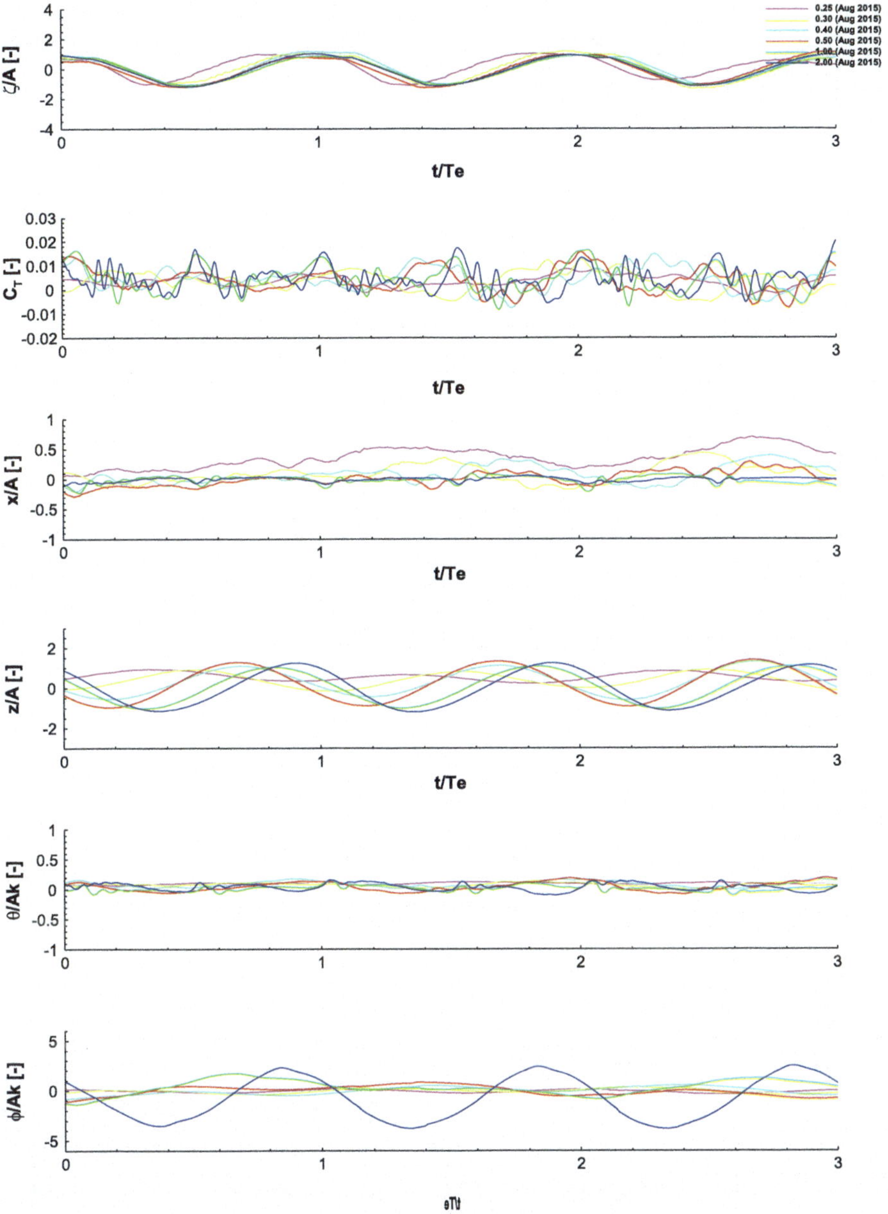

Fig. 20 Time series for force and motions of KCS in regular oblique waves at $\chi = 90°$ (IIHR Aug-15)

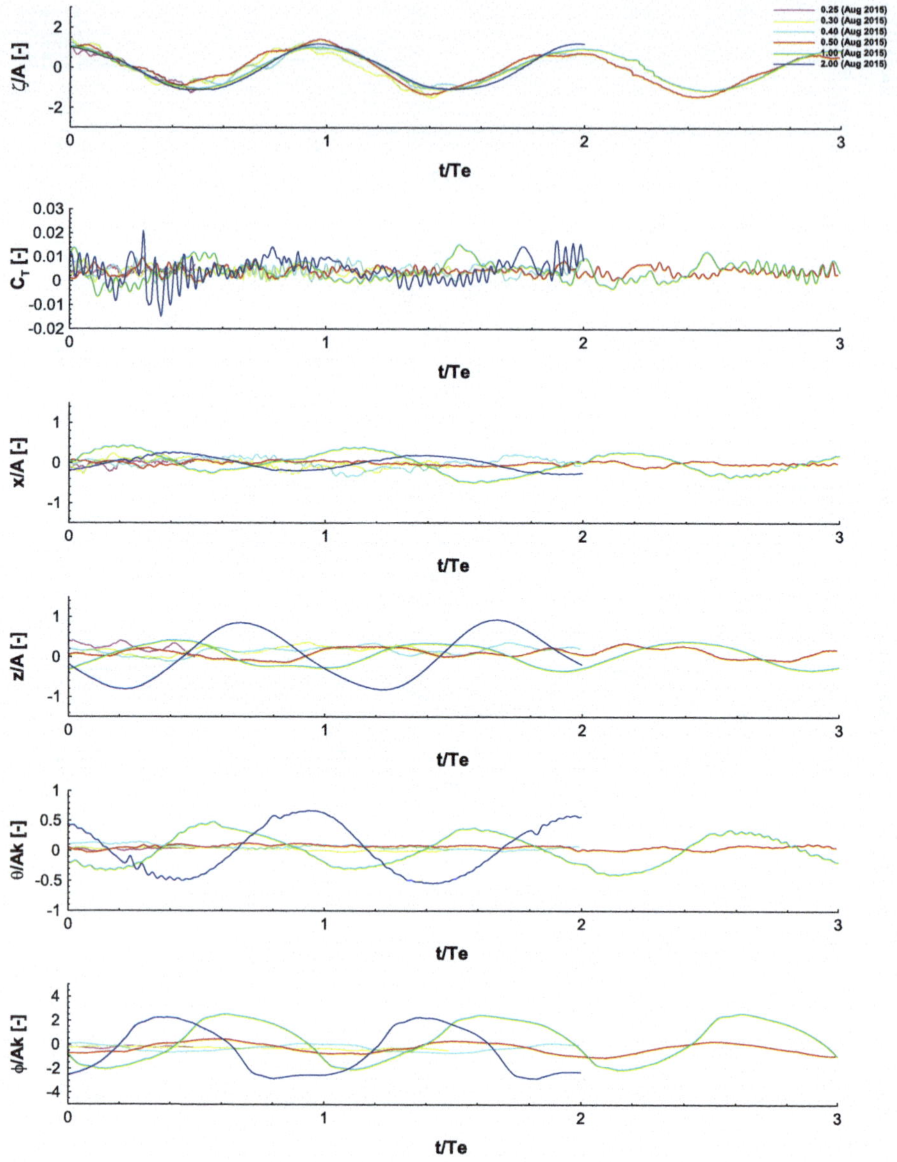

Fig. 21 Time series for force and motions of KCS in regular oblique waves at $\chi = 135°$ (IIHR Aug-15)

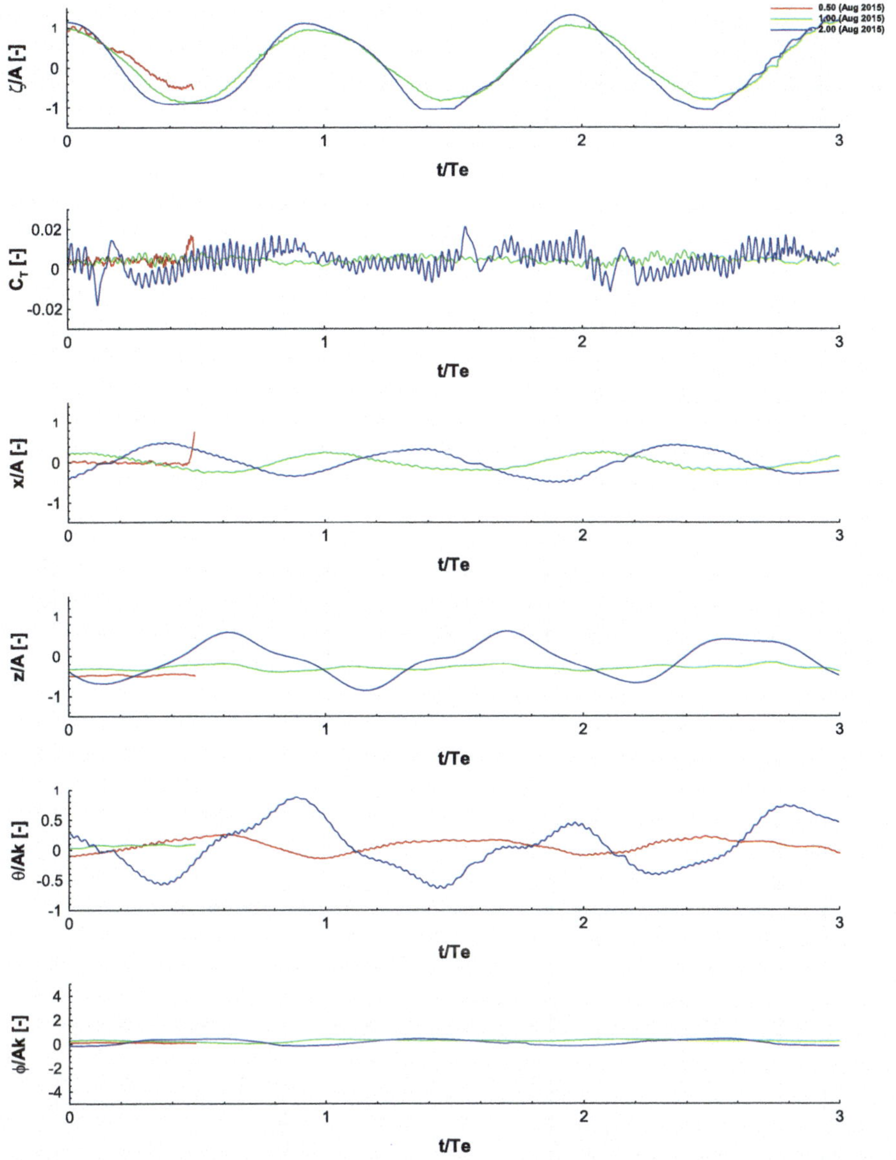

Fig. 22 Time series for force and motions of KCS in regular oblique waves at $\chi = 180°$ (IIHR Aug-15)

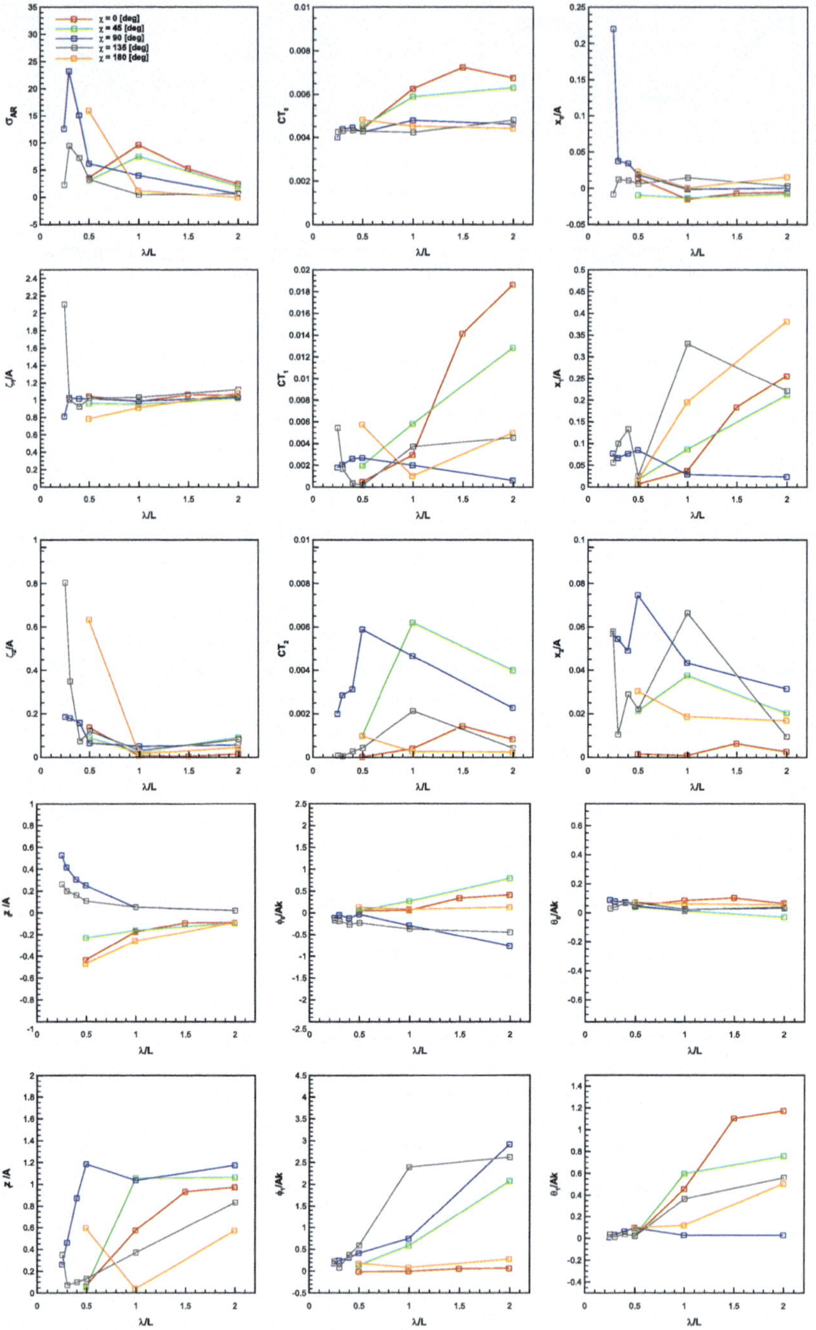

Fig. 23 Harmonic amplitudes of resistance and motions for KCS in regular oblique waves at Fr = 0.26

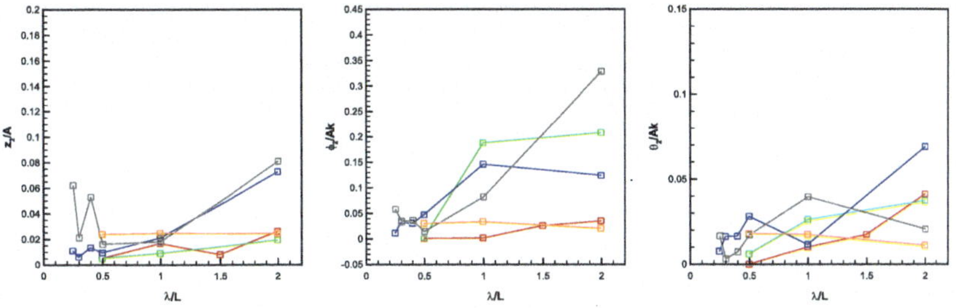

Fig. 23 (continued)

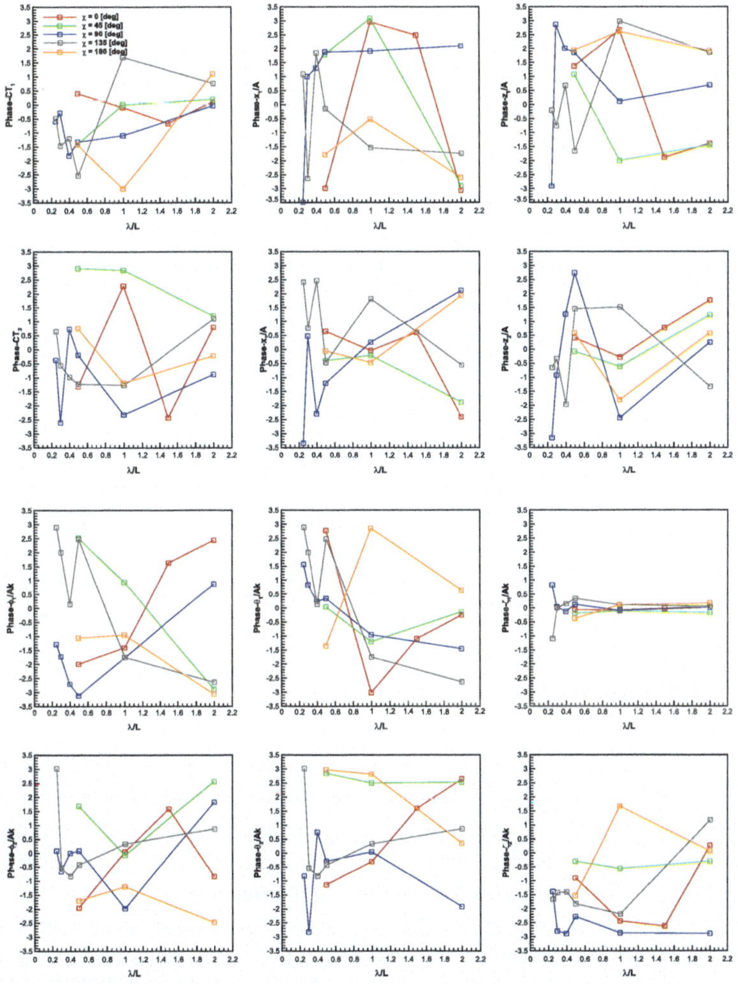

Fig. 24 Harmonic phases of resistance and motions for KCS in regular oblique waves at Fr = 0.26

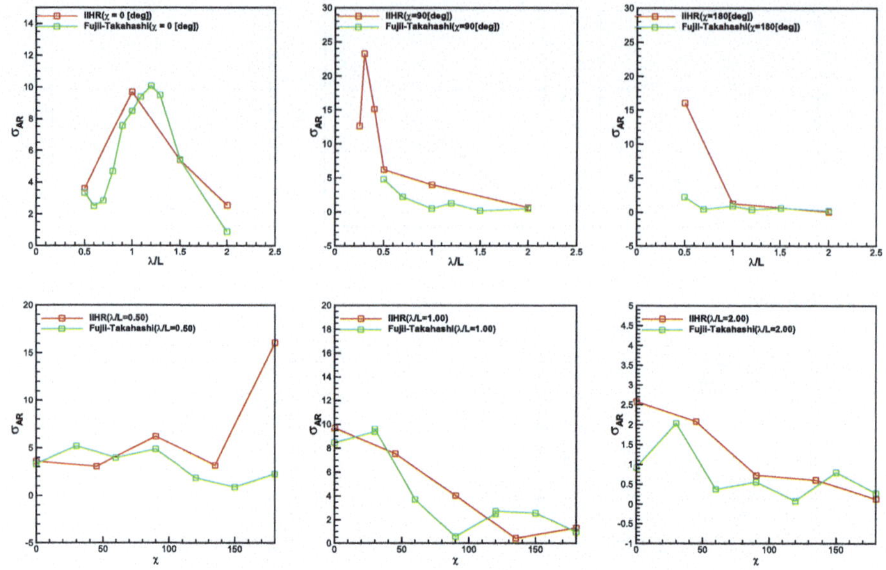

Fig. 25 Added resistance comparison between IIHR and Fujii-Takahashi (1975)

Table 35 Principal-particulars comparison between KCS and S-175

	KCS (IIHR)	S-175 (Fujii-Takahashi 1975)
L [m]	2.700	3.500
B [m]	0.378	0.508
T [m]	0.127	0.190
C_B	0.643	0.572
B/T	2.981	2.674
L/B	7.143	6.890

6 ONRT Calm Water

IIHR has extensive experimental data and uncertainty analysis for ONRT free running calm water self-propulsion and maneuvering; and course keeping/speed loss and maneuvering in regular and irregular waves (Sanada et al. 2013, 2014 and 2018b; Elsheikh 2014; Sadat-Hosseini et al. 2015b; Bottiglieri 2016), as summarized in Table 36. Table 37 lists the particulars of the ship model. The ONRT mass properties adjustment was done similar as for the KCS model as described in Sect. 2 except for k_{xx}. Instead of k_{xx} adjustment, free roll decay tests were performed to adjust the natural roll period to match with the reference value provided by NRIFE (Sanada et al. 2013). The uncertainty analysis is preliminary, and improvements are in progress for

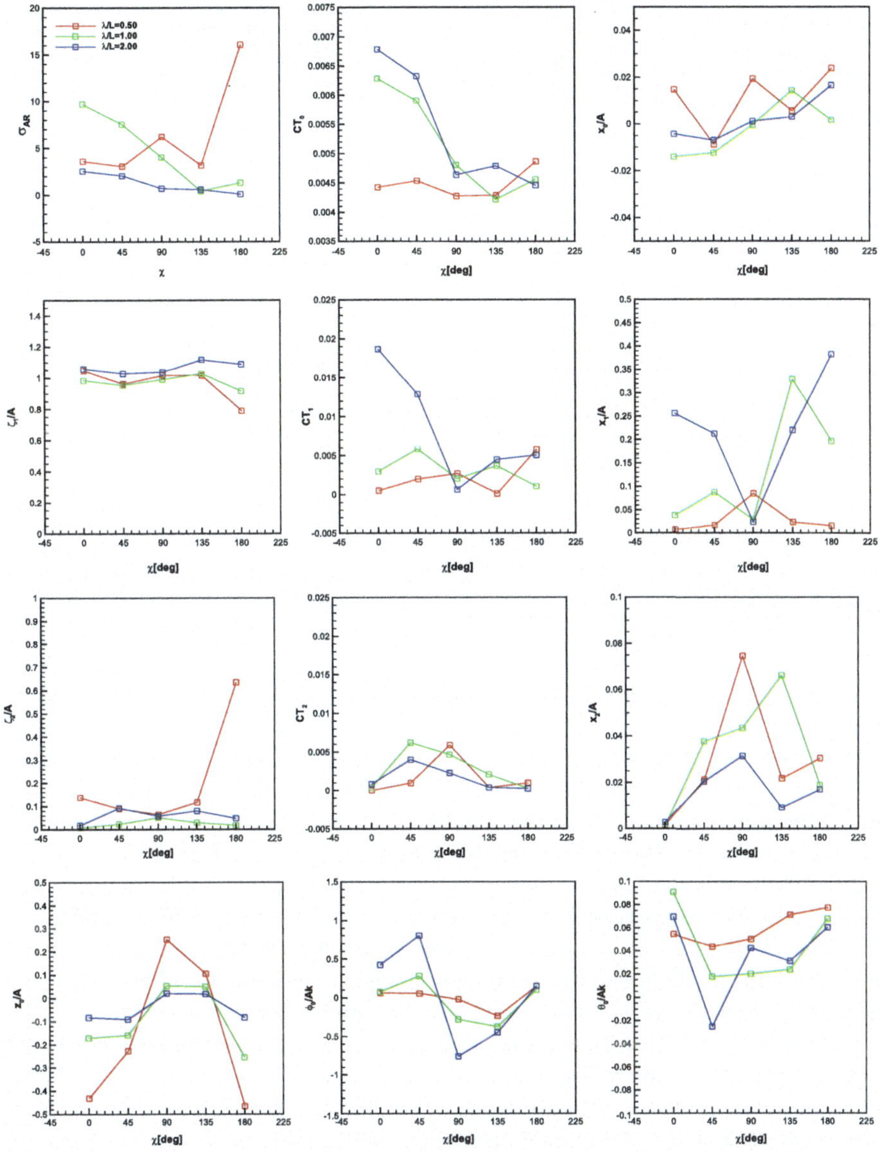

Fig. 26 Harmonic amplitudes of resistance and motions for KCS in regular oblique waves at Fr = 0.26

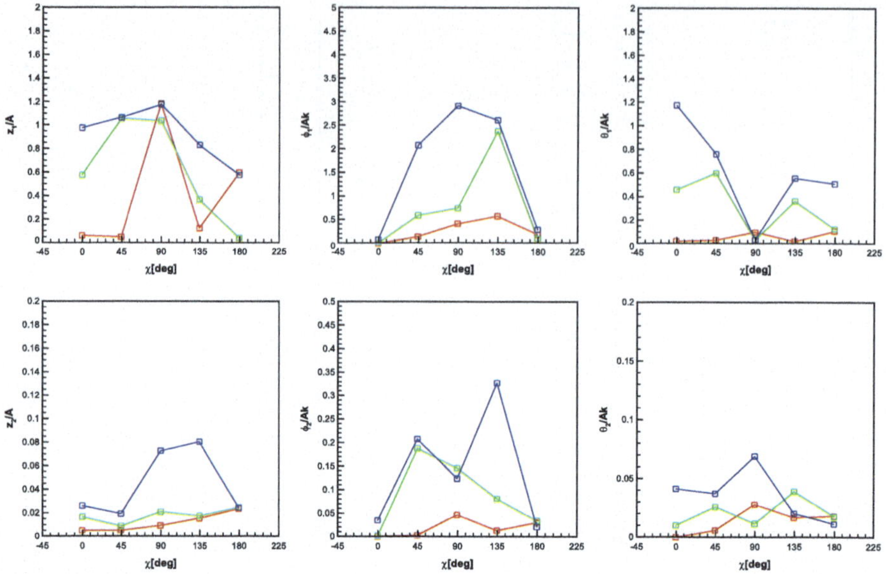

Fig. 26 (continued)

differential filter for surge and sway velocities based on the model position and its uncertainties.

In addition, the time histories display a time shift between the yaw and rudder angle in course keeping/speed loss (the delay is 0.15 s at least) due to the processing time of the free running system. Since the free running system can receive updated yaw angle data every 100 ms from the tracking system and the free-running system requires 50 ms at least to update the rudder after receiving the updated yaw angle. To make an accurate comparison between the time series of the experiments and CFD, the latency of the yaw acquisition should be included as the CFD condition in the future. Both the tracking and free running system data rates are 20 Hz, but they have different internal clocks and it is technically difficult to completely synchronize the current two systems. Time uncertainty due to jittering of the internal clocks should be evaluated in the future. Methods are under investigation to remove this issue in the future; however, it should not have much of an effect on the T2015 V&V and SoA assessment of the CFD. Additional discussion on this issue is provided in Chap. 9, Sect. 4.3.3. Also discussed therein is that a coordinate transformation sign error was discovered post T2015 such that the (u, v) required corrections, which were included in the Chap. 9 CFD assessments and the U_{Xi} values used therein.

Free running calm water self-propulsion and course keeping/speed loss in regular waves with $H/\lambda = 0.02$ and $\lambda/L = 1.0$ are used in T2015. Bottiglieri (2016) provides uncertainty analysis for sinkage, trim, and propeller RPS at $Fr = 0.2$. U_{Xi} for propeller RPS is very small (0.08%D) whereas sinkage and trim showed large uncertainties. The CFD validation studies in Chap. 9 used the U_{Xi} values provided.

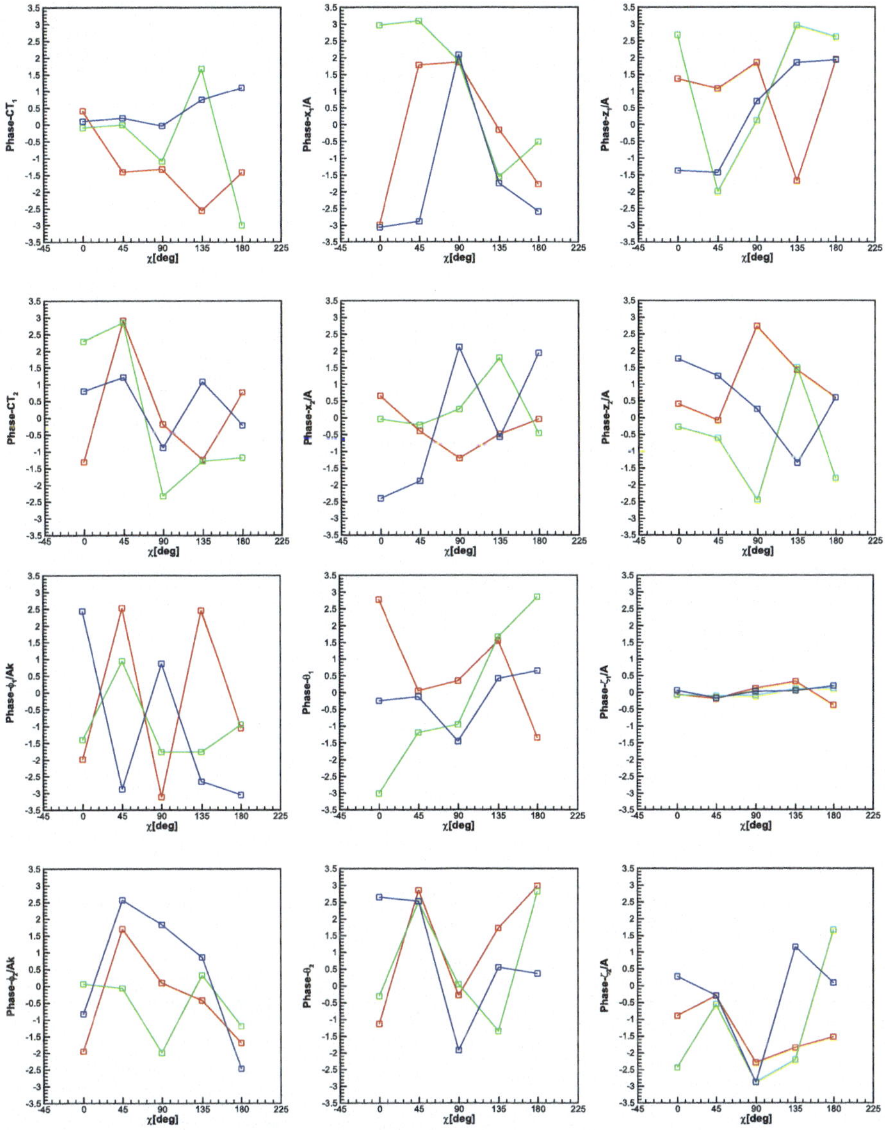

Fig. 27 Harmonic phases of resistance and motions for KCS in regular oblique waves at Fr = 0.26

Table 36 ONRT course keeping and maneuvering tests conditions

	T2015 Case							
	Test	Fr^*	δ [deg]	ψ^{**} [deg]	H/λ	λ/L	χ [deg]	Numbers of runs
Calm water	Course keeping	0.2	N/A	0,180			0,180	3,3
				± 45, ± 135			± 45, ± 135	3/3,3/3
				± 90			± 90	3/3
	Zigzag	0.2	10	10			0,180	3,3
			20	20				21,3
			35	35,90				3,3
	Turning circle	0.1	± 35	N/A			0	5/5
		0.2					0, ± 90,180	6/21,3/3,3/3,3/3
			−5,−10, −15, −20, −25, −30				0	2,2,2,2,3,3
		0.3	± 35				0	5/5
Head Wave	Course keeping	0.2	N/A	0	0.02	0.5	0	3
						1		3
						1.2		3
	Course keeping	0.2	N/A	0	0.04	0.5	0	3
	Zigzag	0.2	10	10	0.02	0.5,1.0,1.2	0	3,3,3
			20	20		0.5,1.0,1.2		3,3,3
			35	35		0.5,1.0,1.2		3,3,3
	Turning circle	0.1	± 35	N/A	0.02	0.5,1.0,1.2	0	5/5,5/5,5/5
		0.2				0.5,1.0,1.2	0, ± 90,180	2 × 3 × 3 × 4 = 72

(continued)

Table 36 (continued)

	T2015 Case							
	Test	Fr^*	δ [deg]	ψ^{**} [deg]	H/λ	λ/L	χ [deg]	Numbers of runs
			−35		0.04	0.5	0	3
		0.3	± 35		0.02	0.5,1.0	0	5/5,5/5
	Long approach Turning circle	0.2	−35		0.02	0.5,1.0,1.2	0	3,3,3
Following waves	Course keeping	0.2	N/A	180	0.02	0.5	180	3
						1		3
						1.2		3
	Zigzag	0.2	10	10	0.02	0.5,1.0,1.2	180	3,3,3
			20	20		0.5,1.0,1.2		3,3,3
			35	35	0.02,0.03	0.5,1.0,1.2		3,3,3/3,3,3
Beam wave	Course keeping	0.2	N/A	± 90	0.02	0.5	± 90	3
						1		3
						1.2		3
Quartering wave	Course keeping	0.2	N/A	± 45, ± 135	0.02	0.5	± 45, ± 135	3/3,3/3
						1		3/3,3/3
						1.2		3/3,3/3

(continued)

Table 36 (continued)

	T2015 Case							
	Test	Fr*	δ [deg]	ψ^{**} [deg]	H/λ	λ/L	χ [deg]	Numbers of runs
Irregular wave	Course keeping	0.2	N/A	0	Sea State 4 (Tp = 1.26 [s], Hs = 0.038 [m]) (Duration: 60[s],Uniform division by frequency) (3 repeated runs with same phase, 4 runs with diferent phases)		0	7
		0.2	N/A	0	Sea State 4 (Tp = 1.26 [s], Hs = 0.038 [m]) (Duration 30[s], Uniform division by frequency)		0	12
	Long approach Turning Circle	0.2	−35	0	Sea State 4 (Tp = 1.26 [s], Hs = 0.038 [m]) (Duration 120 [s],Uniform division by frequency) (3 repeated runs with same phase, 4 runs with diferent phases)		0	7

(continued)

Table 36 (continued)

	T2015 Case							
	Test	Fr^*	δ [deg]	ψ^{**} [deg]	H/λ	λ/L	χ [deg]	Numbers of runs
		0.2	−35	0	Sea State 4 (Tp = 1.26 [s], Hs = 0.038 [m]) (Duration: 120[s],Uniform division by wave energy) (3 repeated runs with same phase, 4 runs with diferent phases)		0	7

Fr^*: the nominal Froude number
ψ^*: the target yaw angle

Table 37 ONRT Particulars

	Model Scale
Length (L)	3.147 m
Breadth	0.384 m
Depth	0.266 m
Draft	0.112 m
Displacement	72.6 kg
Metacentric height	0.0422 m
Natural roll period	1.665 s
Rudder area	$0.012\ m^2 \times 2$
Block coefficient	0.535
Radius of gyration in pitch	$0.246 \times L$
Maximum rudder Angle	±35°

7 ONRT Head and Oblique Wave

Bottiglieri (2016) provides uncertainty analysis for 0th harmonic amplitudes of heave z/A, roll ϕ/Ak, pitch θ/Ak, yaw ψ/Ak, axial velocity u/U, and side velocity v/U, respectively; and uncertainty analysis for 1st and 2nd harmonic phases for ζ/A, z/A, ϕ/Ak, θ/Ak, ψ/Ak, u/U, and v/U, respectively. The harmonic analysis window sizes for $\mu = 0, 45, 90, 135, 180$ degrees are 10Te, 4Te, 5Te, 3Te, 5Te, respectively.

For 0th harmonic amplitudes, the U_{Xi} values are small for u/U (1 ~ 3%D) whereas other values show large uncertainties. The U_{Xi} values for 1st harmonic amplitudes are reasonably small for most variables except for ϕ/Ak and ψ/Ak for $\mu = 0$ and 180 deg, and u/U for $\mu \neq 0$ and v/U for $\mu \neq 90$. 1st harmonic phases mostly have small uncertainties. The uncertainty values are often larger for 2nd harmonic amplitudes and phases. For CFD validation studies in Chap. 9 used the U_D values provided.

8 Overall Conclusions

The experimental data used for T2015 test cases for KCS added resistance in regular head (2.10) and oblique (2.11) waves and ONRT free running course keeping in regular head (3.12) and oblique (3.13) waves evaluation, including in the former case calm water resistance, sinkage and trim and in the latter case calm water self-propulsion (3.9), is shown to be reasonably reliable. However, clearly it is desirable to have data from more facilities including uncertainty analysis for assessment of facility biases, which will provide more robust data and U_{Di} for CFD validation/SoA assessment.

Acknowledgements The Office of Naval Research supports the IIHR research under grants administered by Drs. Thomas Fu, Woei-Min Lin, Ki-Han Kim, and Robert Brizzolara. Mark Stocker and Michael Bottiglieri master's theses contributed to the IIHR experiments.

References

ASME. (2013). Test Uncertainty: Instruments and Apparatus, PTC 19.1.
Bottiglieri, M. (2016). Uncertainty Assessment For Free-Running Model Cases At The IIHR Wave Basin, M.S. Thesis, University of Iowa, August.
Elsheikh, H. (2014). Maneuvering Characteristics in Calm Water and Regular Waves for ONR Tumblehome, M.S. Thesis, University of Iowa, May.
Fujii, H., Takahashi, T. (1975). Experimental Study on the Resistance Increase of a ship in Regular Oblique Waves, 14th ITTC, Ottawa.
ITTC, 7.5-02-01-01. (2008). Guide to the Expression of Uncertainty in Experimental Hydrodynamics, Specialist Committee on Uncertainty Analysis of 25th ITTC.
Otzen, J. F. (2015a). Uncertainty Assessment for KCS Added Resistance in Waves, FORCE Technology Report 112-31859-1C/ JFO/ 22-May.
Otzen, J. F. (2015b). Uncertainty Assessment for KCS Added Resistance in Waves, FORCE Technology Report 112-31859-2A/ JFO/ 11-November.
Sadat-Hosseini, H. et al. (2015a). Experiments and Computations for KCS Added Resistance for Variable Heading, World Maritime Technology Conference, Providence, Rhode Island, USA.
Sadat-Hosseini, H., Sanada, Y., Stern, F. (2015b). Experiments and CFD for ONRT Course Keeping in Regular Waves, 5th World Maritime Technology Conference November 3–7, Rhode Island Convention & Omni Hotel Providence, Rhode Island, USA.
Sanada,Y., Tanimoto, K., Takagi, K., Toda, Y., & Stern, F. (2013). Trajectories and Local Flow Field Measurements around ONR Tumblehome in Maneuvering Motion. *Ocean Engineering, 72,* 45–65.
Sanada, Y., Elshiekh, H., Toda, Y., & Stern, F. (2014). Effects of Waves on Course Keeping and Maneuvering for Surface Combatant ONR Tumblehome, 30th Symposium on Naval Hydrodynamics Hobart, Tasmania, Australia, 2–7 November.
Sanada, Y., Elshiekh, H., Toda, Y., & Stern, F. (2018b). Effects of Waves on Course Keeping and Manoeuvring for Surface Combatant ONR Tumblehome, online JMST October.
Simonsen, C. D., Otzen, J. F., Joncquez, S., & Stern, F. (2013). EFD and CFD for KCS Heaving and Pitching in Regular Head Waves. *Journal of Marine Science and Technology, 18,* 435–459.
Simonsen, C. D., Otzen, J. F., Nielson, C., & Stern F. (2014). CFD Prediction of Added Resistance of the KCS in Regular Head and Oblique Waves, 30th SNH, Hobart, Australia, 2–7 November.
Stocker, M. (2016). Surge Free Added Resistance Tests In Oblique Wave Headings For The KRISO Container Ship Model, M.S. Thesis, University of Iowa, August.
Toda, Y. (2018). Private communication.
Yugo, S., Htike, A. K., Ayaka, M., Yasuyuki, T., Janne, F. O. (2018a). Claus Simonsen, Frederick Stern, Added Resistance/Powering and Motion Measurements of KRISO Container Ship in Regular Head Waves, Proceedings of 8th PAAMES and AMEC 2018, 9–12 October, Busan.
Zou, L., & Larsson, L. (2014). Additional Data for resistance, sinkage and trim. In Larsson et al. (Ed.), Numerical Ship Hydrodynamics—An Assessment of the Gothenburg 2010 Workshop, Springer Business Media, Dordrecht.

Evaluation of Resistance, Sinkage, Trim and Wave Pattern Predictions for JBC

Lars Larsson

Abstract JBC predictions are presented for four cases: towed and self-propulsion for the bare hull and a hull with an Energy Saving Device (ESD). A statistical evaluation is made based on the 88 resistance predictions submitted. The comparison error is defined as the difference between the measured data and the numerically predicted value. This error is analyzed in different ways. It is seen that the mean signed error is as small as the measurement accuracy, while the mean absolute error is about twice as large. This represents the typical error in the prediction. A similar accuracy was found in the 2010 Workshop. The number of grid cells has increased since 2010 and the required grid size for a given uncertainty of 4% has increased from 3 M cells to 10 M cells. Reasons for this are discussed. As in the previous workshop the two-equation turbulence models produce more accurate results than the more advanced models, although the more and more popular Explicit Algebraic Stress Model (EASM) is close. A surprising result of the analysis is that methods with wall-functions give significantly smaller resistance errors than those with a wall-resolved flow. Grid convergence is discussed and it is shown that the vast majority of results converge with grid refinement, but that the achieved order of accuracy is often far from the theoretical one. Sinkage is much better predicted than in 2010 and the trim results are also improved. Finally, the wave pattern prediction is discussed. The best methods in 2010 predicted the waves extremely well, in fact better than in the present workshop. A consistent shift in the predicted wave phase relative to the data could indicate a measurement error.

L. Larsson (✉)
Chalmers University of Technology, Gothenburg, Sweden
e-mail: lars.larsson@chalmers.se

T. Hino et al. (eds.), *Numerical Ship Hydrodynamics*, Lecture Notes in Applied and Computational Mechanics 94, https://doi.org/10.1007/978-3-030-47572-7_5

1 Resistance

1.1 Statistics

The resistance cases of the Workshop were 1.1 for the bare hull and 1.2 for the hull equipped with an ESD. Resistance results are however also reported for the self-propulsion cases 1.5, without, and 1.6 with an ESD. Altogether, there were 88 fine grid resistance solutions reported. This is a large data set, suitable for statistical evaluation. The particular questions to be answered are:

- *How accurately can resistance be predicted for a bluff hull, like the JBC? And how close are we to experimental data? Is there any improvement compared to the previous Workshops?*

In Table 1 the four cases with reported resistance data are listed together with the corresponding results for the bluff hulls used in the two previous Workshops: the KVLCC2 in Gothenburg (Larsson et al. 2014) and the KVLCC2m in Tokyo (Hino 2005). The former is a VLCC designed and tested at KRISO in Korea in the late 1990s (Van et al. 1998) and the latter is a slightly modified version tested at NMRI in Japan a few years later (Hino 2005).

The table shows the comparison error E defined as $E = D - S$, where D and S are the measured and computed values, respectively. In the third column, the mean error E_{mean}, i.e. the mean value of all reported errors with sign, are presented in percent of D. Column four gives $|E|_{mean}$, i.e. the mean of the absolute value of the error. This is followed by the standard deviation of the error σ, the mean numerical uncertainty U_{SN} and the reported uncertainty of the measured data U_D. Finally, the number of entries is shown.

Perhaps the most striking results of Table 1 are the small mean errors. These are around 1%, which is the same as the measurement accuracy in most cases. There is thus practically no systematic error in the CFD predictions, although there is a scatter between the different methods. This scatter is represented by the mean absolute error

Table 1 Resistance statistics. Blue: towing, yellow: self-propulsion

Hull	Year/Case	E_{mean} %D	$/E/_{mean}$ %D	σ %D	U_{SNmean} %D	U_D %D	No. of Entries
KVLCC2m	2005/1.4	0.0		6.2		0.7	13
KVLCC2	2010/1.2a	-1.7		1.3		1.0	5
JBC	2015/1.1a	1.0	2.1	2.4	3.8	1.0	22
	2015/1.2a	1.0	1.5	2.0	5.1	1.0	20
	2015/1.5a	-0.1	3.2	4.0	4.6	1.0	23
	2015/1.6a	0.7	3.1	4.0	6.0	1.0	23

$|E|_{mean}$, which represents the average deviation from the measured data. Note that the numerical uncertainty U_{SN} reported by the participants is considerably larger than $|E|_{mean}$. Unfortunately neither of these two quantities are available from the two previous Workshops, so we cannot draw any conclusions on their development. However, the standard deviation is available, and there was a significant improvement between 2005 and 2010, but no further improvement after that. Now we should compare only the blue fields, representing towed results, since the self-propelled results in the yellow field are more scattered. Also, we have to keep in mind that there were only 5 entries for KVLCC2 at Gothenburg 2010. It should be noted that the mean signed error for *all* hulls in 2010 was very small, $0.1\%D$, while the standard deviation was $2.1\%D$. We will return to the lack of progress between 2010 and 2015 below.

The resistance statistics may be compared with that of the Fifth AIAA Computational Fluid Dynamics Drag Prediction Workshop (Vassberg 2016), DPW. The standard test case was a wing-body configuration, which was tested in two wind tunnels, where the drag differed by 3.2%. 57 drag predictions were submitted, and the mean drag differed from the mean of the two wind tunnel results by 2.4%. This is a larger comparison error than those of Table 1, and a probable explanation is the larger data uncertainty. The standard deviation was 2.1%, which is similar to the results for the towed case in Table 1. An interesting difference compared with the present workshop was the use of extrapolated predictions. All participants had to carry out a grid dependence study and the drag reported was the continuum solution, i.e. the solution extrapolated to zero cell size.

Table 2 shows the reduction in resistance due to the ESD. Although an increase in resistance might be expected when adding the ESD, both computations and measurements exhibit a resistance reduction. For the towed cases the computed and measured reductions are extremely close, while for the self-propelled case the difference is slightly larger. However, all reductions are within the reported measurement accuracy (see Table 1), so their significance is questionable.

An objective of the series of Workshops is the possibility to study the performance of the different components of the CFD methods. For this reason participants have to report the details of their methods in a questionnaire. The results are collected in a spreadsheet, which may be found in the supplementary information on the web site Springer Extras (for address, se book cover). By linking the performance of a method to its basic components, guidelines may be obtained for future developments. While rather detailed comparisons were possible 35 years ago, when the first workshop was held (Larsson 1981), this is not possible anymore due to the very much increased complexity of the methods. Like at Gothenburg 2010, we will therefore make the

Table 2 Resistance reduction due to ESD

	Computed (mean)	Measured
Towing (1.2a-1.1a)%	−0.7	−0.6
Self-propulsion (1.6a-1.5a)%	−0.2	−1.0

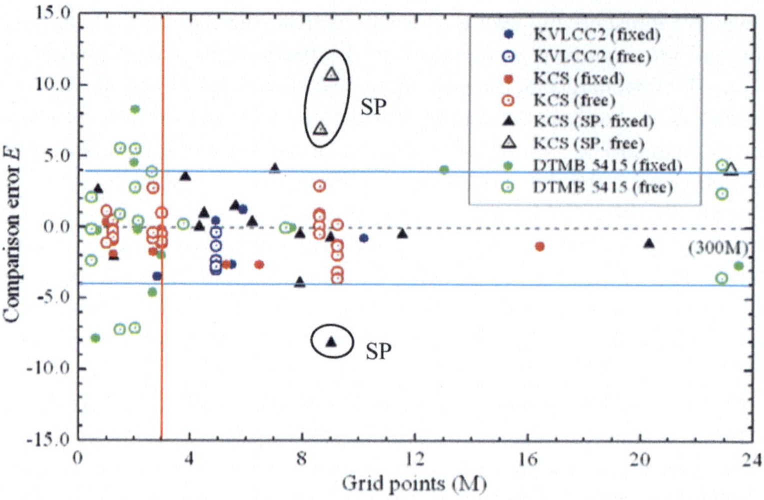

Fig. 1 Gothenburg 2010. Comparison error for all resistance submissions versus grid size. Blue lines represent ±4%. SP stands for self-propulsion

comparison for the two main components: grid and turbulence model. Here we will also add the wall boundary condition.

- *How important is grid size?*

Figure 1 shows the comparison errors for all resistance cases of Gothenburg 2010. It is seen that for grid sizes above 3 million (red line) virtually all errors are within ±4%. The outliers are three self-propulsion cases. The corresponding plot for JBC in the present Workshop is presented in Fig. 2.

Apparently, the ±4% level is now achieved at a larger cell count, approximately 10 million (red line). This is a disappointing result, indicating that more grid cells are needed in 2015 than five years earlier. The reasons for that are not obvious, but one explanation could be general CFD experience of the participants. Over the years there has been a shift from only developers in the first workshops to more and more users in the most recent ones. In the present workshop there were considerably more users than developers. This reflects the maturity of the codes that are now used routinely in hydrodynamic development work, but a regular user may need more grid cells to obtain the same accuracy as an experienced developer. Another reason may be the increasing speed of computers. The manual labor to create a perfect grid is considerable, and it may be more efficient to bear with a less perfect, but larger grid to get a certain accuracy

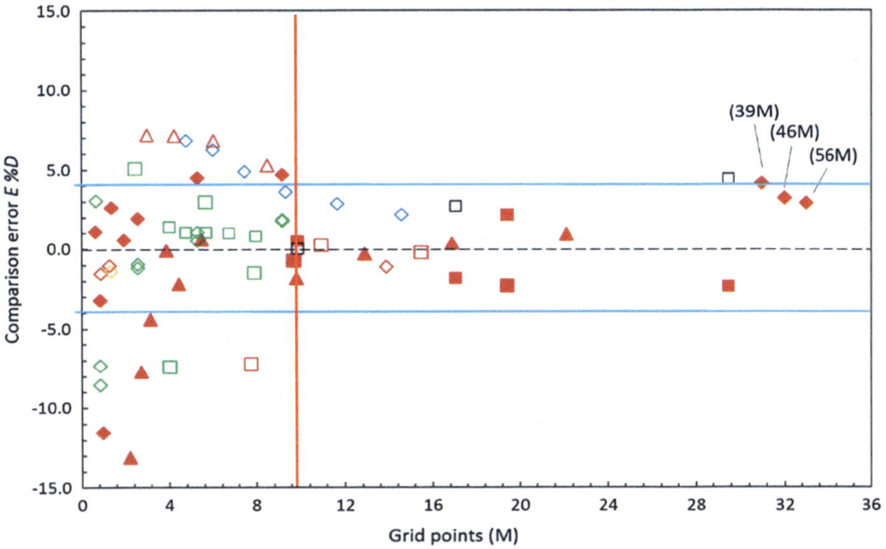

a. Bare hull, Case 1.1

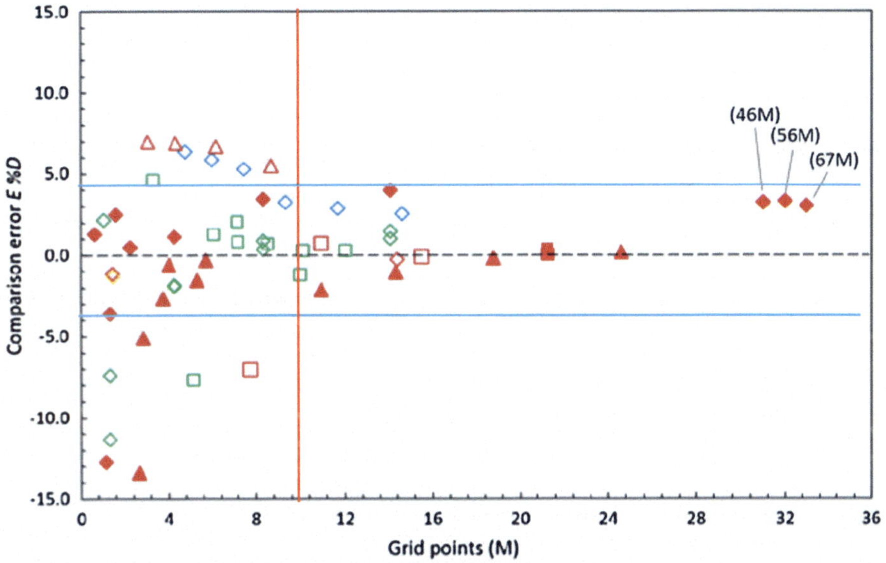

b. Hull with ESD, Case 1.2

Fig. 2 Present JBC submissions. Comparison error for resistance versus grid size. Blue lines represent ±4%

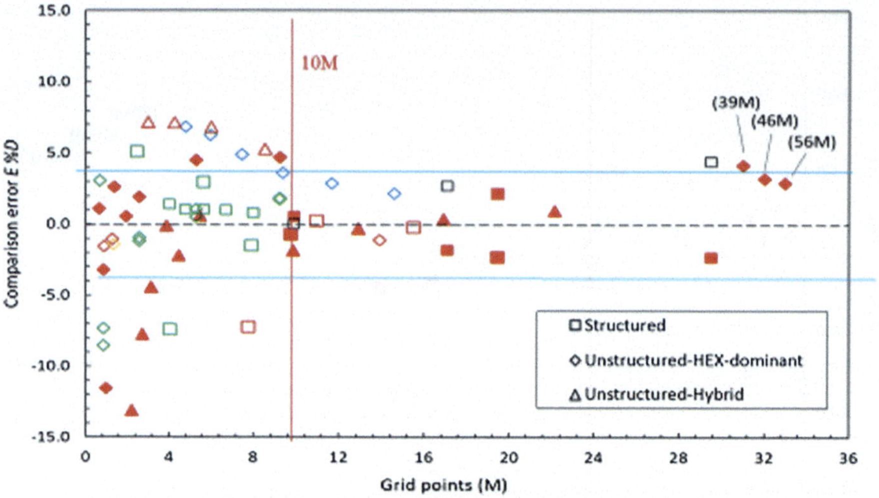

a. Bare hull

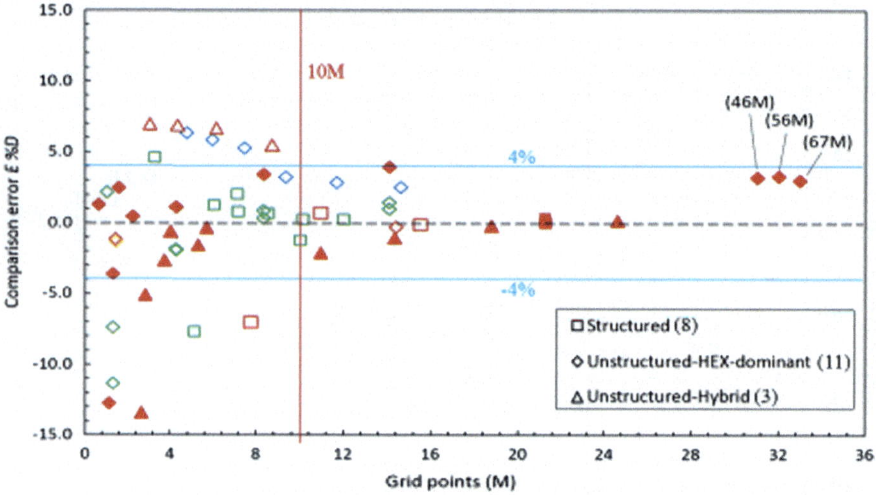

b. Hull with ESD

Fig. 3 Comparison of JBC submissions, resistance versus grid type

The next grid-related question is:

- *Is there any influence of grid type?*

Table 3 Resistance comparison error for different turbulence models

Turbulence model	E_{mean} %D		$\|E\|_{mean}$ %D		No. of entries	
	2015 1.1,1.2	2010 All	2015 1.1,1.2	2010 All	2015 1.1,1.2	2010 All
1-E Spalart Allmaras	-	3.6		3.6	-	1
1-E Menter	-	0.1		0.1	-	3
2-E k-e	0.8	0.3	1.8	2.1	8	37
2-E k-w	0.7	-0.5	1.6	1.8	19	37
EASM	1.6	1.8	2.0	5.9	12	3
RS, ARS	-1.3	-8.0	1.3	-8.0	2	2
DES	4.4	0.6	4.4	3.1	1	6

Figure 3 is the same plot as Fig. 2, but with an emphasis on grid type. Square symbols represent structured grids, diamonds unstructured grids dominated by hexahedral cells and triangles unstructured cells of various shapes. There seems to be a slight advantage of structured grids, particularly in Fig. 3b, but even in Fig. 3a the maximum errors for the structured grids are smaller than for the other two. The largest errors in both figures come from the completely unstructured grids, but the difference compared to the hex-dominated ones is small and may well be insignificant.

- *What is the influence of the turbulence model?*

In Table 3 the comparison error for different turbulence models is presented. Results from Gothenburg 2010 are also included. These are for all three hulls at that workshop. For Tokyo 2015 the results are for cases 1.1 and 1.2. There were no one-equation (1-E) models in the present Workshop, but many two-equation (2-E) models, like in 2010. The Explicit Algebraic Stress Model (EASM) has gained in popularity and the submissions have increased from 3 to 12.

Like in 2010, the two-equation models perform better than the more advanced models. Both the mean signed error and the mean absolute errors are smaller, even though the EASM model is now rather close. The very few submissions for the Reynolds stress models (RS, ARS) makes it difficult to draw any firm conclusions on the accuracy. This holds also for the Detached Eddy Simulation models (DES) in the present Workshop. It should be stressed that these results are for resistance. Wake predictions are a completely different matter.

The final component of the methods to be compared is the boundary condition applied on the hull surface.

- *Are wall-resolved solutions better than those using wall-functions?*

Table 4 shows a comparison between wall-function (WF), and low Reynolds number (LRN), wall resolved solutions for cases 1.1 and 1.2. Most methods use LRN boundary conditions, but it is clear from the table that the wall function approach yields the smallest errors for these two cases. The mean error, the mean absolute error and the standard deviation are all smaller for the wall functions.

Table 4 Resistance comparison error for different wall treatments

Case/BC	E_{mean} %D	$\|E\|_{mean}$ %D	σ % D	No. of entries
1.1, WF	0.3	1.2	1.4	8
1.1, LRN	1.4	2.6	2.7	14
1.2, WF	0.3	1.1	1.3	8
1.2, LRN	1.4	1.8	2.2	12

1.2 Uncertainty Estimation

According to the ITTC (ITTC 2017) a solution has been validated if the comparison error $|E|$ is smaller than the validation uncertainty U_{val}, i.e. $|E| \leq U_{val}$, where $E = D - S$ and $U_{val}^2 = U_{SN}^2 + U_D^2$. U_{SN} is the numerical uncertainty computed from $U_{SN}^2 = U_G^2 + U_I^2$ and U_D is the data uncertainty. U_G and U_I are the grid and iterative uncertainties, respectively. Validation is then achieved at the U_{val} level, meaning that the error is within the "noise level". If on the other hand $|E| \gg U_{val}$, validation has not been achieved and the sign and magnitude of E can be used to improve the CFD modeling.

Information on the procedure for estimating iterative uncertainty U_I was provided by the participants and most of them used the variation of the quantity of interest (here total resistance) during the latter part of the convergence history. To obtain $U_{G,}$ participants were asked to provide solutions for at least three grids. The screening criterion adopted by the organizers for iterative convergence was $|U_I/\varepsilon_{12}| < 0.1$, where ε_{12} is the difference between the solutions on the two finest grids. 25% of the submissions did not satisfy this criterion and these are not included in the statistics below. As shown in Zou and Larsson (2014) a larger U_I will cause an unacceptably large error in U_G.

The error in a numerical solution may be expressed as a series expansion of the typical step size h. When the step size tends to zero the first term in the expansion will become dominant. Thus for sufficiently small h the discretization error δ_{RE} of a solution[1] may be approximated as αh^p, where α is a constant and p is the formal order of accuracy of the method. This error may also be written as $S_1 - S_0$ where S_1 is the known solution and S_0 is the solution for zero step size. We thus have $\delta_{RE} = S_1 - S_0 = \alpha h^p$. For a given step size h the only unknowns are S_0 and α.

In industrial applications the grids are never fine enough for higher order terms in the error expansion to be neglected. However, rather than incorporating more terms in the expression for the error, the same expression is used, but with a variable p. There are then three unknowns of the problem: S_0, α and p, which may be determined from three solutions with systematically varied grids. This is the approach by Xing and Stern (2010). Another technique is advocated by Eca and Hoekstra (2014), where the three unknowns are determined by a least squares fit to more than three solutions. The reason for this more elaborate procedure is to reduce the problem of "scatter" in the

[1]The index stands for Richardson Extrapolation, the basis of this error estimation technique.

solutions caused by, for example, not exactly scaled grids and limiters of different kinds in turbulence models and numerical schemes. The main error estimator in both procedures is the discretization error δ_{RE}. This is used in more or less complex expressions to estimate the grid uncertainty U_G.[2] The general idea is to base the uncertainty estimate on the deviation from the theoretical value of p. The larger the difference the more uncertain the solution.

The Gothenburg 2010 Workshop and the present one offer a unique opportunity to test the uncertainty estimation procedures, and detailed studies of the 2010 results are presented in Zou and Larsson (2014). Here we will make a more limited study of the question:

- *How well does systematic grid variation work for uncertainty estimation in industrial problems?*

Most participants followed the instruction to submit solutions for at least three grids. The largest set was for 1.1.a and 1.2 a and contained 6 grids. There were a few submissions with 5 and slightly more with 4, but the majority reported 3 grids. The total number of grid sets with at least 3 grids was 60. This is a good data base for answering the question above.

First, the convergence of the grid sequencing may be investigated. Figure 4 presents the comparison errors in percent of D for Cases 1.1a and 1.2a. The error is presented as a function of the normalized step size h/h_l, where h_l represents the finest grid. Visual inspection shows that most computations converge. The difference between two successive solutions becomes smaller and smaller, i.e. $0 < R < 1$, where $R = (S_{n+1} - S_n)/(S_n - S_{n-1})$ and S_n represents the solution on the n:th grid. However, there are also solutions that diverge ($|R| > 1$) and with oscillatory convergence ($-1 < R < 0$).

[2]Eça and Hoekstra also use other error estimators.

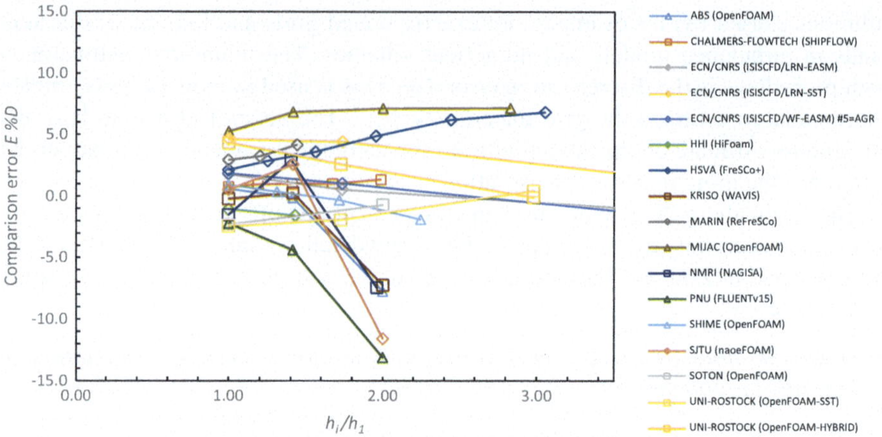

a. Bare hull, Case 1.1a

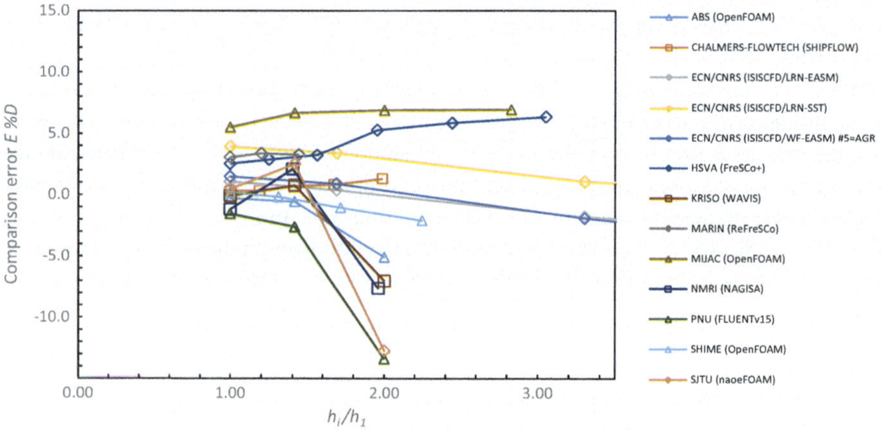

b. Hull with ESD, Case 1.2a

Fig. 4 Resistance versus grid step size

In Table 5 the grid convergence is reported more quantitatively, and a comparison is made with Gothenburg 2010, all cases. In the present Workshop there were 28 submissions with more than 3 grids for cases 1.1a and 1.2a, compared with 42 for all cases in 2010. Of the 28 submissions, 38 grid triplets could be formed (since some contained more than 3 grids). The table is split between structured and unstructured grids. This is an important distinction, since the theory above assumes systematically refined grids, i.e. all cells should be scaled down geometrically, a requirement that is very hard to satisfy for unstructured grids.

There is a significant shift from structured to unstructured grids between the two Workshops. While in 2010 only 7 out of 40 triplets were unstructured, in 2015 this was the case for 25 out of 38. This reflects the larger influence of the general purpose commercial codes, of which practically all use unstructured grids. Another interesting conclusion from Table 5 is that the unstructured grids converge surprisingly well, considering the lack of similarity in the grid refinement. Convergence is obtained for 5 out of 7 in 2010 and 16 out of 25 in 2015. 64% were thus convergent at the present workshop, which is not too far from the structured grids, which converged in 77% of the cases.

Another aspect of the grid refinement is the achieved value of the exponent p, i.e. the achieved order of accuracy. The more different from the theoretical order of accuracy the more uncertain the computed value. In Table 6 statistics are given for the ratio p/p_{th}, where p_{th} is the theoretical order of accuracy. Ideally, this ratio should be 1. This is however not the case at all. For most submissions p/p_{th} is larger than 1 and he average value for these is 2.3 (2.7 in 2010). Those below 1 have an average of 0.5 (0.7 in 2010). While a smaller order of accuracy than the theoretical one is explainable by limiters of different kinds, reducing the order in regions with large gradients, and by less strict boundary conditions, a higher value cannot be explained by such shortcomings. It must be caused by the influence of higher order terms and/or scatter, as explained above. The large values are a matter of concern, since the empirical relations used for the uncertainty estimation are based on smaller values (maximum 2 for Xing and Stern (2010)).

Table 5 Grid convergence

	G2010 all	T2015 1.1a, 1.2a
Submissions >2 grids	42	28
Total # triplets	?	38
Structured grids	33	13
– Convergent	32	10
– Divergent	1	3
Unstructured grids	7	25
– Convergent	5	16
– Divergent	2	9

Table 6 Achieved order of accuracy

	G2010 all	T2015 1.1a, 1.2a
# submissions with p/p_{th}	29	27
# p/p_{th} >1	18	21
– Mean value	2.7	2.3
# p/p_{th} <1	11	6
– Mean value	0.7	0.5

Knowing the grid uncertainty U_G and the iterative uncertainty U_I, the numerical uncertainty U_{SN} may be obtained as explained above. Together with the experimental accuracy U_D this yields the validation uncertainty U_{val}. This can now be compared with the comparison error E. It turns out that $|E| < U_{val}$ in 82% of the reported validations. In 2010 the evaluation was made by the organizers and both uncertainty assessment methods were applied to the reported grid variations. For the Xing and Stern method, $|E| < U_{val}$ in 60% of the submissions, while for the Eca and Hoekstra method this was the case for 91% of the submissions. In fact, |E| is normally considerably smaller than U_{val}. In the present workshop (for JBC) the mean value of |E| was 2.0%D, while the mean value of U_{val} was 5.0%D. Thus, the codes may be considered validated in a mean sense!

2 Sinkage and Trim

Table 7 presents a statistical evaluation of the sinkage predictions for JBC, Cases 1.1a and 1.2a together with results from Gothenburg 2010. Unfortunately, there is only one submission for the bluff hull, KVLCC2 (for varying Froude numbers). Therefore the statistics from all hulls are also included. Apparently, the comparison errors are very large in the 2010 results, 20–30%D. This fact was very surprising and was discussed at length at that workshop, Larsson and Zou (2014). The conclusion was that it is extremely difficult to measure accurately the very small sinkages, of the order of 1 mm for the slow speeds. No uncertainty estimate for the measured data was available, except for the high speed results of the frigate used in that workshop. The conclusion was that the large comparison errors are mainly due to errors in the experimental data. However, a huge improvement is seen in the present computations for JBC in the table. Neither in this case were any uncertainty estimates of the measured data provided.

The corresponding trim statistics are presented in Table 8. For the trim the measured uncertainty was reported in 2010 and it is seen in the table that the computed mean errors are of the same order as this uncertainty. However, there is a large standard deviation in the collected results for all cases. Again, the comparison error is

Table 7 Sinkage statistics

Hull	Case no.	E_{mean} %D	$\|E\|_{mean}$ %D	σ %D	U_D %D	No. of entries
All hulls	**2010**/1.2b, 2.2b, 2.3b 3.2	−21.0		11.2	–	51
KVLCC2 All Fr	**2010**/1.2b	−33.3	33.3	–	–	1
KVLCC2 Fr = 0.142	**2010**/1.2b	−18.2	18.2	–	–	1
JBC	**2015**/1.1a, 1.2a	−1.9	4.7	5.6	–	26

Table 8 Trim statistics

Hull	Case no.	E_{mean} %D	$\|E\|_{mean}$ %D	σ %D	U_D %D	No. of entries
All hulls	**2010/**1.2b, 2.2b, 2.3b 3.2	8.9		36.9	6	51
KVLCC2 All Fr	**2010/**1.2b	7.4	7.4	–	7	1
KVLCC2 Fr = 0.142	**2010/**1.2b	3.2	3.2	–	7	1
JBC	**2015/**1.1a, 1.2a	0.6	3.1	3.5	–	26

very small in the results of the present Workshop. This may well be a result of a more accurate trim measurement, but no uncertainty estimation was presented for the experimental data.

3 Wave Pattern

The wave pattern submissions to the Workshop are listed in Table 9. Three results were submitted by ECN CNRS with different turbulence models and wall treatments. The Volume of Fluid (VOF) method is by far the most popular free-surface method with 9 out of 11 submissions. The remaining two use Level Set. All VOF submissions use hex-dominated, unstructured grids, while the Level Set entries use structured grids. The total number of grid points used in the simulations varies considerably, from 0.7 to 55 million.

Table 9 Wave pattern submissions

Code identifier	Free surface method	Grid type	No. of grid points [M]
ECN CNRS-ISISCFD-LRN EASM	V	U, hex	9.2
ECN CNRS-ISISCFD-LRN SST	V	U, hex	5.5
ECN CNRS-ISISCFD-WF EASM	V	U, hex	2.2
HSVA-FreSCo+	V	U, hex	14.6?
KRISO-WAVIS	LS	S	15.5
MARIC-FINEMarine	V	U, hex	0.7
MARIN-ReFRESCO	V	U, hex	55
MHI-FLUENTv14	V	U, hex	13.9
NMRI-NAGISA	LS	S	7.8
PNU-FLUENTv15	V	U, hex	4.5
SJTU-naoeFOAM	V	U, hex	?

V VOF; *LS* Level Set; *U, hex* Unstructured, hex dominant; *S* Structured

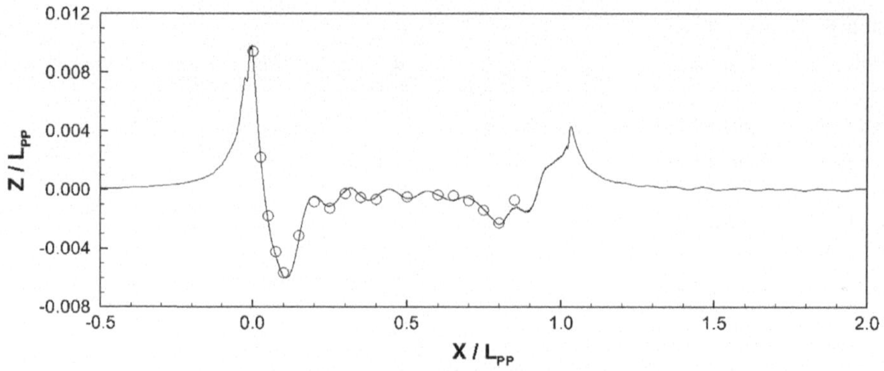

Fig. 5 Typical wave profile along the hull (HSVA-FresCo+)

All methods predict the wave profile along the hull very well. A typical example is shown in Fig. 5, where the circles represent measurements and the line computations. Larger differences are seen in the two wave cut plots to be presented, but first the best result for the KVLCC2 in 2010 is shown in Fig. 6. Here results are presented for three wave cuts and there is a remarkable correspondence with the measured data.

Results for the two wave cuts requested in the present workshop are shown in Fig. 7. These are from the same code as in Fig. 6 (ECN CNRS-ISISCFD) and with the same number of grid points (5.5 million). Although these new results are not bad, they are significantly worse than those of Fig. 6. Starting with the innermost cut ($Y/Lpp = -0.1043$) there is an over prediction of the wave in front of the bow, particularly at the crest, and there is a constant phase shift over the entire length of the hull. However, with the exception of the bow wave, the amplitudes are well predicted back to $X/Lpp = 1.2$, where the grid is coarsened. The over prediction of the bow wave and the phase shift is common to all submissions, indicating that there might be a problem with the reported data. In fact most submissions are very similar to that of Fig. 7. This holds for ECN CNRS-ISISCFD-LRN EASM, ECN CNRS-ISISCFD-WF EASM, HSVA-FreSCo+, KRISO-WAVIS, MARIC-FINEMarine, MARIN-ReFRESCO and MHI-FLUENTv14, while the waves are more damped in NMRI-NAGISA and PNU-FLUENTv15. See the Workshop Proceedings for the other plots!

Very similar features are noticed for the outer cut at $Y/Lpp = -0.1900$. All submissions exhibit an over predicted bow wave and a constant phase shift over the length of the hull. At this *Y*-position, the differences between the methods are larger, however. MARIN-ReFRESCO is very similar to ECN CNRS-ISISCFD-LRN SST shown in Fig. 7, while most of the others under predict wave amplitudes to a smaller or larger extent. HSVA-FresCo+ over predicts the amplitudes and for SJTU-naoeFOAM there is a problem with the level of the surface both in front of and behind the hull. In fact, none of the solutions beats the potential flow solution shown in Fig. 8 for the outer cut. This is along the hull; at the stern and in the wake this solution over predicts the amplitude, as expected.

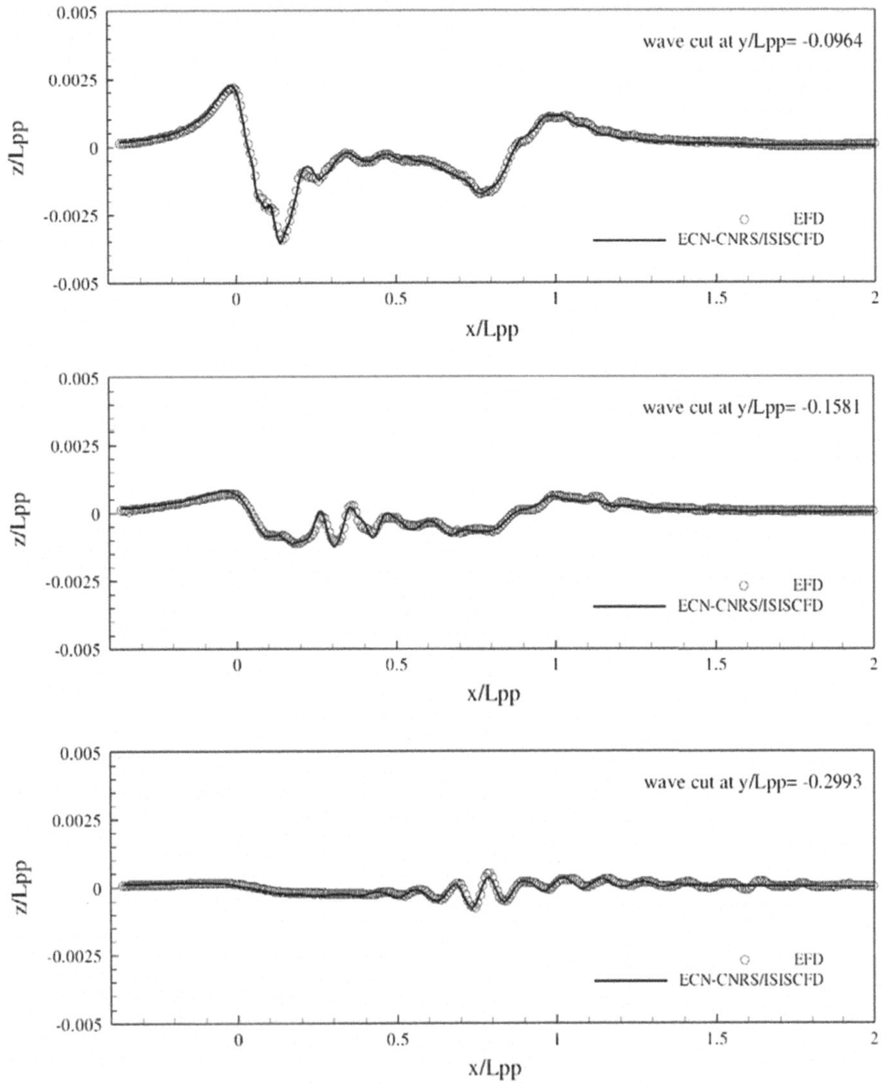

Fig. 6 Three wave cuts (Y/Lpp = −0.0964, −0.1581 and −0.2993) from Gothenburg 2010, KVLCC2

At the Gothenburg 2010 Workshop a diagram for reporting the grid density near the free surface was introduced. This was requested also at the present Workshop. It is presented in Fig. 9 for the submission shown in Fig. 7. The horizontal axis represents a distance either in the X-direction along the symmetry plane/hull surface or in the Y-direction at midship. The vertical scale to the left shows the number of grid points per fundamental wave length (*ppwl*) both in the X-direction (full line)

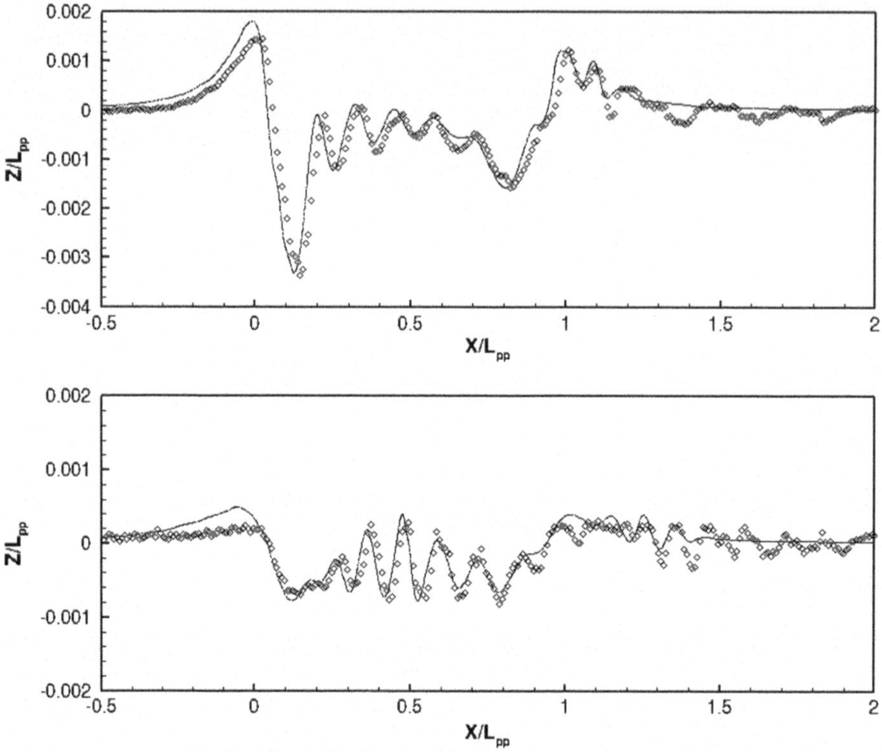

Fig. 7 Two wave cuts ($Y/Lpp = -0.1043$ and -0.1900), JBC. ECN CNRS-ISISCFD-LRN SST

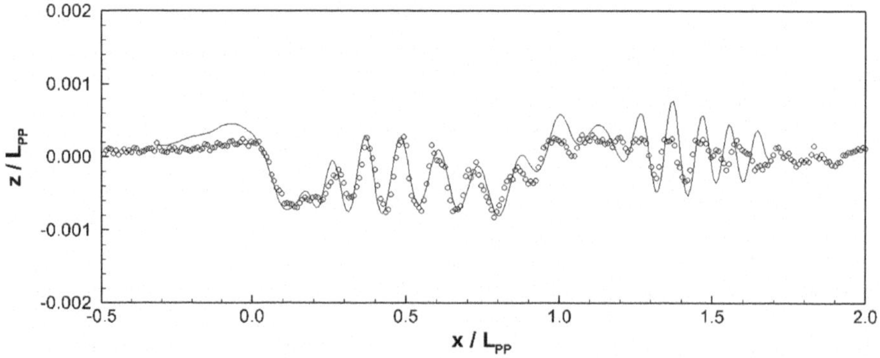

Fig. 8 Potential flow solution at $Y/Lpp = -0.1900$, JBC. CHALMERS-FLOWTECH-SHIPFLOW

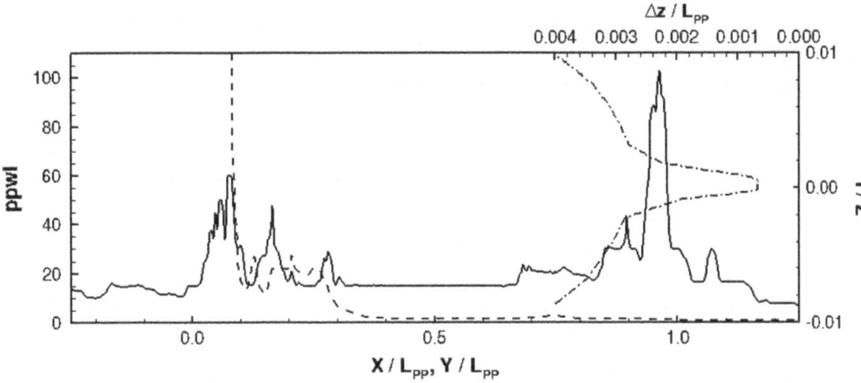

Fig. 9 Grid density. Full line: grid density in the X-direction, Dashed line: grid density in the Y-direction, Dot-dashed line: cell height near the waterline. ECN CNRS-ISISCFD-LRN SST

and the Y-direction (dashed line). To the right, there is a scale for the thickness (Δz) of the cells close to the water line (at $Z/Lpp = 0.00$).

There are peaks in the longitudinal distribution at bow and stern reaching 60–100 ppwl, while along the main part of the hull the density is about 20 ppwl. In the transverse direction (at midship) the density drops rapidly from a very high value in the boundary layer, but stays around 20 ppwl out to about $Y/Lpp = 0.25$. This is the lateral extension of the wave system at midship. There is thus a high density where required, but a rapid drop further out. The cell thickness goes down to about 0.0006 *Lpp* around the undisturbed surface level.

This method (ISIS-CFD) has adaptive grid generation which automatically concentrates cells in regions of large gradients. Most other submissions exhibit larger grid densities along the hull, typically 50 ppwl, and the peaks at the hull ends are normally higher. Also, the density in the transverse direction is mostly larger than in Fig. 9. The cell height around the waterline varies significantly between the methods and are in the range 0.0002–0.002. A surprising plot is that of PNU-FLUENTv15, which exhibits the largest peak longitudinally (800 ppwl), a good density transversely at midship (around 30 ppwl) and a very small cell height at the surface (0.0002 Lpp). Still, it has the largest damping of the waves along the hull. Presumably, this is an effect of the VOF discretization scheme.

4 Conclusions

The Workshop submissions for JBC contained four cases: the hull with or without an ESD both in towing and self-propulsion. Altogether 88 predictions were provided. In this chapter, resistance, sinkage and trim, as well as wave pattern results are evaluated. Conclusions are as follows:

- Like at the previous workshops the mean signed comparison error for resistance is very small, around 1%, which is also the accuracy reported for the experimental data. However, more interesting is the mean absolute comparison error, since this represents the average deviation from the data. This is around 2% for the towed cases and 3% for self-propulsion. The scatter is represented by the standard deviation and is about 2% and 4% respectively for the two conditions. In the 2005 and 2010 Workshops the standard deviations were 6 and 1%. There was thus a significant reduction in scatter between 2005 and 2010, but no further reduction in 2015.
- A resistance reduction around 1% is predicted with the ESD fitted. This is confirmed by the measurements.
- For the comparison error to be smaller than 4% the grids have to contain more than 10 M cells. In 2010 the corresponding limit was 3 M cells. Thus, more grid cells were required for the same accuracy in 2015.
- There is hardly any influence of the grid type on the predicted resistance. If any, there is a slight advantage of the structured grids, compared to hex-dominated and completely unstructured grids.
- Like in 2010 the two-equation turbulence models predict better results than the more advanced models. The k-ε and k-ω models have mean signed comparison errors less than 1% and mean absolute errors just below 2%. EASM has gained in popularity and its mean absolute error is now 2%.
- Resistance predictions using wall-functions are more accurate than those with wall resolved flows. Both the absolute error and the standard deviation are about twice as large with wall resolution.
- Grid convergence is demonstrated for 64% of the unstructured grids and 77% of the structured grids. The achieved order of accuracy in the grid refinement studies is in many cases far from the theoretical accuracy. Most methods predict a too high order of accuracy, 2.3 times the theoretical one, on the average. For those with a too low order of accuracy the mean value is 0.7 times the nominal one.
- As stated above, the mean absolute comparison error in all resistance predictions is 2%, which is considerably smaller than the mean validation uncertainty, 5%, composed of the mean numerical uncertainty of 4.9% and the stated experimental accuracy of 1%. The methods may thus be considered validated in a mean sense.
- The comparison errors for sinkage were very large in 2010, but are now reduced considerably. The mean signed error, absolute error and standard deviation are now −1.9%, 4.7% and 5.6% respectively.
- For trim the corresponding errors are also reduced considerably and are now 0.6%, 3.1% and 3.5%, respectively.
- Very accurate predictions are reported for the wave profile along the hull, while less correspondence with measurements is demonstrated for two wave cuts. There is a consistent phase shift in all results, which may indicate an error in the data. A small over prediction of the bow wave height is also noted in all results.

References

Eca, L., & Hoekstra, M. (2014). A procedure for the estimation of the numerical uncertainty of CFD calculations based on grid refinement studies. *Journal of Computational Physics, 262,* 104–130.

Hino, T. (Ed.). (2005). *Proceedings of CFD Workshop Tokyo 2005*. NMRI Report.

ITTC. (2017). ITTC—Recommended Procedures and Guidelines. Uncertainty Analysis in CFD. Verification and Validation Methodology and Procedures.

Larsson, L. (Ed.). (1981). *SSPA-ITTC Workshop on Ship Boundary Layers*. SSPA Report 90, Gothenburg, Sweden.

Larsson, L., Stern, F., & Visonneau, M. (Eds.). (2014). *An assessment of the Gothenburg 2010 workshop*. Dordrecht: Springer Science and Business Media.

Larsson, L., & Zou, L. (2014). Evaluation of resistance, sinkage and trim, self-propulsion and wave pattern predictions. In *Numerical ship hydrodynamics—An assessment of the Gothenburg 2010 workshop*. Dordrecht: Springer Science and Business Media.

Van, S. H., Kim, W. J., Yim, D. H., Kim, G. T., Lee, C. J., & Eom, J. Y. (1998). Flow measurement around a 300 K VLCC model. In *Proceedings of the Annual Spring Meeting, SNAK*, Ulsan (pp. 185–188).

Vassberg, J. C. (2016). Challenges and accomplishments of the AIAA CFD drag prediction workshop series. In *AVT-246 Progress and Challenges in Validation Testing for Computational Fluid Dynamics*, Avila, Spain.

Xing, T., & Stern, F. (2010). Factors of safety for Richardson extrapolation. *Journal of Fluids Engineering, 132*(6).

Zou, L., & Larsson, L. (2014). CFD verification and validation in practice—A study based on resistance submissions to the Gothenburg 2010 workshop on numerical ship hydrodynamics. In *30th Symposium on Naval Hydrodynamics*, Hobart, Australia.

Analysis of the Local Flow around JBC

Michel Visonneau

Abstract This chapter is devoted to an analysis of the local flow field for test cases 1-3, 1-4, 1-7 and 1-8. The general objective of this analysis is to assess the present computational approaches according to their ability to represent the influence of a duct on the local flow field in presence or absence of a propeller, thanks to the existence of local flow measurements mainly performed by NMRI. General conclusions based on the numerous contributions, are drawn concerning the reliability of turbulence closures in these peculiar flow configurations, the influence of the propeller representation on the local flow field and the ability of CFD to represent the local influence of a duct on the flow at the stern region and in the wake of a ship.

1 Introduction

This chapter is devoted to an analysis of the local flow field for test cases 1-3, 1-4, 1-7 and 1-8. The general objective of this analysis is to assess the present computational approaches according to their ability to represent the influence of a duct on the local flow field in presence or absence of a propeller, thanks to the existence of local flow measurements mainly performed by NMRI. During the previous G2010 workshop, the role played by the turbulence models was systematically assessed for several ships and it was established that the flow in the stern region of the KVLCC2 was strongly dependent on the turbulence closures, once a grid fine enough was used. The objective of this chapter is to conduct a very similar analysis in a more complex framework involving the interaction of the stern flow with an energy-saving device with or without a propeller. General conclusions based on the numerous contributions, have to be drawn concerning the reliability of turbulence closures in these peculiar flow configurations, the influence of the propeller representation on the local flow field and the ability of CFD to represent the local influence of a duct on the flow at the stern region and in the wake of a ship.

M. Visonneau (✉)
LHEEA Laboratory, ECN/CNRS, 1, rue de la Noe, 44321 Nantes, Cedex 3, France
e-mail: michel.visonneau@ec-nantes.fr

T. Hino et al. (eds.), *Numerical Ship Hydrodynamics*, Lecture Notes in Applied and Computational Mechanics 94, https://doi.org/10.1007/978-3-030-47572-7_6

2 Experiments

The Japan Bulk Carrier (JBC) is a Capesize bulk carrier equipped with a stern duct as an energy saving device (ESD). National Maritime Research Institute (NMRI), Yokohama National University and Ship Building Research Center of Japan (SRC) were jointly involved in the design of this ship hull, duct and propeller. Its length between perpendiculars is $L_{pp} = 280$ m. Its service speed is 14.5 knots, leading to a Froude number $Fn = 0.142$ and a Reynolds number at model scale of $Re = 7.46\ 10^6$ with $L_{pp} = 7.00$ m at model scale. Its main particulars are described in Table 1. This ship was selected as one of the testcases of the Tokyo2015 workshop on numerical ship hydrodynamics, the latest edition in a series of workshops which also includes Gothenburg 2010 (Larsson et al. 2014). Towing tank experiments, including resistance tests, self-propulsion tests and PIV measurements of stern flow fields, were performed at NMRI, SRC and Osaka University. Several test cases were considered, all with free sinkage and trim; test cases 1-3a (resp. 1-4) are for towing test without (resp. with) ESD, cases 1-7 (resp. 1-8) for self-propulsion tests without (resp. with) ESD (see Table 2). Global force measurements and local LDV velocity profiles at three sections named S2, S4 and S7 (i.e. $X/L_{pp} = 0.9625$, $X/L_{pp} = 0.9843$ and $X/L_{pp} = 1.0000$) before and after the propeller and duct were also provided by the organizers. Figure 1a, b show a view of the stern without and with ESD with the location of the local measurement sections.

Table 1 JBC—main particulars

Main particulars		Full scale
Length between perpendiculars	L_{pp} (m)	280.0
Length of waterline	L_{WL} (m)	285.0
Maximum beam of waterline	B_{WL} (m)	45.0
Depth	D (m)	25.0
Draft	T (m)	16.5
Displacement volume	∇ (m^3)	178369.9
Wetted surface area (without ESD)	$S_{0-w/oESD}$ (m^2)	19556.1
Wetted surface area (with ESD)	S_{0-wESD} (m^2)	19633.9
Block coefficient	$\nabla/(L_{pp}B_{WL}T)$	0.8580
Midship section coefficient	CM	0.9981
LCB (%L_{PP}), fwd+		2.5475
Vertical center of gravity (from keel)	KG (m)	NA
Metacentric height	GM (m)	NA
Moment of inertia	K_{xx}/B	NA
Moment of inertia	K_{yy}/L_{pp}, K_{zz}/L_{pp}	NA
Propeller center, long. location (from FP)	x/L_{pp}	0.985714
Propeller center, vert. location (below WL)	$-z/L_{pp}$	−0.0404214
Propeller rotation direction (view from stern)		Clockwise

Table 2 Definition of the test cases

Test case	1-3a	1-3b	1-4	1-7	1-8
Experiments	NMRI (towing tank)	TUHH (wind tunnel)	NMRI (towing tank)	NMRI (towing tank)	NMRI (towing tank)
Propeller (self-propulsion)	No	No	No	Yes	Yes
Energy saving device	No	No	Yes	No	Yes

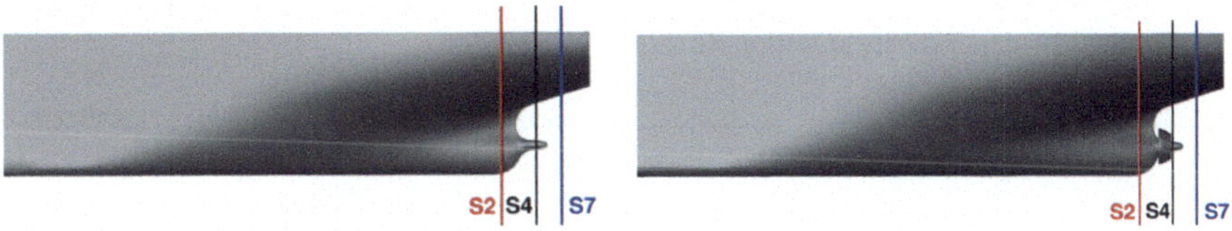

(a) Without ESD *(b) With ESD*

Fig. 1 Side views of the JBC hull

Moreover, TUHH performed wind-tunnel LDV experiments around a double-body model of the JBC without rudder nor propeller at a slightly lower Reynolds number Re = 2.74 10^6 with L_{pp} = 3.513 m and U = 11.8 m/s. These experiments and the computations performed in the same conditions are described in the test case 1-3b.

3 Test Case 1-3a

3.1 Description of Experimental Results

Test case 1-3a concerns the local flow analysis around the JBC without duct nor propeller towed in the NMRI towing tank, in free trim and sinkage conditions. Figures 2, 3 and 4 show the experimental iso wakes and iso contours of the two crossflow velocity components (v, w) at three sections S2, S4 and S7 already defined before.

From Fig. 2a taken at station S2, one can notice the presence of an intense longitudinal vortex which is characterised by an island of isowakes comprised between 0.2 and 0.4. A second closed isowake contour is visible closer to the wall but at the boundary of a region where PIV experiments are likely less reliable. One notices also that it is not possible to get reliable experimental information very close to the hull, in the boundary layer. The lateral v component (Fig. 2b) is always positive with closed contours associated to level 0.0 to 0.2 in the core of the main longitudinal

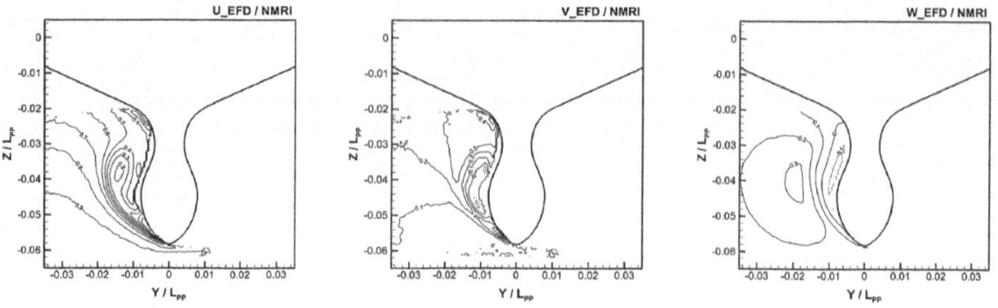

(a) Iso-u contours *(b) Iso-v contours* *(c) Iso-w contours*

Fig. 2 Case 1-3a—experiments at station S2

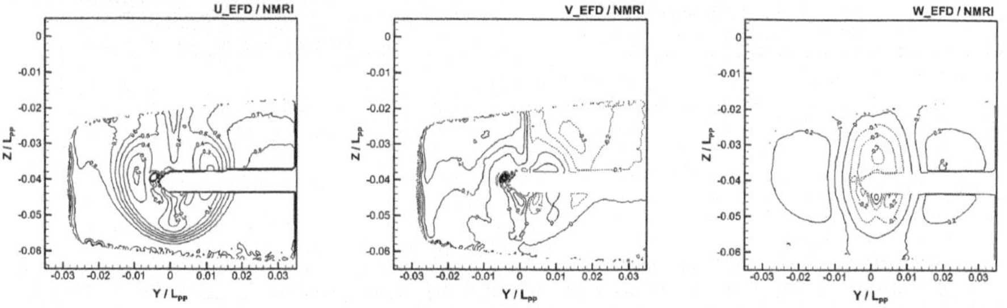

(a) Iso-u contours *(b) Iso-v contours* *(c) Iso-w contours*

Fig. 3 Case 1-3a—experiments at station S4

vortex. For the vertical component w (Fig. 2c), one can notice a well defined line (w = 0.0) dividing a region close to the wall where w reaches a minimum value of −0.1 and a region outside of the core of the vortex where w reaches the maximum value of 0.3.

At station S4, Fig. 3a shows the development of the longitudinal vortex with well developed hook-shaped isowake contours. Two secondary vortices are also visible at the boundary of the PIV experiments which are maybe related with the one visible at station S2 (isocontour 0.4) mentioned previously. However, this can be also an experimental artefact. It is interesting to notice for further comparisons with computations, the typical shape of the isowake contours (iso u contours between 0.2 and 0.3) in the lower part of the vertical plane of symmetry. Figure 3b shows the contours of v which are uniformly very small around 0.1–0.2. Figure 3c provides the same information for the vertical component W at station S4. Here, one should notice a minimum value for w of −0.3 in the vertical plane of symmetry and a maximum value of 0.2 outside of the core of the main vortex.

Finally, Fig. 4a shows the isowakes distribution at station S7. An intense longitudinal vortex is still visible with a minimum isowake value of 0.4. It is interesting to mention that the experimental averaged flow field is not fully symmetric (iso contours 0.5 and 0.4, for instance). This asymmetry might be related to the unsteadiness

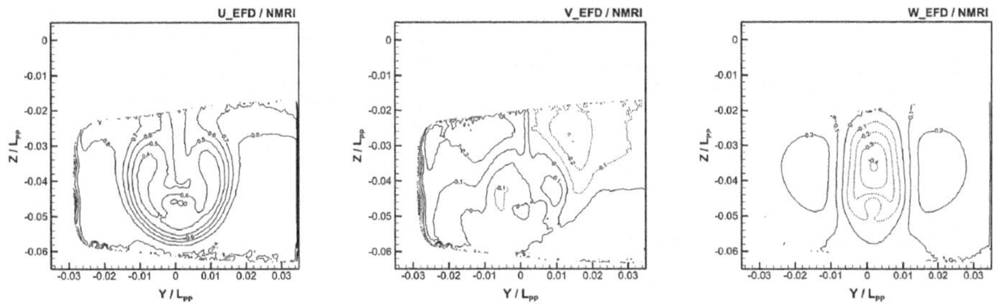

(a) *Iso-u contours* (b) *Iso-v contours* (c) *Iso-w contours*

Fig. 4 Case 1-3a—experiments at station S7

of the stern flow or experimental uncertainties close to the vertical plane of symmetry. Figure 4b shows the contours of v which are again uniformly very small around 0.1–0.2. Figure 4c provides the same information for the vertical component w at station S7. Here, one notices a minimum value for w of −0.4 in the vertical plane of symmetry and a maximum value of 0.2 outside of the core of the main vortex.

3.2 *Review of Contributions*

Eighteen participants provided results for the tescase 1-3a. The main characteristics of their contributions are listed in Table 3 in terms of turbulence closures, wall treatments and discretization. Some participants provided several different contributions which makes possible a comparison of various turbulence closures or wall treatments for the same grid and same code. Moreover, SOTON performed a comparison of codes (OpenFoam vs. Star-CCM+) with the same grid and same turbulence closure.

Among the eighteen participants, eleven used unstructured grids while seven computed on structured grids sometimes locally enriched with overlapping patches (Chalmers, CNRS/INSEAN). It is well known that the prediction of the nominal velocity in the stern region depends strongly on the turbulence closure, as soon as a reasonable dense grid is used. Like during the Gothenburg 2010 workshop, the turbulence models employed by the participants can be classified into three groups: (i) Unsteady Hybrid RANS-LES model (University of Rostock (URO)), (ii) anisotropic non-linear statistical turbulence modeling: Reynolds stress transport models (HHI), Explicit Algebraic Reynolds stress models (Chalmers, ECN/CNRS, NMRI, YNU), (ii) isotropic linear eddy-viscosity model (ECN/CNRS, HSVA, MARIN, URO, …). Therefore, in five years, we can observe a slight increased use of more sophisticated turbulence closures for ship flow computations, probably related with the number of publications illustrating the crucial role of turbulence modeling for the prediction of local flow in the stern region. All the discretization methods are formally second-order accurate and based on multi-block structured or unstructured grids which are all body-fitted. Only Chalmers and CNRS/INSEAN are using an over-

Table 3 Case 1-3a—main characteristics of the computations

Organization code name	Turbulence model	Wall model	Grid characteristics
ABS OpenFOAM	k-ω SST	Wall function	Unstructured grid 5 482 000 cells
Chalmers-Flowtech Shipflow	EARSM	Wall resolved	Structured grid with overlapping 8M cells
CNR-INSEAN Xnavis	Spalart-Allmaras		Structured grid with overlapping 7.7M cells
ECN/CNRS ISIS-CFD aka FINE^{TM}/Marine	SST/EARSM	Wall function Wall resolved	Unstructured grid 5.7/9.2M cells
HHI HiFOAM/StarCCM+	Realizable k-ϵ/RSTM	Wall function	Unstructured grid 1.48M cells
HSVA FreSCo+	2 equations	Wall function	Unstructured grid from 2.1 to 14.6M cells
KRISO Wavis	Realizable k-ϵ/EARSM	Wall function	Structured grid 15.5M cells
MARIC FINE^{TM}/Marine	k-ω SST	Wall function	Unstructured grid 650 000 cells
MARIN ReFRESCO	k-ω SST +DM correction	Wall resolved	Unstructured grid 5.5M cells
MHI Fluentv14	k-ϵ RNG k-ω SST	Wall resolved	Unstructured grid 13.9M cells
MIJAC OpenFOAM	k-ω SST	Wall resolved	Unstructured grid 8.7M cells
NMRI Nagisa	EARSM	Wall resolved	Structured grid 10M cells
PNU Fluentv15	2 equations k-ϵ RNG	Wall resolved Wall function	Unstructured grid 4.45M cells
SHIME OpenFOAM	k-ω SST	Wall resolved	Unstructured grid 22.15M cells
SJTU naoeOpenFOAM	k-ω SST	Wall function	Unstructured grid 2M cells
SOTON OpenFOAM/StarCCM+	k-ω SST	Wall resolved	Structured grid 19.4M cells
UniRostock OpenFOAM	Hybrid RANS-LES k-ω SST	Wall resolved	Structured grid 29.4M cells
YNU SURFv7	EARSM	Wall resolved	Structured grid 2.5M cells

set approach combining structured grids to reach an optimal grid density in the stern region.

3.3 *Grid Sensitivity*

The grids used by the 18 participants can be classified into four categories, based on their sizes:

- Not So Fine: Ncell < 2M cells (MARIC, HHI, SJTU)
- Fine: 2M cells < Ncell < 10M cells (ABS, CHALMERS, CNR/INSEAN, ECN/CNRS, HSVA, KRISO, SHIME, MIJAC, NMRI, PNU, YNU)
- Very Fine: 10M cells < Ncell < 50M cells (SOTON, MHI, MARIN, URO)
- Tremendously Fine: Ncell > 50M cells (None)

During the Gothenburg 2010 workshop, the average size of the grids was 5M cells, excluding the unusually fine grid (305M cells) generated by IIHR for their DES computations. For the Tokyo 2015 workshop, one notices a trend towards the use of significantly finer grids since, five years later, the average size is around 8.4M cells, excluding the very fine grid used by URO for their hybrid RANS-LES computations.

To get an idea of the grid influence, Figs. 5, 6 and 7 show the isowakes at stations S2 to S7 computed by SHIME on two different grids, keeping unchanged the code, discretization and turbulence model. One can hardly see any significant differences between the two results although the number of grid points is multiplied by a factor 2.5.

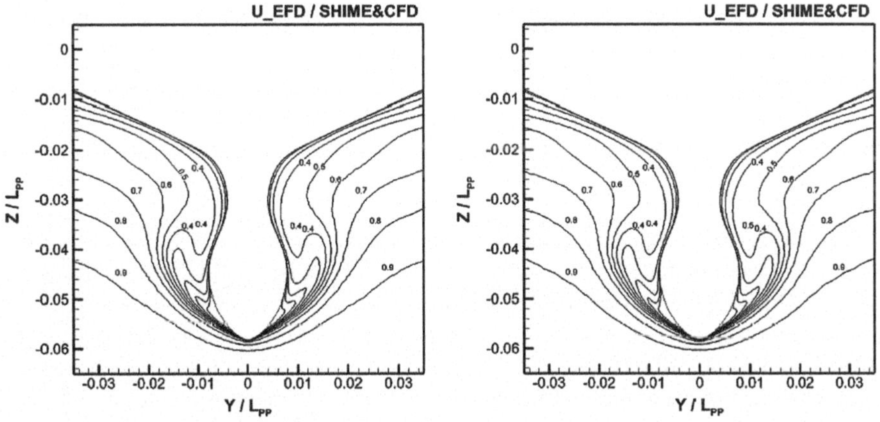

SHIME-OpenFOAM-Fine SHIME-OpenFOAM-VeryFine

Fig. 5 Station S2—computations from SHIME using different grids (left 10M) (right 25M)

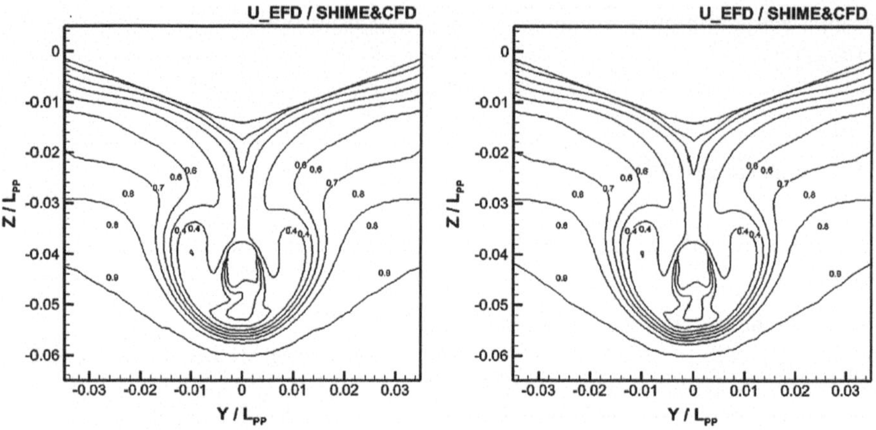

SHIME-OpenFOAM-Fine SHIME-OpenFOAM-VeryFine

Fig. 6 Station S4—computations from SHIME using different grids (left 10M) (right 25M)

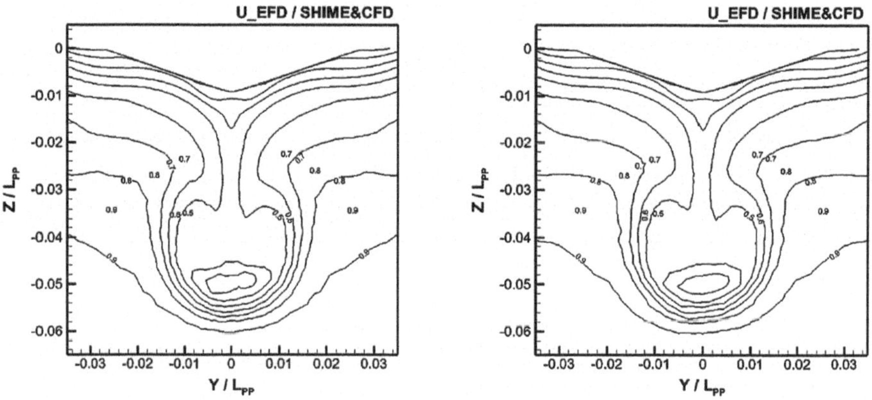

SHIME-OpenFOAM-Fine SHIME-OpenFOAM-VeryFine

Fig. 7 Station S7—computations from SHIME using different grids (left 10M) (right 25M)

Figures 8, 9 and 10 show a comparison of the same isowakes at the same stations as previously, computed by two different teams, SOTON and SHIME, using the same OpenFOAM code, same turbulence model (low Re near wall k-ω SST) and grids of similar density comprised of 20 to 25M cells. It is reassuring to notice that the results are quite similar for all stations, which indicates that some conclusions can be drawn without taking into account specificities related with the user of the computational approach or local details of the grids.

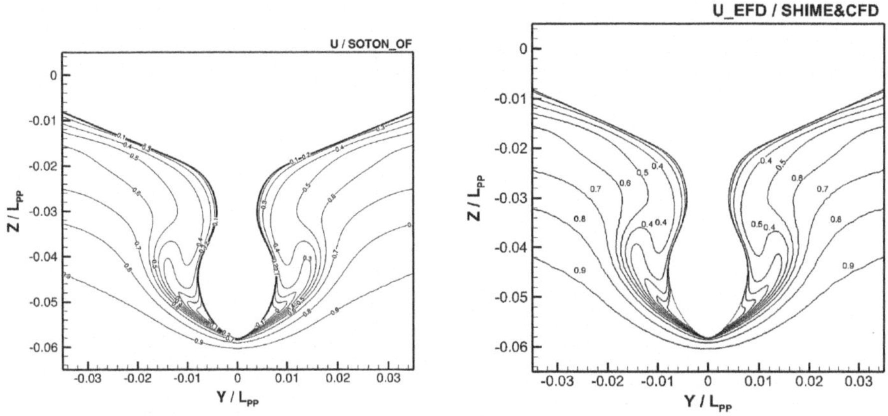

(a) SOTON/OpenFOAM - 20M *(b) SHIME/OpenFOAM - 25M*

Fig. 8 Case 1-3a—comparisons at station S2 between two different contributors using the same code and similar grids

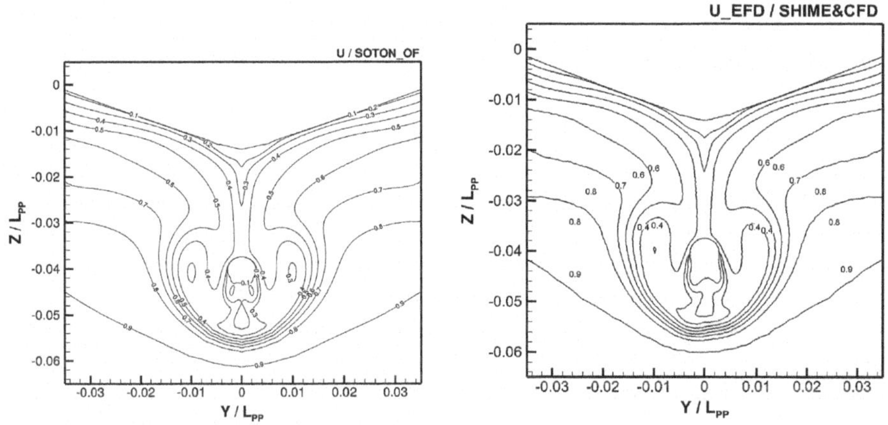

(a) SOTON/OpenFOAM - 20M *(b) SHIME/OpenFOAM - 25M*

Fig. 9 Case 1-3a—comparisons at station S4 between two different contributors using the same code and similar grids

Based on many previous grid sensitivity studies, one can consider that most of the grids used in this workshop (the Fine category) are such that the turbulence modelling error should exceed by large the discretization error, which means that it is meaningful to compare the turbulence closures among computations performed with grids comprised of more than 2M cells.

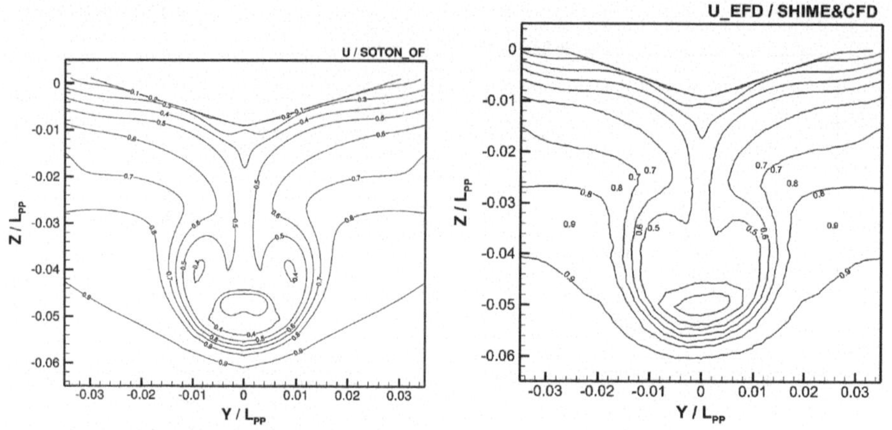

(a) SOTON/OpenFOAM - 20M *(b) SHIME/OpenFOAM - 25M*

Fig. 10 Case 1-3a—comparisons at station S7 between two different contributors using the same code and similar grids

3.4 *Wall Treatment*

Since more and more computations are now performed with a wall function approach, it is interesting to compare wall resolved and wall modeled formulations and see what is the impact of the wall treatment on the development of the flow at the stations analysed in this workshop. Figures 11, 12 and 13 provide such a comparison performed with ISIS-CFD on very similar grids (except below $y^+ = 60$) and with the same anisotropic explicit algebraic Reynolds stress model (EARSM). Globally, differences exist but are not very significant for the three stations used for comparison even if one can notice some local differences affecting the lower values of the isowakes. Based on these comparisons, it is therefore difficult to claim that a wall-function model affects significantly the accuracy of straight ahead resistance viscous computations.

3.5 *Turbulence Closures*

The turbulence models used by the contributors for the case 1-3a can be classified into three groups:

- The isotropic linear closures: INSEAN (Spalart-Allmaras), HSVA (Linear EARSM), MARIN (k-w SST + DM), MARIC (k-ω SST), URO (k-ω SST), PNU (k-ϵ), SHIME (k-ω SST), CSSRC (k-ϵ), ABS (k-ω SST), SOTON (k-ω SST), ECN-CNRS (k-ω SST), MHI (k-ω SST),

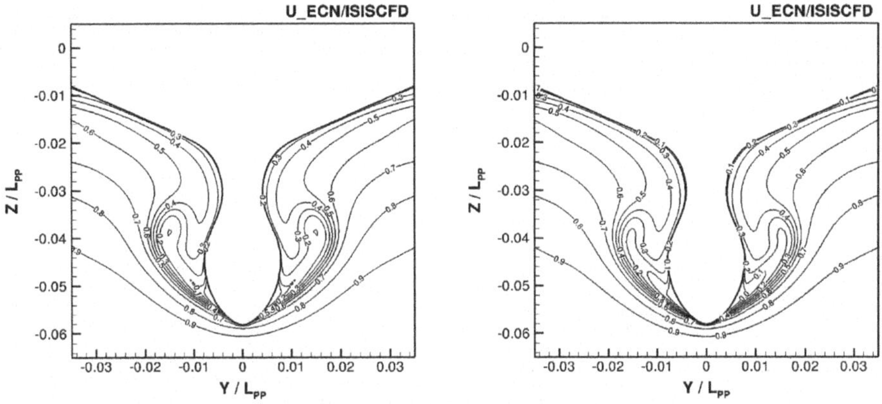

(a) ECN/ISIS-CFD - Wall function *(b) ECN/ISIS-CFD - Wall resolved*

Fig. 11 Case 1-3a—comparisons at station S2 between wall function and wall resolved formulations

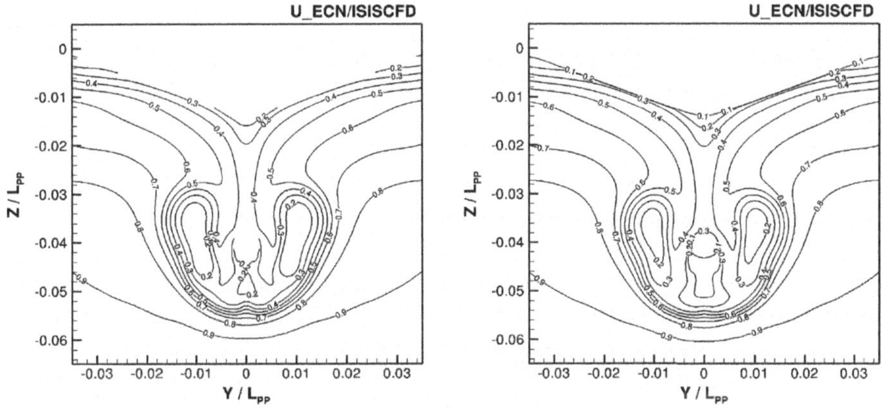

(a) ECN/ISIS-CFD - Wall function *(b) ECN/ISIS-CFD - Wall resolved*

Fig. 12 Case 1-3a—comparisons at station S4 between wall function and wall resolved formulations

- The anisotropic non-linear models: MHI (RSTM), ECN-CNRS (EARSM), NMRI (EARSM), CHALMERS (EARSM), YNU (EARSM), KRISO (EARSM),
- The hybrid RANS-LES models: URO (DES).

In the previous workshop held in Gothenburg in 2010 (see Larsson et al. 2014 for instance), it was noticed that the modelling error induced by the turbulence closures on the velocity distribution in the stern region is the dominating factor of error, greatly surpassing the discretisation error as soon as a reasonably fine grid is used.

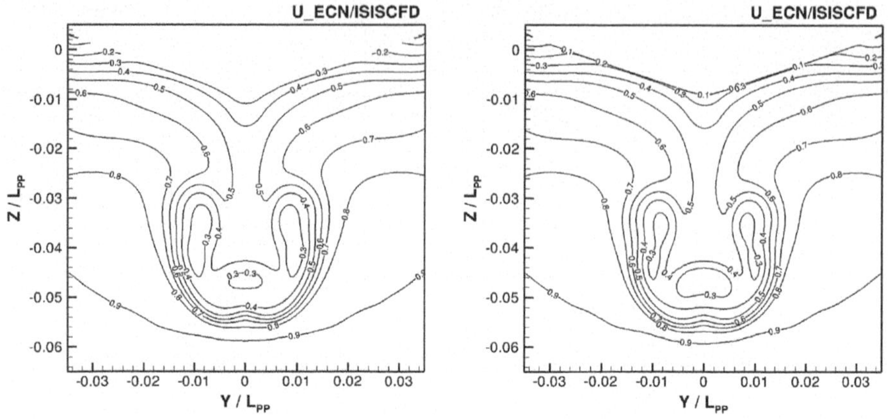

(a) ECN/ISIS-CFD - Wall function *(b) ECN/ISIS-CFD - Wall resolved*

Fig. 13 Case 1-3a—comparisons at station S7 between wall function and wall resolved formulations

In the next subsection, the various turbulence models used by the contributors will be analyzed and compared to see if one can draw common conclusions on the role played by the modelling error in such a flow.

3.5.1 Linear Isotropic Turbulence Closures

Figures 14, 15 and 16 show the iso-contours obtained by the one equation Spalart-Allmaras model implemented in Cnavis by INSEAN. It is clearly unable to provide the right level of longitudinal vorticity for all the experimental cross-sections. It is also true at a lesser degree for the linear EARSM variant implemented by HSVA in Fresco+ (see Figs. 17, 18 and 19). The k-ω SST model appears to behave consistently better in the results shown by University of Rostock with OpenFoam although still far from the experiments (see Figs. 20, 21 and 22). In particular, none of these models is able to predict the characteristic hook shape visible at station S2 with a low level u = 0.2 present in the measurements. The closed contours u = 0.3 and u = 0.2 visible in the experiments at station S4 are also not captured by any of these simulations.

3.5.2 Non-linear Anisotropic Turbulence Closures

The second category of RANSE turbulence models comprises closures taking into account the turbulence anisotropy. Imperfectly when Explicit Algebraic Reynolds Stress models (EARSM) are used or (potentially) more accurately if Full Reynolds Stress Transport Models (RSM) are utilised. Figures 23, 24 and 25 show a comparison of computations performed by ECN/CNRS with the same code (ISIS-CFD) and the

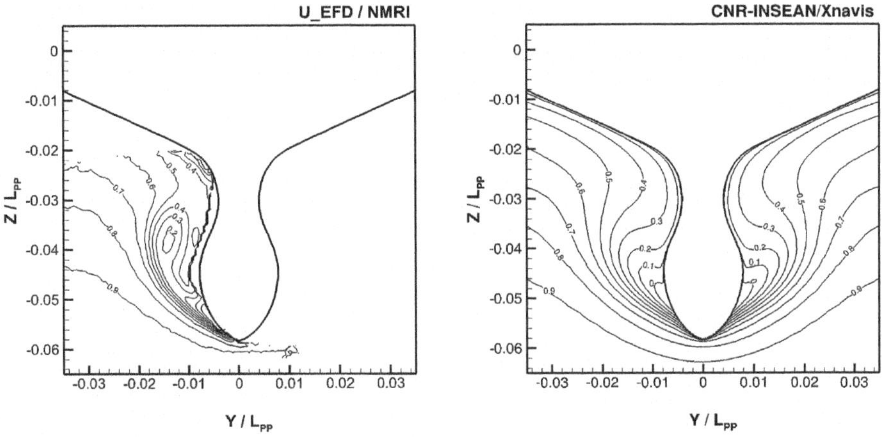

(a) *Experiments* (b) *Spalart-Allmaras*

Fig. 14 Case 1-3a—comparisons at station S2 between INSEAN/Cnavis and experiments

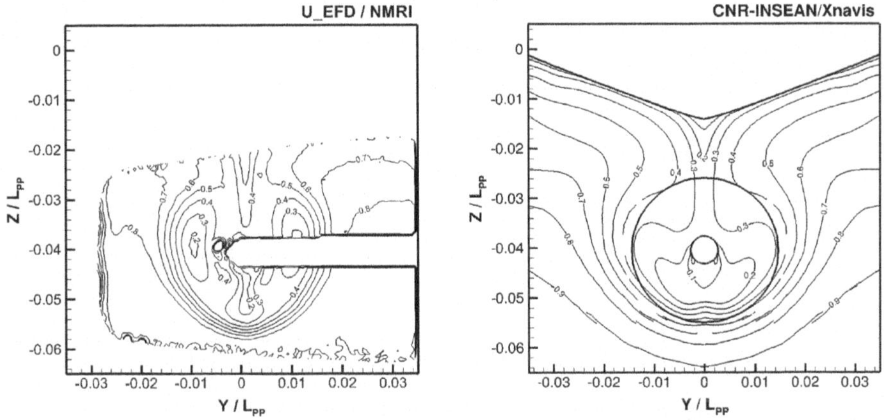

(a) *Experiments* (b) *Spalart-Allmaras*

Fig. 15 Case 1-3a—comparisons at station S4 between INSEAN/Cnavis and experiments

same grid with k-ω SST and EARSM turbulence closures. It appears clearly that EARSM leads to a stern flow with more longitudinal vorticity. Figure 24 shows for instance at station S4 that EARSM accurately captures the isowake 0.2 while k-ω SST fails to reach this result. Low iso wakes are of course related with higher longitudinal vorticity on the core of the main vortex, a result which is associated with the increased production of vorticity related with the turbulence anisotropy.

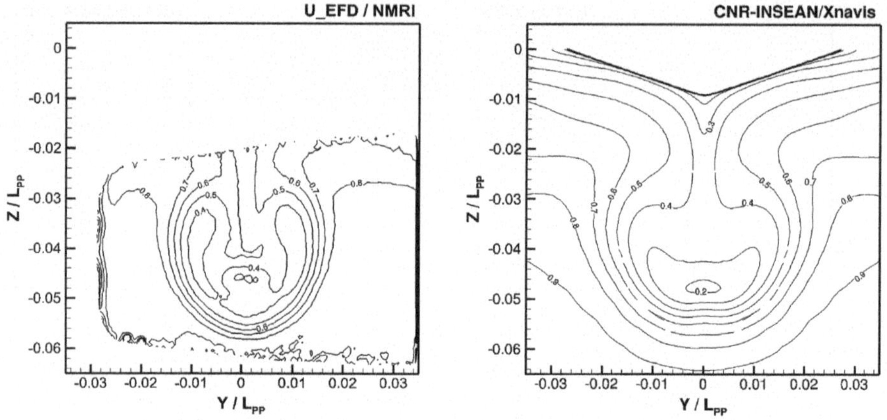

(a) Experiments *(b) Spalart-Allmaras*

Fig. 16 Case 1-3a—comparisons at station S7 between INSEAN/Cnavis and experiments

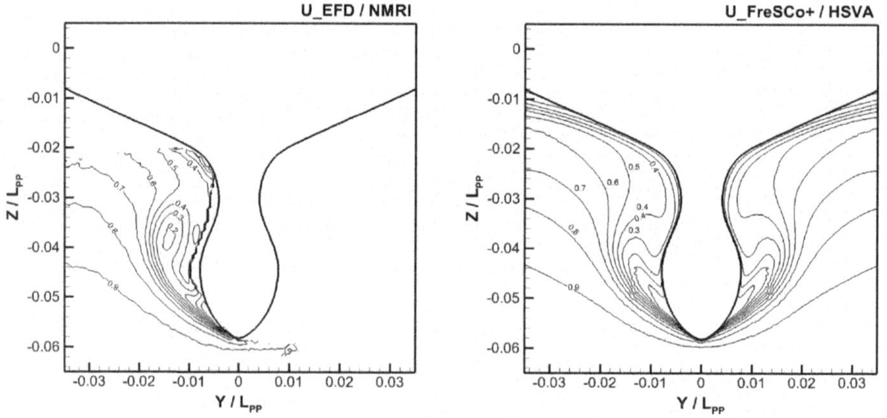

(a) Experiments *(b) Linear EARSM*

Fig. 17 Case 1-3a—comparisons at station S2 between HSVA/Fresco+ and experiments

Figures 26, 27, 28, 29, 30 and 31 show a comparison of the simulations of Chalmers University (resp. ECN/CNRS) with the same EARSM closure implemented in ShipFlow (resp. ISIS-CFD). Although these computations are performed by Chalmers on a slightly finer grid than the one used by ECN/CNRS, one can observe a very similar result. The lower isowake levels in the core of the main averaged vortex are better captured at every experimental sections and the trend provided by both codes are almost identical although completely different codes based on

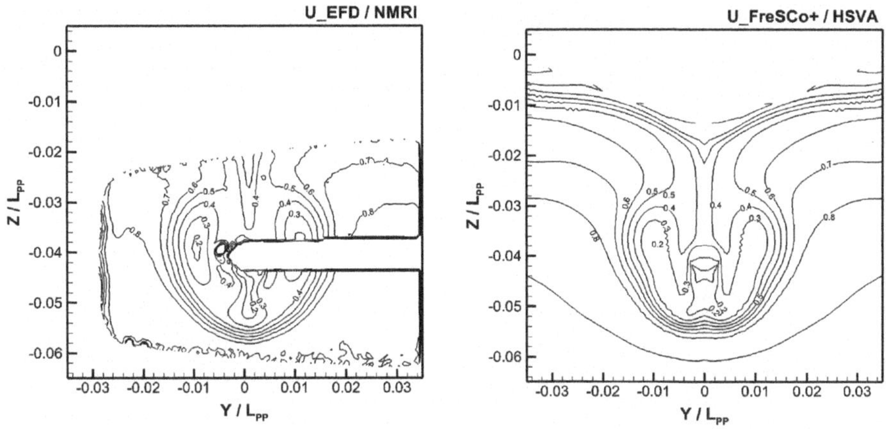

(a) Experiments *(b) Linear EARSM*

Fig. 18 Case 1-3a—comparisons at station S4 between HSVA/Fresco+ and experiments

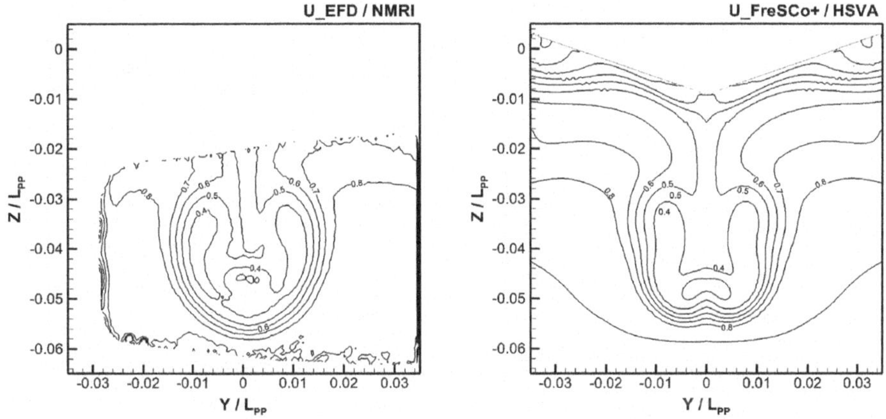

(a) Experiments *(b) Linear EARSM*

Fig. 19 Case 1-3a—comparisons at station S7 between HSVA/Fresco+ and experiments

totally different numerical methodologies are compared. However, for both results, one should notice that the island corresponding to the iso-level u = 0.2 is captured by none of these simulations. However, the improvement brought by the EARSM closure for both contributors, underlines the crucial role played by the turbulence closure as soon as the simulation is carried out on a fine enough grid.

It is also very interesting to assess the predictive capabilities of Full RSM Transport Turbulence Closures for this typical flow. Full RSM transport equations model is based on the discretisation of six transport equations for the Reynolds stresses and

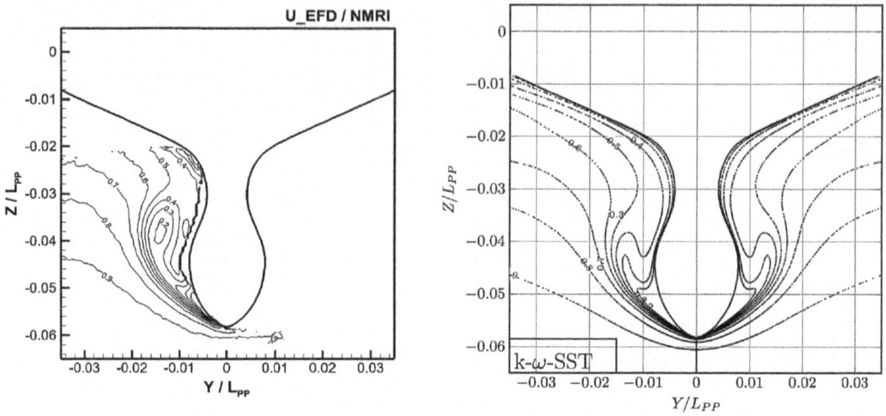

(a) Experiments *(b) k-ω SST*

Fig. 20 Case 1-3a—comparisons at station S2 between URO/Lemos and experiments

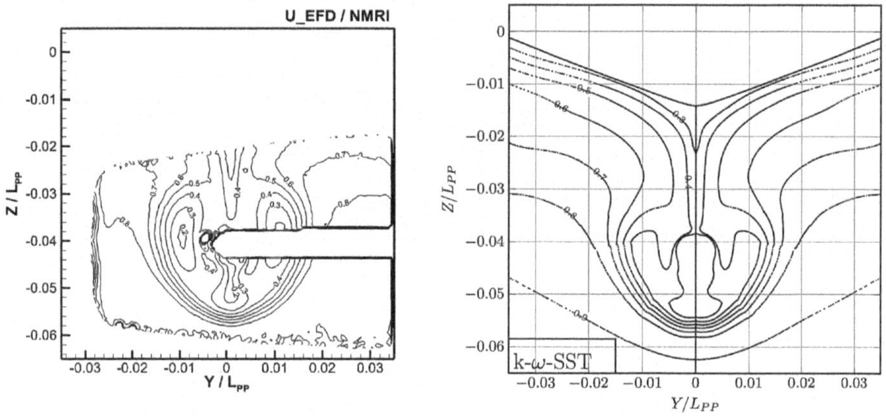

(a) Experiments *(b) k-ω SST*

Fig. 21 Case 1-3a—comparisons at station S4 between URO/Lemos and experiments

one for the turbulence dissipation or frequency. They are supposed to be superior to EARSM because they are exact for convection and production but the other terms like dissipation, diffusion and pressure velocity correlations are crudely modelled. Moreover, due to the absence of explicit turbulence viscosity, this class of models is known to be less robust and more sensitive to discretisation errors on distorted grids and high Reynolds flows. Such a RSTM closure is implemented in Star-CCM+ and Figs. 32, 33 and 34 show results obtained by HHI. The agreement with experiments is very good for stations S4 and S7. At station S2, the simulated longitudinal vortex appears to be slightly too large but for the first time, the isowake $u = 0.2$ is present in the computations at station S2.

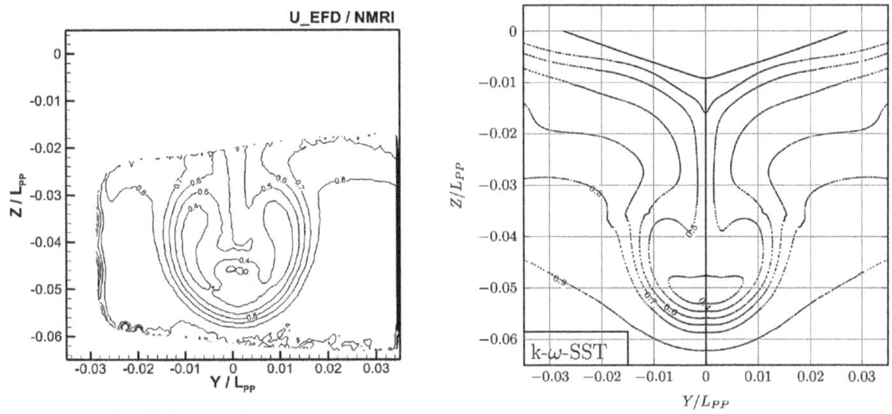

(a) *Experiments* (b) *k-ω SST*

Fig. 22 Case 1-3a—comparisons at station S7 between URO/Lemos and experiments

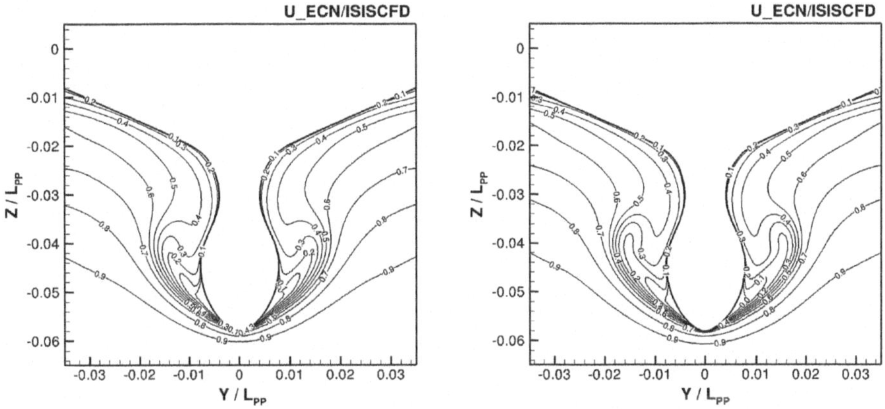

(a) *k-ω SST* (b) *EARSM*

Fig. 23 Case 1-3a—ECN/CNRS—comparisons at station S2 between two turbulence closures (same grid, same code ISIS-CFD)

3.5.3 Hybrid RANS-LES Turbulence Closures

In the previous subsections, one has observed that the turbulence anisotropy should be taken into account by the turbulence models to simulate the right level of longitudinal vorticity in stern flows. EARSM provides a better solution as confirmed independently by two contributors using the same model and RSTM leads to an almost perfect representation of the measurements. All these models are based on statistical

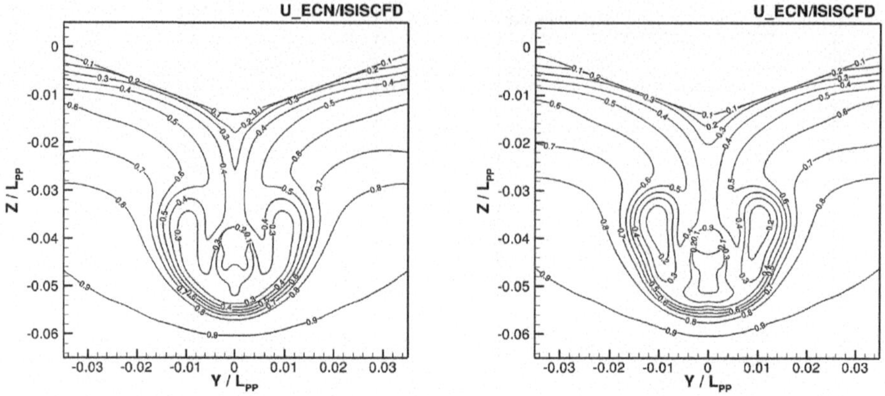

(a) *k-ω SST* (b) *EARSM*

Fig. 24 Case 1-3a—ECN/CNRS—comparisons at station S4 between two turbulence closures (same grid, same code ISIS-CFD)

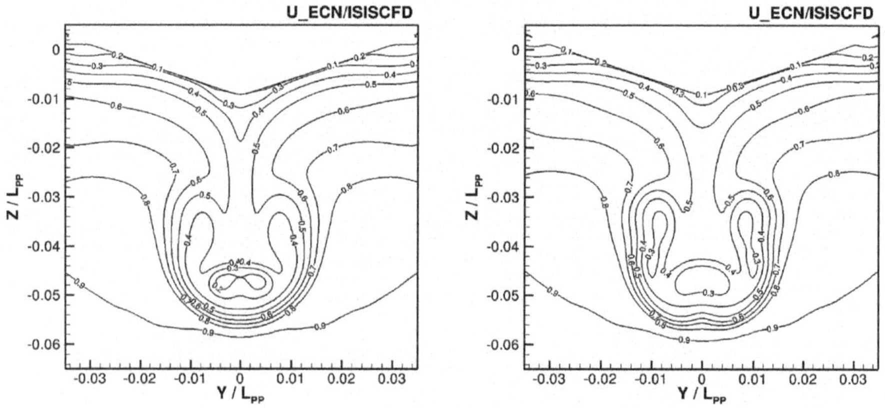

(a) *k-ω SST* (b) *EARSM*

Fig. 25 Case 1-3a—ECN/CNRS—comparisons at station S7 between two turbulence closures (same grid, same code ISIS-CFD)

modelling and are theoretically valid for steady averaged flows since URANSE can not simulate the intrinsic flow instabilities appearing in any turbulent flows. To refine further the model and approach the real physics, it is interesting to have recourse to hybrid RANS-LES unsteady formulations combining RANSE modelling close to the wall and a somewhat under-resolved LES formulations outside of the boundary layer. DES variants proposed by Spalart and colleagues belong to such a class of models. The University of Rostock provides two flow simulations based on a DES and IDDES hybrid RANS-LES closures. The results for the sections S2 to S7 are given by Figs. 35, 36, 37, 38, 39 and 40. Both are based on a double body con-

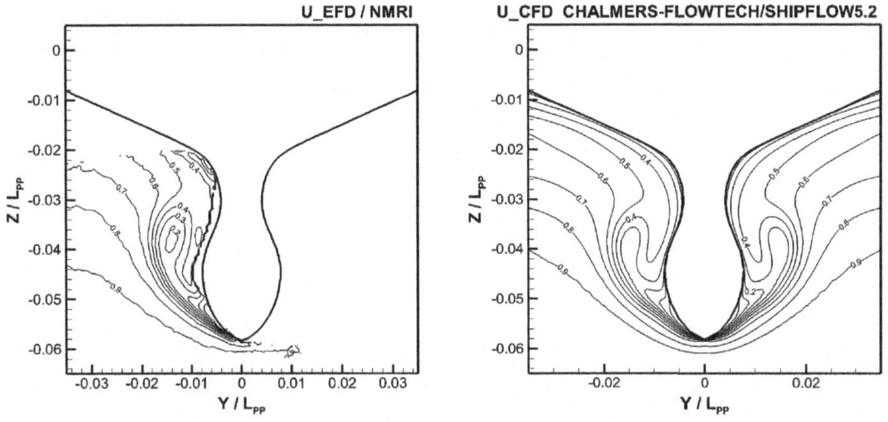

(a) *Experiments* (b) *EARSM*

Fig. 26 Case 1-3a—comparisons at station S2 between CHALMERS/Shipflow and experiments

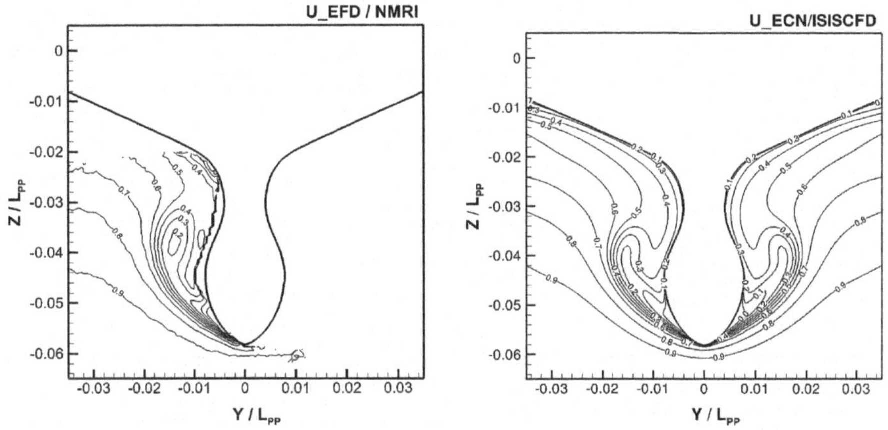

(a) *Experiments* (b) *EARSM*

Fig. 27 Case 1-3a—comparisons at station S2 between ECN/ISIS-CFD and experiments

figuration. DES is able to capture the isowake u = 0.2 at station S2 while IDDES provides a solution with less longitudinal vorticity at this section. However, at stations S4 and S7, the agreement with measurements is degraded, although slightly better with IDDES, when compared to the DES solution. Globally, the agreement on the isowakes is not significantly better for all stations than what can be obtained with EARSM or full RSTM statistical closures.

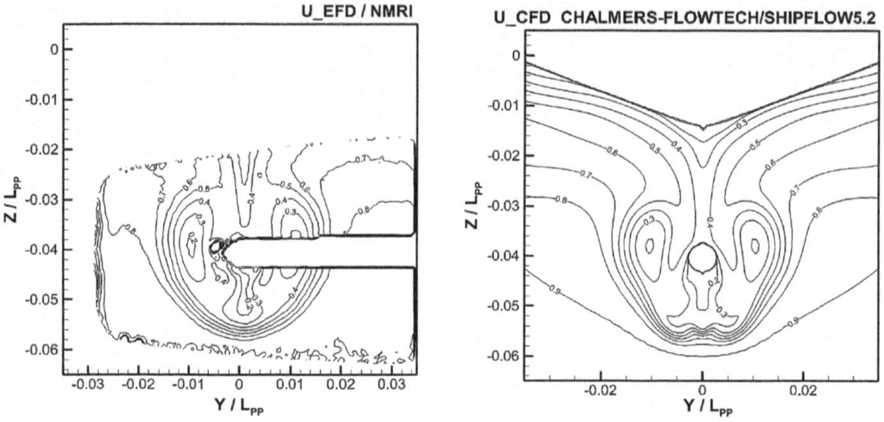

(a) Experiments *(b) EARSM*

Fig. 28 Case 1-3a—comparisons at station S4 between CHALMERS/Shipflow and experiments

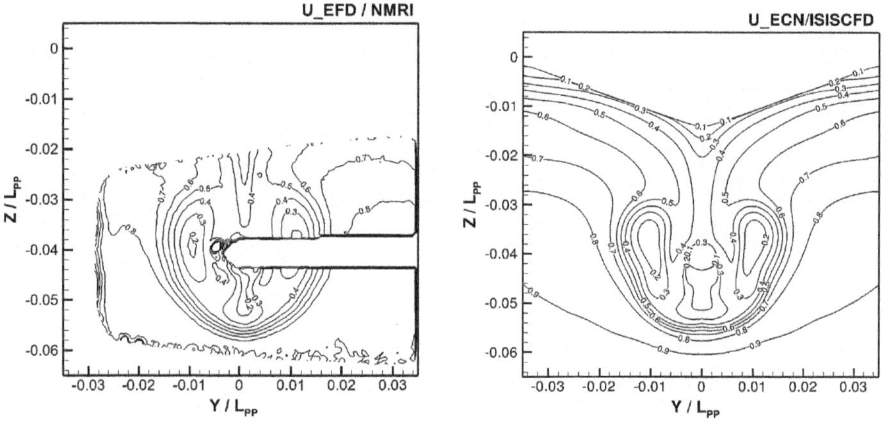

(a) Experiments *(b) EARSM*

Fig. 29 Case 1-3a—comparisons at station S4 between ECN/ISIS-CFD and experiments

3.6 Influence of the Free-Surface and Wall Treatment

The previous hybrid RANS-LES computations were performed on a double body i.e. by replacing the free-surface by a plane of symmetry. It is interesting to assess the influence of the free-surface on the isowakes at the experimental stations. Figures 41, 42 and 43 show computations done by ECN/CNRS with ISIS-CFD with and without free-surface on grids of very similar density. Apart from some minor modifications on the shape of lower isowakes, the influence of the free-surface on the vortical flow at the stern is not significant, which means that the hybrid RANS-LES computa-

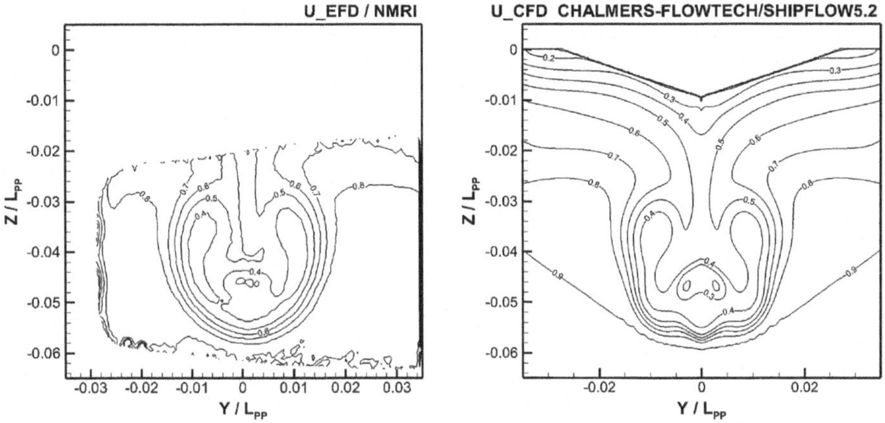

(a) Experiments *(b) EARSM*

Fig. 30 Case 1-3a—comparisons at station S7 between CHALMERS/Shipflow and experiments

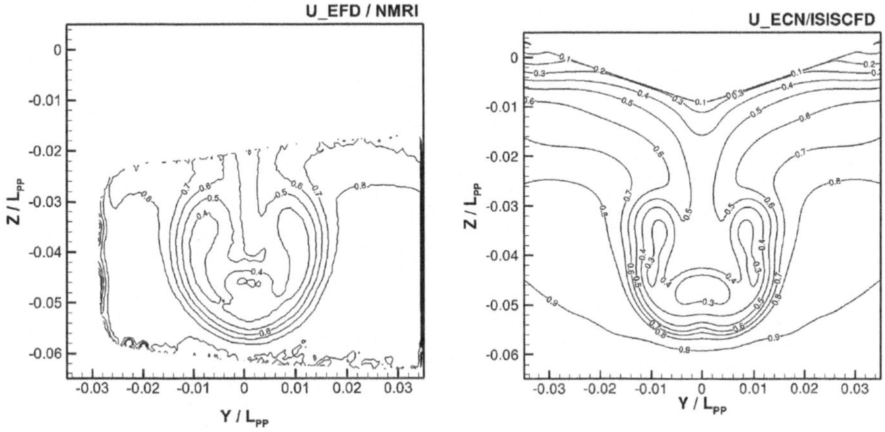

(a) Experiments *(b) EARSM*

Fig. 31 Case 1-3a—comparisons at station S7 between ECN/ISIS-CFD and experiments

tions performed by URO are not jeopardized by the lack of free-surface boundary condition.

The role played by the wall boundary condition is always very controversial and it is worthwhile to assess the results obtained by the same code on grids which only differ by the wall treatment. ECN/CNRS has conducted such computations with ISIS-CFD on two grids built with HEXPRESS^{TM}. Both grids have a viscous layer close to the hull but the wall resolved grid has a first point with a y^+ value below one

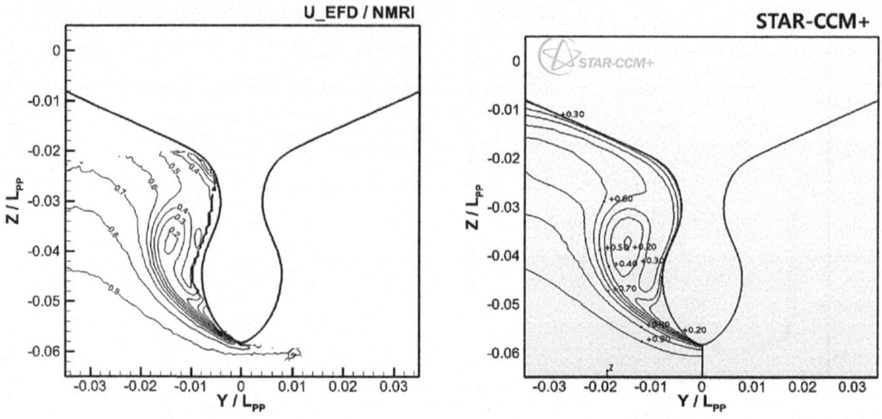

(a) Experiments (b) RSM

Fig. 32 Case 1-3a—comparisons at station S2 between HHI/Star-CCM+ and experiments

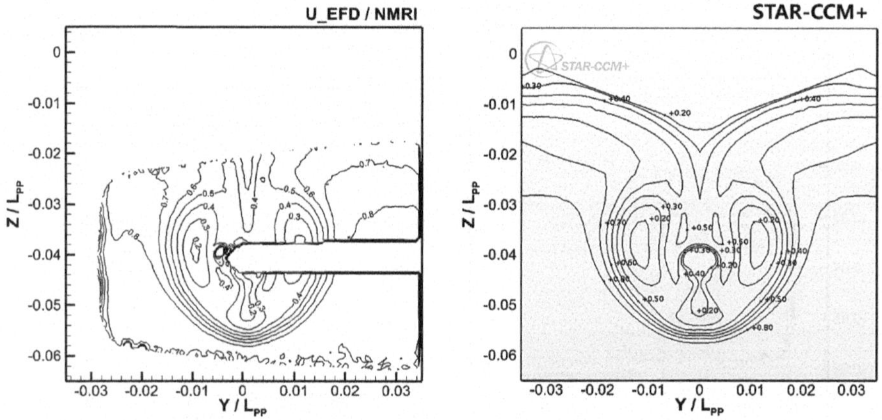

(a) Experiments (b) RSM

Fig. 33 Case 1-3a—comparisons at station S4 between HHI/Star-CCM+ and experiments

in average over the hull while the wall modeled grid has a first point located around $y^+ = 80$. Figures 44, 45 and 46 show a comparison of the isowake distribution at the usual experimental stations. Apart from the shapes of isowakes $u = 0.2$ and 0.3, one does not notice large differences between the results at the experimental stations.

3.6.1 Codes Comparison

Finally, based on the available contributions, it is possible to perform a code comparison in very fair conditions. By fair, one means using exactly the same grid and

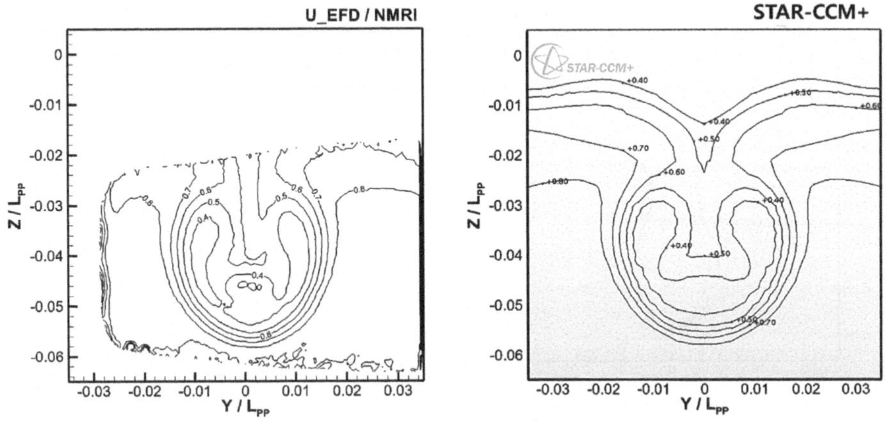

(a) Experiments (b) RSM

Fig. 34 Case 1-3a—comparisons at station S7 between HHI/Star-CCM+ and experiments

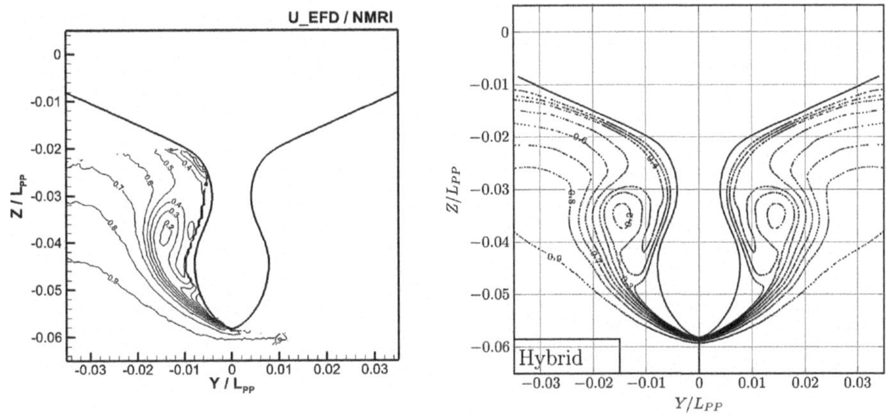

(a) Experiments (b) Hybrid RANS-LES - DES

Fig. 35 Case 1-3a—comparisons at station S2 between URO/Lemos and experiments

turbulence closure. SOTON delivered computations in such conditions with Open-FOAM and StarCCM+ and Figs. 47, 48 and 49 show a comparison of the isowakes at thc usual experimental cross-sections. It is remarkable to see that the figures are almost identical, which underlines the level of maturity reached by computational codes nowadays.

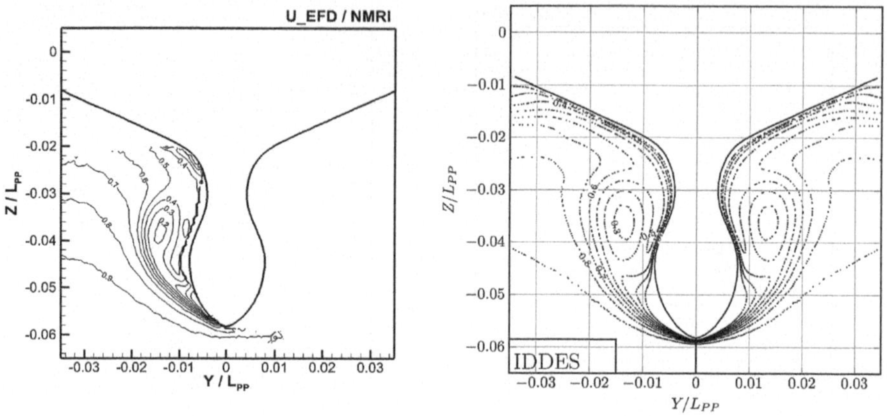

(a) Experiments (b) Hybrid RANS-LES - IDDES

Fig. 36 Case 1-3a—comparisons at station S2 between URO/Lemos and experiments

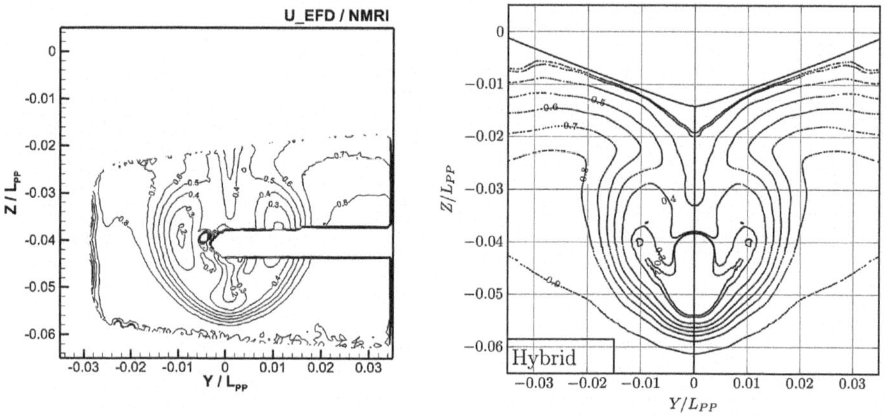

(a) Experiments (b) Hybrid RANS-LES - DES

Fig. 37 Case 1-3a—comparisons at station S4 between URO/Lemos and experiments

3.7 Case 1-3a —Temporary Conclusions

On the basis of this first analysis based on some representative examples taken from the numerous contributions, one can draw the following conclusions:

- With the typical grids which are used by the contributors (5–10 M points), the influence of the grid discretisation is moderate for RANSE, which means that the numerical error appears to be under control.

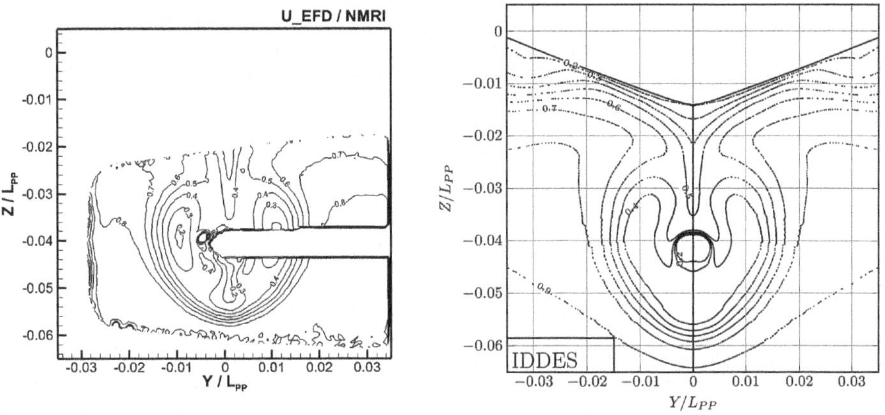

(a) Experiments *(b) Hybrid RANS-LES - IDDES*

Fig. 38 Case 1-3a—comparisons at station S4 between URO/Lemos and experiments

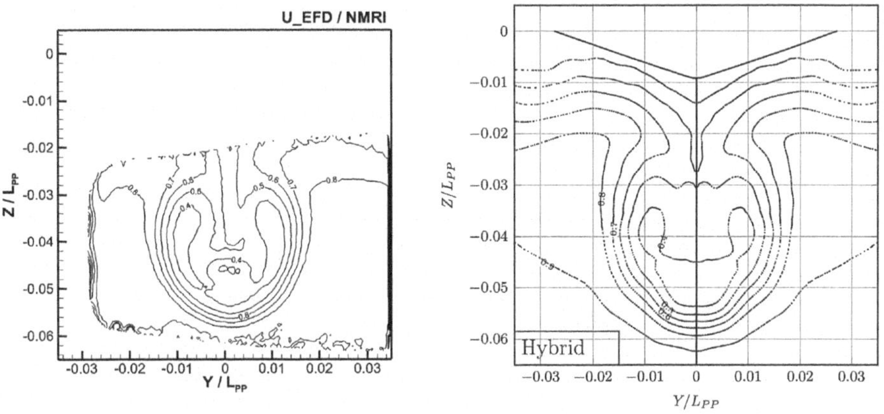

(a) Experiments *(b) Hybrid RANS-LES - DES*

Fig. 39 Case 1-3a—comparisons at station S7 between URO/Lemos and experiments

- As noticed during the Gothenburg2010 workshop for the analysis of the local flow around the KVLCC2, the major influence comes from the turbulence closure. Linear isotropic closures significantly under-predict the longitudinal vorticity at S2 while full RSM closures slightly over-predict it at the same station.
- Non-linear anisotropic closures (EARSM) offer a good compromise from the standpoint of the local flow although they slightly underpredict the vorticity at S2.
- New results from hybrid RANS-LES are promising but IDDES seems to over-predict the vorticity again. The agreement with the isowakes distribution is worse than what is achieved with the best RANS statistical turbulence closures.

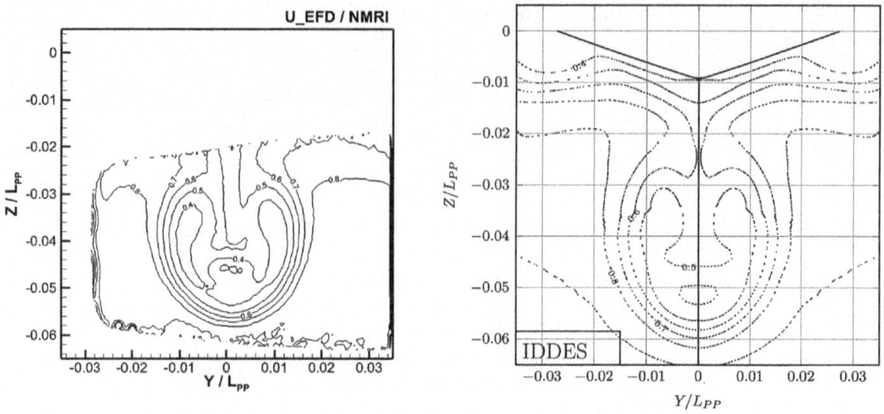

(a) Experiments *(b) Hybrid RANS-LES - IDDES*

Fig. 40 Case 1-3a—comparisons at station S7 between URO/Lemos and experiments

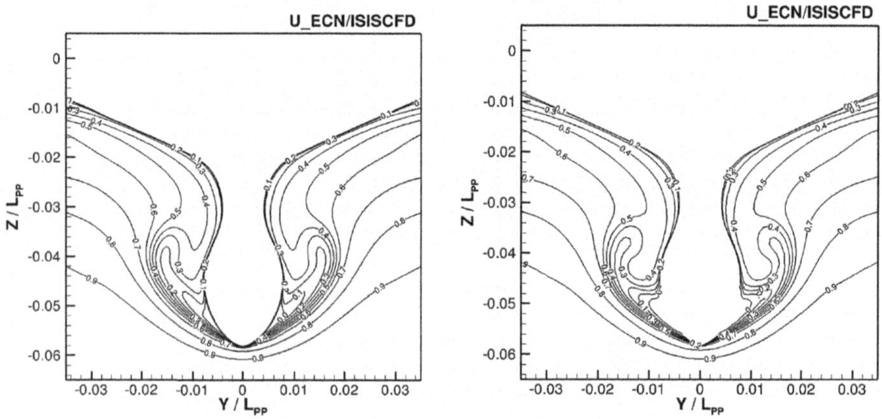

(a) Fr=0.142 *(b) Fr=0.0*

Fig. 41 Case 1-3a—ISIS-CFD—influence of the free-surface at station S2

- Computational codes seem to be mature since one does not observe large discrepancies when the same turbulence closures on similar grids are used.
- The influence of the free-surface on the local flow is not negligible but not large.
- The influence of the wall boundary treatment on the development of the stern flow is weak.

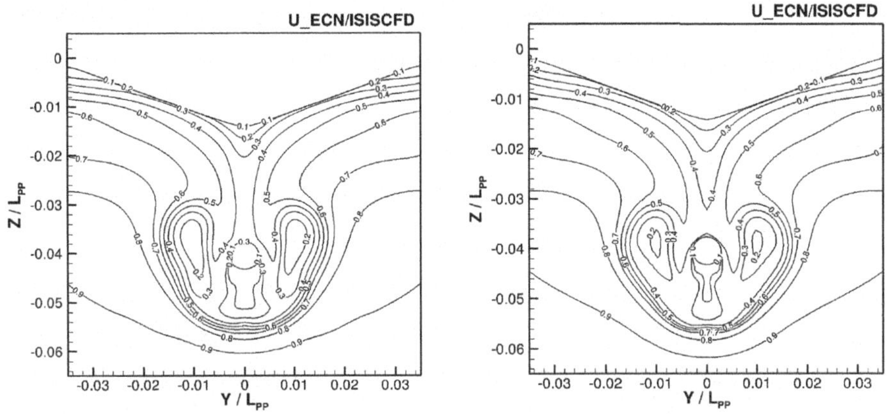

(a) Fr=0.142 *(b) Fr=0.0*

Fig. 42 Case 1-3a—ISIS-CFD—influence of the free-surface at station S4

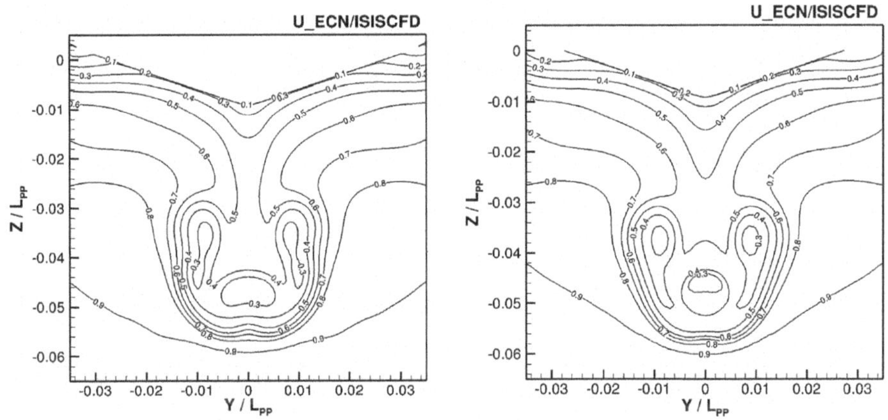

(a) Fr=0.142 *(b) Fr=0.0*

Fig. 43 Case 1-3a—ISIS-CFD—influence of the free-surface at station S7

3.8 Local Vortex Flow Analysis

3.8.1 Objectives of This Local Analysis

In the previous editions of the Gothenburg/Tokyo workshops on numerical ship hydrodynamics, the local flow analysis was uniquely based on the inspection of the flow characteristics at specific cross-sections where experiments were available. Most of the time, this analysis was based on the visual inspection of iso-lines of characteristic quantities like the iso-wakes, iso values of turbulent stresses, etc. Although, this

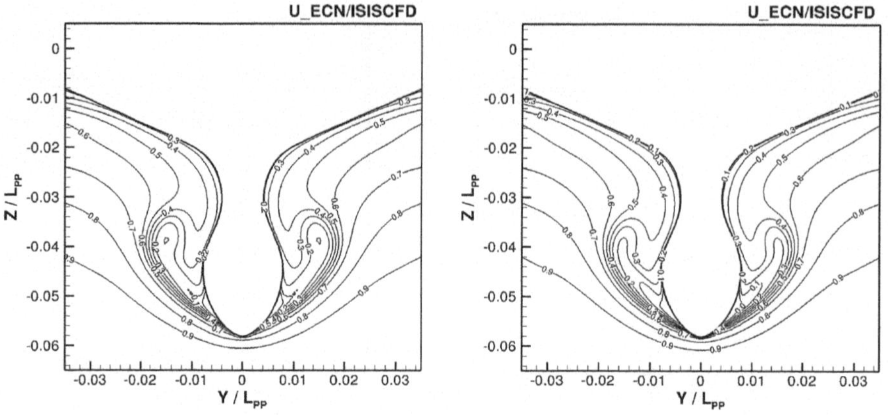

(a) Wall modeled *(b) Wall resolved*

Fig. 44 Case 1-3a—ISIS-CFD—influence of the wall treatment at station S2

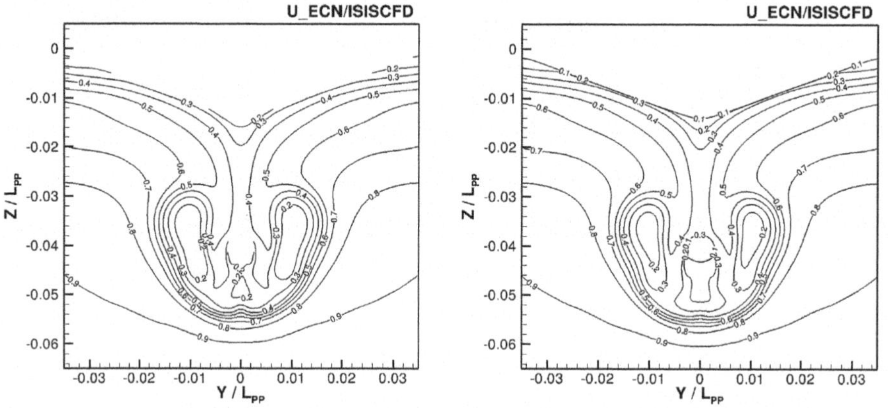

(a) Wall modeled *(b) Wall resolved*

Fig. 45 Case 1-3a—ISIS-CFD—influence of the wall treatment at station S4

approach provides a global analysis of the flow for each experimental cross-section, it can be misleading since it is based on visual inspection of iso-lines for which local gradients are difficult to appreciate. For the first time, during the Tokyo 2015 edition, it was decided to enrich this cross-section based evaluation by a more detailed and local vortex flow analysis in order to draw more elaborate conclusions about the generation and evolution of the longitudinal vortices. What is aimed by this study is to provide a detailed inspection and comparison of the experiments and computational results inside the core of the main bilge vortex.

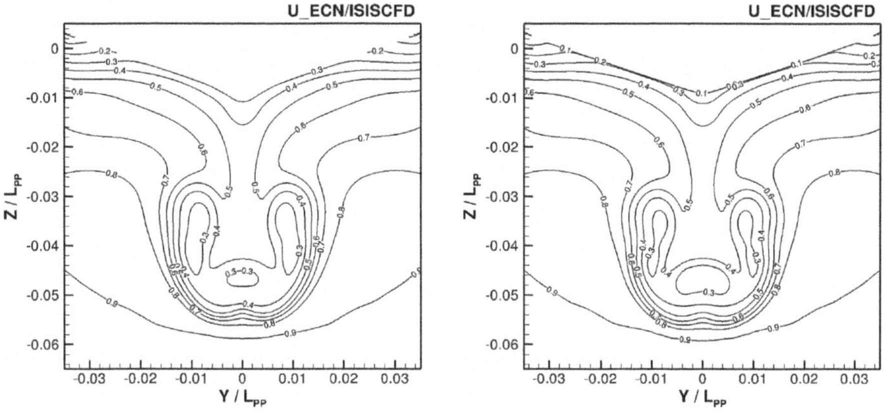

(a) *Wall modeled* (b) *Wall resolved*

Fig. 46 Case 1-3a—ISIS-CFD—influence of the wall treatment at station S7

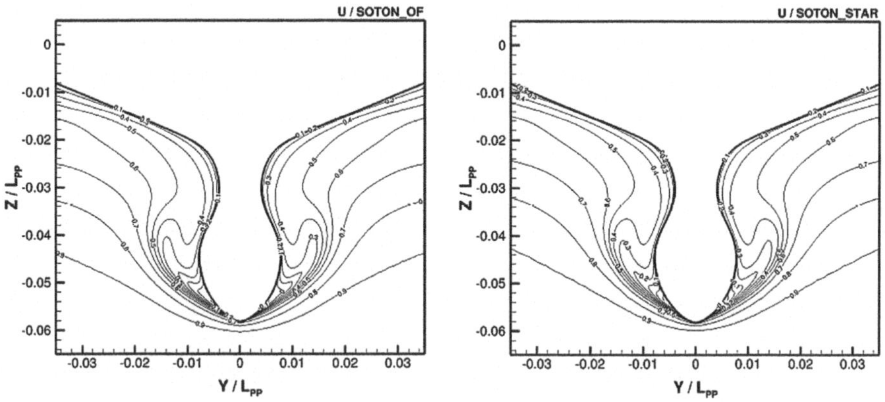

SOTON-OF SOTON-STAR

Fig. 47 Case 1-3a—SOTON—comparison between OpenFOAM and StarCCM+ at station S2

3.8.2 Procedure, Difficulties and Perils of the Exercise

What we wished to compare was the y and z transversal evolutions of characteristic flow data across the vortex center at stations S2, S4 and S7. The main difficulty is to locate the center of the main bilge vortex since (i) it is implicitly stated that there is only one such vortex which can be detected without any ambiguity and (ii) that it is feasible to locate precisely its center at several cross-sections. First of all, it was decided to use the same information in computations and experiments to locate the vortex center. This means that, since we had no three-dimensional information from the measurements (no tomographic PIV), we were obliged to use the local max(ω_x)

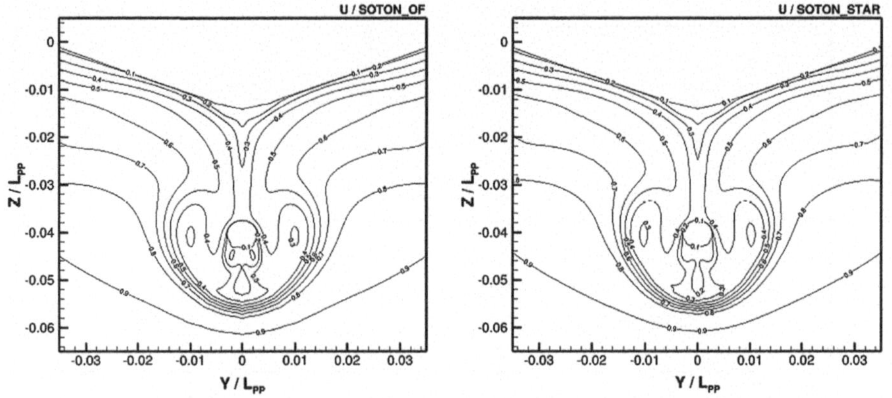

SOTON-OF SOTON-STAR

Fig. 48 Case 1-3a—SOTON—comparison between OpenFOAM and StarCCM+ at station S4

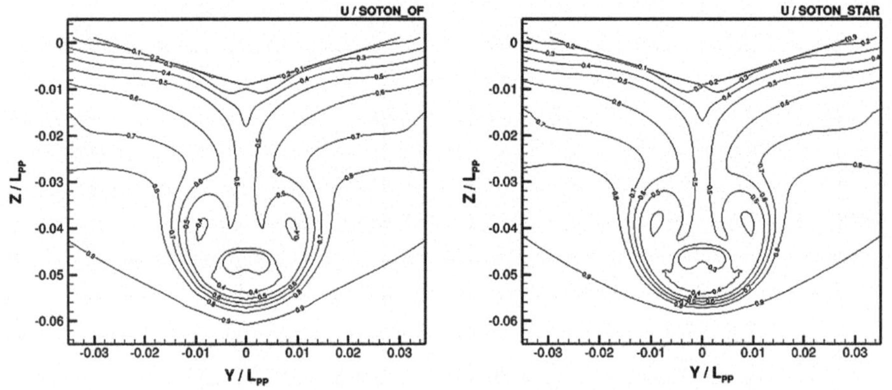

SOTON-OF SOTON-STAR

Fig. 49 Case 1-3a—SOTON—comparison between OpenFOAM and StarCCM+ at station S7

to locate the vortex center in each section instead of max(Q). This procedure is relatively reliable for stations S4 an S7 where the main vortex is roughly aligned with the x direction but it is less justified for station S2 where some computations indicate that the vortex is not aligned with the x axis. This is the reason why some additional figures were also produced based on max(Q) instead of max(ω_x) to locate the center of the vortex at station S2.

3.8.3 Genesis from NATO-AVT183—DTMB5415 at Straight Ahead Condition

The original idea came from the Tomographic PIV experiments performed at IIHR by Prof. Frederick Stern's team on the flow around the US Navy frigate DTMB5415 at straight ahead condition. These remarkable experiments were used in the framework of the NATO/AVT-183 collaborative project to assess the ability of various codes to simulate accurately the flow physics in the core of the longitudinal vortices shed at the sonar dome (Bhushan et al. 2019). Figure 50b gives a global overview of the various vortices emanating from the sonar dome and a typical view of the three-dimensional box where TPIV experiments are performed. Figure 51 shows the longitudinal evolution in the core of the Sonar Dome Vortex of the invariant Q and turbulence kinetic energy TKE for experiments and various computations. Figures 52 and 53 show the radial evolution in the core of the SDV vortex of the longitudinal vorticity, the invariant Q, the turbulence kinetic energy TKE and the longitudinal component of the velocity U. Here, computations performed by IIHR using two different turbulence modeling approaches are compared to the experiments.

From these figures, one can notice on one hand that the longitudinal evolution of Q and TKE were satisfactorily predicted by all the computations. On the other hand, large differences between the statistical and hybrid RANS/LES turbulence models run by IIHR were revealed by the comparison on the radial evolution in the core of the SDV vortex, hybrid RANS/LES DES (resp. statistical) model over (resp. under) predicting the magnitudes of Q and ω_x. It is worthwhile to underline also the large differences observed in the radial distribution of TKE between statistical and hybrid RANS/LES turbulence closures.

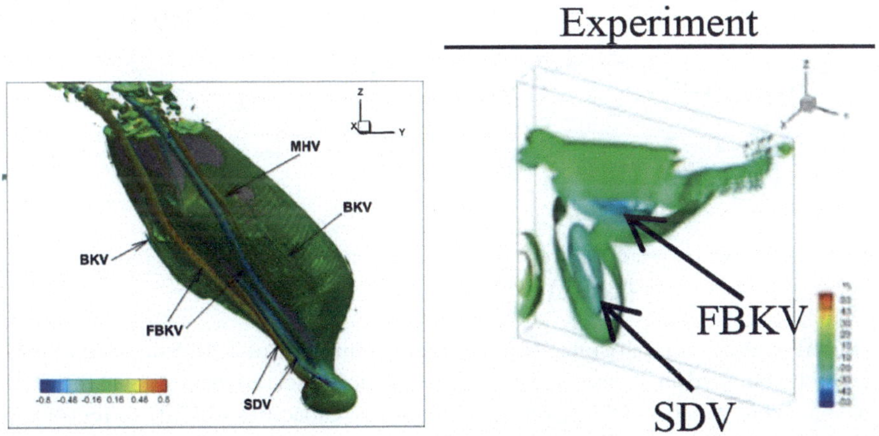

(a) 3D view (SDV and FBKV vortices) *(b) IIHR 3D experiments - ω_x behind the sonar dome*

Fig. 50 NATO AVT183—longitudinal vortices evolution and 3D tomographic PIV around the DTMB5415 at straight ahead condition

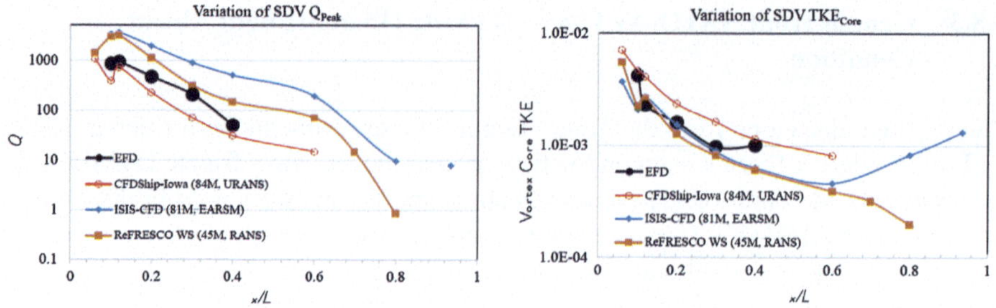

(a) *Invariant Q* (b) *TKE*

Fig. 51 NATO AVT183—longitudinal evolution of Q and TKE in the core of the SDV vortex

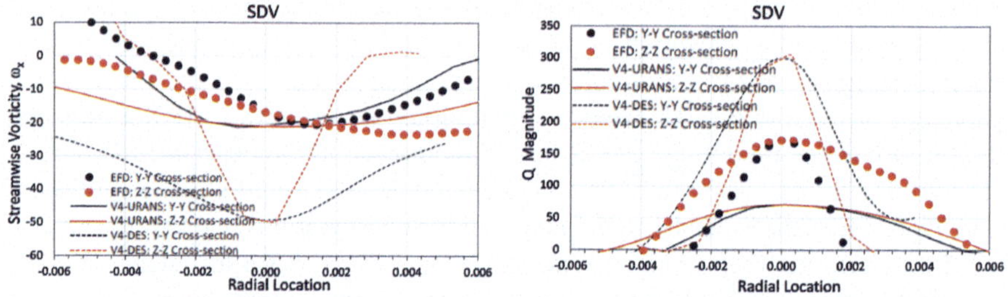

(a) ω_x (b) *Q*

Fig. 52 NATO AVT183—radial evolution of ω_x and Q in the core of the SDV vortex

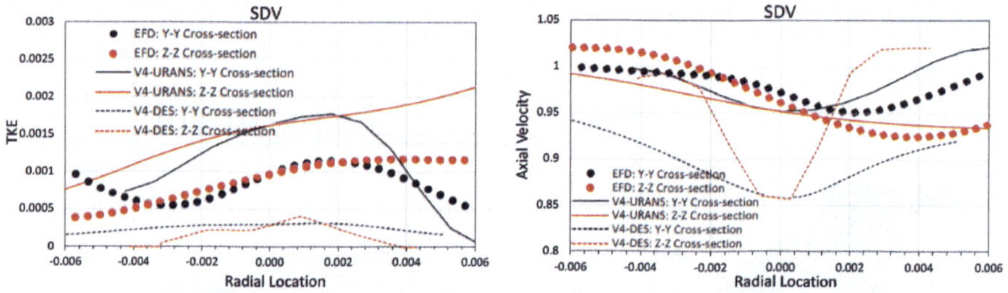

(a) *TKE* (b) *U*

Fig. 53 NATO AVT183—radial evolution of TKE and U in the core of the SDV vortex

Based on these comparisons, the organizing committee of T2015 decided to propose such a detailed analysis to see if the flow physics observed in the core of the SDV was somewhat generic and to assess the predictive capacities of CFD to reproduce it in the case of the main bilge vortex developing at the aft part of the JBC.

3.8.4 Local Vortex Flow Analysis—Longitudinal Evolution

As indicated previously, the core of the main bilge vortex was detected by looking for the point where ω_x is maximum since only 2D experiments at sections S2, S4 and S7 were available. In the computations, a downstream streamline was shed from the vortex core at S2 and the data was extracted along this streamline. The data obtained in that way were compared with that extracted along an upstream streamline originating from the vortex core at S7 and no major differences were observed. Figure 54 showing the longitudinal evolution of the Y and Z coordinates of the core, establishes that the trajectory of this vortex is reasonably well predicted by most of the contributors. A better agreement with the experiments is observed on the vertical location than on the horizontal one. All the computations predict a wavy longitudinal evolution of the Y and Z, which means that the core of the vortex follows a kind of helicoidal trajectory. Globally, all the computations indicate the same trend from upwind to downwind, although a relatively large dispersion can also be noticed.

Figure 55 shows the longitudinal evolution of the second invariant Q and the longitudinal component of the vorticity. Apart from Chalmers which predicts an

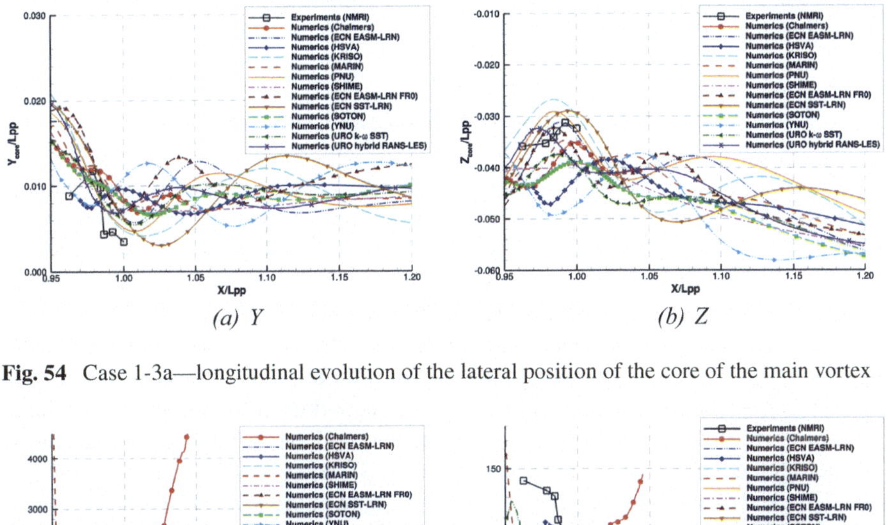

(a) Y *(b) Z*

Fig. 54 Case 1-3a—longitudinal evolution of the lateral position of the core of the main vortex

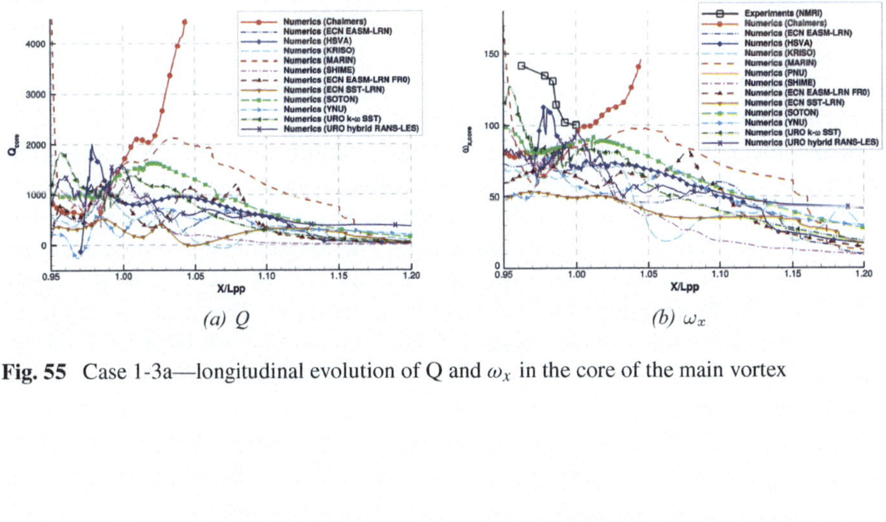

(a) Q *(b) ω_x*

Fig. 55 Case 1-3a—longitudinal evolution of Q and ω_x in the core of the main vortex

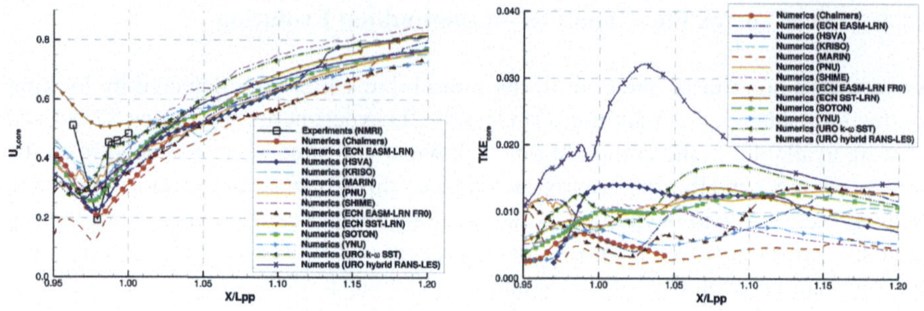

(a) U *(b) Tke*

Fig. 56 Case 1-3a—longitudinal evolution of U and TKE in the core of the main vortex

erroneous longitudinal evolution of Q, all the contributions appear globally similar in terms of magnitude. The magnitude of the longitudinal vorticity measured by NMRI is initially under-evaluated in the near-wake by all the computations. After $X/L_{pp} = 1.0$, the agreement between measurement and computations looks improved but it is difficult to draw any solid conclusions because of the lack of experiments after $X/L_{pp} = 1.00$ (Fig. 56).

3.8.5 Local Vortex Flow Analysis—Transversal Evolution

In order to better understand the local flow physics in the core of this bilge vortex, it was also decided to report the transversal evolution across the core along y and z directions. The core was determined by locating the point where ω_x was maximum to stick with the experiments in the absence of any measurement of the second invariant Q. This is an important limitation which should be underlined since the point where ω_x is maximum coincides with the core of the vortex only if the vortex is aligned with the x direction, which is not always true for this validation exercise. Figures 57, 58, 59 and 60 show the transversal evolution of the second invariant Q, the longitudinal vorticity ω_x, the longitudinal velocity component U and the turbulence kinetic energy TKE across the core of the main vortex at station S2. First of all, one notices that the typical marked Gaussian shape of Q found in the core of the SDV DTMB5415 vortex core is hardly visible here in most of the computations (except Marin) although more marked in the Y than in Z directions. This trend is confirmed by Fig. 58 showing the transversal evolution of ω_x. The agreement between measured and computed ω_x and U is reasonable for most of the computations if one takes into account the underlying uncertainties of such a comparison. The longitudinal component of the velocity appears to be smooth as it is expected in the core of a vortex and this trend is reasonably well represented by most of the contributors. This analysis on Q, ω_x and U applies to the other stations S4 and S7 as well (see Figs. 61, 62, 63, 64, 65, 66 and 67).

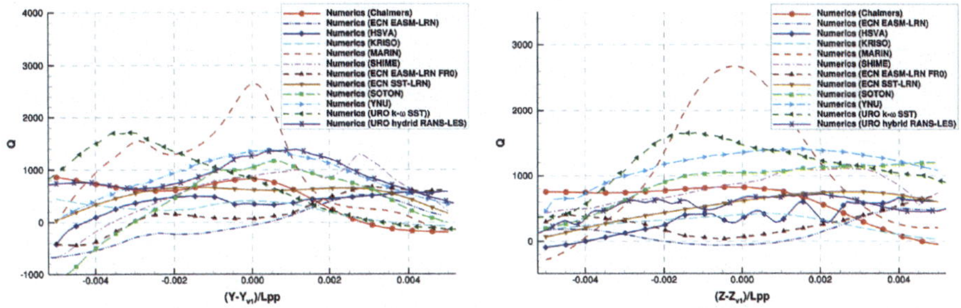

(a) Along y *(b) Along z*

Fig. 57 Case 1-3a—transversal evolution along y and z of Q across the core of the main vortex at station S2

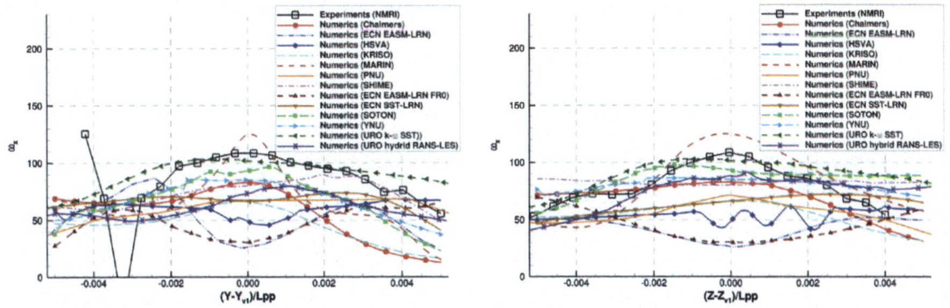

(a) Along y *(b) Along z*

Fig. 58 Case 1-3a—transversal evolution along y and z of ω_x across the core of the main vortex at station S2

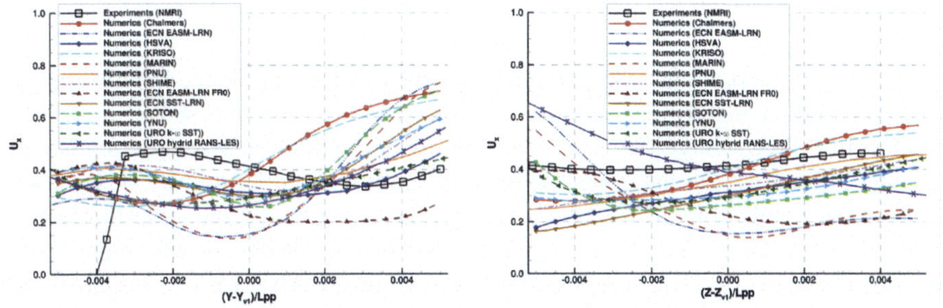

(a) Along y *(b) Along z*

Fig. 59 Case 1-3a—transversal evolution along y and z of U across the core of the main vortex at station S2

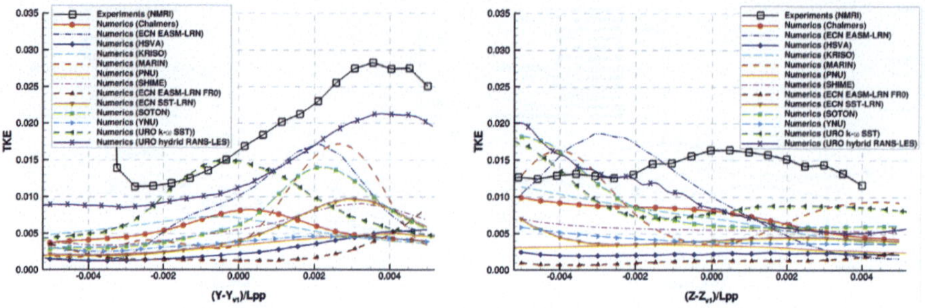

(a) Along y *(b) Along z*

Fig. 60 Case 1-3a—transversal evolution along y and z of TKE across the core of the main vortex at station S2

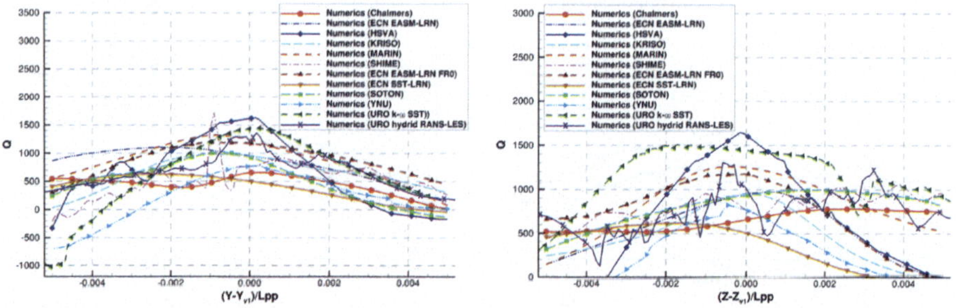

(a) Along y *(b) Along z*

Fig. 61 Case 1-3a—transversal evolution along y and z of Q across the core of the main vortex at station S4

Contrary to the previous comparisons, the study of the turbulence kinetic energy in the core of the vortex reveals unexpected trends. The experiments show a relatively high level of TKE which is not predicted by the computations based on statistical turbulence closures. On the contrary, the contribution from University of Rostock using a time-resolved hybrid RANS-LES approach is the only one able to capture the level of turbulence kinetic energy indicated by the NMRI measurements (see Fig. 60). The level of TKE computed with this hybrid RANS-LES formulation, in very good agreement with NMRI measurements, is three to ten times higher than what is simulated by the isotropic or anisotropic RANSE models. This trend is fully confirmed by the comparisons at stations S4 and S7 (see Figs. 64 and 68) where the agreement between NMRI measurements and the hybrid RANS-LES computations of URO appears even better.

Figures 61, 62, 63 and 64 show the transversal evolution of the second invariant Q, the longitudinal vorticity ω_x, the longitudinal velocity component U and the turbulence kinetic energy TKE across the core of the main vortex at the second experimental station S4.

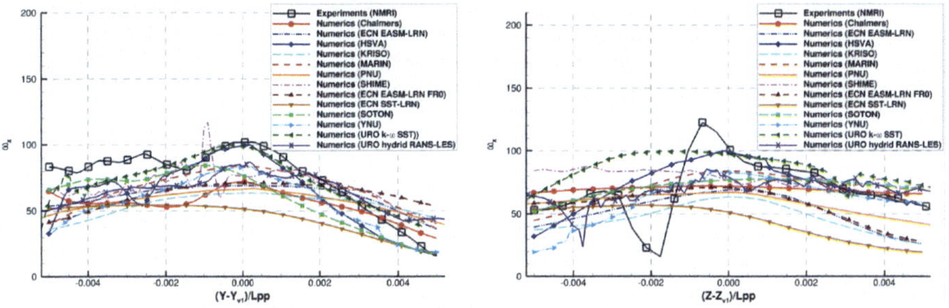

(a) Along y *(b) Along z*

Fig. 62 Case 1-3a—transversal evolution along y and z of ω_x across the core of the main vortex at station S4

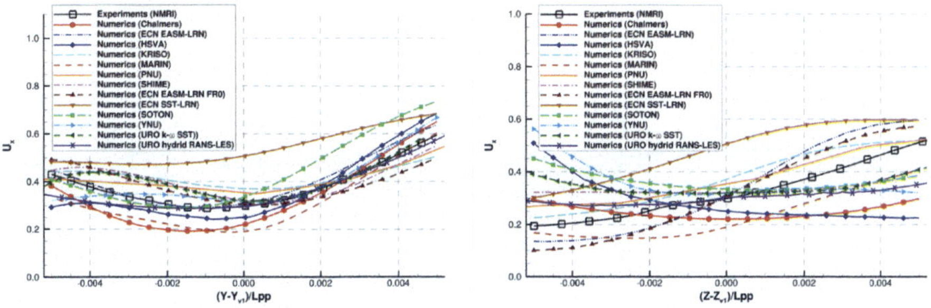

(a) Along y *(b) Along z*

Fig. 63 Case 1-3a—transversal evolution along y and z of U across the core of the main vortex at station S4

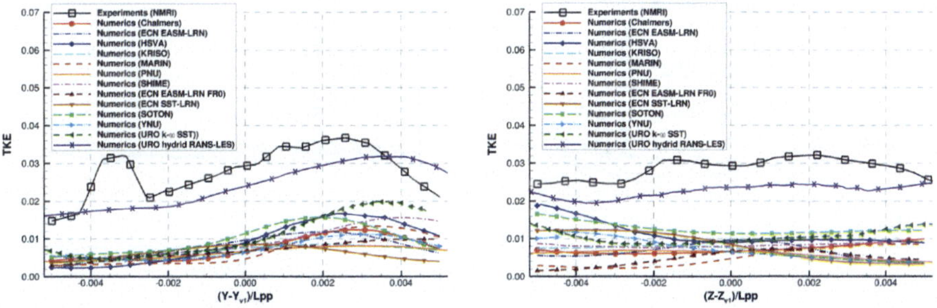

(a) Along y *(b) Along z*

Fig. 64 Case 1-3a—transversal evolution along y and z of TKE across the core of the main vortex at station S4

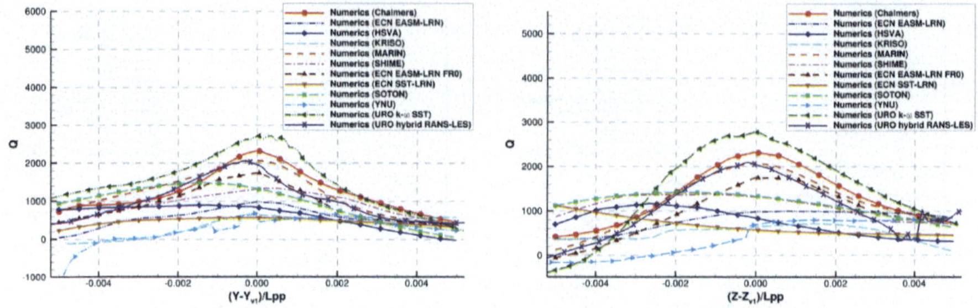

(a) Along y *(b) Along z*

Fig. 65 Case 1-3a—transversal evolution along y and z of Q across the core of the main vortex at station S7

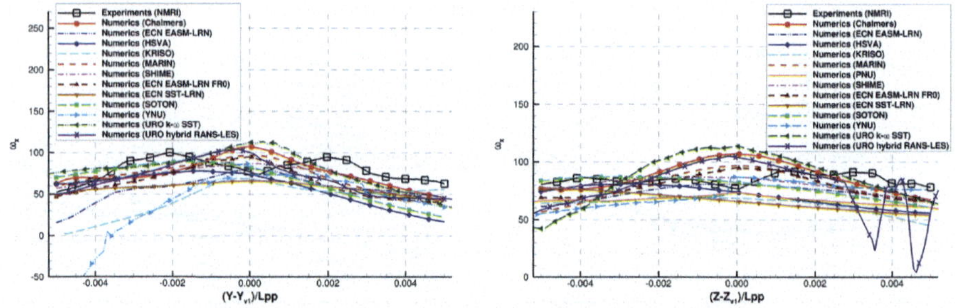

(a) Along y *(b) Along z*

Fig. 66 Case 1-3a—transversal evolution along y and z of ω_x across the core of the main vortex at station S7

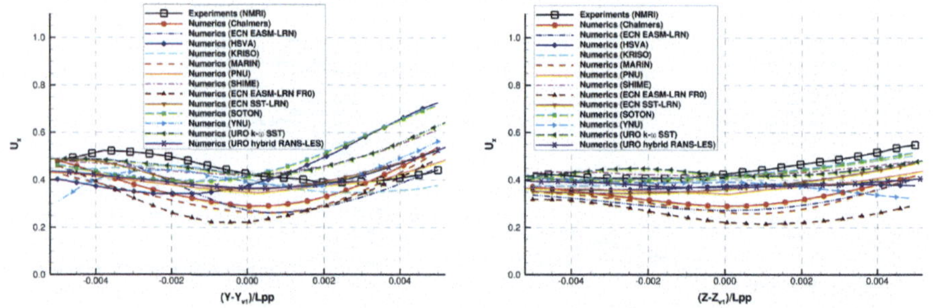

(a) Along y *(b) Along z*

Fig. 67 Case 1-3a—transversal evolution along y and z of U across the core of the main vortex at station S7

Figures 65, 66, 67 and 68 show the transversal evolution of the second invariant Q, the longitudinal vorticity ω_x, the longitudinal velocity component U and the turbulence kinetic energy TKE across the core of the main vortex at the last experimental station S7.

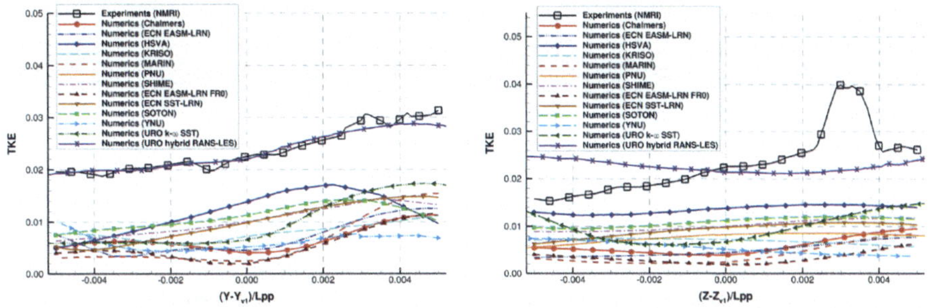

(a) Along y *(b) Along z*

Fig. 68 Case 1-3a—transversal evolution along y and z of TKE across the core of the main vortex at station S7

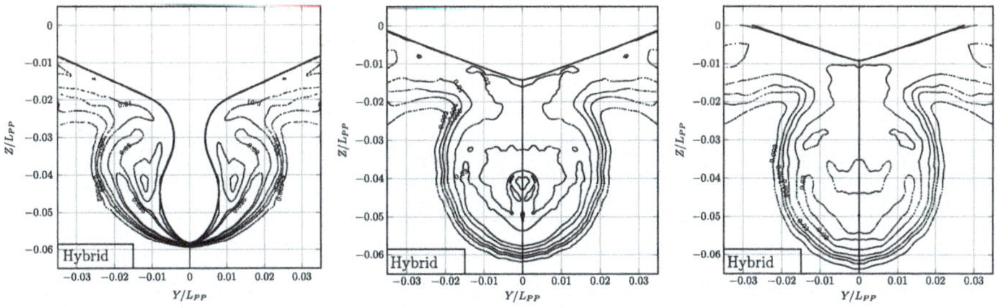

(a) S2 *(b) S4* *(c) S7*

Fig. 69 Case 1-3a—total TKE distribution at stations S2, S4 and S7—hybrid RANS-LES computations from URO

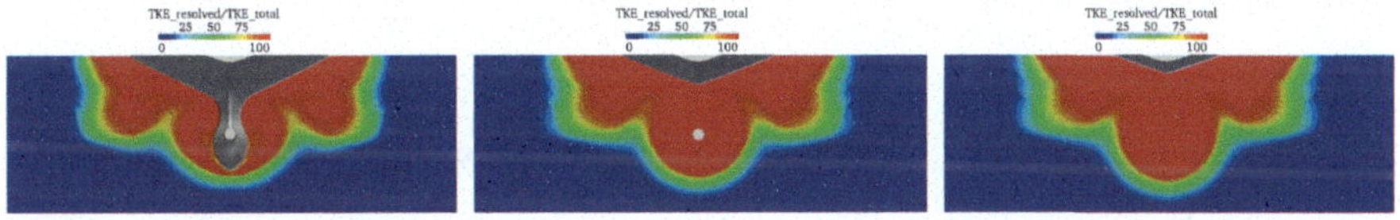

(a) S2 *(b) S4* *(c) S7*

Fig. 70 Case 1-3a—ratio between resolved and total TKE distributions at stations S2, S4 and S7—hybrid RANS-LES computations from URO

To analyze the balance between modeled and resolved TKE in URO's hybrid RANS-LES computations, Figs. 69 and 70 show the TKE distributions at S2, S4 and S7 and the ratio between total and resolved TKE. It is clear that most of the turbulence kinetic energy computed by URO comes from the unsteady resolved part, while the modeled part is negligible in the core of the bilge vortex.

It is of course unexpected to find a relatively high level of TKE in the core of longitudinal vortex since this high level is often associated with a high level of turbulent dissipation if one traditionally thinks in terms of linear eddy-viscosity

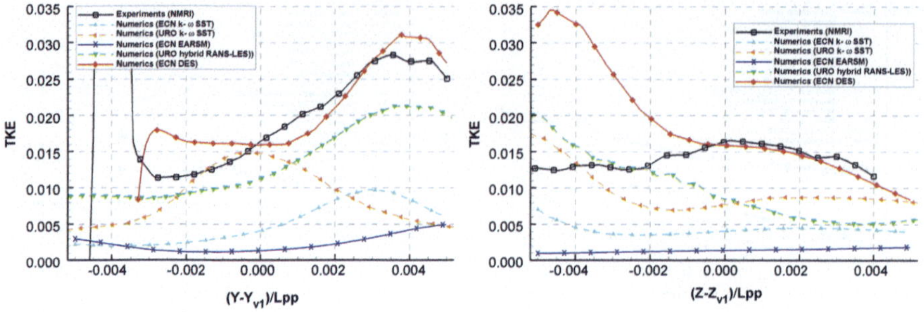

(a) Along y *(b) Along z*

Fig. 71 Case 1-3a—horizontal and vertical evolutions of TKE around the vortex center at station S2 (from Visonneau et al. 2016)

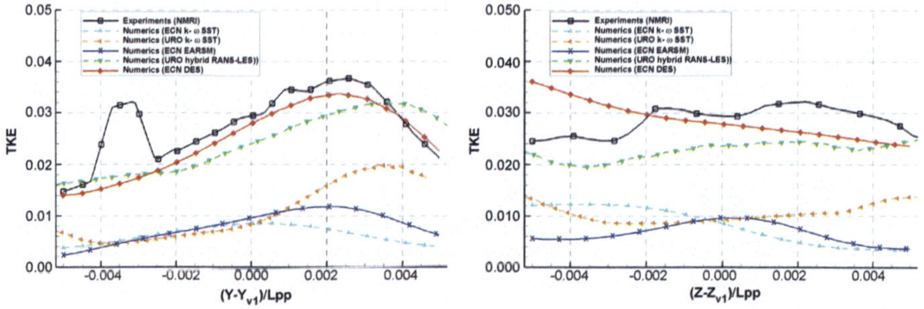

(a) Along y *(b) Along z*

Fig. 72 Case 1-3a—horizontal and vertical evolutions of TKE around the vortex center at station S4 (from Visonneau et al. 2016)

turbulence models. Whether the NMRI TKE measurements are reliable or not was debated during the workshop. NMRI, which realized the experiments, was (and still is) skeptical about the reliability of their TKE measurements because of a seemingly too low PIV frequency. Chapter "Experimental Data of Resistance, Sinkage, Trim, Self-propulsion Factors, Longitudinal Wave Cut and Detailed Flow for JBC With and Without an Energy Saving Circular Duct" of the present book addresses this topic by comparing NMRI, OU and TUHH measurements of TKE but, up to now, no solid conclusion can be drawn by comparing experiments which were not performed at the same Reynolds number. To shed some light on this enigma, ECN/CNRS decided to carry out additional unsteady post-workshop computations with a similar hybrid RANS-LES closure based on a DES-SST turbulence model to check if similar trends were observed independently of the flow solver. These new DES-SST computations were performed on a grid around the complete double-body hull comprised of 66 million points locally refined in the core of the bilge vortex and a time step $\Delta t = 0.006$ s. Figures 71, 72 and 73 showing these new results for TKE, fully confirm the results obtained by University of Rostock during the Tokyo2015 workshop.

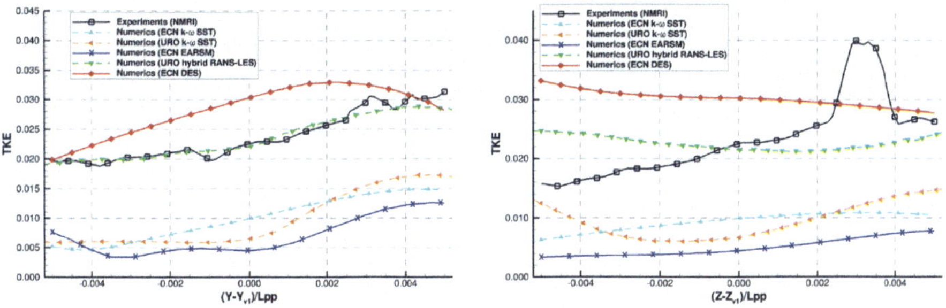

(a) *Along y* (b) *Along z*

Fig. 73 Case 1-3a—horizontal and vertical evolutions of TKE around the vortex center at station S7 (from Visonneau et al. 2016)

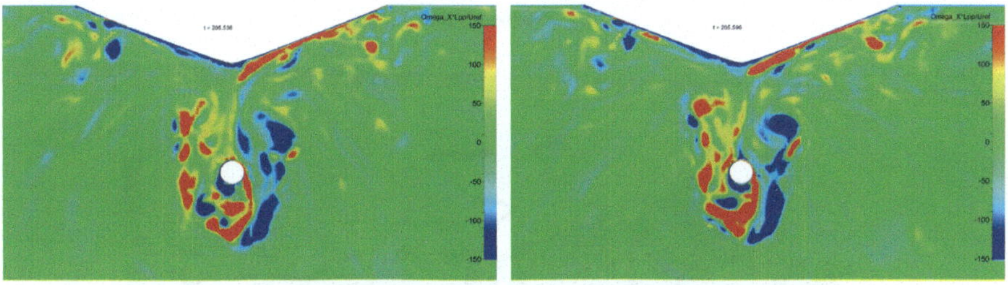

(a) *Time 205.536s* (b) *Time 205.596s*

Fig. 74 Case 1-3a—instantaneous views of the longitudinal vorticity at section S2 (from Visonneau et al. 2016)

To support this interpretation, Fig. 74 provides also two instantaneous views of the longitudinal vorticity at section S2 separated by ten time steps i.e. 0.06s, extracted from the DES-SST computations. These figures show that the unsteady flow predicted by a hybrid RANS-LES closure exhibits large scale unsteadiness in the core of the vortex associated with ring-like vortices which play a major role in the local flow physics. These intense and strongly unsteady smaller vortical structures contribute to the high level of TKE through the strong unsteady velocity fluctuations at the point of measurement located in the core of the time-averaged longitudinal vortex, making the resolved part of TKE far much important than the modelled one. More informations could be found in (Visonneau et al., 2016).

It is believed that this is the fundamental reason which can explain concomitant large levels of averaged turbulence kinetic energy and longitudinal vorticity. The unsteady motion of these smaller scale vortical structures contributes to a high level of TKE which is associated with relatively low frequency macroscopic fluctuations. This level of temporal fluctuations is probably correctly measured by NMRI since the frequency of this evolution is clearly lower than the experimental measurement frequency of 6 Hz reported by NMRI. For instance, Fig. 75 showing a FFT decomposition of TKE at point (X = −3.391 m (i.e. 0.984428 L_{pp}), Y = −0.065949 m, Z

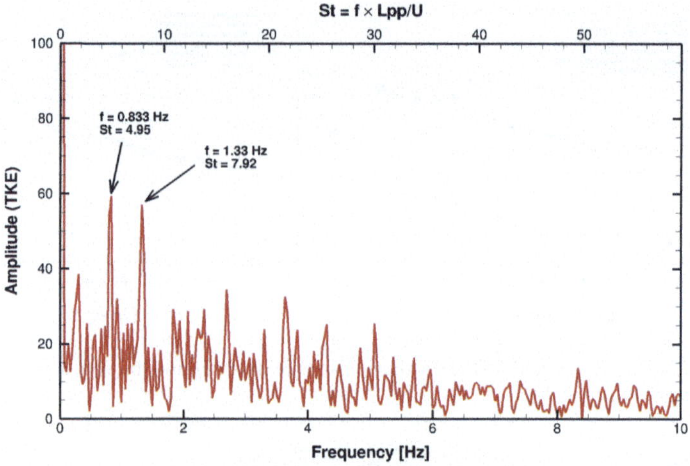

Fig. 75 Case 1-3a—FFT decomposition of TKE (from Visonneau et al. 2016)

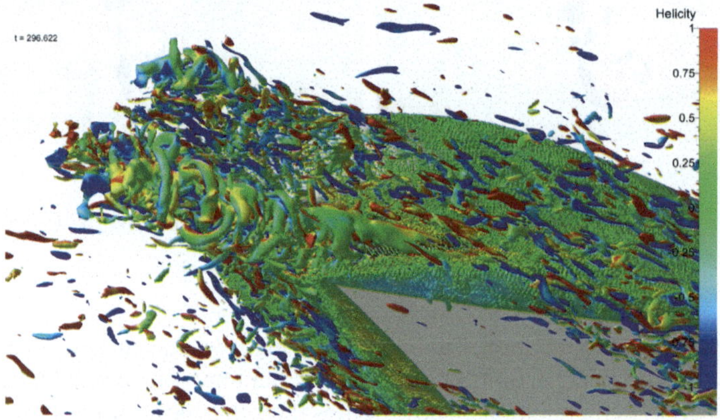

Fig. 76 Case 1-3a—instantaneous view of Q invariant colored by the helicity (from Visonneau et al. 2016)

= 0.102806 m), exhibits two peaks at 0.833 and 1.18 Hz, peaks which could have been captured by NMRI's experiments (6 Hz).

To understand the origins of this large scale unsteadiness, one should refer to Fig. 76 which gives an instantaneous view of the iso-surfaces of the Q invariant colored by the helicity. The figure clearly shows a succession of ring vortices which are created after the onset of an open separation linked with the initial thickening of the boundary layer. This large scale unsteadiness is likely to be due to the peculiar design of JBC ($C_B = 0.858$). The rapid reduction of the hull sections at the stern, reflected by the high value of C_B, creates the condition of an open separation followed by a flow reversal and a strong unsteadiness revealed by the shedding of ring vortices.

In chapter "Experimental Data of Resistance, Sinkage, Trim, Self-propulsion Factors, Longitudinal Wave Cut and Detailed Flow for JBC With and Without an Energy Saving Circular Duct", several additional experiments are compared in order to assess the existence of this high level of turbulence kinetic energy in the core of this longitudinal vortex. Experiments conducted on the same hull by Osaka University and TUHH (in a wind tunnel) can be compared to check if similar observations are made concerning the value reached by TKE in the core of the bilge vortex. Unfortunately, these measurements are not made at the same Reynolds number. Figure 77 taken from chapter "Experimental Data of Resistance, Sinkage, Trim, Self-propulsion Factors, Longitudinal Wave Cut and Detailed Flow for JBC With and Without an Energy Saving Circular Duct" shows the distribution of the squared root of TKE normalized by the friction velocity extracted along the same horizontal line at the shaft height at the same cross-section S4 for the three experiments mentioned above (see Fig. 78 for the location of the line of extraction). Moreover, data coming from the measurements on the KVLCC2 (Lee et al. 2003) approximately at the propeller plane are added to see if one can observe a generic behaviour. Figure 78 shows the location of the line of extraction (in red) for each experiment. The distributions are plotted as $\frac{\sqrt{tke}}{u_\tau}$ versus y^+. Also plotted is the same scaled data from the propeller plane of KVLCC2. The data is taken from the wind tunnel hot-wire measurements (Lee et al. 2003) where Reynolds number is $Re = 4.610^6$. The wall distance y is measured from the shaft edge and the friction velocity u_τ is estimated from the local c_f of a turbulent boundary layer of a flat plate using the formula of Prandtl-Schlichting as below:

$$c_f = (2.0log_{10}(Re_x) - 0.65)^{-2.3} \tag{1}$$

with $u_\tau/U = \sqrt{c_f/2}$.

One clearly observes an increase of the peak value of normalised TKE when the Reynolds number increases. This trend is correctly captured by a posteriori computations performed on the JBC by ECN/CNRS with the ISIS-CFD code (see Fig. 79). This figure shows a comparison between JBC experiments and two computations based either on the anisotropic EARSM turbulence closure or on the hybrid RANS/LES DES-SST turbulence model. Although the location of the peak is slightly shifted outwards compared to the experiments, it is worthwhile to notice that only the unsteady DES-SST closure is able to capture the right peak level of TKE, thanks to the resolved large scale unsteadiness present at this location, while the statistical EARSM model fails to capture this local peak of TKE. Figure 80 shows the isowake contours at the same S4 section used to generate the previous figures, both for NMRI experiments and ECN/CNRS computations using EARSM and DES-SST turbulence models. Black color is used for the NMRI experiments while blue (resp. red) color is used for the EARSM (resp. DES-SST) simulations. First of all, one can notice that the line of extraction does not pass through the center of the vortex for the computations, a location where the peak value of TKE is reached in NMRI experiments. Moreover, the agreement between NMRI measurements and both simulations is good,

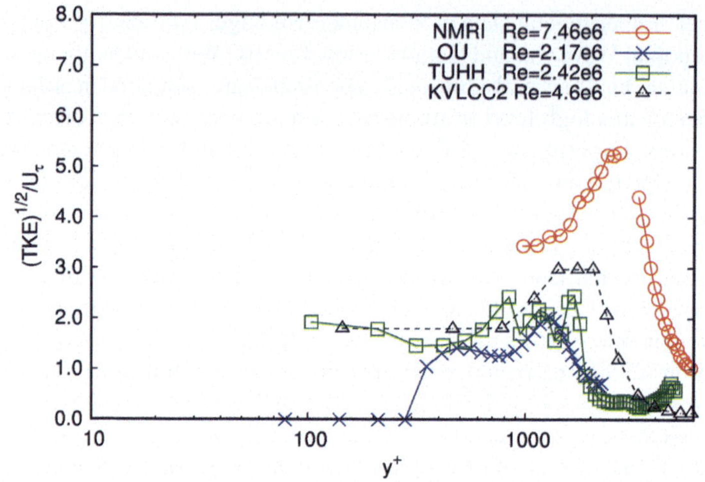

Fig. 77 Case 1-3a—normalized transversal distribution of the turbulent kinetic energy for various experiments (JBC and KVLCC2)

Fig. 78 Case 1-3a—cross-sections used to extract the TKE transversal distribution for NMRI, TUHH, OU and KVLCC2 experiments

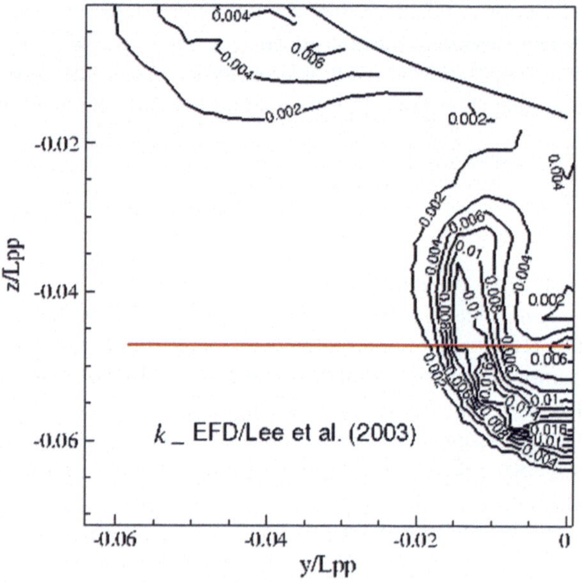

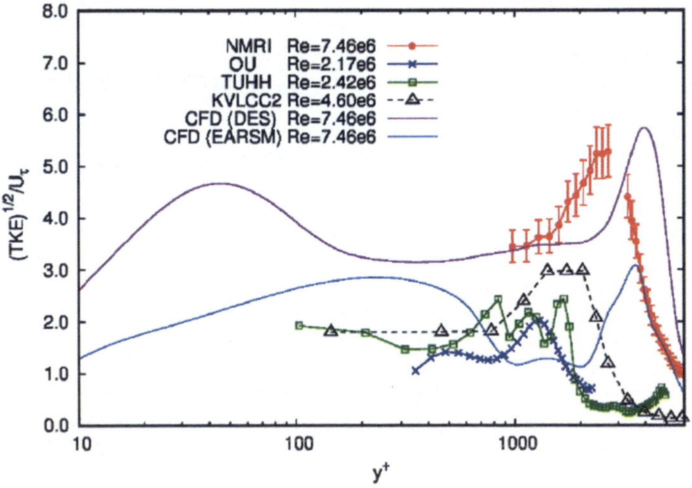

Fig. 79 Case 1-3a—normalized transversal distribution of the turbulent kinetic energy for various experiments (JBC and KVLCC2)—comparisons with ECN/CNRS-ISIS-CFD computations for the JBC test case using EARSM and DES-SST

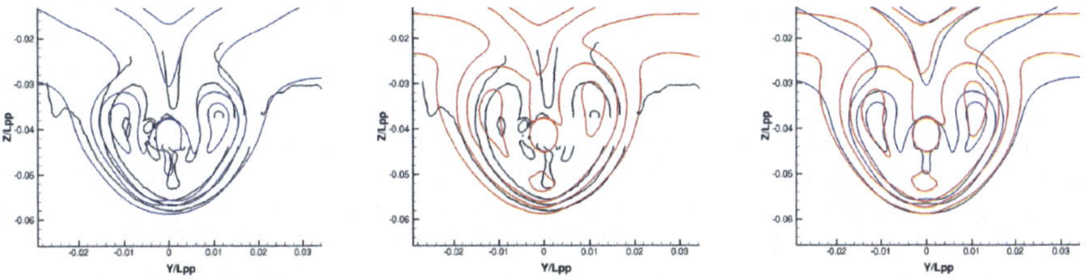

(a) NMRI experiments vs EARSM simulation

(b) NMRI experiments vs DES-SST simulation

(c) EARSM vs DES-SST simulations

Fig. 80 Case 1-3a—station S4—isowakes from NMRI experiments (black) and ECN/CNRS computations—EARSM (blue) and DES-SST (red)

which means that the two turbulence closures considered here provide a satisfactory evaluation of the velocity field. Figure 81 shows the distribution of TKE measured by NMRI with labels identifying the maximum and average values at section S4. One can also observe that the maximum value of TKE (0.05) is reached in small islands in the NMRI experiments while the first well-defined and regular iso-level surrounding these irregular islands is 0.03. Then, Fig. 82 shows the distribution of turbulence kinetic energy normalised by the square of the ship velocity at the same S4 section with the same conventions of color, as described previously. The maximum normalised TKE level is 0.05 while the minimum level is 0.01. Five isolevels are drawn. As observed previously in the transversal extraction, the TKE distribution depends strongly on the turbulence closures. The maximum TKE value reached by EARSM is 0.027 while it is 0.050 for DES-SST like in the NMRI experiments.

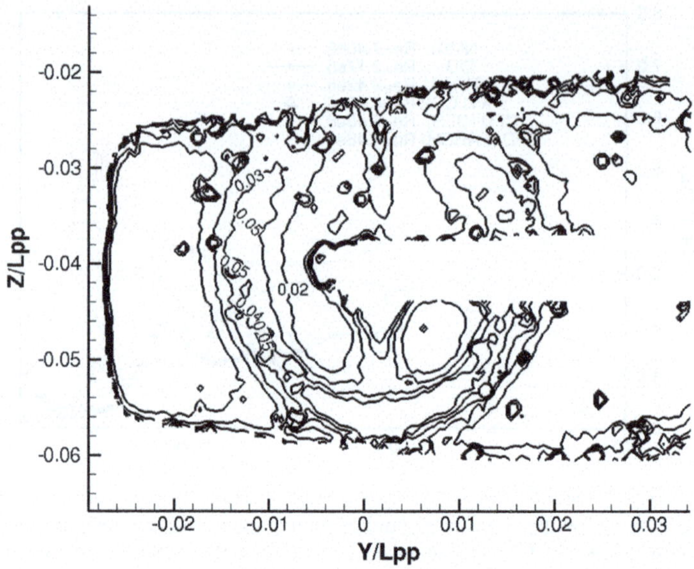

Fig. 81 Case 1-3a—station S4—isowakes from from NMRI experiments

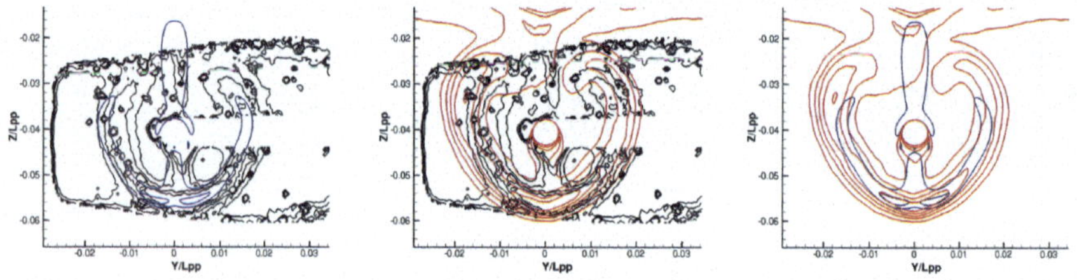

(a) NMRI experiments vs EARSM simulation *(b) NMRI experiments vs DES-SST simulation* *(c) EARSM vs DES-SST simulations*

Fig. 82 Case 1-3a—station S4—isowakes from from NMRI experiments (black) and ECN/CNRS computations—EARSM (blue) and DES-SST (red)

3.8.6 Temporary Conclusions on the Core Vortex Analysis

This first analysis of the flow in the core of the main bilge vortex was surprisingly interesting, despite the known limitations due to the absence of 3D local PIV measurements. First of all, it appears necessary to use locally refined grid in the center of the bilge vortex to capture more accurately the flow physics. The grids employed in the computations were probably locally too coarse to capture accurately the detailed physics. Despite these limitations, the agreement between most of the computations and the rather coarse experiments was satisfactory in terms of global trends. The longitudinal vorticity was somewhat underestimated and the longitudinal velocity distribution was fairly reproduced. These experiments pointed out very large differences on the turbulence kinetic energy and this was the most striking (and unexpected) result of this comparison. The linear or non-linear statistical models were unable to reproduce this result, providing TKE results which were three to ten times smaller than what was measured. Only the unsteady hybrid RANS-LES model from University of Rostock produced results in remarkably good agreement with the NMRI measurements, thanks to the major contribution of the resolved turbulence kinetic energy. Whether the NMRI TKE measurements are reliable or not is still a matter of intense debate but the remarkable agreement shown by the blind URO hybrid RANS-LES computations (confirmed one year later by ECN/CNRS computations) and the typical frequency of occurence of the large scale structures contained in the core of the bilge vortex, seem to indicate that NMRI's measurements correctly represented the local large scale unsteadiness associated with the onset and progression of a vortex along a hull with such a large block-coefficient hull. The additional experiments provided by Osaka University and Technical University of Hamburg-Harburg compared with NMRI results seem to plead in favor of a Reynolds number influence on the peak value of TKE. However, further much more detailed measurements will be necessary to draw safer conclusions.

3.9 *Case 1-3a —Concluding Remarks*

On the basis of this first analysis based on some representative examples taken from the numerous contributions, one can draw the following conclusions:

- With the typical grids which are used by the contributors (5–10 M points), the influence of the grid discretisation is moderate for RANSE, which means that the numerical error appears to be under control.
- As noticed during the Gothenburg2010 workshop for the analysis of the local flow around the KVLCC2, the major influence comes from the turbulence closure. Linear isotropic closures significantly under-predict the longitudinal vorticity at S2 while full RSM closures slightly over-predict it at the same station.
- Non-linear anisotropic closures offer a good compromise from the standpoint of the local flow although they slightly underpredict the vorticity at S2.
- New results from hybrid RANS-LES are promising but IDDES seems to over-predict the vorticity again. The agreement with the isowakes distribution is worse than what is achieved with the best RANS statistical turbulence closures.
- Computational codes seem to be mature since one does not observe large discrepancies when the same turbulence closures on similar grids are used.
- The influence of the free-surface on the local flow is not negligible but not large.
- The influence of the wall boundary treatment on the development of the stern flow is weak.

This first analysis of the flow in the core of the main bilge vortex was surprisingly interesting, despite the known limitations due to the absence of 3D local PIV measurements. First of all, it appears necessary to use locally refined grid in the center of the bilge vortex to capture more accurately the flow physics. The grids employed in the computations were probably locally too coarse to capture accurately the detailed physics. Despite these limitations, the agreement between most of the computations and the rather coarse experiments was satisfactory in terms of global trends. The longitudinal vorticity was somewhat underestimated and the longitudinal velocity distribution was fairly reproduced. These experiments pointed out very large differences on the turbulence kinetic energy and this was the most striking (and unexpected) result of this comparison. The linear or non-linear statistical models were unable to reproduce this result, providing TKE results which were three to ten times smaller than what was measured. Only the unsteady hybrid RANS-LES model from University of Rostock produced results in remarkably good agreement with the NMRI measurements, thanks to the major contribution of the resolved turbulence kinetic energy. Whether the NMRI TKE measurements are reliable or not is still a matter of intense debate but the remarkable agreement shown by the blind URO hybrid RANS-LES computations (confirmed one year later by ECN/CNRS computations) and the typical frequency of occurence of the large scale structures contained in the core of the bilge vortex, seems to indicate that NMRI's measurements correctly represented the local large scale unsteadiness associated with the onset and progression of a vortex along a hull with such a large block-coefficient. However, further much more detailed measurements will be necessary to draw safer conclusions. The reader can refer to chapter "Experimental Data of Resistance, Sinkage, Trim, Self-propulsion Factors, Longitudinal Wave Cut and Detailed Flow for JBC With and Without an Energy Saving Circular Duct" for a more detailed discussion and comparison between NMRI experiments and two other experiments conducted later on in a towing tank on a smaller model at Osaka University and in a wind tunnel on a double body JBC by Technical University of Hamburg-Harburg (TUHH).

4 Test Case 1-3b

4.1 Description of Experimental Results

Table 4 provides a comparison between the two experimental conditions, namely cases 1-3a and 1-3b. The wind tunnel experiments are conducted on a double body while the free-surface is taken into account in the towing tank. But the Froude number is low and one does not think that the free-surface deformation influences the flow at the experimental locations. More significant is the difference in terms of Reynolds number, the TUUH experiments having been performed at a Re which is 2.72 times smaller.

Figure 83 shows the locations of the experimental stations and a global view of the longitudinal vortex developing in the stern part of the ship, represented by an iso Q surface colored by the normalized helicity.

Table 4 Case 1-3b—flow conditions for the towing tank experiments (NMRI) and wind-tunnel tests (TUHH)

	NMRI	TUHH
Type	Towing tank (calm water)	Wind tunnel
Model	Single body	Double body
Scale	1:40	1:80
Lpp (m)	7.00	3.513
V0 (m/s)	1.179	11.8
Re	$7.46\ 10^6$	$2.74\ 10^6$
Fr	0.142	–
ρ (kg/m^3)	998.2	1.2
ν (m^2/s)	$1.107\ 10^6$	$1.5\ 10^5$

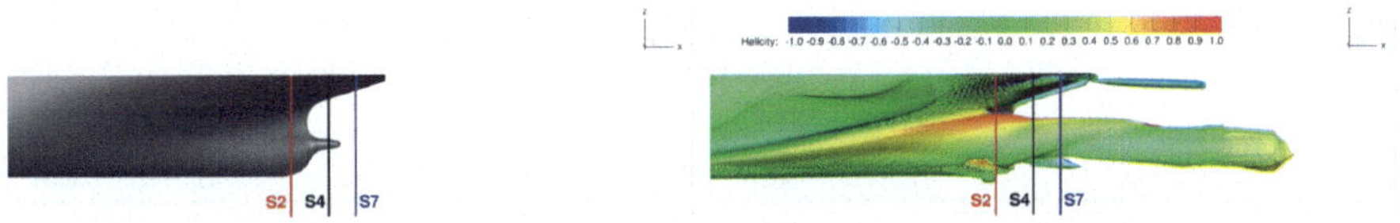

(a) Hull (b) Flow

Fig. 83 Case 1-3b—view of the locations of the three experimental stations S2, S4 and S7 and of the main longitudinal vortex visualized with the second invariant iso-surface

To understand the difference between NMRI and TUHH experiments, Figs. 84, 85 and 86 show the experimental isowakes measured by NMRI and TUHH at stations S2, S4 and S7. Globally, the same averaged intense bilge vortex is visible in both experiments. However, the experiments conducted at TUHH reveal a flow characterised by a slightly less intense longitudinal vortex since the isowakes 0.3 and 0.2 are not visible at station S2 in TUHH measurements. Similarly, at station S7, the isowake 0.4 is way less developed in TUHH than in NMRI measurements.

4.2 *Review of Contributions*

Four participants provided results for the test case 1-3b. The main characteristics of their contributions are listed in Table 5 in terms of turbulence closures, wall treatments and discretization. ECN/CNRS provided two different contributions with and without the automatic grid refinement functionality while FOI showed a comparison between RANS and LES with different wall treatments. Finally, SRC provided a reference wall-resolved LES computation on a grid composed of 4.9G cells.

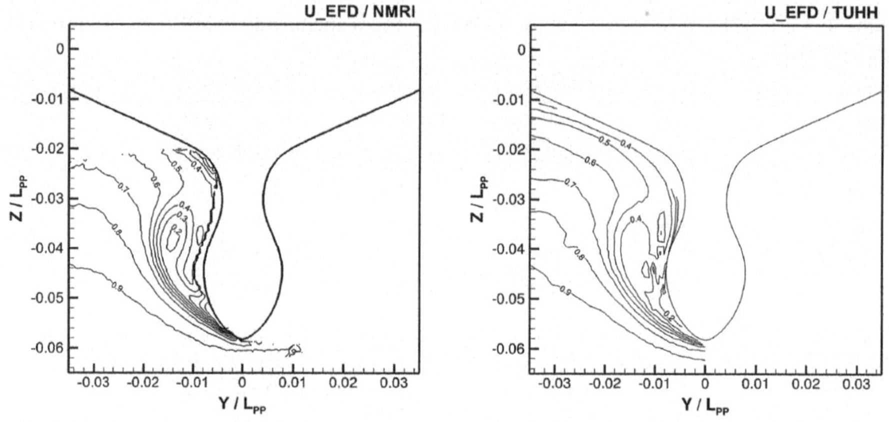

(a) *NMRI* (b) *TUHH*

Fig. 84 Case 1-3b—station S2—case without duct—experimental isowakes from NMRI and TUHH

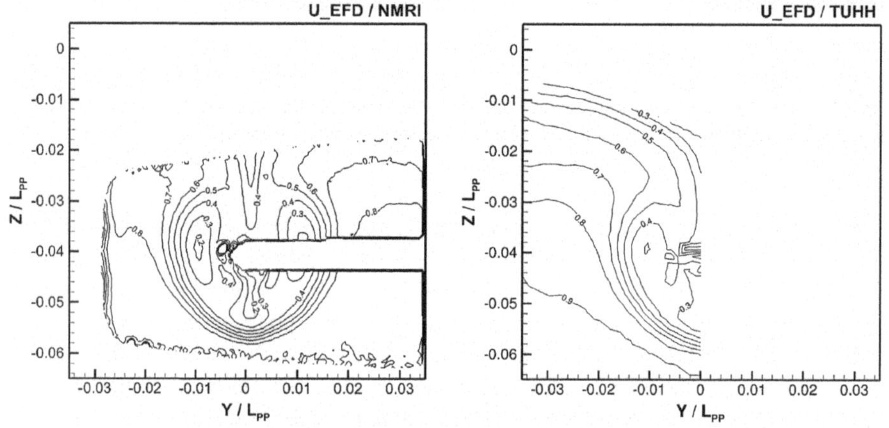

(a) *NMRI* (b) *TUHH*

Fig. 85 Case 1-3b—station S4—case without duct—experimental isowakes from NMRI and TUHH

4.3 Grid Sensitivity

As for the case 1-3a, the grids used by the three participants can be classified into four categories, based on their sizes:

- Not So Fine: Ncell < 2M cells (None)
- Fine: 2M cells < Ncell < 10M cells (ECN without AGR)
- Very Fine: 10M cells < Ncell < 50M cells (ECN with AGR)
- Tremendously Fine: Ncell > 50M cells (FOI (150M cells), SRC (4.9G cells !!!))

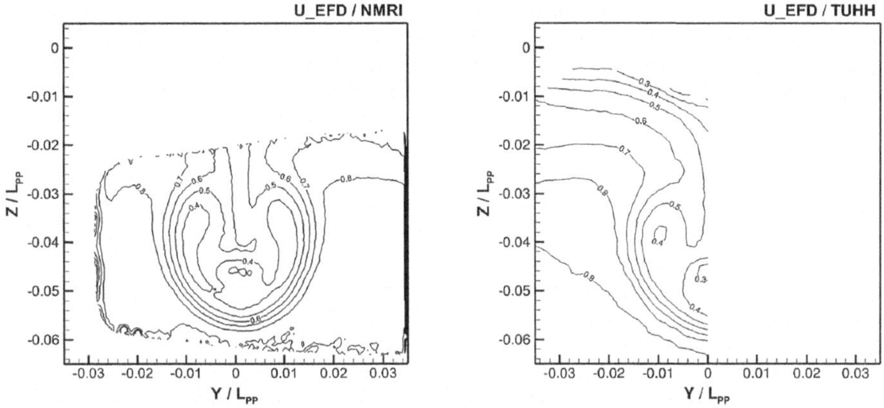

(a) NMRI (b) TUHH

Fig. 86 Case 1-3b—station S7—case without duct—experimental isowakes from NMRI and TUHH

Table 5 Case 1-3b—main characteristics of the computations

Organization code name	Turbulence model	Wall model	Grid characteristics
ECN-CNRS ISIS-CFD aka FINE^{TM}/Marine	SST/EARSM	Wall function Wall resolved	Unstructured grid 5.7/9.2M cells
FOI OpenFOAM	LES RANSE/LES	Wall modeled Wall resolved	Unstructured grid 143M cells
SRC FrontFlow/blu	LES	Wall resolved	Unstructured grid 4.9B cells

First of all, one can start to study the grid sensitivity by comparing the results provided by ECN/CNRS with or without local grid adaptation. Since the same code and same turbulence model is used, such an internal comparison should provide unbiased informations on the grid sensitivity which are therefore easier to interprete. The automatic grid refinement (AGR) used in these computations is based on the Hessian of the convective fluxes (see Wackers et al. 2014), which increases significantly the grid density in the regions of high shear stress. Figures 87, 88 and 89 show the isowake distribution at stations S2 to S7 with and without activating the automatic grid refinement. With AGR, at station S2, the intensity of the longitudinal is clearly increased as it is visible through the addition of the closed isowake 0.2 in the core of the averaged bilge vortex. The same trend is observed at station S4 with the addition of an isowake 0.1 when AGR is activated. It is also worthwhile to notice that the flow close to the bottom part of the vertical plane of symmetry is significantly modified by the addition of grid points.

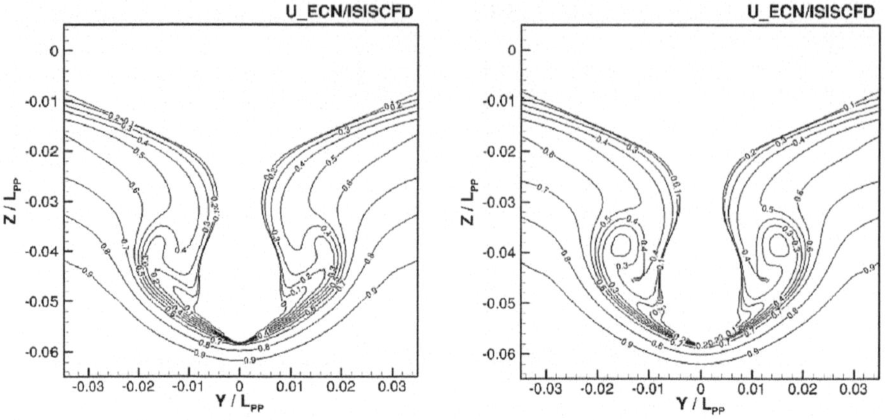

ECN_CNRS-ISISCFD-noAGR-1.3b

(a) Without AGR

ECN_CNRS-ISISCFD-AGR-1.3b

(b) With AGR

Fig. 87 Case 1-3b—grid influence—station S2

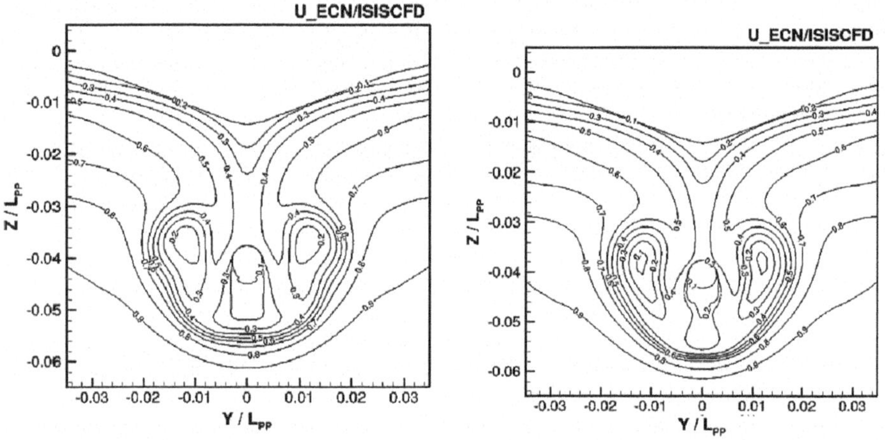

ECN_CNRS-ISISCFD-noAGR-1.3b

(a) Without AGR

ECN_CNRS-ISISCFD-AGR-1.3b

(b) With AGR

Fig. 88 Case 1-3b—grid influence—station S4

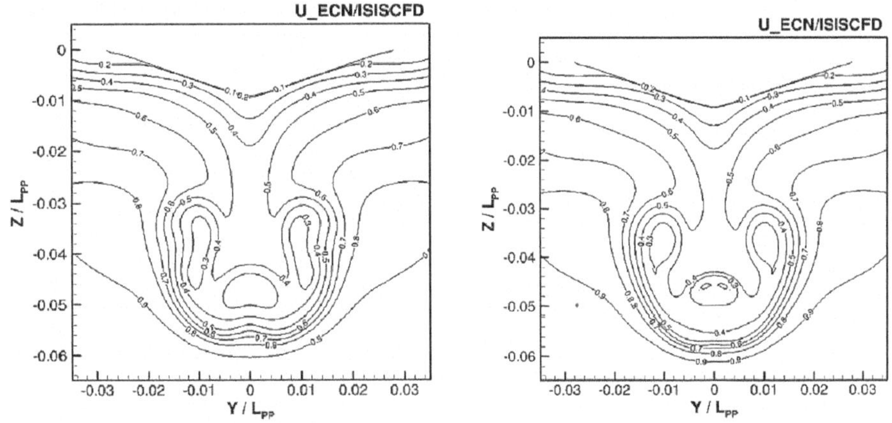

ECN_CNRS-ISISCFD-noAGR-1.3b

(a) *Without AGR*

ECN_CNRS-ISISCFD-AGR-1.3b

(b) *With AGR*

Fig. 89 Case 1-3b—grid influence—station S7

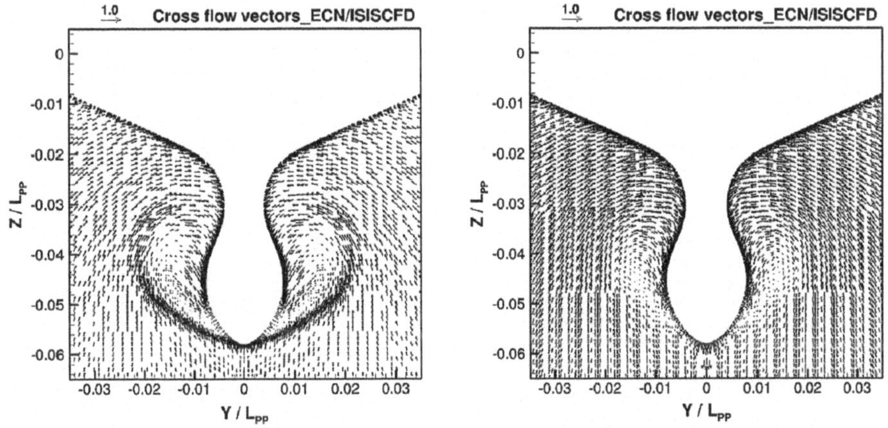

ECN_CNRS-ISISCFD-AGR-1.3b

ECN_CNRS-ISISCFD-noAGR-1.3b

Fig. 90 Case 1-3b—grid influence—station S2

Figures 90, 91 and 92 show the distribution of the secondary velocity components (v and w) with and without automatic grid refinement. One can indirectly sees the regions with high shear-stress where points are automatically added.

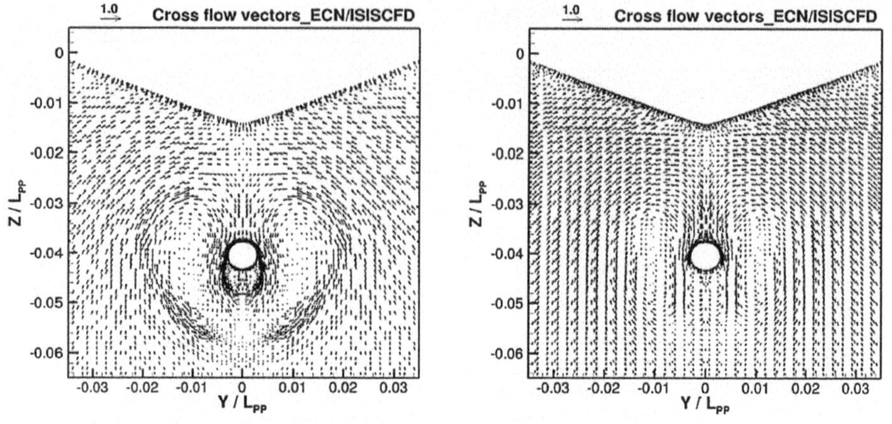

ECN_CNRS-ISISCFD-AGR-1.3b ECN_CNRS-ISISCFD-noAGR-1.3b

Fig. 91 Case 1-3b—grid influence—station S4

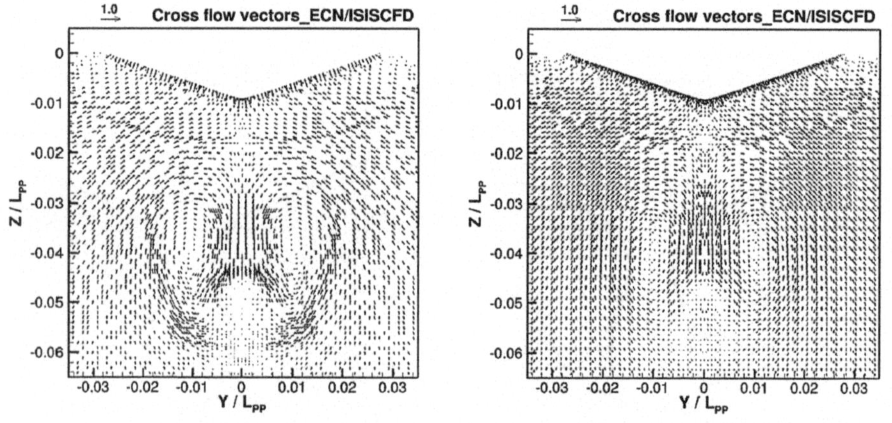

ECN_CNRS-ISISCFD-AGR-1.3b ECN_CNRS-ISISCFD-noAGR-1.3b

Fig. 92 Case 1-3b—grid influence—station S7

4.4 Wall Treatment

Figures 93, 94 and 95 show a comparison between a wall-resolved and a wall-modelled RANS (k-ω SST) solution performed by FOI. Some minor differences are visible but one can not conclude that the wall treatment has a crucial influence on the computed isowakes. One might notice a slight asymmetry in the results indicating that the computations are not fully converged, probably because of the existence of large scale unsteadiness near the stern of the ship.

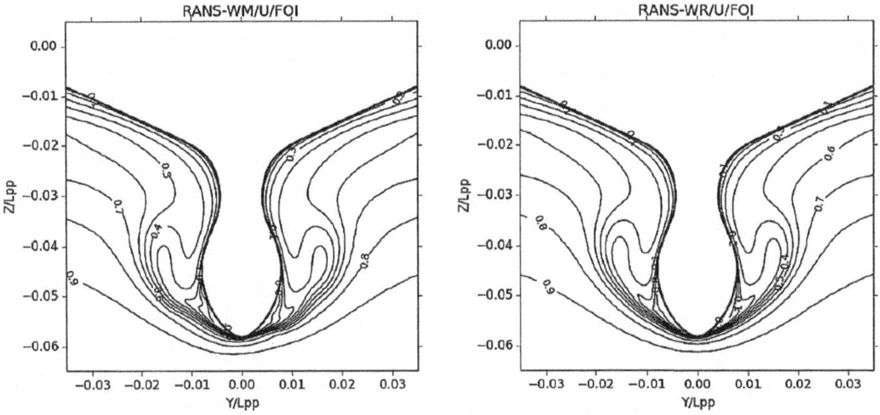

FOI-RANSWM-1.3b FOI-RANSWR-1.3b

Fig. 93 Case 1-3b—station S2—FOI RANS wall-modeled and wall-resolved solutions

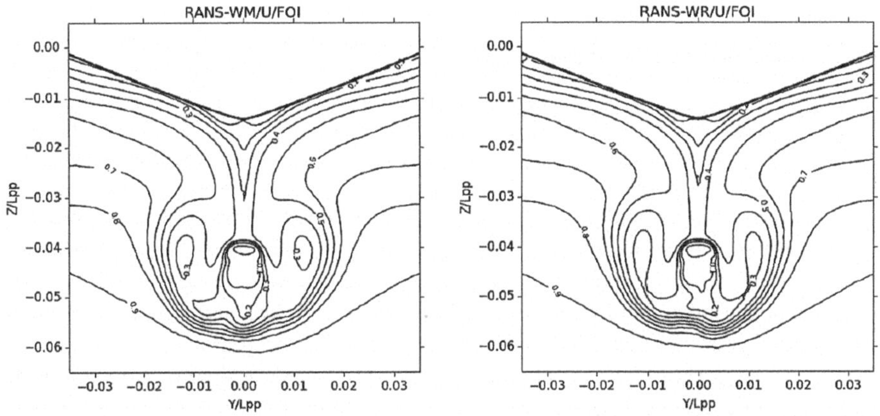

FOI-RANSWM-1.3b FOI-RANSWR-1.3b

Fig. 94 Case 1-3b—station S4—FOI RANS wall-modeled and wall-resolved solutions

4.5 Turbulence Closures

For Case-1.3b, the turbulence models can be organized into three groups:

- the isotropic linear closures: FOI (k-ω SST),
- the anisotropic non-linear models: ECN-CNRS (EARSM),
- the LES models: FOI (NWM-LES), SRC (LES).

The next subsections will be devoted to a comparison of these various solutions and turbulence models mainly by assessing two different flow characteristics, the distributions of isowakes and turbulence kinetic energy at the three experimental stations S2, S4 and S7.

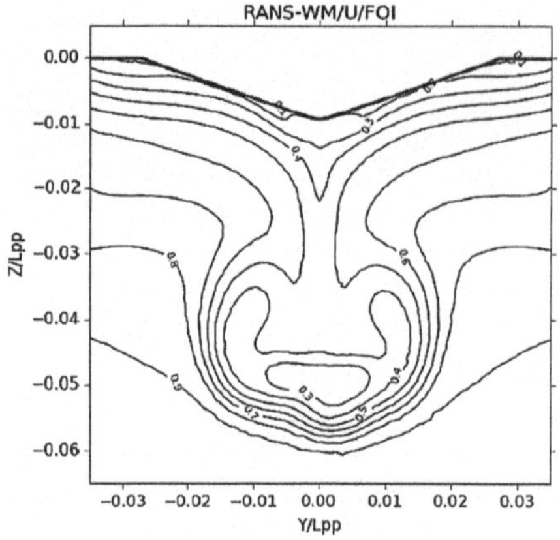

FOI-RANSWM-1.3b

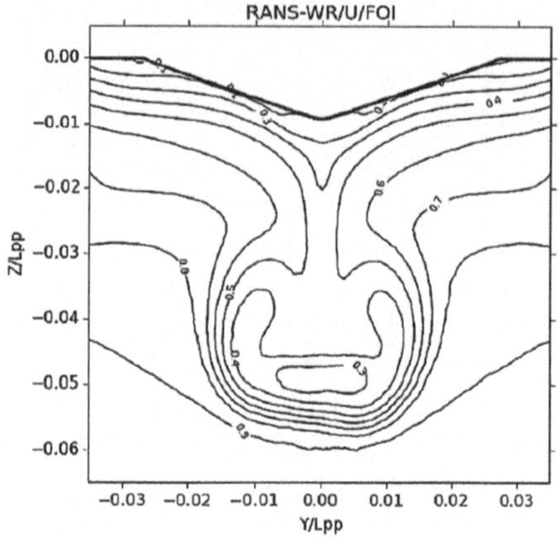

FOI-RANSWR-1.3b

Fig. 95 Case 1-3b—station S7—FOI RANS wall-modeled and wall-resolved solutions

4.5.1 Assessment of Turbulence Closures

First of all, results provided by FOI using the code Open FOAM are considered. The turbulence model is the linear isotropic k-ω SST closure equipped with a low Re near-wall formulation. Figures 96, 97 and 98 show comparisons on the isowakes between TUHH experiments and the wall-resolved FOI RANS solution. This solution appears similar to the one obtained by ECN/CNRS (see Figs. 99, 100 and 101) with EARSM and local grid adaptation. No closed contours are observed in the core of the vortex at station S2 and at station S4, U = 0.3 is the lowest iso-contour which is captured in the core while ECN/CNRS captures U = 0.2. A similar trend is observed at station S7. Based on the TUHH measurements, it is not possible to draw any firm conclusion about the best agreement in the core of this vortex. Despite the use of a locally finer grid generated by AGR for ECN/CNRS, it is believed that the difference between FOI and ECN/CNRS RANS simulations is to be mainly attributed to the use of the non-linear anisotropic EARSM turbulence closure by the French team, a closure which is known to lead to more intense longitudinal vortices.

Interestingly, several teams proposed results based on various LES turbulence models. FOI provided wall-modeled LES computations on a grid composed of 150M cells while SRC proposed a remarkable wall-resolved LES simulation computed on a grid made of 4.9 G cells. Figures 102, 103, 104, 105, 106 and 107 show comparisons on the iso-wake contours between these two results at stations S2 to S7. First of all, one may notice that both computations predict a more intense longitudinal vortex, as indicated by the lower values of the isowake in the core. At S2, the vortex predicted by FOI (with a closed contour U = 0.1 in the core) appears way too large compared to the measurements and the SRC LES result. It is impossible to know whether this might be attributed to a mesh which is too coarse, too large a time step, too small an averaging period or to the influence of the wall model. Reassuringly, the wall-

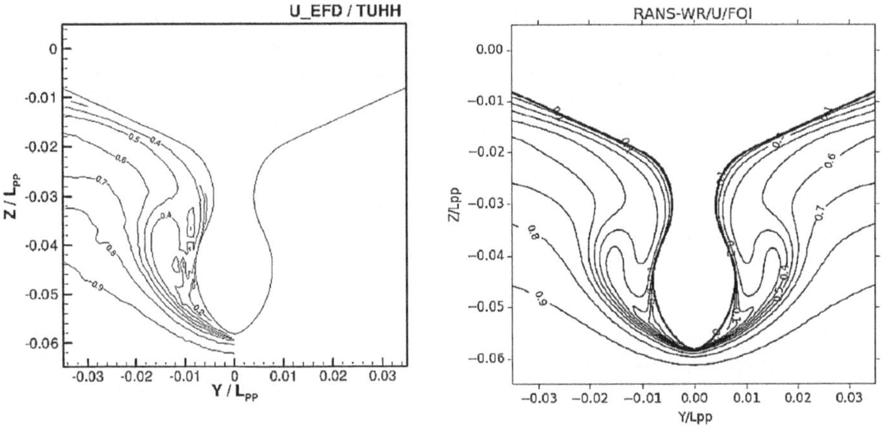

EFD(TUHH) FOI-RANSWR-1.3b

(a) Experiments *(b) k-ω SST*

Fig. 96 Case 1-3b—station S2—results from FOI-Open Foam

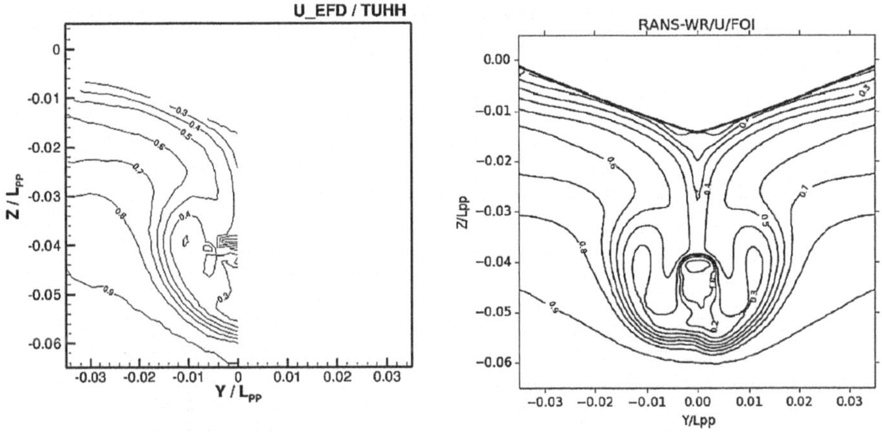

EFD(TUHH) FOI-RANSWR-1.3b

(a) Experiments *(b) k-ω SST*

Fig. 97 Case 1-3b—station S4—results from FOI-Open Foam

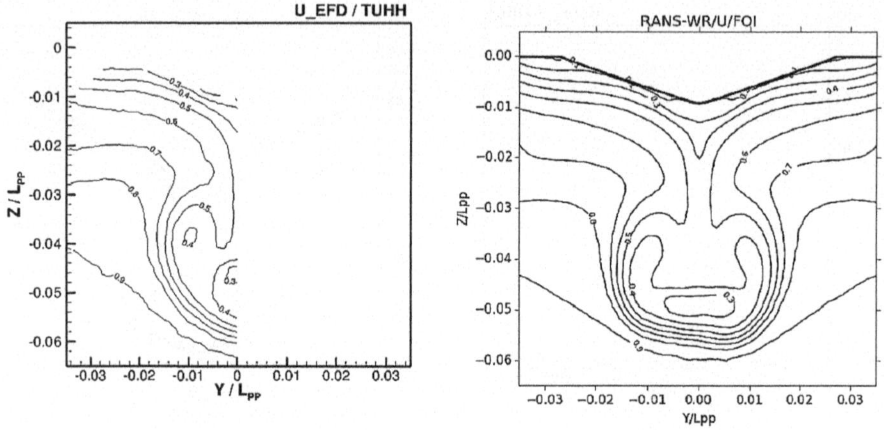

EFD(TUHH) FOI-RANSWR-1.3b

(a) Experiments *(b) k-ω SST*

Fig. 98 Case 1-3b—station S7—results from FOI-Open Foam

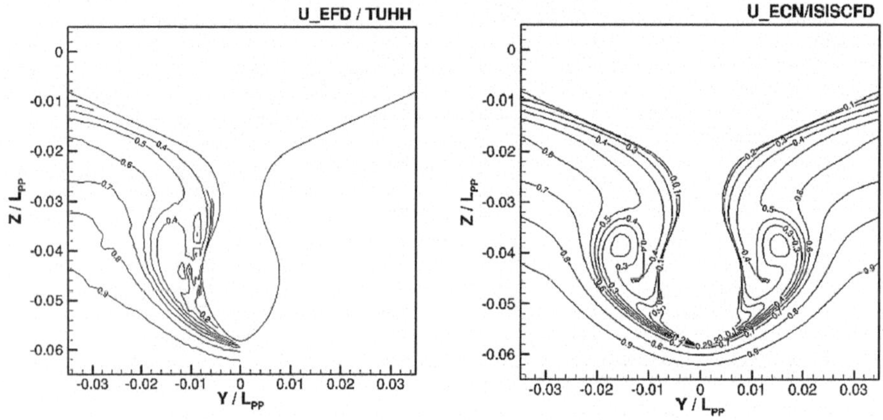

EFD(TUHH) ECN_CNRS-ISISCFD-AGR-1.3b

(a) Experiments *(b) EARSM*

Fig. 99 Case 1-3b—station S2—results from ECN-ISIS-CFD

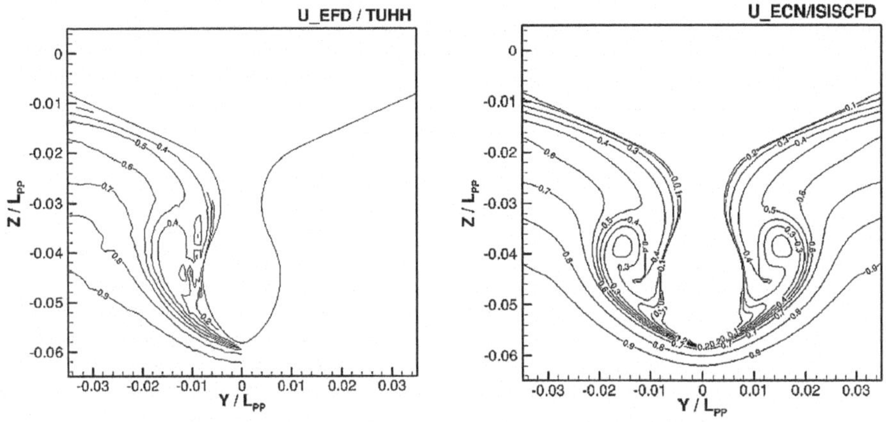

EFD(TUHH) ECN_CNRS-ISISCFD-AGR-1.3b

(a) Experiments *(b) EARSM*

Fig. 100 Case 1-3b—station S4—results from ECN-ISIS-CFD

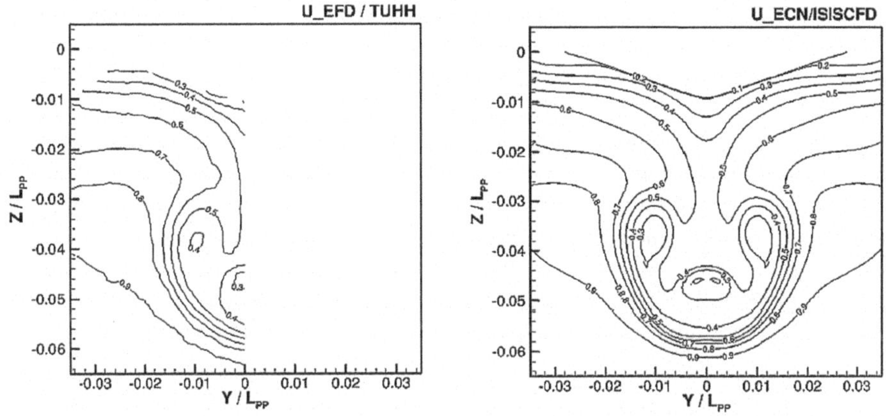

EFD(TUHH) ECN_CNRS-ISISCFD-AGR-1.3b

(a) Experiments *(b) EARSM*

Fig. 101 Case 1-3b—station S7—results from ECN-ISIS-CFD

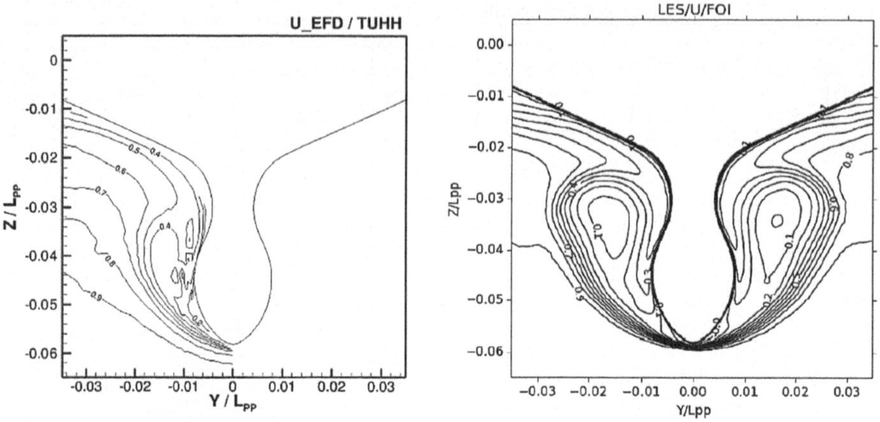

EFD(TUHH) FOI-LESWM-1.3b

(a) Experiments *(b) LES*

Fig. 102 Case 1-3b—station S2—results from FOI-OpenFoam

resolved LES SRC simulation appears in good agreement with TUHH measurements and other RANS computations at station S2. The same remarks hold for the stations S4 and S7, although LES SRC results provide a vortex which appears more intense than what is measured at station S7 as indicated by the comparison on the isowake U = 0.4. However, the solution provided by the wall resolved LES SRC computation appears to agree globally very well with both TUHH measurements and RANS computations. A comparison limited to the iso-U contours does not do justice to the LES simulations since this flow around the JBC is not characterised by a strong unsteady detached flow. However, the SRC simulation should provide a kind of reference prediction of the Reynolds stresses and turbulence kinetic energy far much accurate than the statistical turbulence models and the current experiments.

4.5.2 Focus on the Turbulence Kinetic Energy in the Core of the Main Longitudinal Vortex

In the case 1-3a, the local core vortex analysis revealed large differences concerning the prediction of the turbulence kinetic energy between the various teams and consistently, between (U)RANS and hybrid RANS-LES turbulence closures. In particular, hybrid RANS-LES models computed significantly higher levels of TKE in the core of the main vortex, due to the contribution of the resolved velocity fluctuations. The validity of the NMRI measurements was also questioned because of a too low frequency of acquisition, although the resolved velocity fluctuations captured by every hybrid RANS-LES models led to values of TKE in good agreement with NMRI measurements.

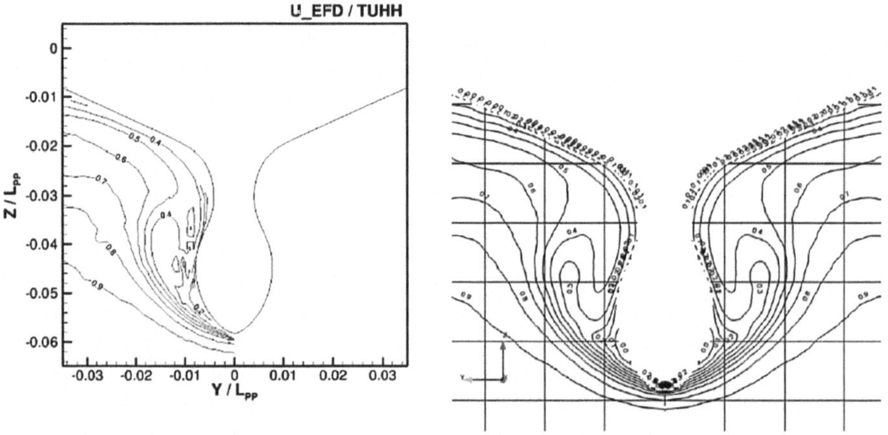

EFD(TUHH) SRC-LES-1.3b

(a) Experiments *(b) LES*

Fig. 103 Case 1-3b—station S2—results from SRC

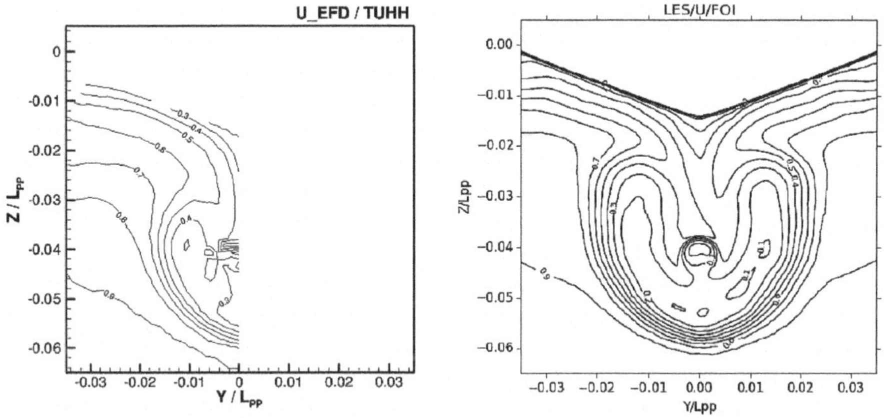

EFD(TUHH) FOI-LESWM-1.3b

(a) Experiments *(b) LES*

Fig. 104 Case 1-3b—station S4—results from FOI-OpenFoam

Although the Reynolds number is 2.72 times smaller for the case 1-3b, it is interesting to examine the levels of TKE predicted by the various simulations, and particularly, by the wall resolved LES simulation provided by SRC.

Figure 108 shows a comparison between the TKE contours computed by a statistical closure (here EARSM) and the wall-resolved LES at station S2. Globally, the iso-contour maps look similar but the extension of the region where TKE is not negligible is larger with LES than with EARSM. This is the first noticeable difference. If one looks at the maximum values of TKE, one sees that they occur for both closures

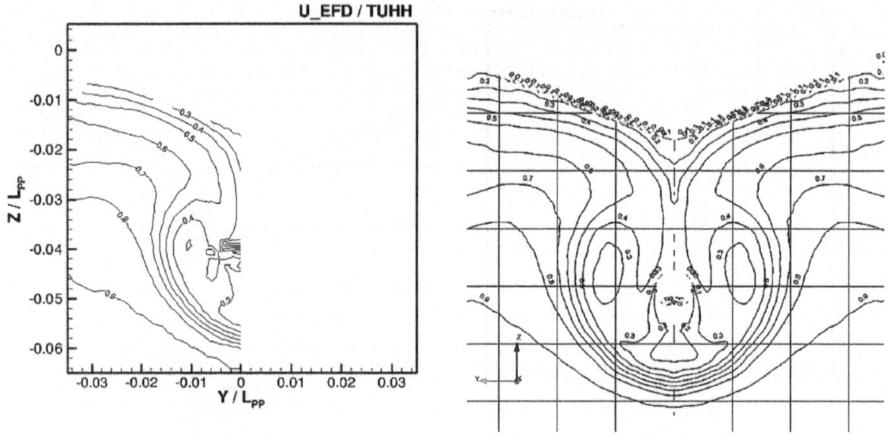

EFD(TUHH) SRC-LES-1.3b

(a) Experiments *(b) LES*

Fig. 105 Case 1-3b—station S4—results from SRC

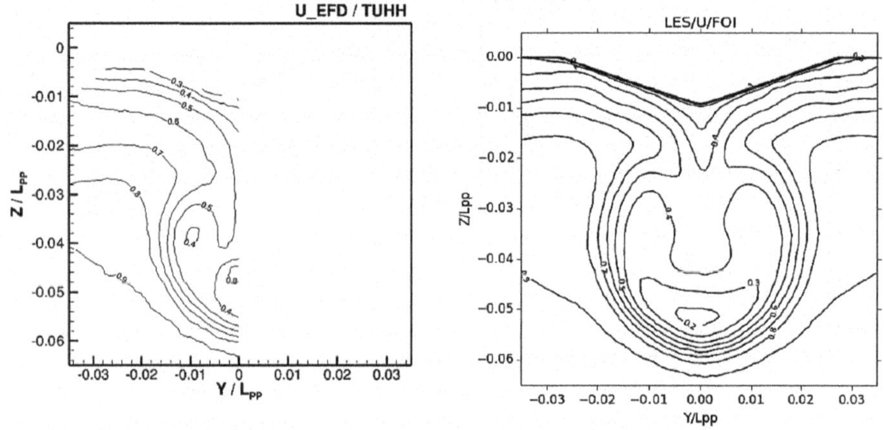

EFD(TUHH) FOI-LESWM-1.3b

(a) Experiments *(b) NWM-LES*

Fig. 106 Case 1-3b—station S7—results from FOI-OpenFoam

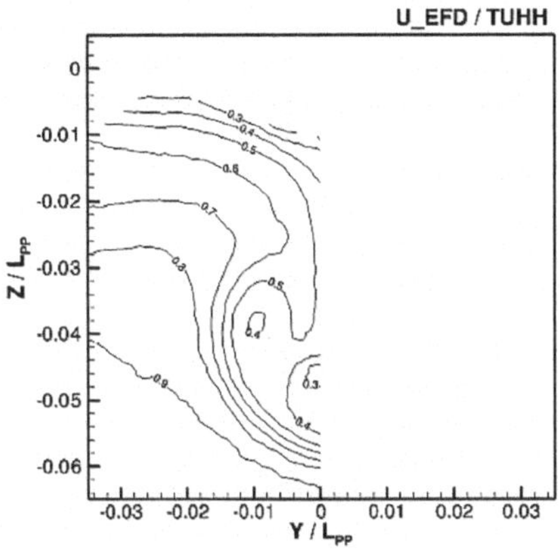

EFD(TUHH)

(a) Experiments

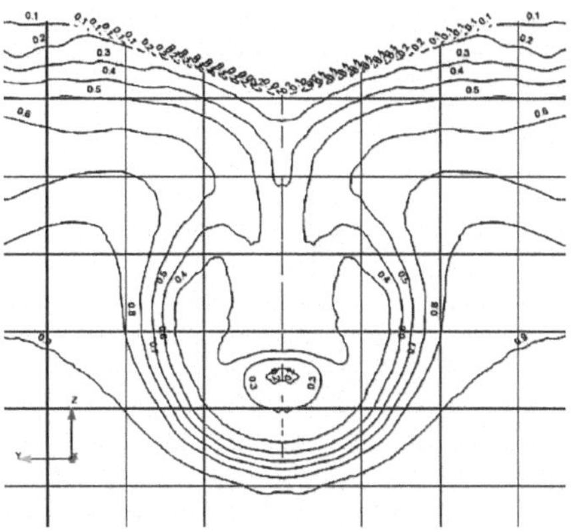

SRC-LES-1.3b

(b) LES

Fig. 107 Case 1-3b—station S7—results from SRC

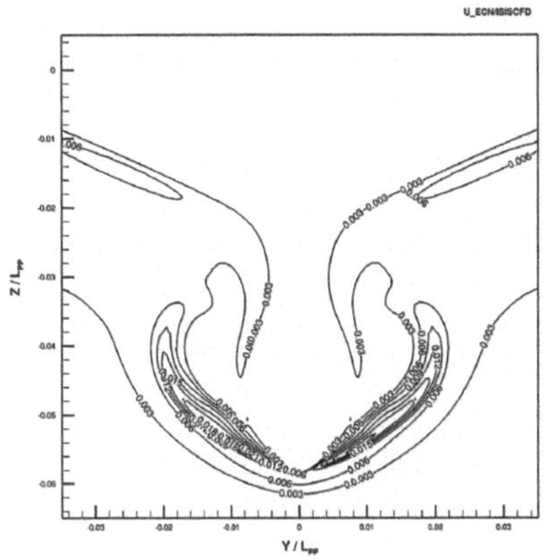

(a) ECN-CNRS/EARSM

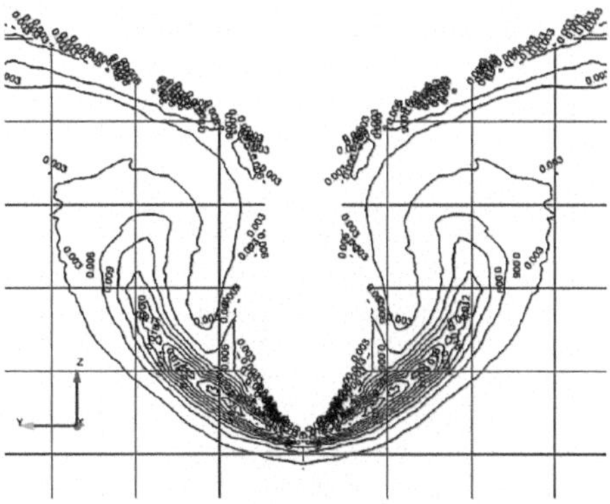

SRC-LES-1.3b(S2)

(b) SRC/LES

Fig. 108 Case 1-3b—TKE contours—station S2

at the bottom of the region a bit away from the vertical plane of symmetry. Here, the maximum value predicted by LES is around 0.022 while it is around 0.018 for EARSM. Moreover, in the region where the center of the vortex should be located, the values of TKE predicted by LES are significantly higher than what is predicted by EARSM.

Figure 109 shows a comparison between the TKE contours computed by a statistical closure (here EARSM) and the wall-resolved LES at station S4. The differences on the TKE contours which were already visible at station S2 are more marked at station S4. The region with high TKE is much more developed for LES than for RANS closures with maximum values ranging from 0.03 for LES to 0.02 for RANS.

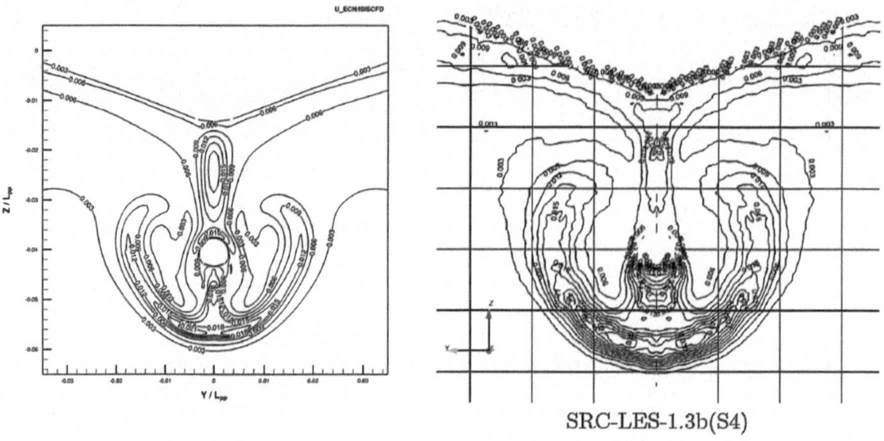

(a) *ECN-CNRS/EARSM* (b) *SRC/LES*

Fig. 109 Case 1-3b—TKE contours—station S4

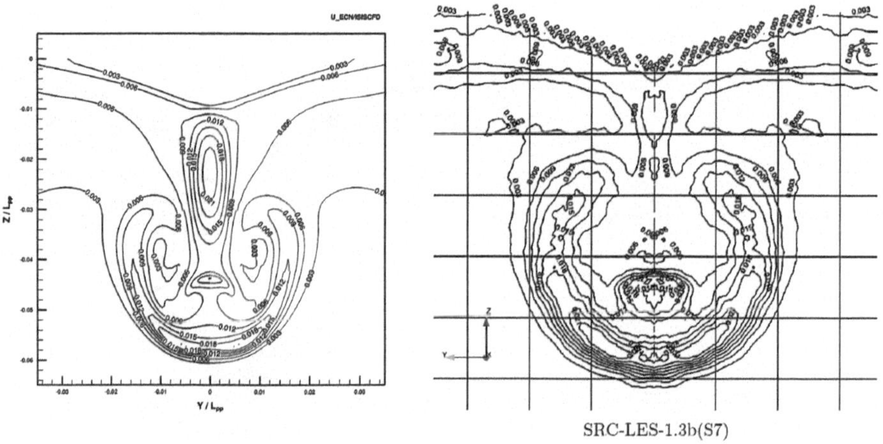

(a) *ECN-CNRS/EARSM* (b) *SRC/LES*

Fig. 110 Case 1-3b—TKE contours—station S7

Here again, in the core of the vortex, the turbulence kinetic energy computed by LES is significantly higher than what is modeled by the EARSM RANSE closure.

Finally, Fig. 110 shows a comparison between the TKE contours computed by a statistical closure (here EARSM) and the wall-resolved LES at station S7. The analysis made for station S4 can be repeated for station S7 without any alteration. In the core of the vortex, TKE predicted by LES appears to be about two times higher than what is modeled by EARSM.

Globally, the comparisons of the TKE maps at stations S2 to S7 computed by LES and RANSE closures confirm what was observed in the local vortex analysis conducted for the case 1-3a. The levels of turbulence kinetic energy simulated by LES are significantly higher than what is modeled by RANS. This suggests that, even if this flow does not exhibit any strong coherent structures as it is found for separation around bluff bodies, there is a significant amount of large scale velocity fluctuation which is resolved by LES in the core of the vortex and is maintained during its progression in the wake, that is not modelled by the RANSE closures. It is believed that this is a major flow characteristic which is associated to the large scale unsteadiness of the flow in the stern region of JBC.

4.6 Case 1-3b—Concluding Remarks

On the basis of the analysis presented above, a few conclusions can be drawn:

- A sensitivity to the local grid density is observed, which contradicts somewhat the conclusions drawn for the previous case 1-3a,
- However, the major influence comes from the turbulence closure, as expected. Linear isotropic closures under-predict the longitudinal vorticity at S2 while full RSM closures tend to over-predict it at the same station.
- For the first time, two LES computations are presented. The (almost) wall-resolved LES results from SRC predict a level of TKE in the core of the bilge vortex in agreement with NMRI's measurements and higher than what is modeled by EARSM. Results from NWM-LES are promising but this model seems to over-predict the vorticity. It would have been interesting to check TKE in the bilge vortex but this information was not available at the time of the workshop.

5 Test Case 1-4

5.1 Description of Experimental Results—Influence of the Duct

Test case 1-4 concerns the local flow analysis around the JBC with duct but without propeller towed in the NMRI towing tank, in free trim and sinkage conditions. The location of the experimental stations with respect to the duct are shown in Figs. 111 and 112 shows the stern equipped with its duct. Figures 113, 114, 115, 116, 117 and 118 show the influence of the duct on the experimental iso-wakes and iso-contours of the two crossflow velocity components at three sections S2, S4 and S7 already defined before.

At station S2, no significant influence of the duct can be noticed since the slight differences observed on the isowakes close to hull might be attributed to measurement uncertainties. This indicates that the upwind influence of the duct is confined to a close neighborhood since no modification is visible on the flow velocity distribution at station S2. Station S4 is located in the close wake of the duct and consequently,

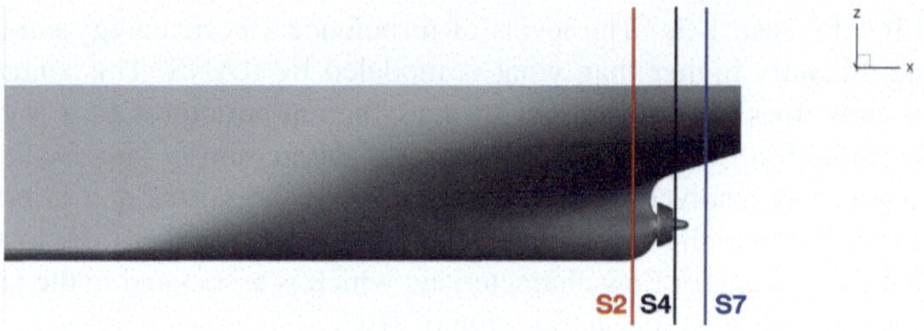

Fig. 111 Case 1-4—appended hull geometry and experimental stations

Fig. 112 Case 1-4—view of the stern equipped with its duct

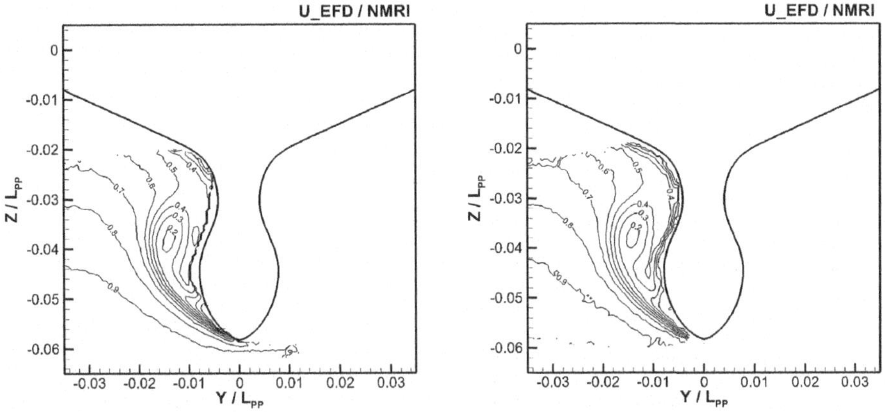

(a) *Without duct* (b) *With duct*

Fig. 113 Case 1-4—experiments—influence of the duct on the isowakes distribution at station S2

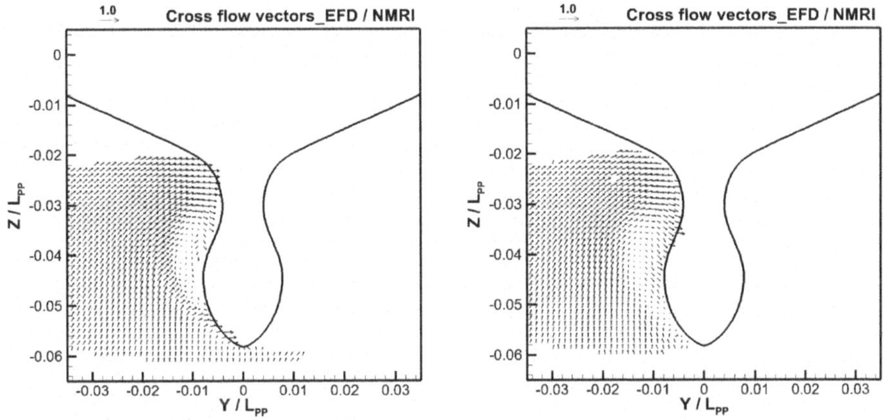

(a) *Without duct* (b) *With duct*

Fig. 114 Case 1-4—experiments—influence of the duct on the secondary velocities distribution at station S2

one can notice large differences on the flow velocity distribution due to the presence of the duct. First of all, the duct does not remove the longitudinal vortex as indicated by the hook shape of the isowakes behind the duct. The location of the core of the longitudinal vortex is moved down and closer to the vertical plane of symmetry but the vortical intensity does not seem to have been significantly modified by the duct, if one looks at the distortion of the isowakes. Correlated with the vortex translation, the flow appears slightly accelerated in the upper part of duct's wake where the isowake $U = 0.4$ is replaced by $U = 0.5$. In the bottom part of the duct's wake, the flow appears decelerated with an extended region characterized by the new isowakes $U = 0.1$ and $U = 0.0$, which means that one can not exclude the existence of a local

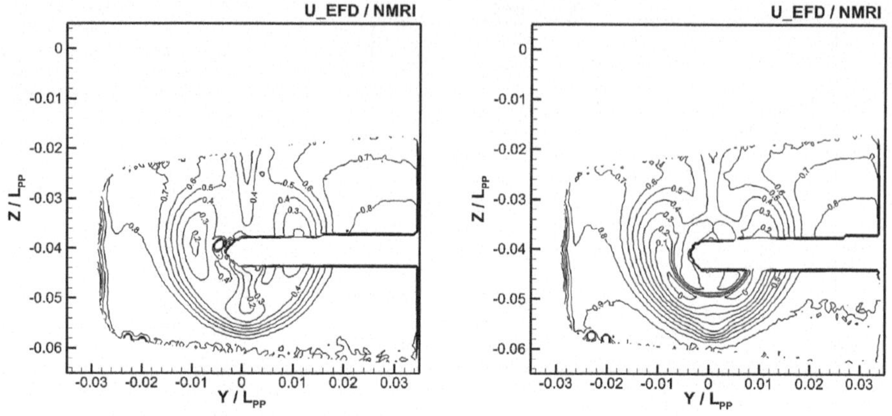

(a) Without duct (b) With duct

Fig. 115 Case 1-4—experiments—influence of the duct on the isowakes distribution at station S4

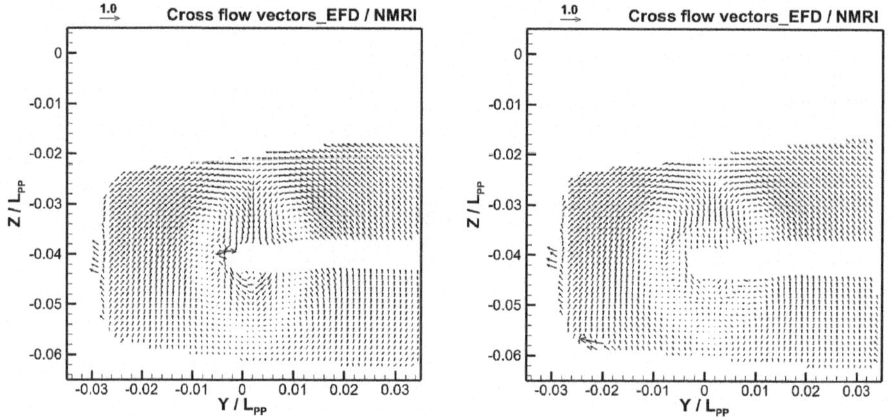

(a) Without duct (b) With duct

Fig. 116 Case 1-4—experiments—influence of the duct on the secondary velocities distribution at station S4

flow separation in the core of this vortex. Likewise, one can notice the existence of closed contour U = 0.4 in the center of the duct's wake, close to the vertical plane of symmetry. This peculiar flow region is confirmed by the distribution of the secondary velocity components. The plots at station S7 confirm the flow topology detected at station S4. Instead of a unique distorted U = 0.4 isowake, one can notice one additional iso-contour U = 0.3, which indicates that the flow in the core of the main vortex is slightly decelerated at station S7 when the duct is mounted on the hull. Accordingly, the vortex is slightly shifted down in presence of a duct.

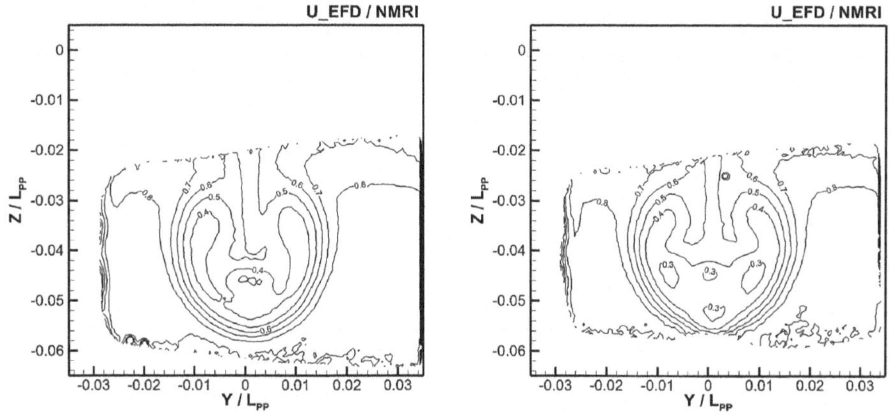

(a) Without duct *(b) With duct*

Fig. 117 Case 1-4—experiments—influence of the duct on the isowakes distribution at station S7

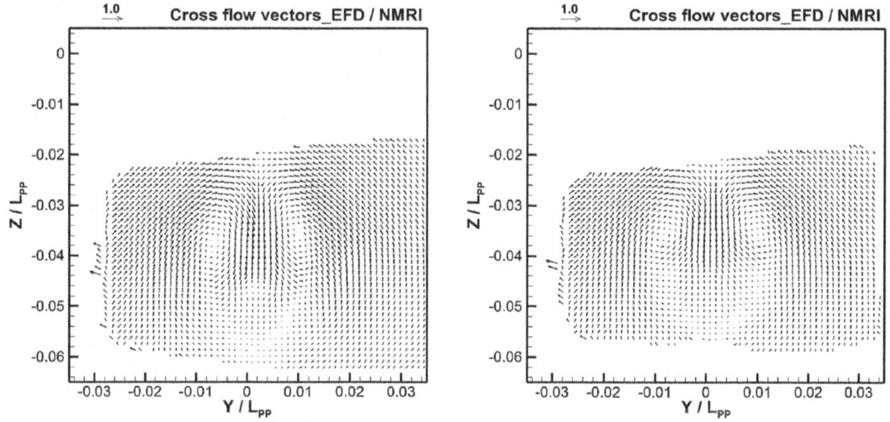

(a) Without duct *(b) With duct*

Fig. 118 Case 1-4—experiments—influence of the duct on the secondary velocities distribution at station S7

5.2 Review of Contributions

Seventeen participants provided results for the test case 1-4. The main characteristics of their contributions are listed in Table 6 in terms of turbulence closures, wall treatments and discretizations. Some participants provided several different contributions, which makes possible a comparison of various turbulence closures or wall treatments for the same grid and same code. Moreover, SOTON performed a comparison of codes (OpenFoam vs. Star-CCM+) with the same grid and same turbulence closure.

Table 6 Case 1-4—main characteristics of the computations

Organization code name	Turbulence model	Wall model	Grid characteristics
ABS OpenFOAM	k-ω SST	Wall function	Unstructured grid 5 726 752 cells
Chalmers-Flowtech Shipflow	EARSM	Wall resolved	Structured grid with overlapping 12M cells
CNR-INSEAN Xnavis	Spalart-Allmaras	Wall resolved	Structured grid with overlapping 8.6M cells
ECN/CNRS ISIS-CFD aka FINE^{TM}/Marine	SST/EARSM	Wall function Wall resolved	Unstructured grid 8.7/14.1M cells
HHI HiFOAM/StarCCM+	Realizable k-ϵ/RSTM	Wall function	Unstructured grid 1.4M cells
HSVA FreSCo+	2 equations	Wall function	Unstructured grid from 2.1 to 14.6M cells
KRISO Wavis	Realizable k-ϵ/EARSM	Wall function	Structured grid 15.5M cells
MARIC FINE^{TM}/Marine	k-ω SST	Wall function	Unstructured grid 650 000 cells
MARIN ReFRESCO	k-ω SST +DM correction	Wall resolved	Unstructured grid 6.67M cells
MHI Fluentv14	k-ϵ RNG k-ω SST	Wall resolved	Unstructured grid 14.4M cells
MIJAC OpenFOAM	k-ω SST	Wall resolved	Unstructured grid 8.5M cells
NMRI Nagisa	EARSM	Wall resolved	Structured grid 10 056 704 cells
PNU Fluentv15	2 equations k-ϵ RNG	Wall resolved Wall function	Unstructured grid 5.31M cells
SHIME OpenFOAM	k-ω SST	Wall resolved	Unstructured grid 24.6M cells
SJTU naoeOpenFOAM	k-ω SST	Wall function	Unstructured grid 2.25M cells
SOTON OpenFOAM/StarCCM+	k-ω SST	Wall resolved	Structured grid 21.3M cells
YNU SURFv7	EARSM	Wall resolved	Structured grid 3 267 584 cells

Among the seventeen participants, eleven used unstructured grids while six computed on structured grids sometimes locally enriched with overlapping patches. It is well known that the prediction of the nominal velocity in the stern region depends strongly on the turbulence closure, as soon as a reasonable dense grid is used. The turbulence models employed by the participants computing this test case can be classified into two groups: (i) anisotropic non-linear statistical turbulence modeling: Reynolds stress transport models (HHI), Explicit Algebraic Reynolds Stress Models (Chalmers, ECN/CNRS, NMRI, YNU), (ii) isotropic linear eddy-viscosity model (ECN/CNRS, HSVA, MARIN, URO, ...). Concerning the linear isotropic turbulence models, most of the contributors use the k-ω SST model but HHI, MHI and KRISO use also the realizable k-ε closure while CNR-INSEAN and Marin prefer to use a one equation model (Spalart-Allmaras or 1 eq. Menter with Dacles-Mariani modification). Therefore, in five years, we can observe a slight increase of more sophisticated turbulence closures used for ship flow computations, probably related with the number of publications illustrating the crucial role of turbulence modeling for the prediction of local flow in the stern region. All the discretization methods are formally second-order accurate and based on multi-block structured or unstructured grids which are all body-fitted, with the exception of Chalmers and CNR-INSEAN which are using an overset approach combining structured grids to reach an optimal grid density in the stern region.

5.3 *Influence of the Turbulence Closures*

5.3.1 Linear Isotropic Turbulence Closures

Starting the analysis of turbulence closures by the one equation models, Figs. 119, 120 and 121 provide the results with the Spalart-Allmaras model. Clearly, the longitudinal vorticity is strongly under-estimated as illustrated by the open iso-contour $u = 0.2$ and the wrong shapes of $u = 0.3$ and $u = 0.4$ at station S2. These trends are confirmed at stations S4 and S7. The results provided by MARIN which employs also a k-ω SST model with Dacles-Mariani correction, are very different as illustrated by Fig. 122 for station S2. Here, the longitudinal vortex is far more intense, even if the iso-contour $u = 0.2$ appears to be too extended, which indicates an over-estimated longitudinal vorticity. The same trend is recognizable at station S4 (resp. S7) where the iso-contour $u = 0$ and the reversed flow region (resp. $u = 0.3$) appear also significantly over-estimated (see Figs. 123 and 124). It is interesting to notice that these two results obtained with a one equation model do not show at all the same trend, which perhaps indicates that the Dacles-Mariani correction plays a crucial role here.

To illustrate the behaviour of two equation linear isotropic models, the results from HSVA using a linear EARSM model and SHIME using a more classical k-ω SST closure have been retained to illustrate the general trend observed with these turbulence closures. Globally, the intensity of the longitudinal vortex is under-estimated as indicated by the iso-contours in the core of this vortex at station S2 (see Figs. 125

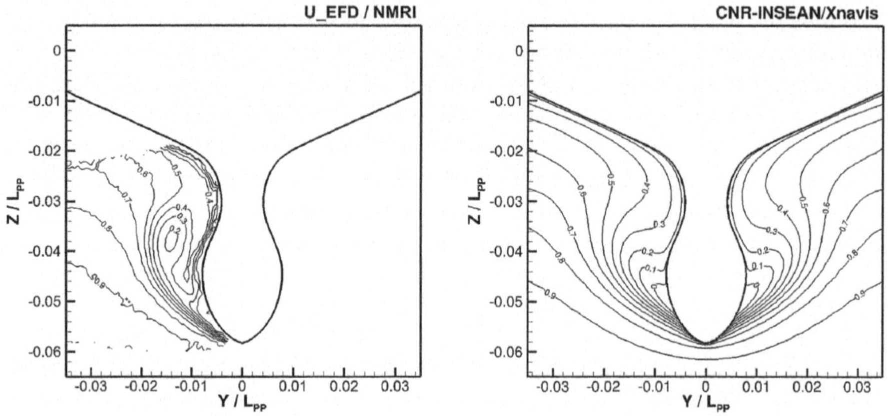

(a) Experiments *(b) INSEAN/Spalart-Allmaras*

Fig. 119 Case 1-4—linear isotropic turbulence closures—station S2

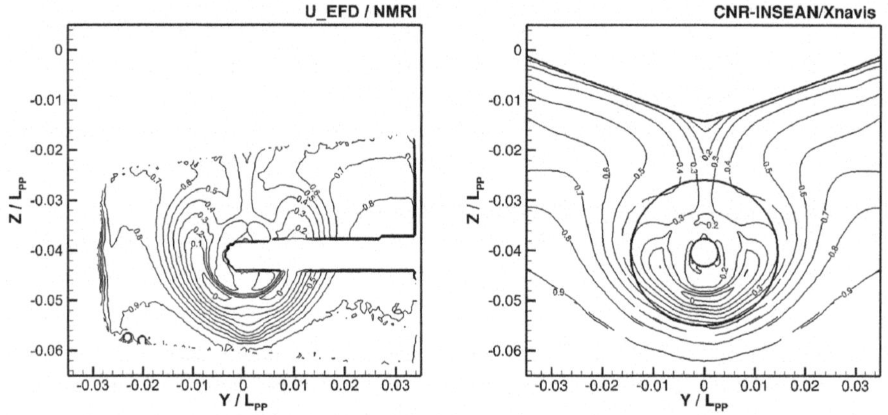

(a) Experiments *(b) INSEAN/Spalart-Allmaras*

Fig. 120 Case 1-4—linear isotropic turbulence closures—station S4

and 128). At station S4 just behind the duct, the results from HSVA are in very good agreement with the experiments while those provided by SHIME show a noticeable asymmetry and a deteriorated agreement with the measurements as shown by Figs. 126 and 129. It is interesting to notice that the same asymmetry is present at the same station in ABS and SJTU results using the same code (OpenFoam) and the same turbulence model. Most the computations based on k-ω SST are able to predict reasonably well the flow at station S4 (Figs. 127 and 130).

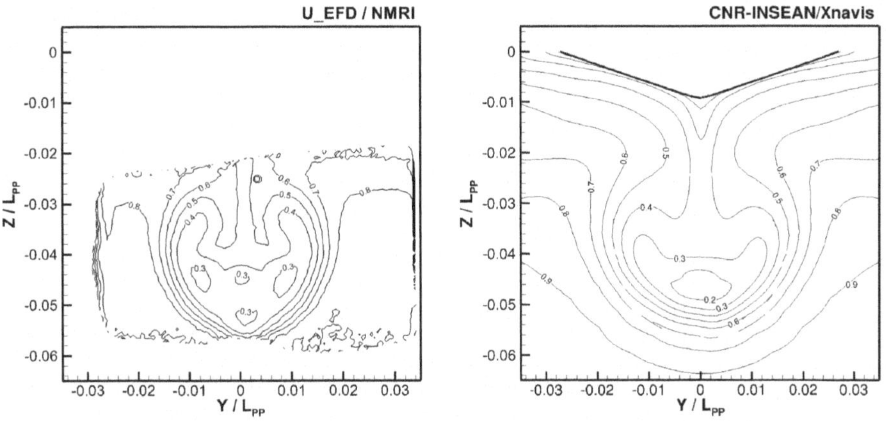

(a) Experiments (b) INSEAN/Spalart-Allmaras

Fig. 121 Case 1-4—linear isotropic turbulence closures—station S7

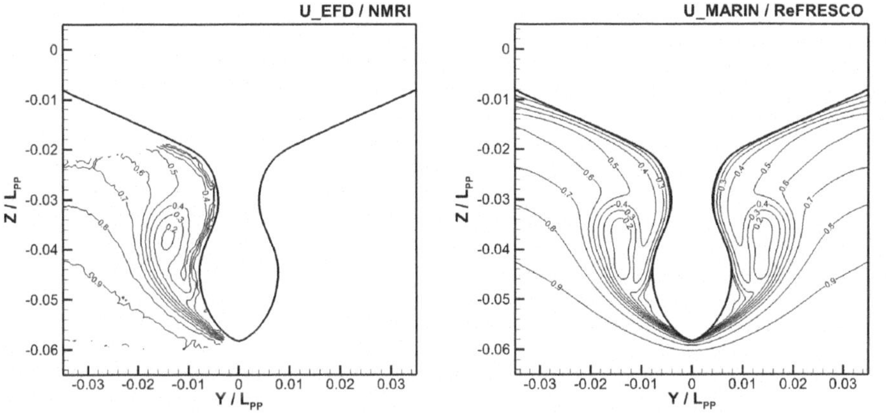

(a) Experiments (b) MARIN/k-ω SST + DM correction

Fig. 122 Case 1-4—MARIN—k-ω SST + DM correction—station S2

5.3.2 Non-linear Anisotropic Turbulence Closures

Reynolds Stress Transport and non-linear anisotropic EARSM models are used by HHI, ECN/CNRS, Chalmers, NMRI and YNU. All these models exhibit very similar trends which are now summarized. At station S2, they are the only models able to predict, without any ad-hoc corrections, the right shape of the iso-contour $u = 0.2$ with a trend towards a slight over-estimation for the RSTM closure shown in Fig. 140. EARSM models perform reasonably well as illustrated by Chalmers results (see Fig. 131). However, as confirmed by ECN/CNRS (not shown here), the location of the core of the vortex is located slightly further from the hull, compared to the

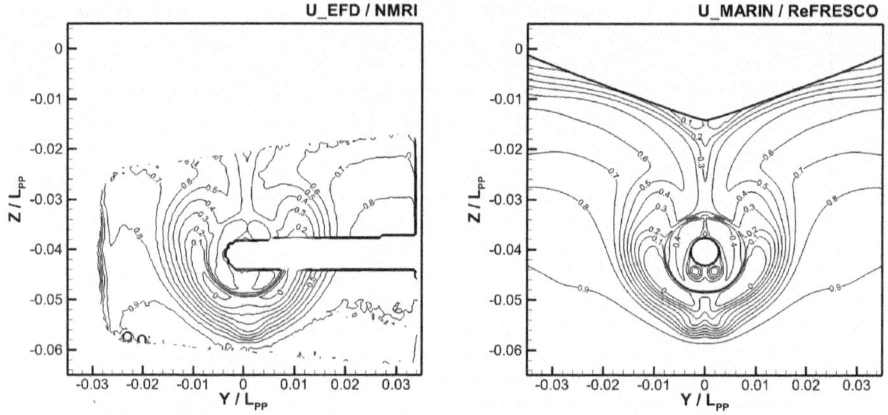

(a) Experiments *(b) MARIN/k-ω SST + DM correction*

Fig. 123 Case 1-4—MARIN—k-ω SST + DM correction—station S4

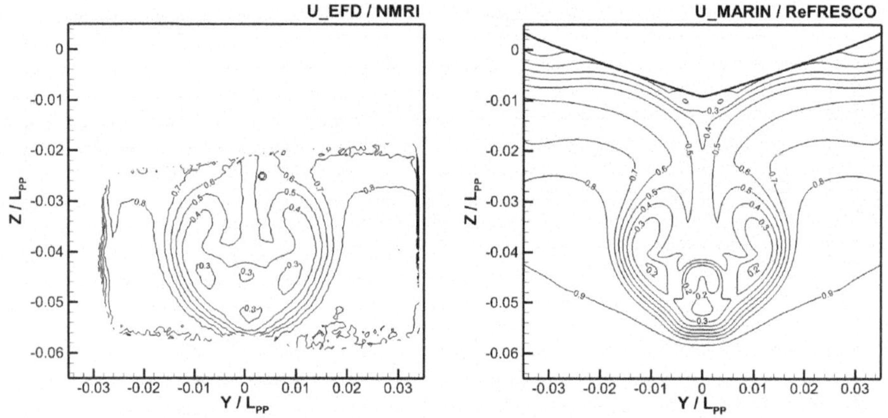

(a) Experiments *(b) MARIN/k-ω SST + DM correction*

Fig. 124 Case 1-4—MARIN—k-ω SST + DM correction—station S7

experiments, which indicates that the computed vortex is likely to be thicker than the measured one. At station S4, the distorsion of the outer iso-contours ($u = 0.4$ and 0.5) is better represented by the non-linear anisotropic turbulence closures but there is a common trend to overestimate the size of the reversed flow region in the center of the vortex (see for instance EARSM (Fig. 132) or RSTM (Fig. 141) based computations (Figs. 133, 134, 135, 136, 137, 138, 139 and 142).

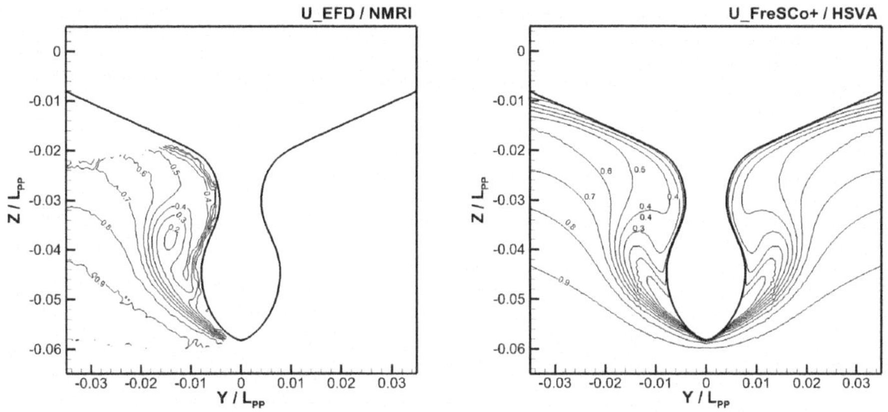

(a) Experiments *(b) HSVA/Linear EARSM*

Fig. 125 Case 1-4—linear EARSM from HSVA—station S2

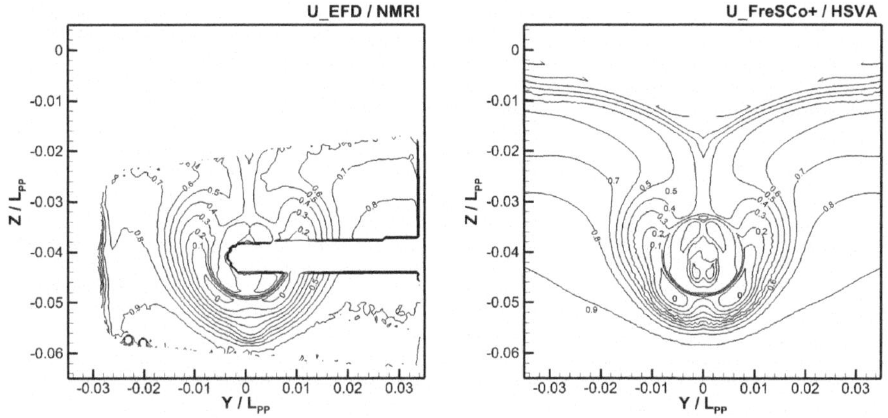

(a) Experiments *(b) HSVA/Linear EARSM*

Fig. 126 Case 1-4—linear EARSM from HSVA—station S4

What is clear one more time is that the turbulence closure plays a central role in the prediction of the vortex core flow physics as indicated by Figs. 143, 144 and 145 which compare on the same grid the k-ω SST and EARSM turbulence models implemented in the ECN/CNRS code.

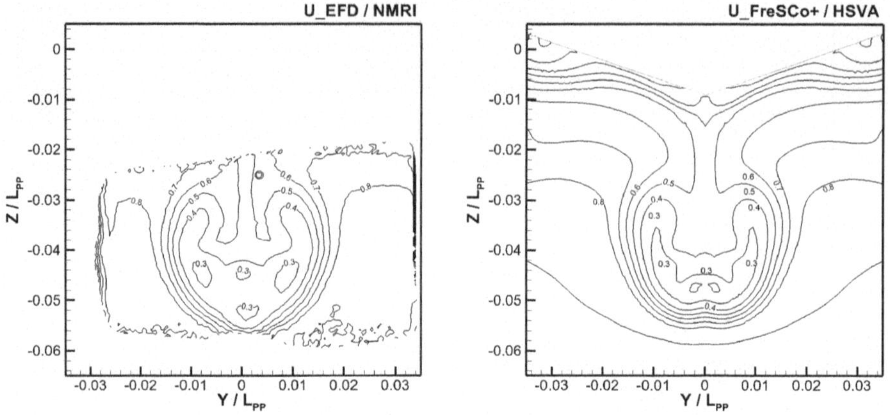

(a) *Experiments* (b) *HSVA/Linear EARSM*

Fig. 127 Case 1-4—linear EARSM from HSVA—station S7

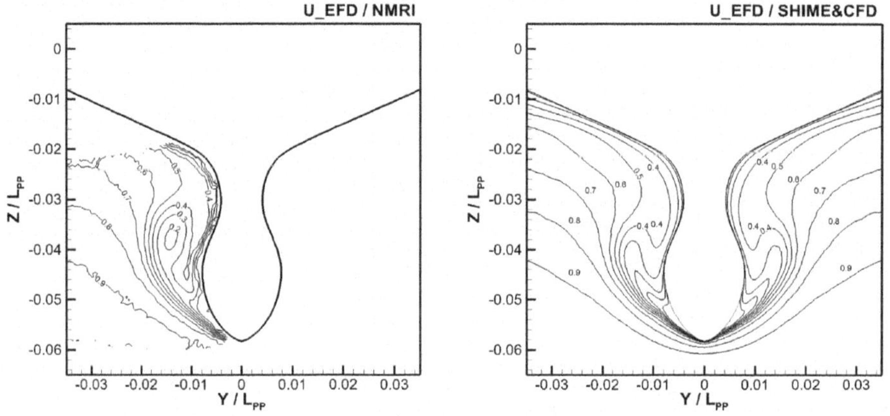

(a) *Experiments* (b) *SHIME/k-ω SST*

Fig. 128 Case 1-4—SHIME—k-ω SST—station S2

5.4 Case 1-4a—Concluding Remarks

Based on the comparisons performed during this workshop, one can say that the turbulence models have a critical role to play once the grid is fine enough. The main modifications of the flow created by the presence of the duct are correctly reproduced by linear isotropic models which provide the correct qualitative trend with the remarkable exception of the linear model implemented by HSVA in Fresco+.

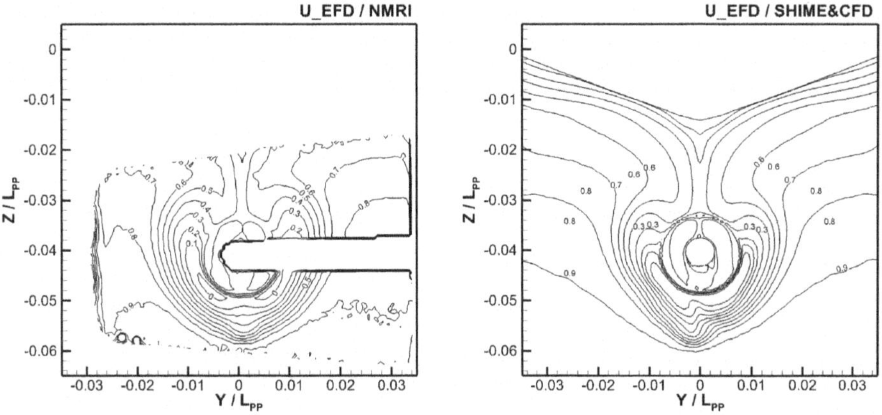

(a) Experiments *(b) SHIME/k-ω SST*

Fig. 129 Case 1-4—SHIME—k-ω SST—station S4

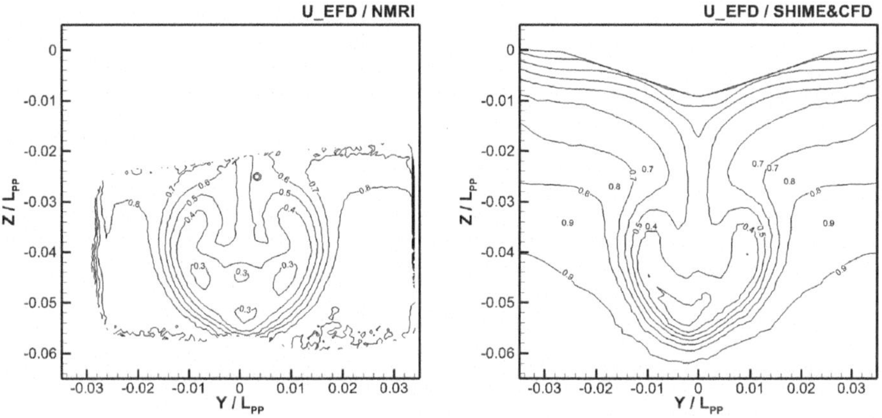

(a) Experiments *(b) SHIME/k-ω SST*

Fig. 130 Case 1-4—SHIME—k-ω SST—station S7

To get a more quantitative agreement, moving from isotropy to anisotropy appears necessary although the inner part of the vortex is not perfectly reproduced as shown by the occurence of a reversed flow region which is over-estimated by both RSTM and EARSM turbulence closures.

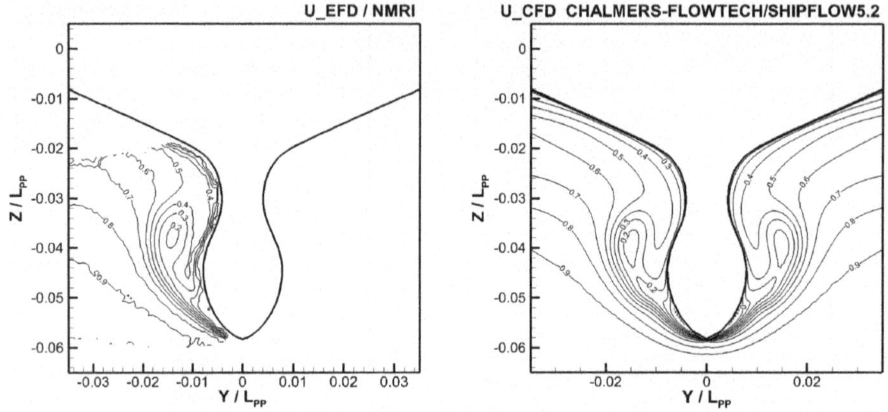

(a) Experiments *(b) CHALMERS/EARSM*

Fig. 131 Case 1-4—CHALMERS—EARSM turbulence closure—station S2

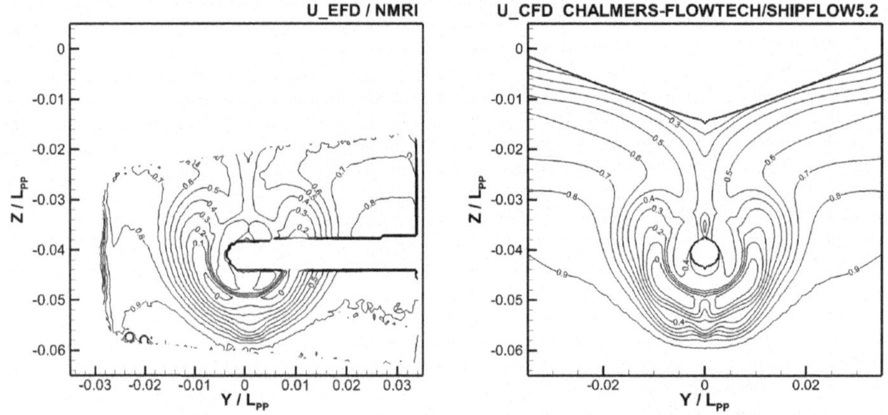

(a) Experiments *(b) CHALMERS/EARSM*

Fig. 132 Case 1-4—CHALMERS—EARSM turbulence closure—station S4

6 Test Case 1-7

6.1 Description of the Experiments

Test case 1-7 concerns the local flow analysis around the JBC without duct but with propeller towed in the NMRI towing tank, in free trim and sinkage conditions. The location of the experimental stations are indicated in Figs. 146 and 147 shows the stern with its five blades propeller.

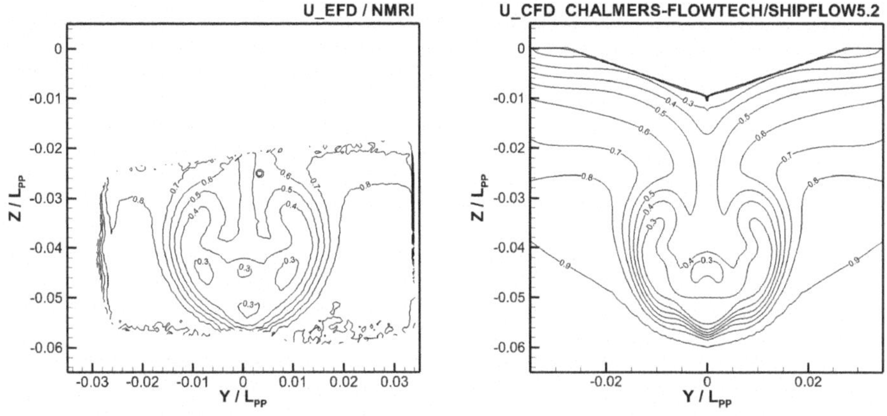

(a) Experiments (b) CHALMERS/EARSM

Fig. 133 Case 1-4—CHALMERS—EARSM turbulence closure—station S7

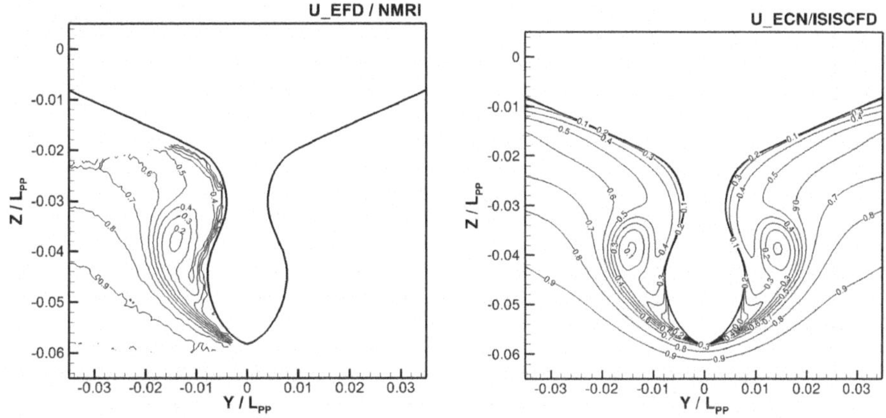

(a) Experiments (b) ECN/CNRS - EARSM

Fig. 134 Case 1-4—ECN/CNRS—EARSM turbulence closure—station S2

Figures 148, 149, 150, 151, 152 and 153 show the influence of the propeller on the experimental iso-wakes and averaged crossflow velocity components at three sections S2, S4 and S7 already defined before. At station S2, the propeller has already an influence on the flow as shown by Fig. 148. With propeller, the isowakes $u = 0.3$ and $u = 0.4$ are no more visible, which means that the core of the longitudinal vortex is accelerated, probably due to the succion effect associated with the presence of the propeller. Interestingly, the iso-contour $u = 0.5$ has evolved from a distorted hook shape (without propeller) to a closed contour in the vicinity of the hull (with propeller), which indicates also a significant acceleration of the flow close to the hull. At station S4, the experiments are noisy and hard to interprete (see Fig. 150). One can

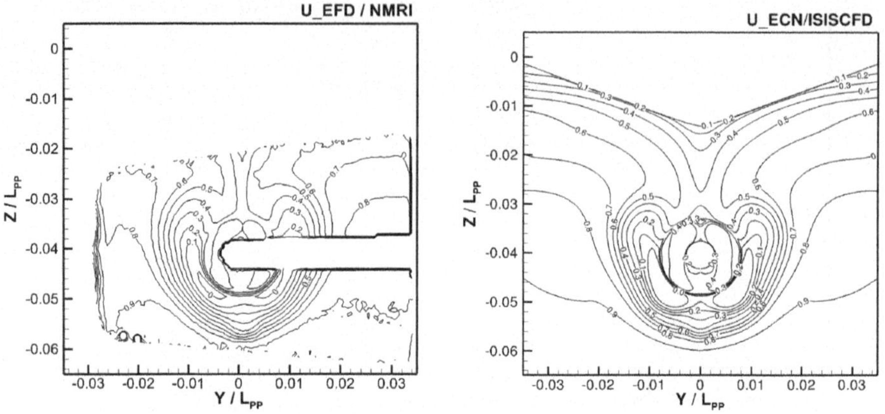

(a) *Experiments* (b) *ECN/CNRS - EARSM*

Fig. 135 Case 1-4—ECN/CNRS—EARSM turbulence closure—station S4

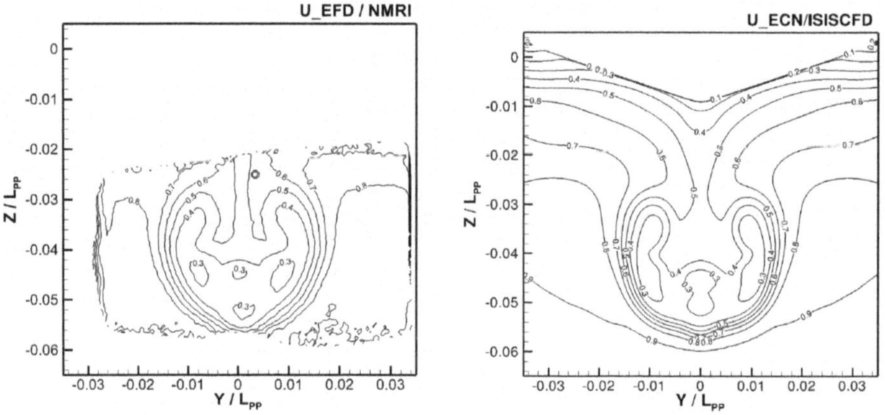

(a) *Experiments* (b) *ECN/CNRS - EARSM*

Fig. 136 Case 1-4—ECN/CNRS—EARSM turbulence closure—station S7

at least observe the strong succion effect illustrated by the respective locations of the isowakes u = 0.7 and 0.8. At station S7 (see Fig. 152), the influence of the propeller is way more obvious and easier to analyze. One notices two circular isowakes (u = 0.9 and 1.0) bounding an inner region in the wake of the propeller. This wake is asymmetric and characterized by the presence of large zone u = 1.2 with a shape of Moon's crescent mainly located on the right part of the wake but also present in the left part. On the left side, one can see isocontours with high levels (u = 1.3) while on the right side, one notices the existence of lower level iso-contours (u = 1.1 to 0.7).

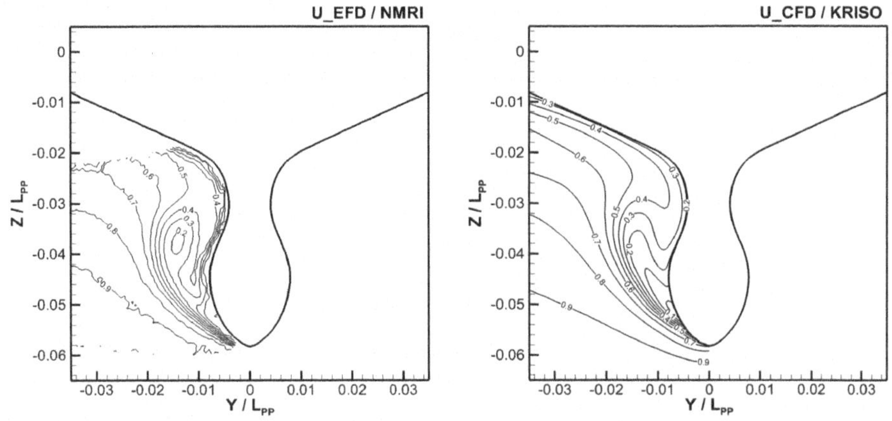

(a) Experiments *(b) KRISO - EARSM*

Fig. 137 Case 1-4—KRISO—EARSM turbulence closure—station S2

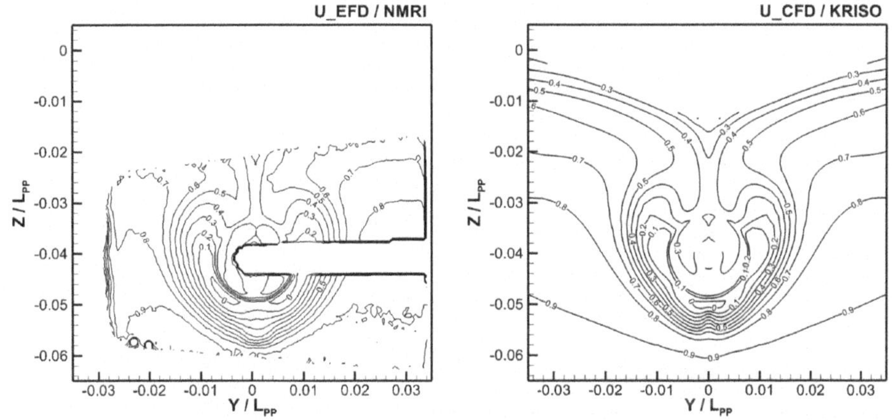

(a) Experiments *(b) KRISO - EARSM*

Fig. 138 Case 1-4—KRISO—EARSM turbulence closure—station S4

Finally, phase-averaged isowakes and secondary velocities from station S4 to S7 for 0°, 24° and 48° were also measured by NMRI. At station S4, these results are difficult to exploit because of the absence of measurements in the main part of the wake. This is the reason why we will restrict the analysis to the isowakes and secondary velocities measured at station S7. These experimental results are shown in Figs. 154 and 155. The main interest of these phase-averaged experiments is to show the structure of the wake flow behind the rotating propeller for several positions of its five blades. In particular, we expect to be able to detect the phase-averaged position of the tip vortices which are shed behind each rotating blade. The locations of these five tip vortices should be indicated by the presence of several closed isowakes (one per

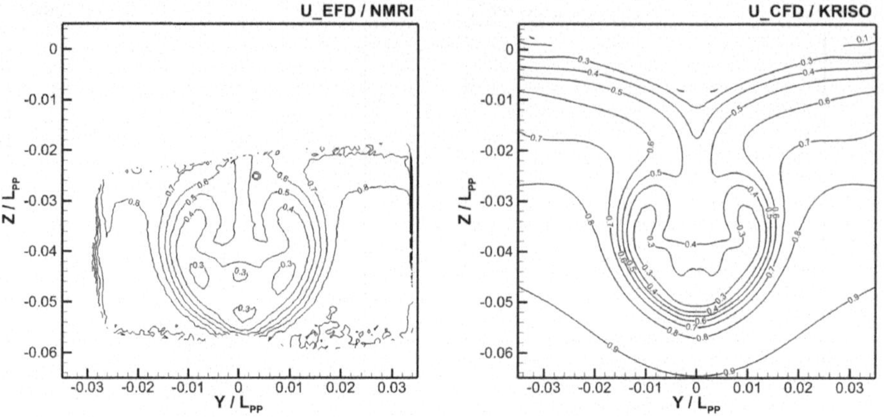

(a) *Experiments* (b) *KRISO - EARSM*

Fig. 139 Case 1-4—KRISO—EARSM turbulence closure—station S7

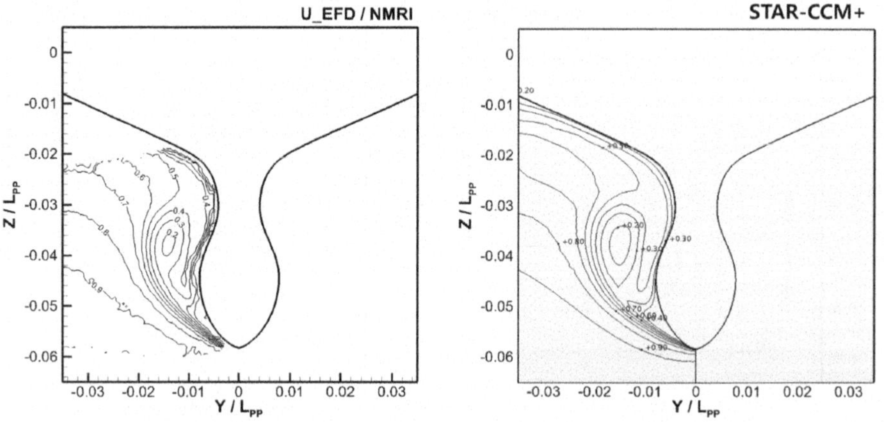

(a) *Experiments* (b) *HHI - RSM*

Fig. 140 Case 1-4—HHI—RSM turbulence closure—station S2

blade) associated to a dimensionless longitudinal velocity higher than one. However, one must remember that these tip vortices are progressing around an helicoidal path, which means that they are never strictly orthogonal to the plane of measurements. This means that the measured Cartesian secondary velocity components (v, w) will probably not bring any significant information regarding the local rotational motion inside each tip vortex since the plane of measurement is not orthogonal to the local axis of this vortex.

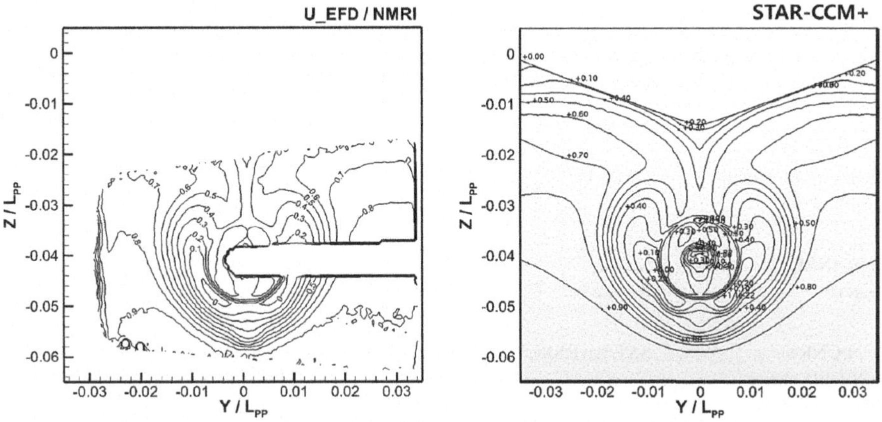

(a) Experiments *(b) IIIII - RSM*

Fig. 141 Case 1-4—HHI—RSM turbulence closure—station S4

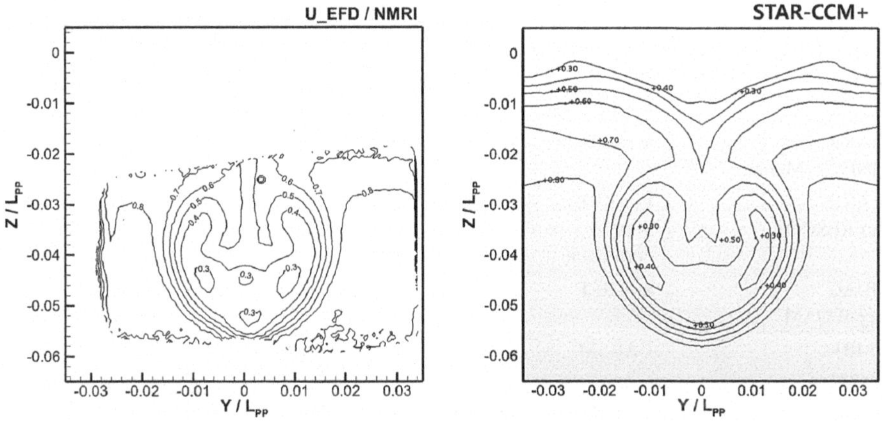

(a) Experiments *(b) HHI - RSM*

Fig. 142 Case 1-4—HHI—RSM turbulence closure—station S7

6.2 Review of Contributions

Eighteen participants provided results for the test case 1-7. The main characteristics of their contributions are listed in Table 7 in terms of turbulence closures, wall treatments, discretization and propeller treatment. Some participants provided several different contributions, which makes possible a comparison of various turbulence closures or wall treatments for the same grid and same code. Moreover, SOTON performed a comparison of codes (OpenFoam vs. Star-CCM+) with the same grid and same turbulence closure, as previously.

Table 7 Case 1-7—main characteristics of the computations

Organization Code name	Turbulence model	Wall model	Grid characteristics	Propeller treatment
ABS OpenFOAM	k-ω SST	Wall function	Unstructured grid 7 3710 43 cells	Actual propeller
Chalmers-Flowtech Shipflow	EARSM	Wall resolved	Structured grid with overlapping 18.4M cells	Lifting line
CNR-INSEAN Xnavis	Spalart-Allmaras	Wall resolved	Structured grid with overlapping 10.6M cells	Body-force/actual propeller
ECN/CNRS ISIS-CFD aka FINE^{TM}/Marine	SST/EARSM	Wall function Wall resolved	Unstructured grid 18.7M cells	Body-force/Actual propeller
HHI HiFOAM/StarCCM+	Realizable k-ϵ/RSTM	Wall function	Unstructured grid 3.5M cells	Moving reference frame
HSVA FreSCo+	2 equations	Wall function	Unstructured grid 14.6M cells	Vortex lattice
IIHR REX	DDES	Wall resolved	Unstructured grid 67.7M cells	Actual propeller
KRISO Wavis	Realizable k-ϵ/EARSM	Wall function	Structured grid 15.5M cells	Lifting surface
MARIC FINE^{TM}/Marine	k-ω SST	Wall function	Unstructured grid 2M cells	Body force
MARIN ReFRESCO	k-ω SST + DM correction	Wall resolved	Unstructured grid 17M cells	Actual propeller
MIJAC OpenFOAM	k-ω SST	Wall resolved	Unstructured grid 8.7M cells	Simplified propeller theory
NMRI Nagisa	EARSM	Wall resolved	Structured grid 7.8M cells	Simplified propeller theory
PNU Fluentv15	2 equations k-ϵ RNG	Wall resolved Wall function	Unstructured grid 17.1M cells	Actual propeller
SHIME OpenFOAM	k-ω SST	Wall resolved	Unstructured grid 22.15M cells	N/A
SJTU naoeOpenFOAM	k-ω SST	Wall function	Unstructured grid 4.7M cells	Simplified propeller theory
SOTON OpenFOAM/StarCCM+	k-ω SST	Wall resolved	Structured grid 19.4M cells	Body force prescribed
UniRostock OpenFOAM	Hybrid LES k-ω SST	Wall resolved	Structured grid 22.8M cells	Actual propeller
YNU SURFv7	EARSM	Wall resolved	Structured grid 2.5M cells	Simplified propeller theory

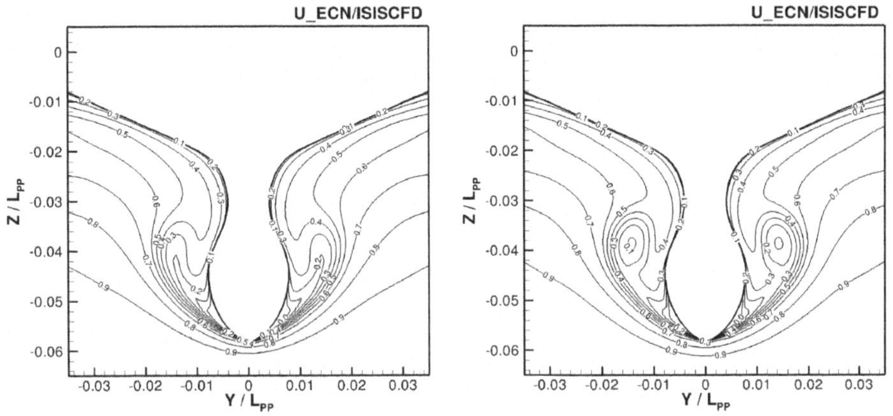

(a) *ECN/CNRS - k-ω SST* (b) *ECN/CNRS - EARSM*

Fig. 143 Case 1-4—ECN/CNRS—influence of turbulence anisotropy—station S2

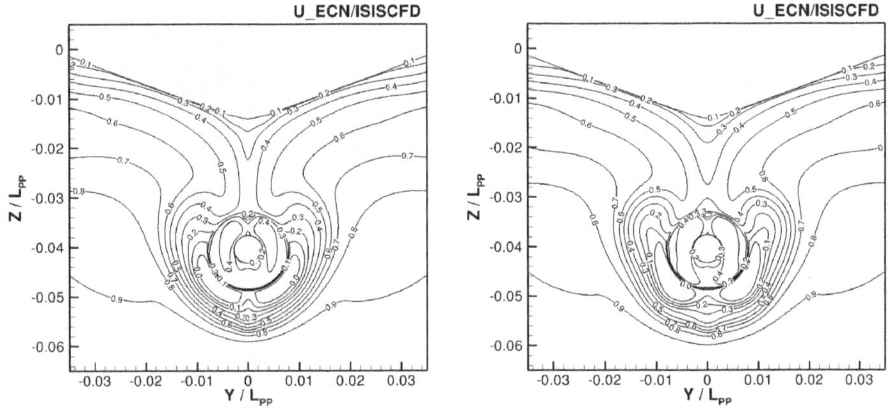

(a) *ECN/CNRS - k-ω SST* (b) *ECN/CNRS - EARSM*

Fig. 144 Case 1-4—ECN/CNRS—influence of turbulence anisotropy—station S4

Among the eighteen participants, twelve used unstructured grids while six computed on structured grids sometimes locally enriched with overlapping patches. It is well known that the prediction of the nominal velocity in the stern region depends strongly on the turbulence closure, as soon as a reasonably dense grid is used. The turbulence models employed by the participants computing this test case can be classified into three groups: (i) hybrid RANS/LES turbulence closures (University of Rostock, IIHR), (ii) anisotropic non-linear statistical turbulence modeling: Reynolds stress transport models (HHI), Explicit Algebraic Reynolds stress models (Chalmers, ECN/CNRS, NMRI, YNU), (iii) isotropic linear eddy-viscosity model (ECN/CNRS, HSVA, MARIN, URO, …). Concerning the linear isotropic turbulence models, most

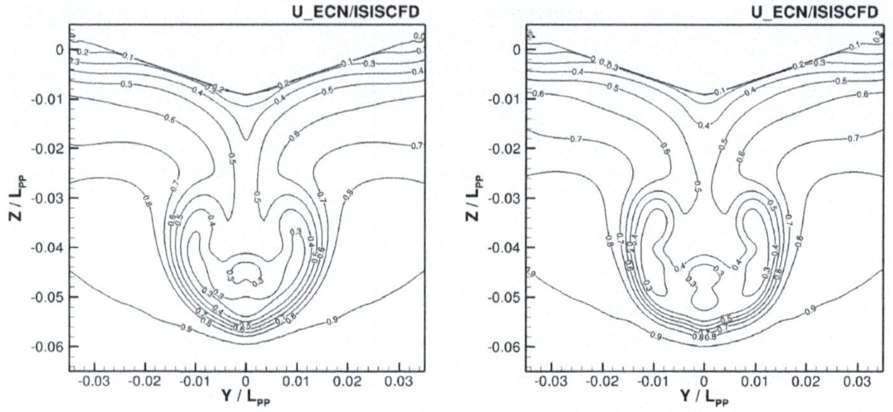

(*a*) *ECN/CNRS - k-ω SST* (*b*) *ECN/CNRS - EARSM*

Fig. 145 Case 1-4—ECN/CNRS—influence of turbulence anisotropy—station S7

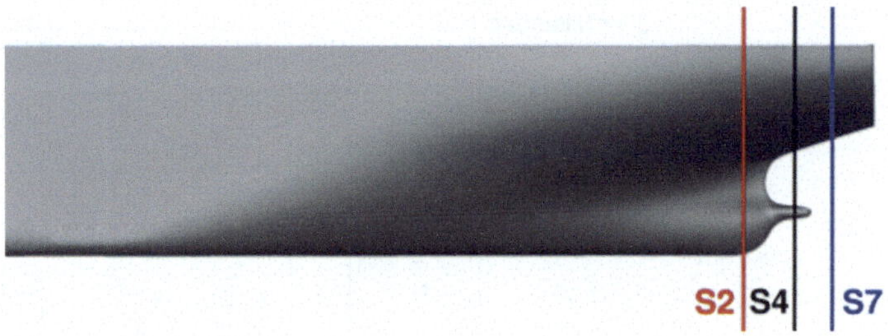

Fig. 146 Case 1-7—hull geometry and experimental stations

of the contributors use the k-ω SST model but HHI, MHI and KRISO employ also the realizable k-ε closure while CNR-INSEAN and MARIN prefer to use a one equation model (Spalart-Allmaras or one eq. Menter with Dacles-Mariani modification). Although the influence of turbulence models was assessed by various authors at model scale without propeller, it will be interesting to check if this is also true in presence of a rotating propeller. All the discretization methods are formally second-order accurate and based on multi-block structured or unstructured grids which are all body-fitted, with the exception of Chalmers and CNR-INSEAN which are using an overset approach combining structured grids to reach an optimal grid density in the stern region.

The propeller is treated in various ways. Seven contributors (ABS, ECN/CNRS, CNR/INSEAN, IIHR, MARIN, PNU, University of Rostock) implemented an actual propeller while seven other organizations (Chalmers, KRISO, MARIC, MIJAC, NMRI, SJTU, SOTON and YNU) used various simplified propeller models described in Table 7. Moreover, HHI used a propeller computational model based on a Mov-

Fig. 147 Case 1-7—view of the stern with its propeller

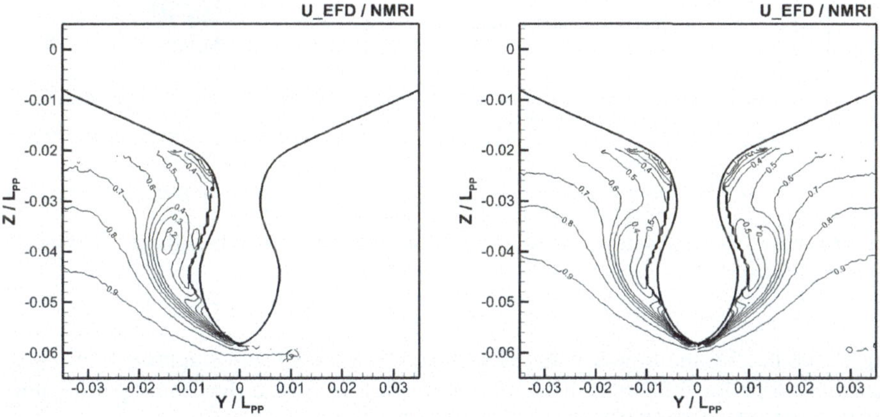

(a) Without propeller *(b) With propeller*

Fig. 148 Case 1-7—experiments—influence of the propeller on the isowakes distribution at station S2 in the absence of duct

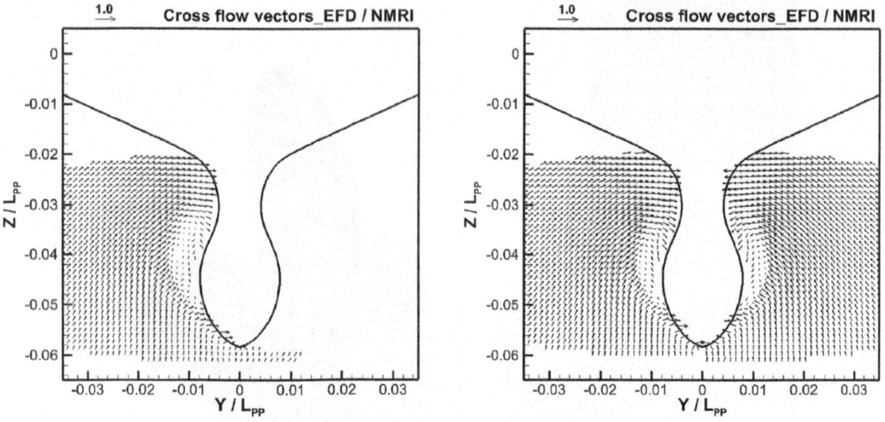

(a) *Without propeller* (b) *With propeller*

Fig. 149 Case 1-7—experiments—influence of the propeller on the distribution of mean secondary velocities at station S2

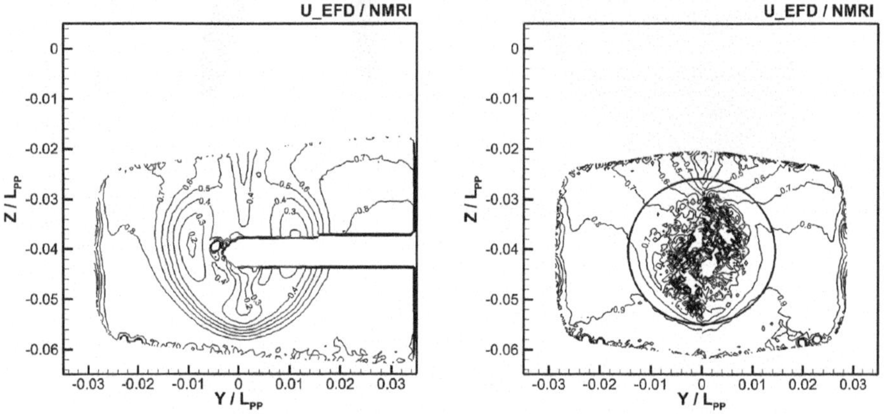

(a) *Without propeller* (b) *With propeller*

Fig. 150 Case 1-7—experiments—influence of the propeller on the isowakes distribution at station S4 in the absence of duct

ing Reference Frame model while HSVA used a Vortex Lattice approach to represent the propeller influence. It is also interesting to notice that some participants (CNR/INSEAN, ECN/CNRS) provided results with both an actual and modelled propeller, which can be used to quantify the impact of the propeller model on the local flow.

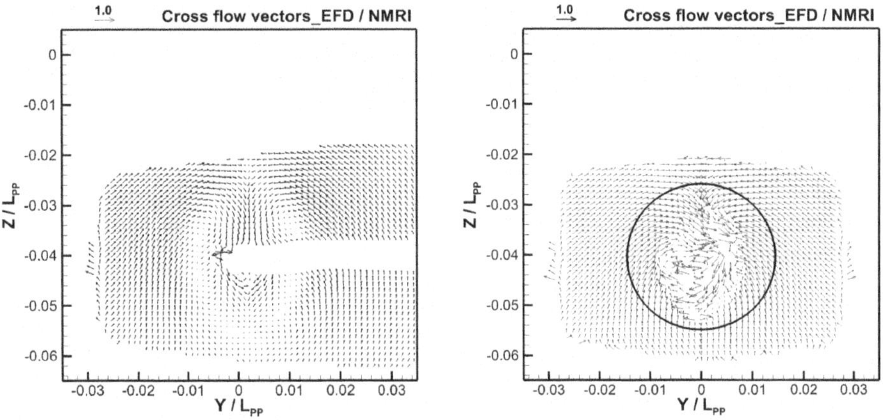

(a) *Without propeller* (b) *With propeller*

Fig. 151 Case 1-7—experiments—influence of the propeller on the distribution of mean secondary velocities at station S4

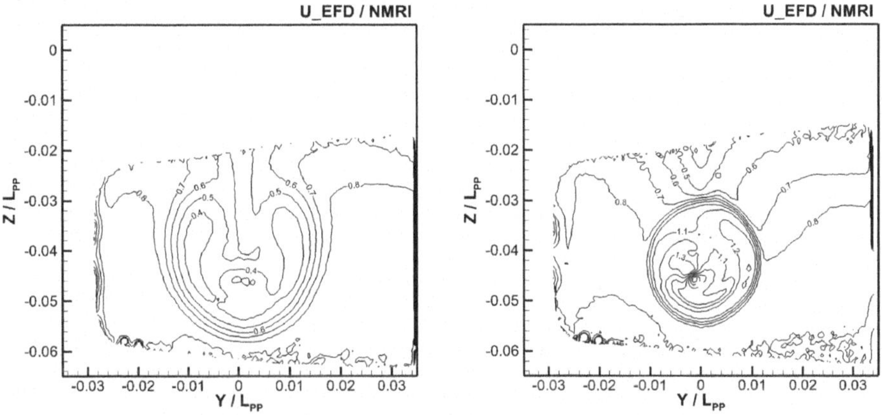

(a) *Without propeller* (b) *With propeller*

Fig. 152 Case 1-7—experiments—influence of the propeller on the isowakes distribution at station S7 in the absence of duct

6.3 *Influence of Propeller Treatment*

Figures 156 and 157 show a comparison between a body force representation of the propeller and the computation of the rotating propeller based on sliding grids. Both computations were performed by ECN/CNRS. It is interesting to notice that the propeller treatment has already a significant influence on the isowake distribution at station S2 due to a different modelling of the succion effect (see Fig. 156). But, the occurence of the isowake u = 0.5 visible close to the wall in NMRI experiments, is captured by none of these computations. In the wake of the propeller, at station S7,

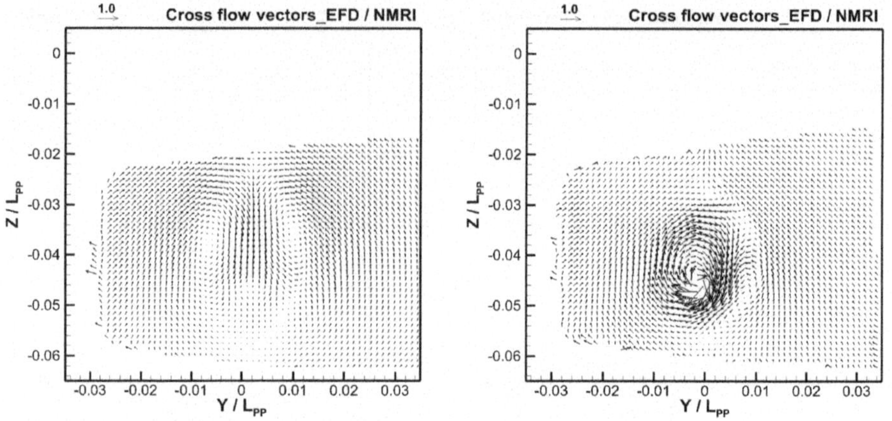

(a) *Without propeller* (b) *With propeller*

Fig. 153 Case 1-7—experiments—influence of the propeller on the distribution of mean secondary velocities at station S7

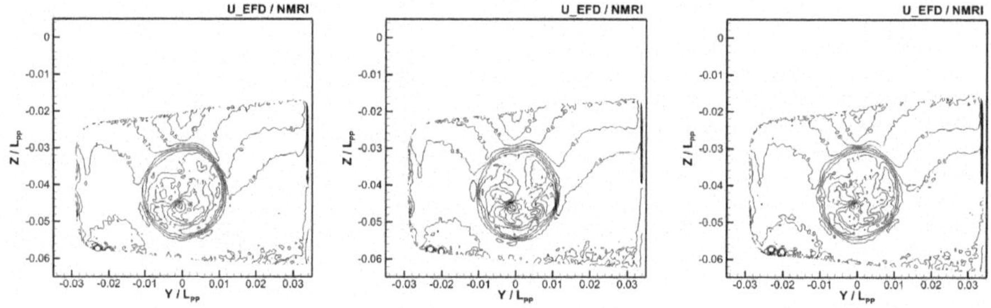

(a) *0°* (b) *24°* (c) *48°*

Fig. 154 Case 1-7—experiments—distribution at station S7 of isowakes at 0°, 24° and 48°

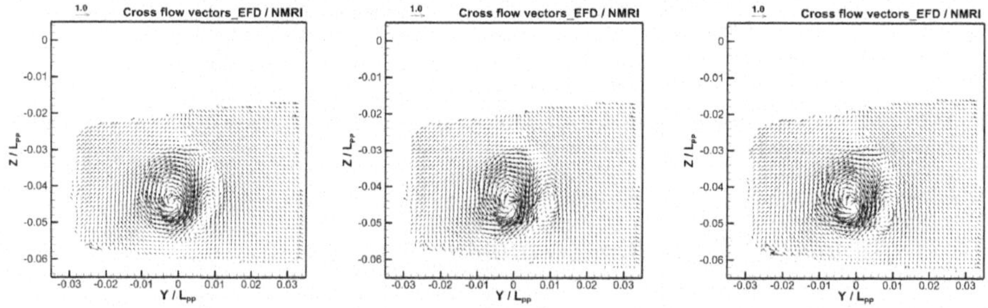

(a) *0°* (b) *24°* (c) *48°*

Fig. 155 Case 1-7—experiments—distribution at station S7 of secondary velocities at 0°, 24° and 48°

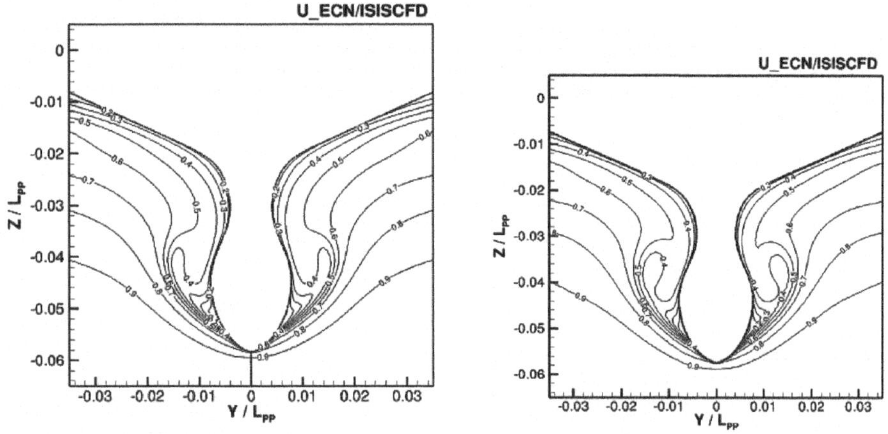

(a) Body Force *(b) Actual Propeller*

Fig. 156 Case 1-7—ECN/CNRS—distribution of isowakes at station S2 for body force and actual propeller

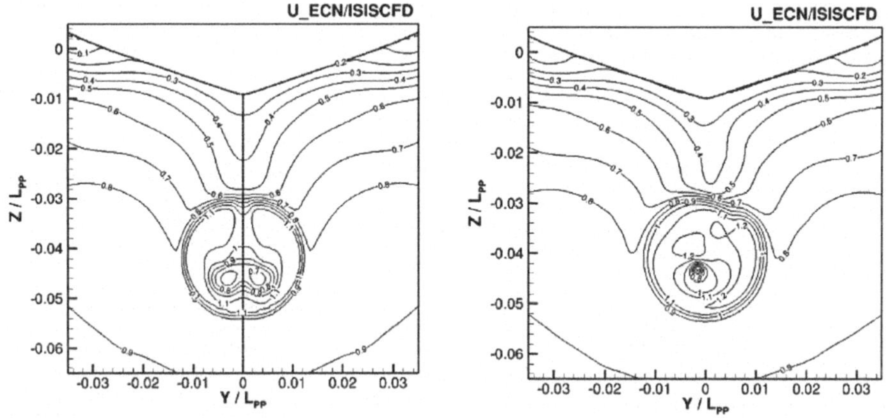

(a) Body Force *(b) Actual Propeller*

Fig. 157 Case 1-7—ECN/CNRS—distribution of isowakes at station S7 for body force and actual propeller

the only valid approach is the one based on the computation of the rotating propeller which provides the correct asymmetry of the mean isowakes (see Fig. 157) since this very simplified body force approach provides a symmetric wake flow. The isowake $u = 1.2$ is correctly captured on the right hand side of the figure but its extent on the left hand side is much more limited than in NMRI's experiments.

Other examples of computations based on an actual propeller are provided by IIHR using Rex and a DDES turbulence model at station S7 (see Fig. 158a), MARIN with ReFRESCO (see Fig. 158b), PNU with Fluent (see Fig. 158c) or CNR/INSEAN with

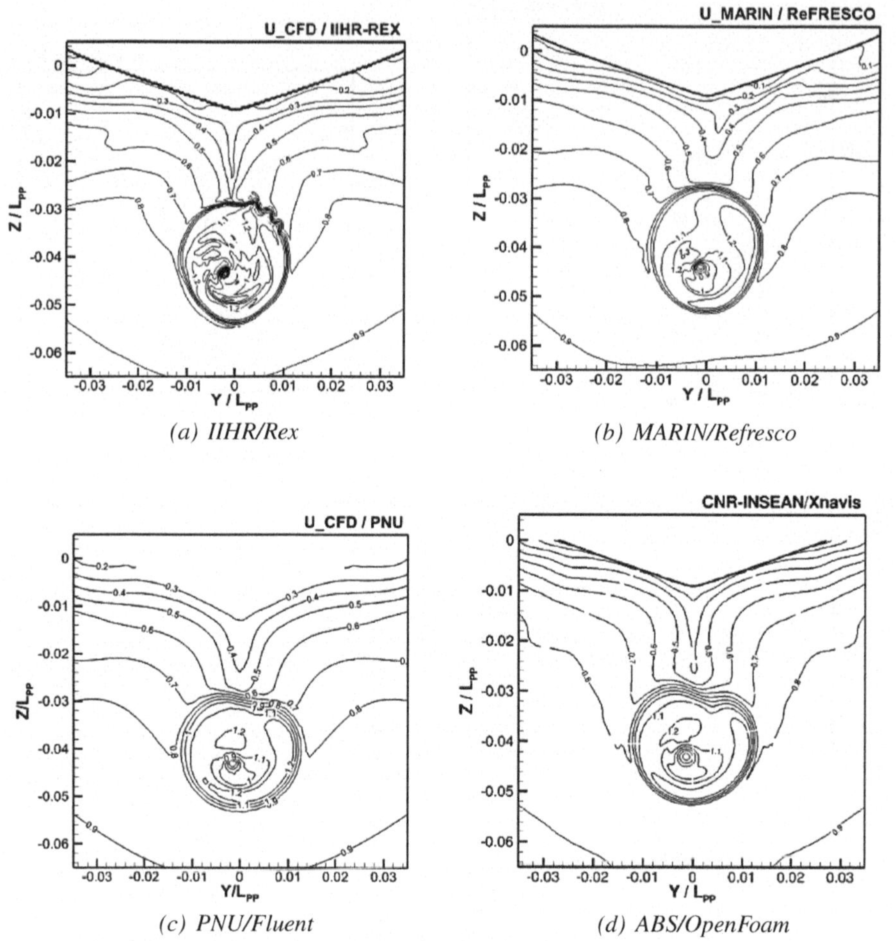

(a) IIHR/Rex

(b) MARIN/Refresco

(c) PNU/Fluent

(d) ABS/OpenFoam

Fig. 158 Case 1-7—distribution of isowakes at station S7 by various contributors discretizing the actual propeller

Xnavis (see Fig. 158d). Globally, all these computations are in very good agreement with NMRI's measurements. The location of the $u = 1.2$ isowake is globally very well captured and the isowake $u = 1.3$ is even present in MARIN contribution. Results from IIHR/Rex employing an IDDES hybrid RANS/LES model are also very promising, although the noisy nature of the isowakes might indicate a too short averaging time.

More sophisticated body force models can be used since they offer a very interesting alternate choice to the very expensive Actual Propeller simulations. A good example is provided by Chalmers which uses a lifting line model to represent the flow around the propeller. Figure 159 shows their computed isowakes at stations S2 and S7, which are in reasonable agreement with the experiments although the isowake

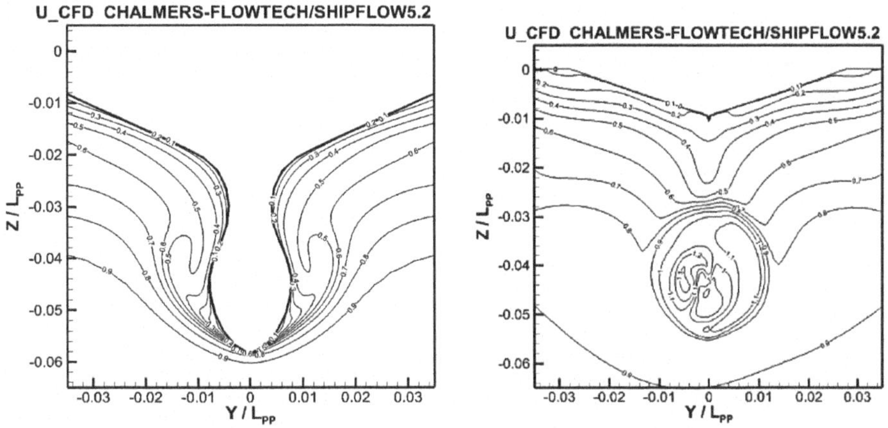

(a) Station S2 *(b) Station S7*

Fig. 159 Case 1-7—Chalmers—distribution of isowakes at station S2 and S7 for a lifting line approach

u = 1.2 is not captured at station S7. HSVA developed a coupling with a vortex lattice approach to represent the action of the propeller. Their results at stations S2 and S7 are shown in Fig. 160. At station S7, it seems that the isowake u = 1.2 is not located on the right side of the propeller's wake, which is difficult to explain. Finally, Fig. 161 shows the isowake distribution at station S7 computed by NMRI and YNU. Both institutions used a simplified propeller theory coupled with their RANS solver to represent the propeller effect. As mentioned previously for Chalmers results, one can observe that the isowake u = 1.2 is hardly captured in these results, NMRI being marginally better than YNU from that standpoint.

In order to complete the analysis, it is interesting to compare the phase-averaged isowake distribution at 0°, 24° and 48° since they were also measured by NMRI (see Fig. 162). Figures 163, 164, 165, 166 and 167 show the isowakes obtained by different teams. At first glance, one notices a very large dispersion of the results, mainly due to the lack of discretisation points in this specific region. It is clear that very fine grids probably adapted to the instantaneous location of the tip vortices would be necessary to capture all these flow details. This can be observed, for instance, in Fig. 165 or Fig. 166 where no isolated vortical structures are visible. The results obtained by ABS with OpenFoam (Fig. 163), and to a lesser degree, by INSEAN using Xnavis (Fig. 164) are more convincing. The grids used by these contributors range from 7M to 17M cells. The most detailed solution is provided by IIHR using Rex (Fig. 167) on a grid comprised of 68 M cells. One can clearly distinguish the location of tip vortices inside the propeller disk characterised by a very large longitudinal velocity gradient. In the center of the wake, a spot of isowakes probably indicates the location of the propeller hub wake and around this structure, on can distinctly detect five islands of isowakes which are associated with the traces of tip vortices. It is well known now, that it is mandatory to use a locally adapted mesh refinement to capture accurately

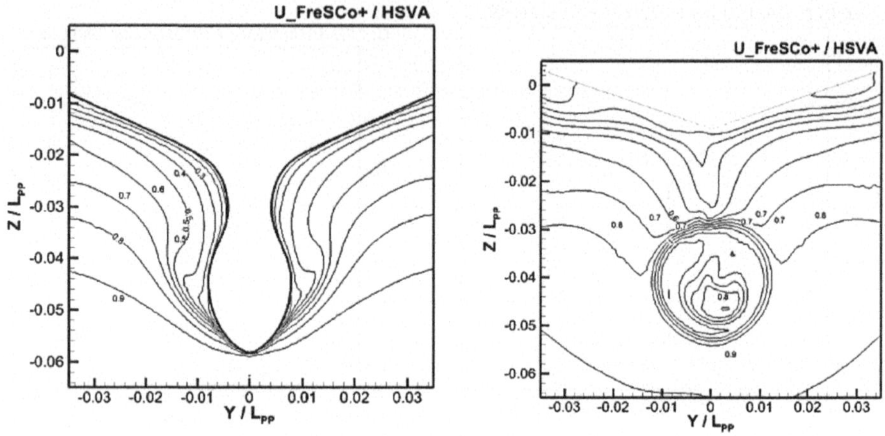

(a) Station S2 (b) Station S7

Fig. 160 Case 1-7—HSVA—distribution of isowakes at station S2 and S7 for a vortex lattice approach

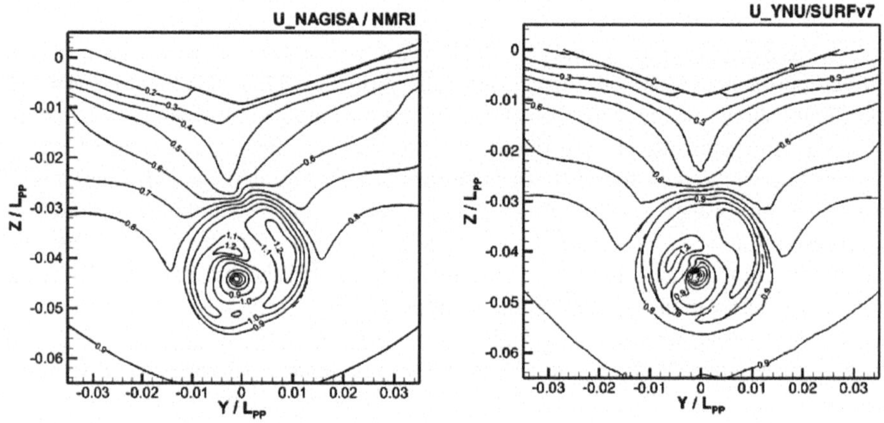

(a) NMRI/Nagisa (b) YNU/Surfv7

Fig. 161 Case 1-7—NMRI and YNU—distribution of isowakes at station S7 using simplified propeller theories

these structures or, by default, an extremely fine grid in the propeller wake. But a very fine discretization is not enough since classical RANS closures dissipate very quickly the vorticity of each of these tip vortex during their progression in the wake, contrary to the hybrid RANS/LES closure which is able to maintain the right level of longitudinal vorticity far from the tip vortex onset (see for instance Guilmineau et al. 2018 for more details on the turbulence model's influence in the wake of a propeller). Based on an hybrid RANS/LES closure, IIHR/Rex fully confirms this observation.

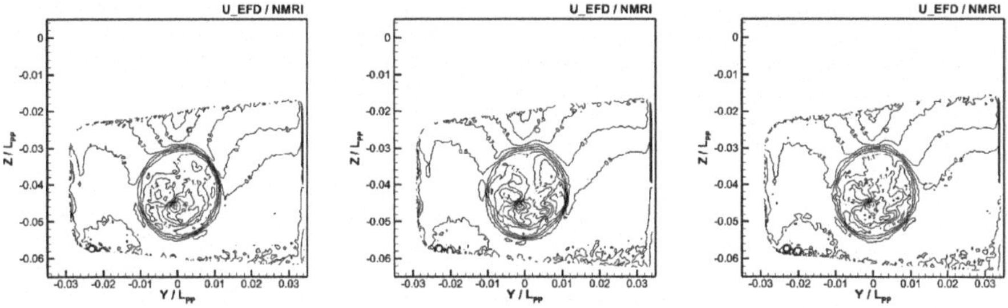

(a) 0° *(b) 24°* *(c) 48°*

Fig. 162 Case 1-7—experiments—distribution at station S7 of isowakes at 0°, 24° and 48°

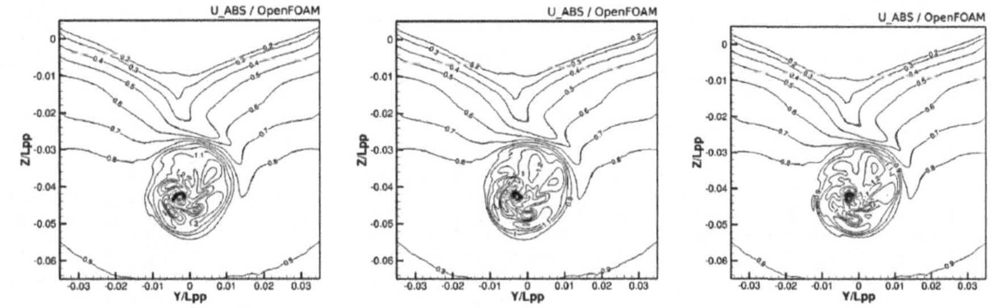

(a) 0° *(b) 24°* *(c) 48°*

Fig. 163 Case 1-7—ABS/OpenFoam—distribution at station S7 of isowakes at 0°, 24° and 48°

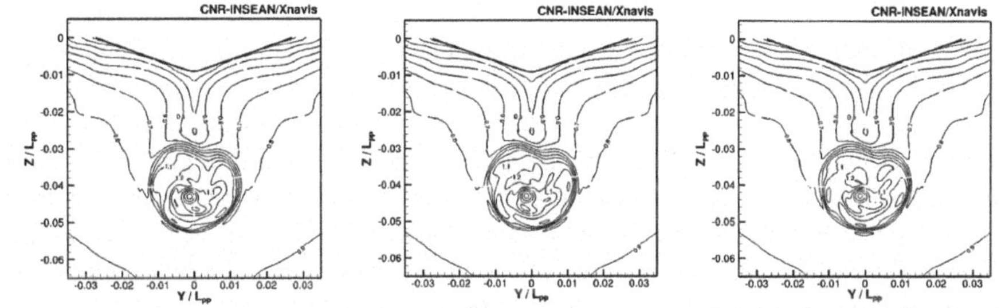

(a) 0° *(b) 24°* *(c) 48°*

Fig. 164 Case 1-7—INSEAN/Xnavis—distribution at station S7 of isowakes at 0°, 24° and 48°

6.4 Influence of Turbulence Closures

It is more difficult to analyze the role played by the turbulence modelling in case of a rotating propeller since one does not have at our disposal results combining the same propeller treatment with different turbulence closures. Moreover, the flow is strongly accelerated after the propeller disk, which means that the global flow is

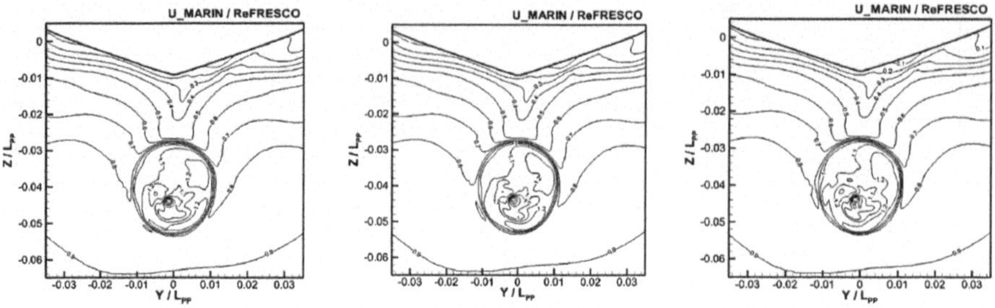

(a) 0° (b) 24° (c) 48°

Fig. 165 Case 1-7—MARIN/ReFresCo—distribution at station S7 of isowakes at 0°, 24° and 48°

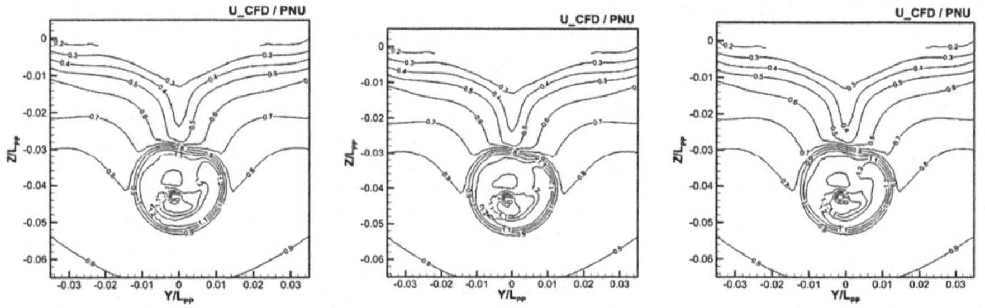

(a) 0° (b) 24° (c) 48°

Fig. 166 Case 1-7—PNU/Fluent—distribution at station S7 of isowakes at 0°, 24° and 48°

rather controlled by the pressure than by turbulence. However, as briefly mentioned in the previous section, the evolution of the individual tip vortices and their local flow structure are strongly dependent on the turbulence closure as long as a very fine local grid is generated to capture these flow details. It was noticed in previous systematic computational studies that RANS models are able to predict the onset of tip vortex but only hybrid RANS/LES can simulate accurately the detailed flow physics in the core of these vortices (accurate value of TKE, for instance) and their progression without dissipation in the wake of the propeller.

In order to get a first idea on the influence of turbulence closures, we are going to select computations performed with the actual propeller and compare RANS with hybrid RANS/LES models. Of course, at this stage, it is impossible to perform a true verification study and evaluate the role played by the discretisation error. However, we will choose on one hand, PNU/Fluent to represent the RANS category using a k-ε and on the other hand, IIHR/Rex to represent the hybrid RANS/LES class. In terms of mesh density, IIHR/Rex has generated a grid comprised of 68 M cells while PNU/Fluent uses a coarser grid comprised of 17M cells. Figure 168 shows the averaged iso-wake distribution at station S7 for both contributors. At first glance, it is obvious that the result from IIHR is richer in terms of vortical structures. The isowake distribution delimitating the propeller disk is also steeper for IIHR than for

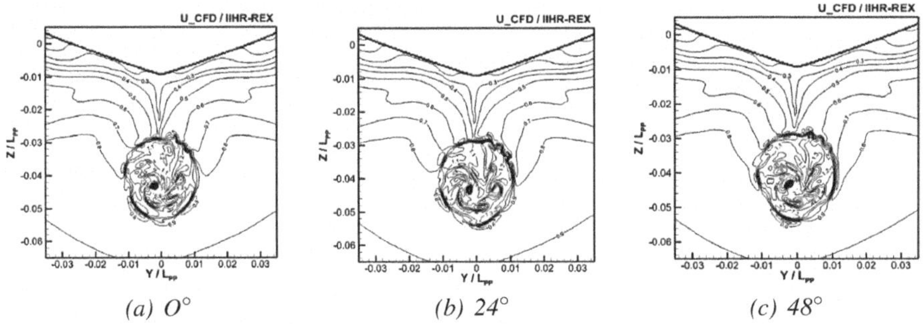

Fig. 167 Case 1-7—IIHR/Rex—distribution at station S7 of isowakes at 0°, 24° and 48°

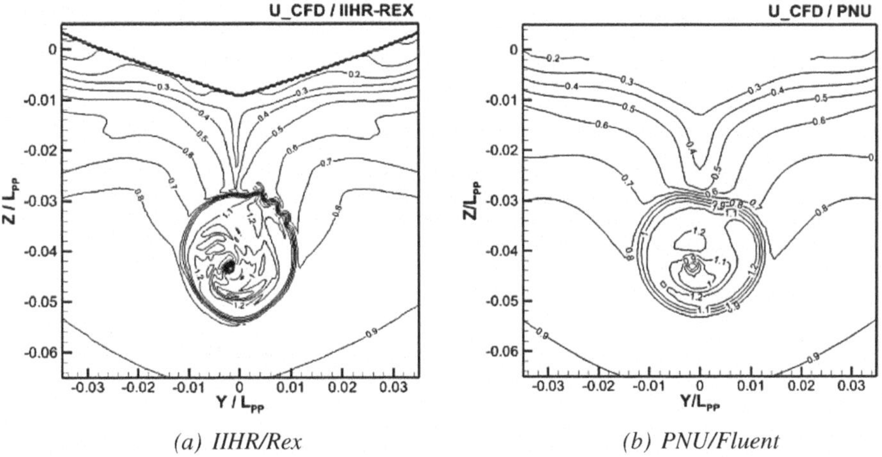

Fig. 168 Case 1-7—distribution of isowakes at station S7 with various turbulence closures

PNU. However, this could be associated with a different grid distribution. In the propeller disk, the island u = 1.2 is significantly more extended in IIHR than in PNU computations. This is a striking difference which might or might not be attributed to the turbulence closure but IIHR results are there in better agreement with the experiments. The iso-line u = 1.3 located in the upper part of the propeller disk is predicted by none of the contributors. PNU predicts a well defined iso-contour u = 1.2 while IIHR gets a set of lower iso-contours u = 1.2 associated probably to various vortical structures. This effect is probably due to an averaging time which is still too short to get a statistically converged solution.

Figures 169 and 170 recall the comparisons of phase-averaged isowakes for PNU and IIHR. Here, as mentioned previously, the two solutions strongly differ. IIHR is able to capture the traces and evolution of the five tip vortices while PNU provides globally identical pictures for the three phase-averaged results which do not exhibit any temporal evolution. It seems that the natural unsteadiness of the flow in the wake of the rotating propeller is not captured by this simulation. The same remark applies

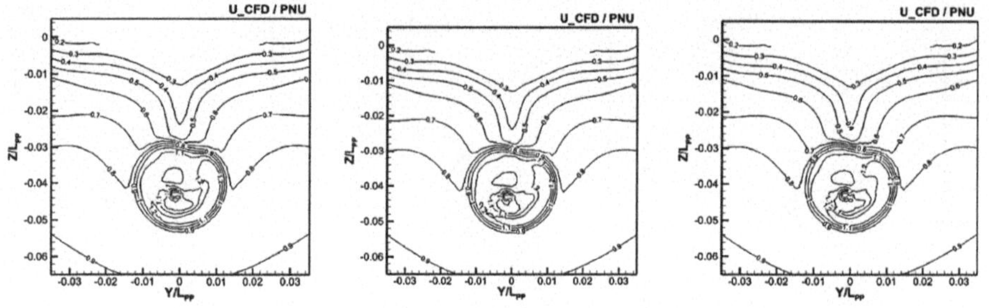

(a) 0° *(b) 24°* *(c) 48°*

Fig. 169 Case 1-7—PNU/Fluent—distribution at station S7 of isowakes at 0°, 24° and 48°

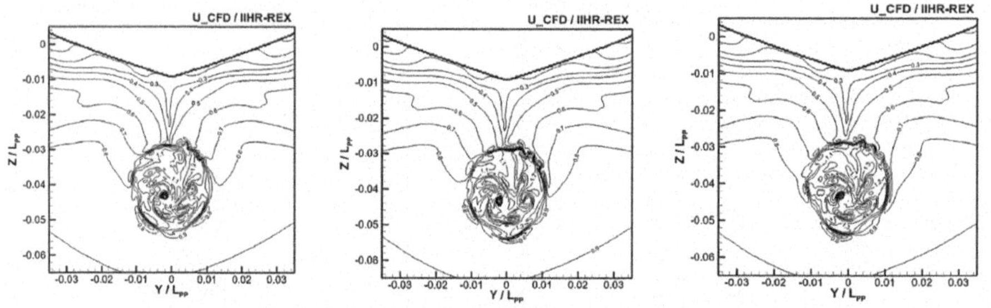

(a) 0° *(b) 24°* *(c) 48°*

Fig. 170 Case 1-7—IIHR/Rex—distribution at station S7 of isowakes at 0°, 24° and 48°

to MARIN's results which computed the flow with a rotating propeller (see Fig. 165). The agreement between IIHR and NMRI's phase-averaged experiments is globally good, taking into account the experimental uncertainty.

6.5 *Case 1-7—Concluding Remarks*

This section provided a short analysis of the local flow around the JBC with a rotating propeller at one station S2 located before the propeller and two stations S4 and S7, located after the propeller. Various meshes, turbulence models and propeller treatments were used by the contributors, which renders a rigourous analysis almost impossible at this stage. However, a few general conclusions based on the comparison of the results can be drawn.

- The computations based on the simulation of the actual propeller are the only ones which are able to represent all the characteristics of the flow in the wake of propeller, specifically at station S7. Various codes using different meshes and different turbulence models yield globally comparable results in terms of isowake distribution when the real rotating propeller is taken into account in the simulation.

In particular, the highest isowake level is systematically captured by the simulations and its extent is very correctly reproduced.

- However, the price to pay to get this level of agreement is very high, due to the use of a very small time step (about one time step per degree of rotation), which makes this approach not realistic for computations in waves for instance. Various alternate coupling strategies based on simplified propeller models are therefore of great practical interest although less accurate in terms of local representation of the isowake distribution. Globally, the local flow produced by these simplified models is quite satisfactory for some of them, even if they are not able to capture all the characteristics of the wake flow. Some coupled models are not satisfactory but at this stage, it is difficult to identify the reasons which explains these disappointing performances.
- A systematic study of the influence of turbulence models is hardly feasible on the basis of these results. It seems that, due to the fact that the flow after the propeller is pressure-controlled, the turbulence model plays a more minor role on the averaged flow field. Classical linear eddy viscosity based closures are doing reasonably well. Hybrid RANS/LES model used by IIHR/Rex yields very good results but not necessarily way superior for the average flow field to what is obtained with usual k-ω-like models to justify the very high cost associated with this time-accurate approach. However, this hybrid RANS/LES model is the only one able to capture most of the details of the flow revealed by the phase-averaged measurements, which makes it probably useful when propeller/rudder interaction during maneuvers are considered, in the absence of any new RANS model.

7 Test Case 1-8

7.1 Description of the Experiments

Test case 1-8 concerns the local flow analysis around the JBC with a ducted propeller towed in the NMRI towing tank, in free trim and sinkage conditions. The location of the experimental stations are shown in Figs. 171 and 172 shows the ducted propeller. With these last experiments, we are going to be able to determine the influence of the duct on the local flow with a rotating propeller behind it while the test case 1-4 provided the same kind of information without the presence of a propeller.

Figures 173, 174, 175, 176, 177 and 178 show the influence of the duct on the experimental time-averaged iso-wakes and transversal velocity components at three sections S2, S4 and S7 already defined before. Already at station S2, the influence of the duct is visible through an intensification of the longitudinal vorticity close to the wall. This is indirectly indicated by the isowake u = 0.4 which is closed when the duct is present. The same influence was noticed previously without the presence of a propeller. It is also worthwhile to notice the presence of the closed isowake u = 0.5 which is more extended when the duct is there. The experiments at station

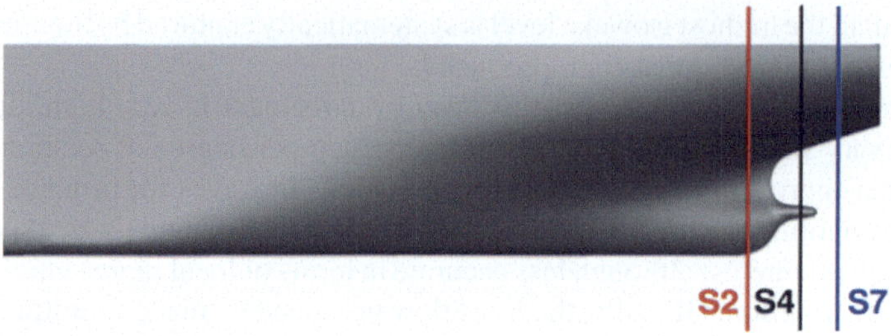

Fig. 171 Case 1-8—hull geometry and experimental stations

Fig. 172 Case 1-8—view of the stern with its ducted propeller

S4 are hardly exploitable but station S7 provides very interesting information on the influence of the duct in the wake of the propeller. The duct leads to a general reduction of the longitudinal velocity. For instance, the isowake u = 1.3 has disappeared when the duct is mounted. The region where u is comprised between 1.1 and 1.2 is considerably extended with a duct, which means that the duct creates a noticeable homogeneization of the longitudinal velocity distribution in the wake of the propeller, even in the central part of the propeller disk. A by-product of this effect is the shape of the isowake u = 1.2 which is strongly reduced to the right part of the propeller disk when the duct is there.

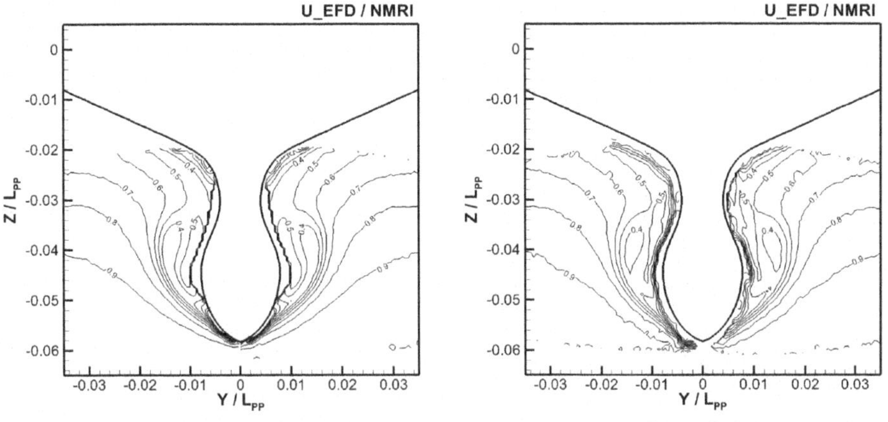

(a) Without duct (b) With duct

Fig. 173 Case 1-8—experiments—influence of the duct on the isowakes distribution at station S2

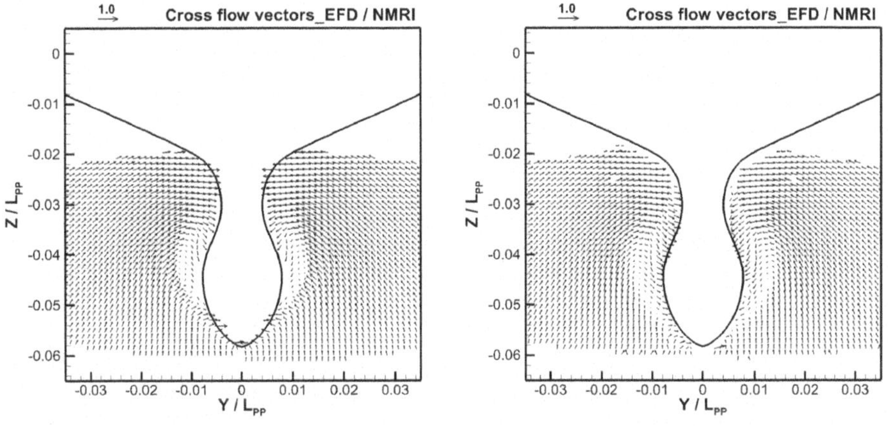

(a) Without duct (b) With duct

Fig. 174 Case 1-8—experiments—influence of the duct on the secondary velocities distribution at station S2

Figures 179, 180, 181, 182, 183 and 184 provide the same information for the phase-averaged iso-wakes at stations S4 and S7 for the same angles as previously, i.e. 0°, 24° and 48°. As mentioned before, experiments at station S4 are hard to analyze but station S7 provides again very interesting information on the influence of the duct on the phase-averaged flow. Figure 182 confirms that the wake is more homogeneous in case of the presence of a duct. The region where u is greater than 1.2 is strongly reduced at 0° in case of duct and the central part of the wake appears less noisy. At 24° (see Fig. 183), the region where u is greater than 1.3 is also strongly reduced when the duct is present. At 48° (see Fig. 184), an opposite influence is

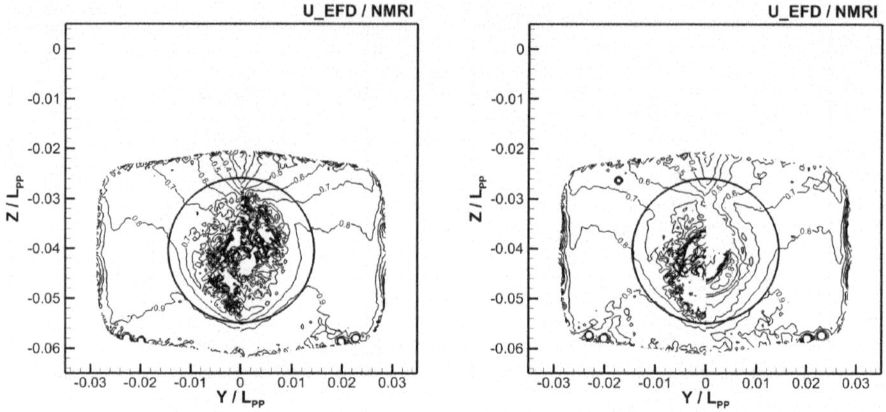

(a) Without duct *(b) With duct*

Fig. 175 Case 1-8—experiments—influence of the duct on the mean isowakes distribution at station S4

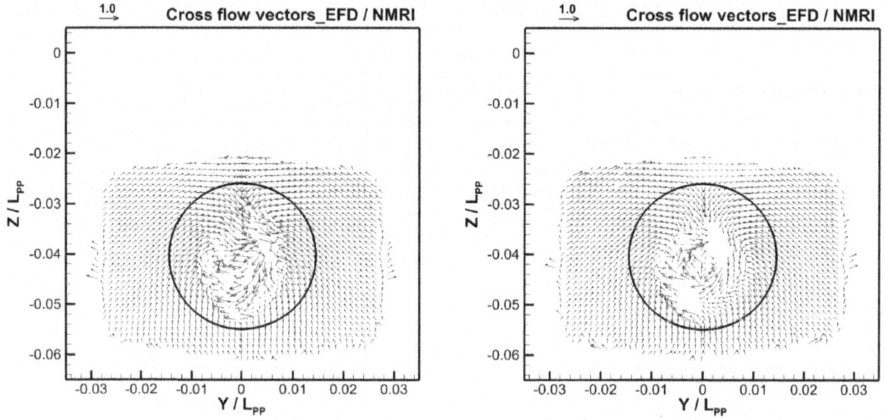

(a) Without duct *(b) With duct*

Fig. 176 Case 1-8—experiments—influence of the duct on the secondary velocities distribution at station S4

noticed: the region where u is greater than 1.2 is slightly more extended when the duct is mounted than without it. This compensation for various blade angle might contribute to the global homogeneization of the averaged flow field observed when the duct is present.

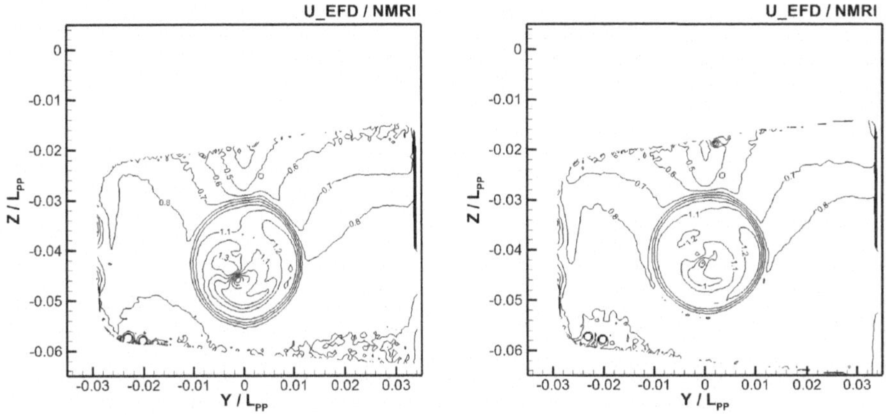

(a) Without duct *(b) With duct*

Fig. 177 Case 1-8—experiments—influence of the duct on the mean isowakes distribution at station S7

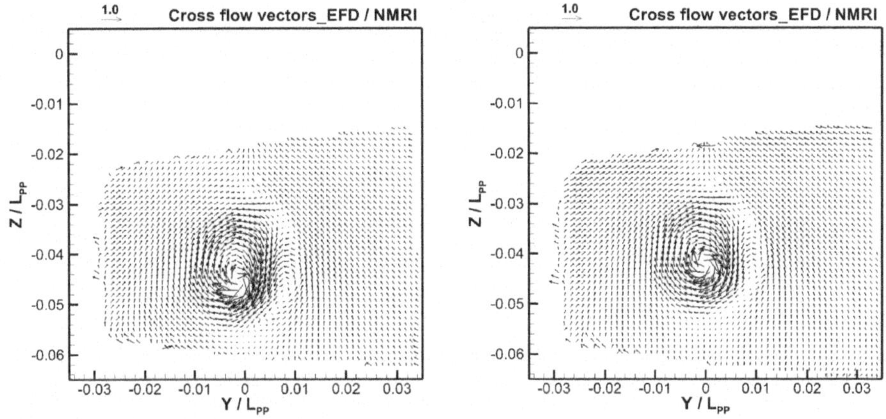

(a) Without duct *(b) With duct*

Fig. 178 Case 1-8—experiments—influence of the duct on the secondary velocities distribution at station S7

7.2 Review of Contributions

Seventeen participants provided results for the test case 1-8. The main characteristics of their contributions are listed in Table 8 in terms of turbulence closures, wall treatments, discretization and propeller treatment. Some participants provided several different contributions, which makes possible a comparison of various turbulence closures or wall treatments for the same grid and same code. Moreover, SOTON

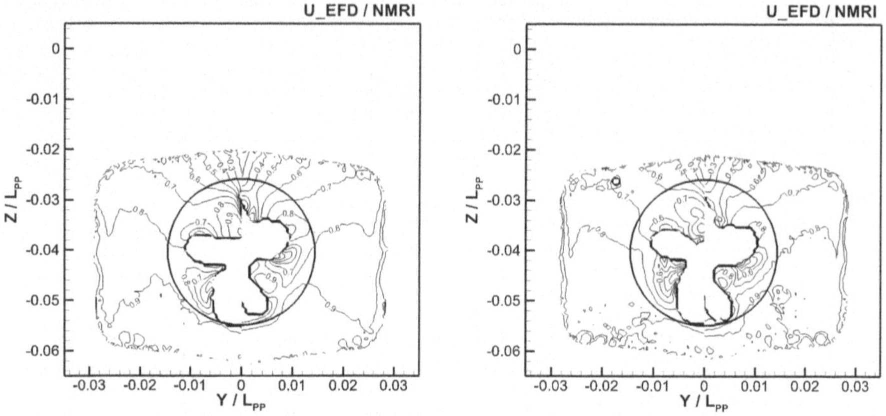

(a) *Without duct* (b) *With duct*

Fig. 179 Case 1-8—experiments—influence of the duct on the isowakes distribution at 0° and station S4

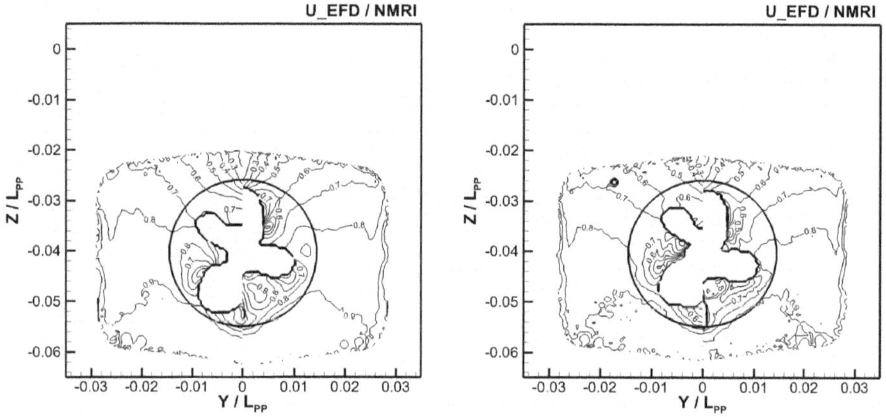

(a) *Without duct* (b) *With duct*

Fig. 180 Case 1-8—experiments—influence of the duct on the isowakes distribution at 24° and station S4

performed a comparison of codes (OpenFoam vs. Star-CCM+) with the same grid and same turbulence closure, as previously.

Except University of Rostock which is no more present in this list, the contributors to case 1-8 were the same as the ones who contributed to case 1-7. For the sake of conciseness, the characteristics of their contributions are not recalled here and the reader could refer to the comments accompanying Table 7 in case of irrepressible need.

Table 8 Case 1-8—main characteristics of the computations

Organization Code name	Turbulence model	Wall model	Grid characteristics	Propeller treatment
ABS OpenFOAM	k-ω SST	Wall function	Unstructured grid 7 452 332 cells	Actual propeller
Chalmers-Flowtech Shipflow	EARSM	Wall resolved	Structured grid with overlapping 22.4M cells	Lifting line
CNR-INSEAN Xnavis	Spalart-Allmaras		Structured grid with overlapping 11.4M cells	Body-force/actual propeller
ECN-CNRS ISIS-CFD aka FINE^{TM}/Marine	SST/EARSM	Wall function Wall resolved	Unstructured grid 25.3M cells	Body-force/Actual propeller
HHI HiFOAM/StarCCM+	Realizable k-ϵ/RSTM	Wall function	Unstructured grid 3.5M cells	Moving Reference Frame
HSVA FreSCo+	2 equations	Wall function	Unstructured grid 14.6M cells	Vortex lattice
IIHR REX	DDES	Wall resolved	Unstructured grid 70.9M cells	Actual propeller
KRISO Wavis	Realizable k-ϵ/EARSM	Wall function	Structured grid 15.5M cells	Lifting surface
MARIC FINE^{TM}/Marine	k-ω SST	Wall function	Unstructured grid 2M cells	Body force
MARIN ReFRESCO	k-ω SST + DM correction	Wall resolved	Unstructured grid 23M cells	Actual propeller
MIJAC OpenFOAM	k-ω SST	Wall resolved	Unstructured grid 8.7M cells	Simplified propeller theory
NMRI Nagisa	EARSM	Wall resolved	Structured grid 10M cells	Simplified propeller theory
PNU Fluentv15	2 equations k-ϵ RNG	Wall resolved Wall function	Unstructured grid 17.5M cells	Actual propeller
SHIME OpenFOAM	k-ω SST	Wall resolved	Unstructured grid 22.15M cells	N/A
SJTU naoeOpenFOAM	k-ω SST	Wall function	Unstructured grid 4.9M cells	Simplified propeller theory
SOTON OpenFOAM/StarCCM+	k-ω SST	Wall resolved	Structured grid 21.3M cells	Body force prescribed
YNU SURFv7	EARSM	Wall resolved	Structured grid 3.3M cells	Simplified propeller theory

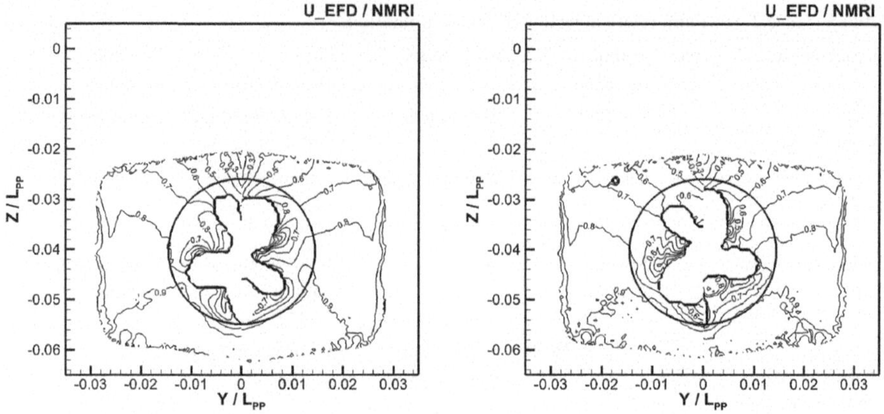

(a) *Without duct* (b) *With duct*

Fig. 181 Case 1-8—experiments—influence of the duct on the isowakes distribution at 48° and station S4

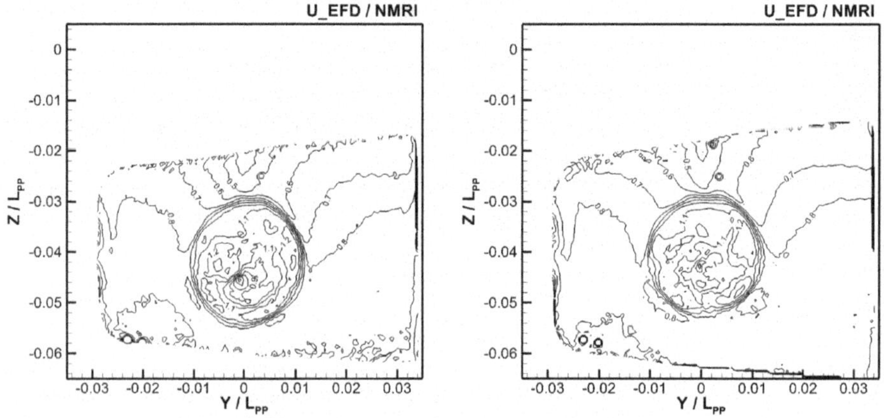

(a) *Without duct* (b) *With duct*

Fig. 182 Case 1-8—experiments—influence of the duct on the isowakes distribution at 0° and station S7

7.3 Influence of the Duct on the Local Flow

In order to see if the computations are able to reproduce the salient features characterising the influence of the duct on the local flow and the unsteady nature of the flow in the wake of the propeller, one will start with the more complete computational approach, i.e. the one based on an actual rotating propeller and an hybrid RANS/LES turbulence model. Then, let us analyse the contribution of IIHR/Rex to check if these computations are able to catch the local influence of the duct on the flow before and after the propeller. Figures 185 and 186 show the isowakes with and without duct

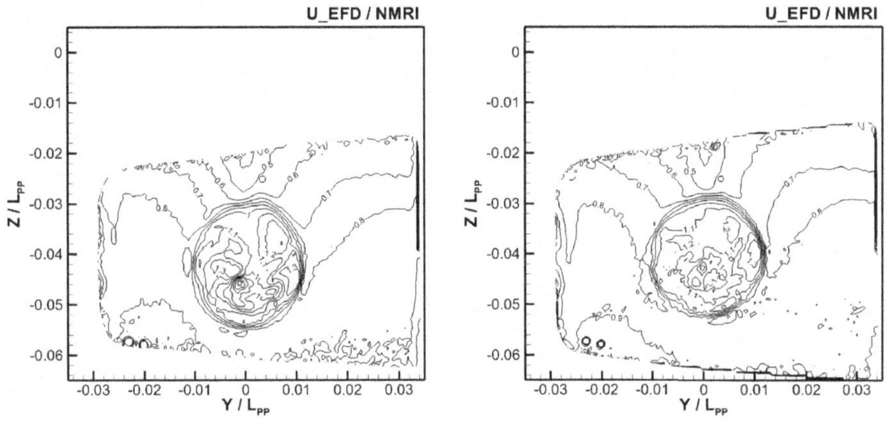

(a) Without duct (b) With duct

Fig. 183 Case 1-8—experiments—influence of the duct on the isowakes distribution at 24° and station S7

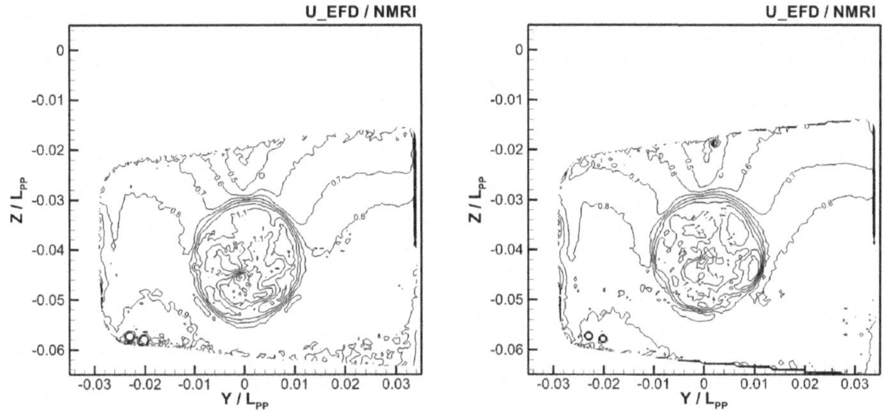

(a) Without duct (b) With duct

Fig. 184 Case 1-8—experiments—influence of the duct on the isowakes distribution at 48° and station S7

computed by IIHR/Rex at stations S2 and S7. These computations are based on an hybrid RANS-LES model (DDES). At station S2 (see Fig. 185), computations do not show any difference between the two configurations, which means that, either the experiments are not accurate enough to assess this level of detailed comparisons, or the computational model is not able to capture the modification of the upstream influence of the duct at station S2. Once again, it should be noticed that the peculiar shape of the closed isowake u = 0.5 in the vicinity of the hull at station S2 is not captured by these computations. We will come back to this feature in the next subsection devoted to the influence of turbulence closures.

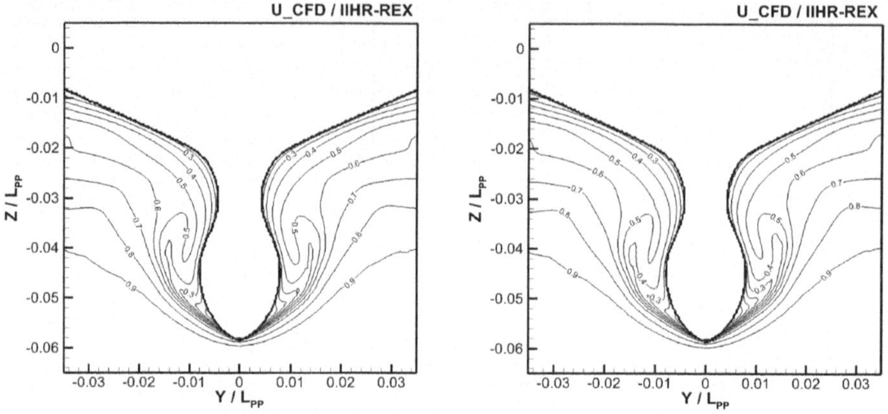

(a) Without duct *(b) With duct*

Fig. 185 Case 1-8—IIHR/Rex—influence of the duct on the distribution of isowakes at station S2 with a propeller

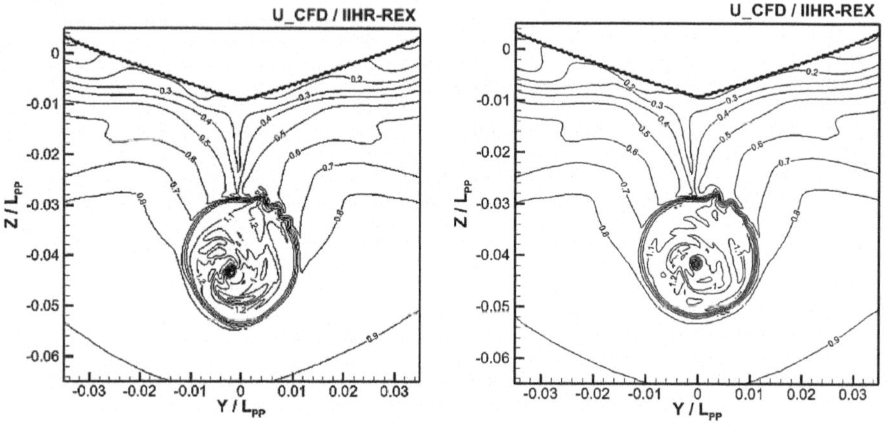

(a) Without duct *(b) With duct*

Fig. 186 Case 1-8—IIHR/Rex—influence of the duct on the distribution of isowakes at station S7 with a propeller

At station S7 (see Fig. 186), the trend towards an homogeneisation of the wake in presence of a duct is well captured by IIHR/Rex. Moreover, the isowake $u = 1.2$ has almost completely disappeared in the simulations with a duct, which is in good agreement with the experimental observation.

Figures 187, 188 and 189 show a comparison of computations performed by IIHR/Rex with the same computational approach as before at station S7 with and without duct for the phase-averaged flow measurements at 0°, 24° and 48°. At 0°, the number of vortices seems reduced (associated with closed isowakes) when the

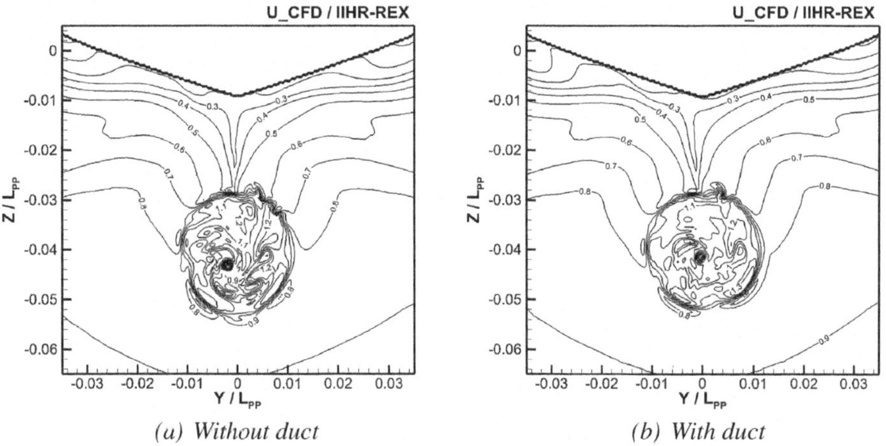

Fig. 187 Case 1-8—IIHR/Rex—influence of the duct on the distribution of isowakes at 0° and station S7 with a propeller

duct is present, which is in line with the measurements (see Fig. 187) but one notices an isowake u = 1.3 which is not visible in the experiments. At 24°, (see Fig. 188), computations with duct look again more homogeneous with less local intense structures. Moreover, now the isowakes u = 1.3 have disappeared when the duct is there, which is in agreement with the experimental observation. At 48°, (see Fig. 189), the isowake distribution with duct appears more homogeneous than without duct as noticed for the other blade positions. It is very difficult to analyze in more details these computational results, due to the complexity of the flow in the wake of the propeller.

The previous analysis was performed by using an hybrid RANS/LES turbulence model. It is interesting to see what can be obtained with a more classical RANS model activated in URANS mode. As previously, Figs. 190 and 191 show the time-averaged isowakes at stations S2 and S7, computed by PNU/Fluent using a RNG k-ε turbulence model and an actual rotating propeller. This specific contribution is chosen because it is computed on a fine grid comprised of 17.5 M cells, which should reduce the discretization errors. It is clear from these figures that the presence of the duct has no influence at station S2, contrary to what is shown by the experiments. At station S7, the main effect of the duct is to reduce the extent of the iso-wake u = 1.2 region and to increase significantly the area associated with the iso-wake u = 1.1. This is in very good agreement with the experimental trend. This leads to think that one can rely on RANS simulations for global predictions of time averaged velocity, even in the wake of the propeller.

Figures 192, 193 and 194 show the phase-averaged distributions of isowake at station S7 for 0°, 24° and 48°. Contrary to the previous results of IIHR/Rex, all these three figures are very similar and do not show any occurence of tip vortices. Probably,

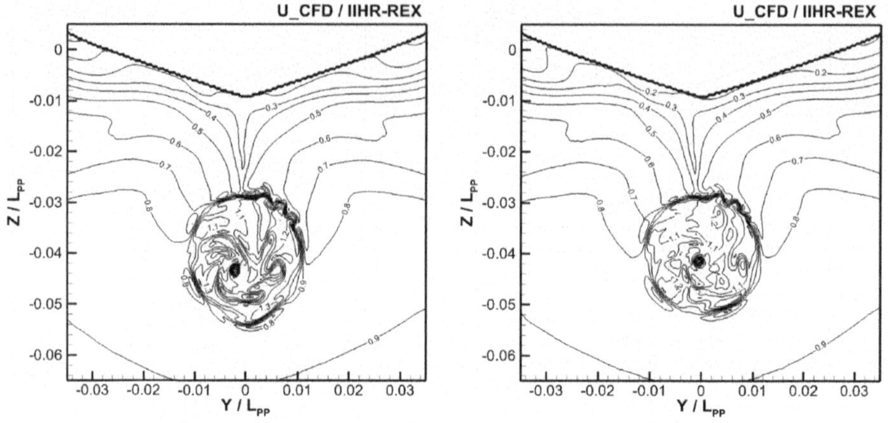

(a) Without duct *(b) With duct*

Fig. 188 Case 1-8—IIHR/Rex—influence of the duct on the distribution of isowakes at 24° and station S7 with a propeller

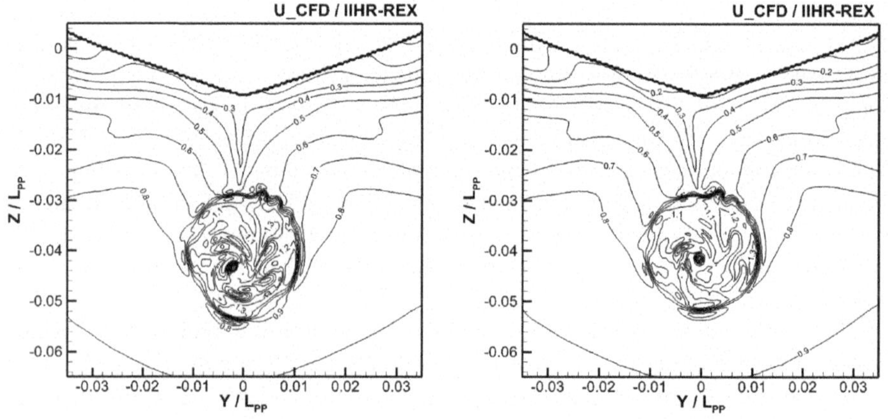

(a) Without duct *(b) With duct*

Fig. 189 Case 1-8—IIHR/Rex—influence of the duct on the distribution of isowakes at 48° and station S7 with a propeller

the tip vortices are already dissipated at this station, due to the combined influence of a grid which is locally too coarse and the use of RANS turbulence model.

All the previous computations were made on the basis of computations taking into account the actual rotating propeller. It is also interesting to see what can be obtained with a modelled propeller without taking into consideration the axisymmetric body-force model which is obviously not able to represent the flow in the wake of a propeller. For instance, Chalmers provided us with a simulation using a propeller modelled on the basis of a lifting line model. Figure 195 shows the time-averaged

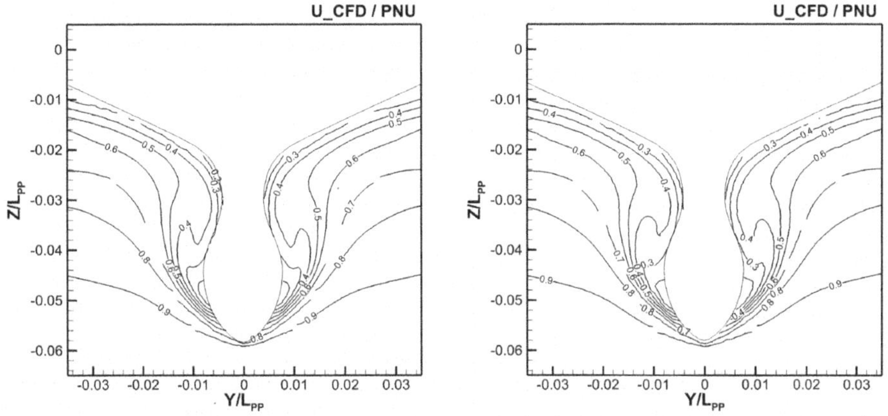

(a) Without duct (b) With duct

Fig. 190 Case 1-8—PNU/Fluent—influence of the duct on the distribution of isowakes at station S2 with a propeller

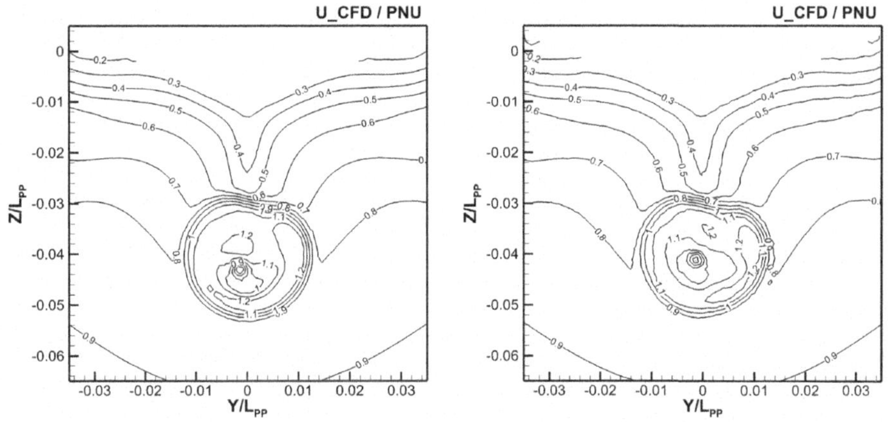

(a) Without duct (b) With duct

Fig. 191 Case 1-8—PNU/Fluent—influence of the duct on the distribution of time-averaged isowakes at station S7 with a propeller

isowakes at station S7 with and without duct as usual. Although the isowake u = 1.2 is not captured by the computations without duct, the trend towards an homogeneization of the wake due to the action of the duct is also well captured by this contribution. This leads us to think that it is not necessary to use an actual propeller to get the trend right, which, if it is confirmed, is a valuable result in the perspective of duct shape optimisation.

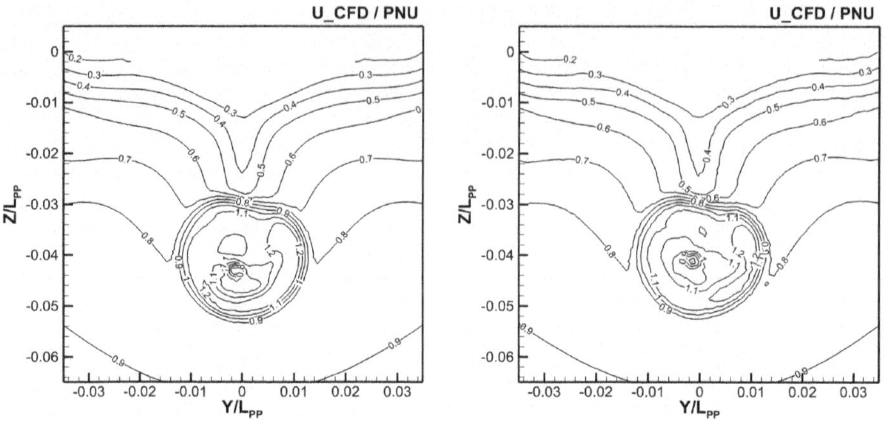

(a) *Without duct* (b) *With duct*

Fig. 192 Case 1-8—PNU/Fluent—influence of the duct on the distribution of isowakes at 0° and station S7 with a propeller

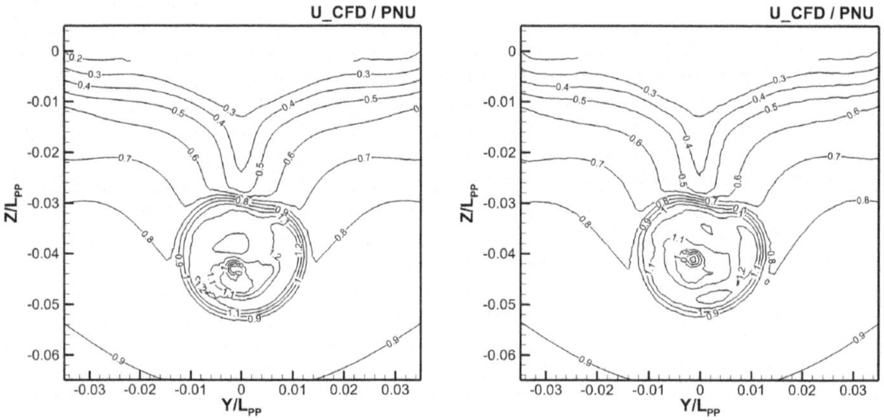

(a) *Without duct* (b) *With duct*

Fig. 193 Case 1-8—PNU/Fluent—influence of the duct on the distribution of isowakes at 24° and station S7 with a propeller

7.4 Influence of Turbulence Closures

As previously, the isowakes at station S2 are better predicted with non-linear EARSM turbulence models as shown by Fig. 196 but this station S2 is not significantly influenced by the presence of the ducted propeller. IIHR/Rex hybrid RANS/LES closure does not behave well at station S2 probably because the LES model is not yet activated there and the flow prediction relies on the linear isotropic k-ω SST which is known to under-predict the longitudinal vorticity.

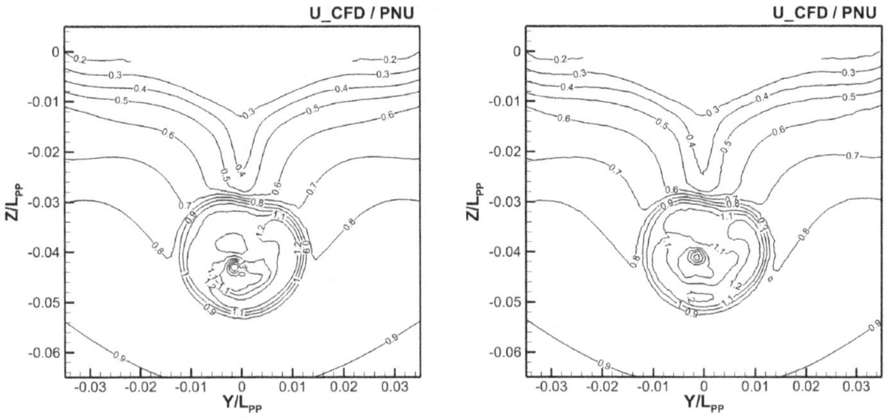

(a) Without duct (b) With duct

Fig. 194 Case 1-8—PNU/Fluent—influence of the duct on the distribution of isowakes at 48° and station S7 with a propeller

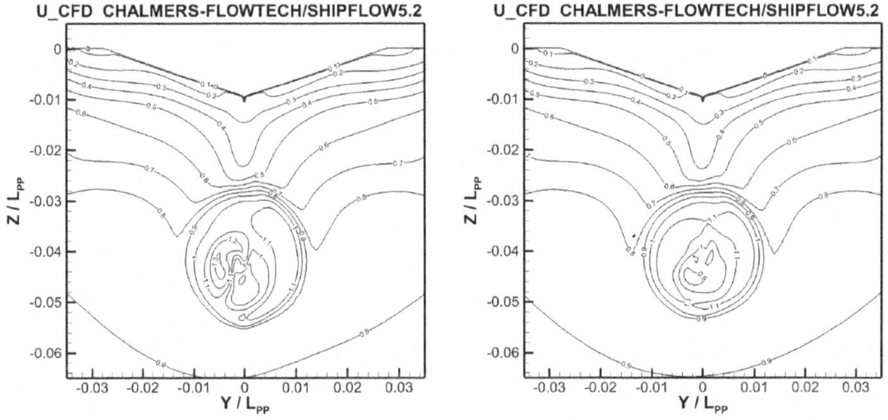

(a) Without duct (b) With duct

Fig. 195 Case 1-8—CHALMERS/ShipFlow—influence of the duct on the distribution of time-averaged isowakes at station S7 with a propeller

Let us now compare a classical RANS model with an hybrid RANS/LES model implemented in the same code, IIHR/Rex, in the wake of the ducted propeller. Figure 197 shows the time-averaged isowakes at station S7 for k-ω SST and DDES turbulence closures with a ducted rotating propeller. At first glance, the k-ω SST model seems to provide a solution in good agreement with the experiments since it is able to predict the existence of the crescent-shaped zone u = 1.2 even if its extent is under evaluated. On the other hand, hybrid RANS/LES results appear much more

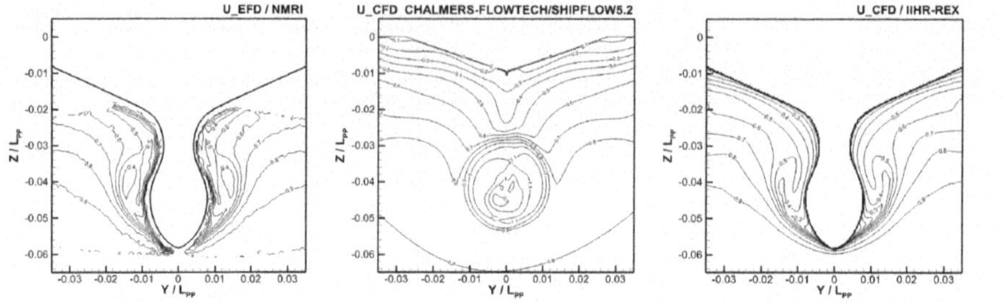

(a) Experiments (b) EARSM Chalmers/ShipFlow (c) Hybrid RANS/LES IIHR/Rex

Fig. 196 Case 1-8—comparison of different turbulence closures at station S2

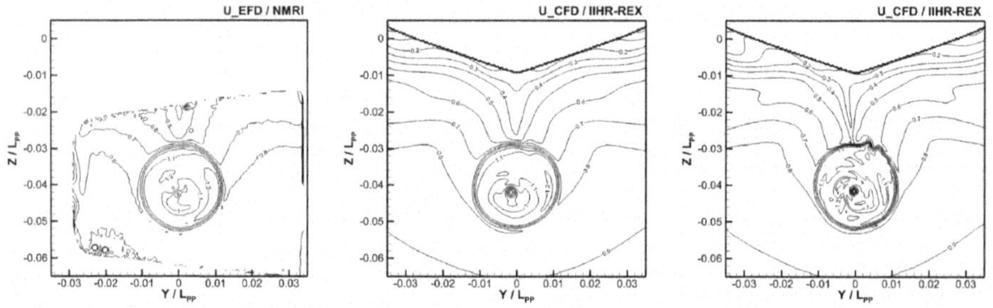

(a) Experiments (b) SST (c) DDES

Fig. 197 Case 1-8—IIHR/Rex—distribution of the averaged isowakes at station S7 for RANS and hybrid RANS/LES turbulence closures

noisy as if the statistical convergence was not yet reached (see the irregular boundary of the isowake u = 1.1 at the top of the propeller disk) and the isowake u = 1.2 is hardly visible. Globally, based on the comparison of isowakes, it is difficult to justify having recourse to the hybrid RANS/LES model which is, at least, ten times more expensive.

Figures 198, 199 and 200 give the same comparison at station S7 for phase-averaged isowakes. Hybrid RANS/LES based simulation provides richer isowakes than what is computed by the URANS approach, which is expected but can be related with a lack of statistical convergence, as well. For each computation, one can follow the evolution of vortical structures with respect to the blade angles, which means that even with a RANS approach, it is possible to capture the traces of the tip vortices at this station, contrary to what was obtained previously by PNU/Fluent using a RNG k-ε turbulence model. It is hard to assess the results in a more detailed way, due to the complexity of the phase-averaged flow.

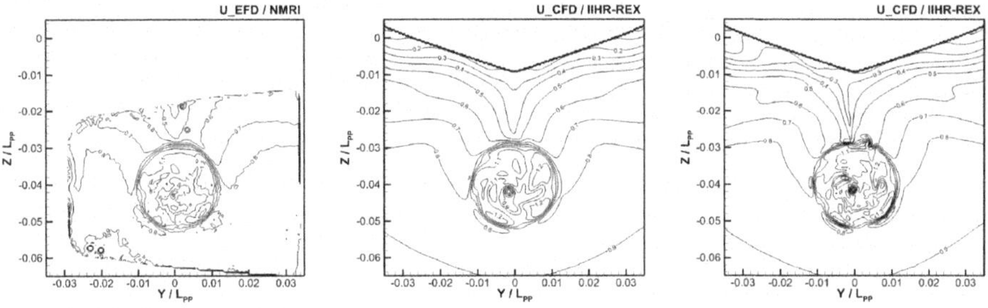

(a) Experiments (b) SST (c) DDES

Fig. 198 Case 1-8—IIHR/Rex—distribution of isowakes at 0° and station S7 for RANS and hybrid RANS/LES turbulence closures

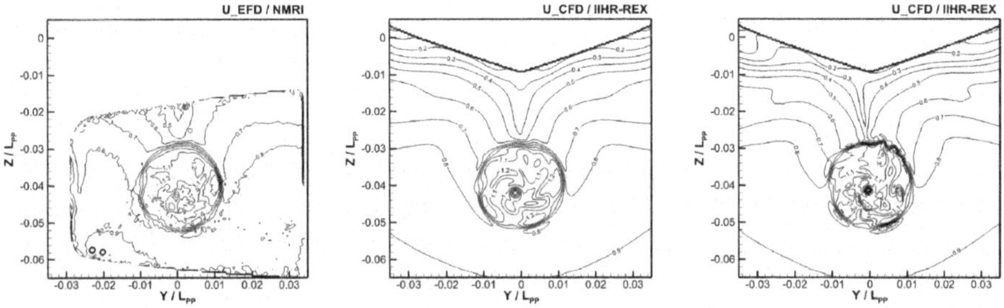

(a) Experiments (b) SST (c) DDES

Fig. 199 Case 1-8—IIHR/Rex—distribution of isowakes at 24° and station S7 for RANS and hybrid RANS/LES turbulence closures

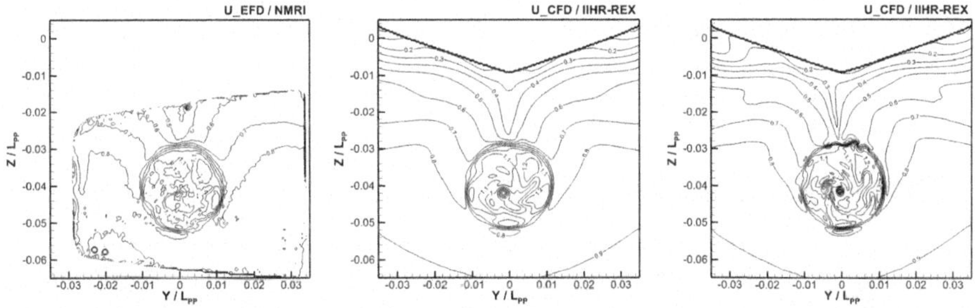

(a) Experiments (b) SST (c) DDES

Fig. 200 Case 1-8—IIHR/Rex—distribution of isowakes at 48° and station S7 for RANS and hybrid RANS/LES turbulence closures

7.5 Case 1-8—Concluding Remarks

This section provided a short analysis of the local flow around the JBC with a ducted rotating propeller at one station S2 located before the propeller and two stations S4 and S7, located after the propeller. Various meshes, turbulence models and propeller treatments were used by the contributors, which renders a rigourous analysis almost impossible at this stage. However, a few general observations that stem from this study are summarized here:

- When a duct is mounted before the propeller, the flow in the wake of the propeller is more homegeneous as indicated by the experimental distribution of isowakes. This influence is correctly captured by most of the contributors, whatever the turbulence model or propeller representation used.
- The isowake distribution behind the ducted propeller is better represented when the actual rotating propeller is accounted for. However, some simplified propeller models work reasonably well and are able to indicate the right trend concerning the influence of the duct.
- RANS models provide a reasonably accurate prediction of the time-averaged flow behind the propeller. For phase-averaged quantities, the analysis of experimental and computational results is hard because of the complexity of the phase-averaged flow field. Some URANS-based computations are in fair agreement with the experiments while some others do not show any unsteadiness in the wake of the propeller, probably due to the combined influence of a grid which is locally too coarse and a turbulence model which is too diffusive. From that standpoint, the hybrid RANS/LES model appears more consistent with the physics of the flow. This is encouraging if one considers the level of details revealed by this computation although the provided contribution from IIHR/Rex might suffer from an unsufficient statistical convergence.

8 General Conclusions

In this chapter, we have reviewed computations of the flow around the Japanese Bulk Carrier (JBC) provided by the contributors to the Tokyo T2015 workshop from the standpoint of the local flow analysis. Five flow configurations were studied: (i) the naked hull without propeller nor duct (case 1-3a), (ii) the double body model (case 1-3b), (iii) the hull with a duct but without propeller (case 1-4), (iv) the hull with a propeller and without duct (case 1-7) and finally, (v) the hull with a ducted propeller (case 1-8). This chapter aimed at understanding the physics of the flow from a local point of view with the help of the flow measurements performed by NMRI at three stations S2, S4 and S7. But several additional objectives were pursued during this analysis. The first one, in the continuity of the G2010 Gothenburg workshop concerned the verification and validation of stern flows over the JBC hull. This hull is characterised by a relatively high block coefficient, which means that the flow is more complex in the stern region. An intense bilge vortex was detected and some questions arose about the steadiness of the flow in the stern region. It was therefore decided to focus the study on the bilge vortex core and try to compare the various contributions in this specific region. The second objective was to understand the mechanisms which might improve the propulsive efficiency when a ship is equipped with a ducted propeller. To separate physical effects, several configurations with and without duct combined with the presence or absence of a propeller were studied with the help of the measurements made by NMRI at stations S2, S4 and S7. This gave us the opportunity (i) to compare the various turbulence closures in presence of a duct or a propeller, (ii) to assess the reliability of the various propeller representations implemented by the contributors, and (iii) to check if the observed influence of the duct on the local flow is accurately captured by the computations.

The main conclusions that stem from all these studies are summarized as follows.

For the case 1-3a, since the typical grids which are used by the contributors (5–10 M points) are fine enough, the influence of the grid discretisation is moderate for RANSE, which means that the numerical error appears to be under control. It was feasible therefore to focus the study on the influence of the turbulence closures and see if it was possible to draw general conclusions form all the contributions. As noticed during the Gothenburg2010 workshop for the analysis of the local flow around the KVLCC2, the major influence comes from the turbulence closure. One noticed that the linear isotropic closures significantly under-predict the longitudinal vorticity while full RSM closures slightly over-predict it. Non-linear anisotropic closures (EARSM) seemed to offer a good compromise from the standpoint of the local flow although they slightly underpredict the vorticity at the key station S2. Contributions employing hybrid RANS-LES were also taken into account. The results appeared promising although IDDES seemed to over-predict the vorticity again. No spectacular advantage was noticed compared to the best RANS models at the measurement stations.

This study was completed by a local core vortex analysis which provided a first interesting attempt to carry out a more local analysis of the time-averaged bilge vortex. Globally, the agreement between most of the computations and the rather coarse experiments was satisfactory in terms of global trends. The longitudinal vorticity was somewhat underestimated and the longitudinal velocity distribution was fairly reproduced. These experiments pointed out very large differences on the turbulence kinetic energy and this was the most striking (and unexpected) result of this comparison. Only the hybrid RANS/LES closures were able to reproduce this characteristic, thanks to the contribution of the resolved turbulence kinetic energy. Whether the NMRI TKE measurements are reliable or not to provide an accurate turbulence kinetic energy at this location is still a matter of intense debate which is partly addressed in chapter "Experimental Data of Resistance, Sinkage, Trim, Self--propulsion Factors, Longitudinal Wave Cut and Detailed Flow for JBC With and Without an Energy Saving Circular Duct" of this book. Further much more detailed measurements will be necessary to draw safer conclusions.

In the case 1-3b, the turbulence closure study led to very similar conclusions; i.e. that the major influence on the computed distribution of isowakes in the stern region comes from the turbulence closure. Linear isotropic closures under-predict the longitudinal vorticity at station S2 while full RSM closures tend to over-predict it at the same station. But what was remarkable and quite unique was that two LES computations were presented and especially, one from SRC on an extremely fine grid (4 billion cells) which was in good agreement with measurements and predicted the level of turbulence kinetic energy measured by NMRI in the core of the bilge vortex (at least three times higher than what was modeled by EASM). Unfortunately, it was not possible to exploit further the results of the (almost) wall-resolved LES simulation, due to the unavailability of this flow database.

For the case 1-4, it was again noticed that the turbulence models have a critical role to play once the grid is fine enough. The main modifications of the flow created by the presence of the duct were correctly reproduced by linear isotropic models which provided the correct qualitative trend. A better quantitative agreement was reached by anisotropic closures although the inner part of the vortex was not perfectly reproduced as illustrated by the occurence of a reversed flow region which was over-estimated by both RSTM and EARSM turbulence closures.

For the test case 1-7, it was observed that the computations based on the simulation of the actual propeller were the only ones able to represent all the characteristics of the flow in the wake of propeller, specifically at station S7. Various codes using different meshes and different turbulence models yielded globally comparable results in terms of isowake distribution when the real rotating propeller is taken into account in the simulation. However, alternate coupling strategies based on simplified propeller models appeared to produce a local flow which was quite satisfactory for some of them, even if they were not able to capture all the characteristics of the wake flow. Hybrid RANS/LES model used by IIHR/Rex yielded very good results but not necessarily way superior for the averaged flow field to what is obtained with usual k-ω-like models to justify the very high cost associated with this time-accurate approach. However, this hybrid RANS/LES model was the only one able to capture most of the details of the flow revealed by the phase-averaged measurements.

For the test case 1-8, the computations confirmed the fact that when a duct is mounted before the propeller, the flow in the wake of the propeller is more homogeneous. This influence was correctly captured by most of the contributors, whatever the turbulence model or propeller representation used. As expected, the isowake distribution behind the ducted propeller was better represented when the actual rotating propeller was accounted for. However, some simplified propeller models worked reasonably well and were able to indicate the right trend concerning the influence of the duct. RANS models provided a reasonably accurate prediction of the time-averaged flow behind the propeller. For phase-averaged quantities, some URANS-based computations were in fair agreement with the experiments while some others did not exhibit any unsteadiness in the wake of the propeller, probably due to the combined influence of a locally too coarse grid and a too diffusive turbulence model. From that standpoint, the hybrid RANS/LES model appeared again more consistent with the physics of the flow.

Finally, some overall considerations can be drawn to conclude this sixth chapter:

(i) Most of the contributors were able to predict the main characteristics of the flows considered in this chapter, provided that the grid is fine enough. And computing on a fine enough grid is no more out of reach, nowadays,

(ii) Once discretization error is under control, the main source of discrepancy is the modelling error, i.e. the turbulence model. From that standpoint, the situation has not much changed compared to the previous edition of this workshop in 2010. Linear isotropic k-ω SST closure remains the usual workhorse because of its robustness and global reliability. However, better results in terms of local flow field like longitudinal vorticity are obtained if one takes into account the turbulence anisotropy. This improvement has an non-negligible impact (a few percents) on the prediction of the resistance.

(iii) The influence of the duct on the flow with and without propeller is correctly captured by a linear isotropic model, at least in terms of trend. This means that it does not seem mandatory to have recourse to sophisticated turbulence models to perform a design study aiming at improving the efficiency of a ducted propeller. However, special care should be used to model the downwind influence of the propeller which is not represented by a simple body force model. Here again, the propeller model appears to be the weak point which has to be improved if one wishes to avoid lengthy computations of the actual rotating propeller.

(iv) Having recourse to more sophisticated and expensive turbulence models like hybrid RANS-LES closures brings much more physics to the local description of flow fields. Illustrations were given in this chapter for instance, by the study of the wake of a rotating propeller or by the fine analysis of the turbulence characteristics in the core of the vortex. Although RANS statistical turbulence closures are able to capture the main flow characteristics, the unsteady hybrid RANS/LES closures provide reliable information on the turbulence characteristics that are out of reach of statistical turbulence models, whatever their complexity.

(v) In several parts of this analysis, it was difficult to draw any firm conclusion because of the lack of reliable data to characterize turbulent fluctuations. There is clearly a need of better experiments in that respect to deepen the analysis of the physical models. I would therefore recommend for the next workshop to complete, as much as possible, the mean flow measurements by reliable experimental information on the velocity fluctuations like the Reynolds Stress tensor or at least the turbulence kinetic energy.

Acknowledgements The study performed by ECN/CNRS was granted access to the HPC resources under the allocation 2015-2a1308 made by GENCI (Grand Equipement National de Calcul Intensif).

References

Bhushan, S., Yoon, H., Stern, F., Guilmineau, E., Visonneau, M., Toxopeus, S., et al. (2019). Assessment of computational fluid dynamics for surface combatant 5415 at straight ahead and static drift β=20°. *Journal of Fluids Engineering, 141*(5), 051101.

Guilmineau, E., Deng, G., Leroyer, A., Queutey, P., Visonneau, M., & Wackers, J. (2018). Numerical simulations for the wake prediction of a marine propeller in straight ahead flow and oblique flow. *Journal of Fluids Engineering, 140*, 1–11.

Larsson, L., Stern, F., & Visonneau, M. (Eds.). (2014). *Numerical ship hydrodynamics: An assessment of the Gothenburg 2010 workshop*. Springer.

Lee, S.-J., Kim, H.-R., Kim, W.-J., & Van, S.-H. (2003). Wind Tunnel tests on flow characteristics of the KRISO 3,600 TEU containership and 300K VLCC double-deck ship models. *Journal of Ship Research, 47*(1), 24–38.

Visonneau, M., Deng, G., Guilmineau, E., Queutey, P., & Wackers, J. (2016, September). Local and global assessment of the flow around the Japan bulk carrier with and without energy saving devices at model and full scale. In *31st Symposium on Naval Hydrodynamics*. CA: Monterey.

Wackers, J., Deng, G. B., Guilmineau, E., Leroyer, A., Queutey, P., & Visonneau, M. (2014). Combined refinement criteria for anisotropic grid refinement in free-surface flow simulation. *Computers & Fluids, 92*, 209–222.

Evaluation of Self-propulsion and Energy Saving Device Performance Predictions for JBC

Takanori Hino

Abstract Self-propulsion cases for the JBC hull with and without an energy saving duct are analyzed in this chapter. Test cases are set up for self-propulsion condition at the ship point. No rudder is fitted in either case. About half of the submissions employ actual propeller models in which a propeller geometry is discretized using a moving mesh and the remainder uses body force models in which propeller effects are considered as a body force computed using external potential-flow based programs. Self-propulsion simulations are carried out in two ways. The first is to follow the self-propulsion test procedure in a towing tank and a propeller revolution rate is adjusted in such a way that propeller thrust and towing force or *SFC* (Skin Friction Correction) for the ship point condition are balanced with ship's resistance. The other way is to fix the propeller revolution rate equal to the experimental value and the force invariance is computed. Thrust and torque coefficients, propeller revolution and ship's resistance components in self-propulsion condition are items to be submitted. Analysis of grid uncertainty is carried out based on the submission data with multiple grids. Average of comparison errors and the standard deviations of propeller thrust and torque together with revolution rates are estimated using the towing tank test data of 7.0 m model. Self-propulsion factors are estimated using the submitted data and compared with the measured data. Finally, model scale delivered powers are used to evaluate overall accuracy of the current CFD analysis in terms of the prediction accuracy of energy saving duct performance.

1 Introduction

In addition to the resistance test cases of 1.1 for the bare hull and 1.2 for the hull with a stern duct as an energy saving device (ESD), the self-prolusion cases are set for JBC without (Case 1.5a) and with (Case 1.6a) ESD. There are 26 submissions for Case 1.5a and 25 submissions for Case 1.6a, respectively. The results are analyzed with

T. Hino (✉)
Yokohama National University, Yokohama, Japan
e-mail: hino-takanori-nf@ynu.ac.jp

T. Hino et al. (eds.), *Numerical Ship Hydrodynamics*, Lecture Notes in Applied and Computational Mechanics 94, https://doi.org/10.1007/978-3-030-47572-7_7

respect to thrust, torque and revolution rate of a propeller and self-propulsion factors. Finally, the performance predictions of the energy saving device are compared and their accuracy is discussed.

2 Overview of Test Cases

Computational conditions are set in accordance with the self-propulsion test conducted in NMRI towing tank (see Chap. 2 for detail).

Both Cases 1.5a and 1.6a are for the self-propulsion simulation at the ship-point condition. No rudder is fitted and flow condition is $Fn = 0.142$ and $Rn = 7.46 \times 10^6$ which corresponds to the experiment with a ship speed of 1.179 [m/s] and ship length (L_{PP}) of 7.000 [m]. Ship motion is free in heave and pitch. Only difference between Cases 1.5a and 1.6a is the presence of ESD.

Geometry of the model propeller and its open water characteristics are provided to participants. Requested procedure for self-propulsion simulation follows the experimental approach, i.e., the revolution rate of a propeller is adjusted to obtain force equilibrium in the longitudinal direction considering the applied towing force (Skin Friction Correction, SFC) for the ship-point condition. Thus, the thrust T is $T = R_{T(SP)} - SFC$, where $R_{T(SP)}$ is the total resistance in self propulsion. SFC is pre-computed as 18.2 [N] for Case 1.5a and 18.1 [N] for Case 1.6a which are the same amounts as in the experiment. The items to be reported are the propeller revolution rate n, the thrust coefficient K_T, the torque coefficient K_Q and the total, pressure and frictional resistance coefficients in the self-propulsion condition, $C_{T(SP)}$, $C_{P(SP)}$ and $C_{F(SP)}$, respectively.

In case the procedure above cannot be carried out, the propeller revolution rate n is set to the measured value of 7.8 [rps] for Case 1.5a and 7.5 [rps] for Case 1.6a and the resulting towing force ($R_{T(SP)} - T$) is to be reported together with, K_T, K_Q, $C_{T(SP)}$, $C_{P(SP)}$, $C_{F(SP)}$ defined above.

3 Submissions

3.1 Bare Hull (Case 1.5a)

Total of 26 solutions are submitted for Case 1.5a as shown in Tables 1 and 2. Table 1 summarizes 14 submissions using actual propeller models in which the propeller geometry is discretized and rotated using moving grids. Table 2 is for 12 submissions with body force propeller models which include lifting line (BL), lifting surface (BS), prescribed (BP), vortex lattice (BV) or simplified propeller (BX) models. Free surface effects are not considered in 10 submissions. The other 16 submissions include free surface effects via VOF model (12), level-set model (3) or potential flow model (1).

Table 1 Submissions with actual propeller model for Case 1.5a

Organization (Code)	Free surface	Prop. M.		C_T × 1000	C_F × 1000	C_{PV} × 1000	K_T	K_Q	n(rps)	R_T-T(N)
EFD (NMRI)	–	–	D	4.811	–	–	0.217	0.0279	7.8	18.2
ABS (OpenFOAM)	V	A	S	4.768	3.205	1.564	0.213	0.0284	7.842	–
			E%D	0.895	–	–	2.057	−1.682	−0.538	–
CNR/INSEAN (XNAVIS)	No	A	S	N/A	N/A	N/A	0.210	0.0304	–	N/A
			E%D	N/A	–	–	3.230	−8.960	–	N/A
CSSRC (FLUENT)	No	A	S	4.757	3.139	1.619	0.223	0.0291	–	17.272
			E%D	−1.094	–	–	2.765	4.194	–	5.100
ECN/CNRS (ISISCFD/WF-EASM-RANSE)	V	A	S	4.661	3.149	1.512	0.214	0.0290	–	17.470
			E%D	3.110	–	–	1.470	−5.550	–	4.010
HHI (HiFoam)	V	A	S	4.841	3.128	1.713	0.217	0.0279	–	18.600
			E%D	−0.630	–	–	−0.140	0.100	–	−2.400
HHI (STRA-CCM+)	V	A	S	4.855	–	–	0.224	0.0294	–	18.100
			E%D	−0.910	–	–	−3.320	−5.380	–	0.770
IIHR (REX-DDES)	LS	A	S	4.812	3.210	1.600	0.219	0.0291	7.818	–

(continued)

Table 1 (continued)

Organization (Code)	Free surface	Prop. M.		C_T × 1000	C_F × 1000	C_{PV} × 1000	K_T	K_Q	n(rps)	R_T-T(N)
			E%D	−0.021	–	–	−0.922	−4.301	−0.231	–
MARIC (FINEMarine)	V	A	S	4.986	–	–	0.228	0.0278	7.800	18.200
			E%D	3.637	–	–	5.069	−0.358	–	
MARIN (ReFreSCo)	V	A	S	4.663	3.065	1.598	0.212	0.0282	–	17.744
			E%D	3.085	–	–	2.517	−1.143	–	2.504
PNU (FLUENTv15)	V	A	S	5.177	3.166	2.011	0.244	0.0285	7.871	18.515
			E%D	−7.608	–	–	−12.442	−2.151	−0.910	−1.732
SHIME (OpenFOAM)	No	A	S	4.852	3.070	1.782	0.216	0.028	–	18.800
			E%D	−0.855	–	–	0.631	−1.853	–	−3.355
UDE (STAR-CCM+)	V	A	S	4.808	2.946	1.862	0.215	0.0264	–	18.600
			E%D	0.062	–	–	0.829	5.376	–	−2.198
UNI-ROSTOCK (OpenFOAM-SST)	No	A	S	5.210	3.554	1.656	0.215	0.027	–	21.450
			E%D	−8.293	–	–	0.921	4.659	–	−17.850
UNI-ROSTOCK (OpenFOAM-HYBRID)	No	A	S	5.075	2.830	2.245	0.232	0.029	–	19.518
			E%D	−5.487	–	–	−6.912	−2.150	–	−7.241

Table 2 Submissions with body force propeller model for Case 1.5a

Organization (Code)	Free surface	Prop. M.		C_T × 1000	C_F × 1000	C_{PV} × 1000	K_T	K_Q	n(rps)	R_T-T(N)
EFD (NMRI)	–	–	D	4.811	–	–	0.217	0.0279	7.8	18.2
CHALMERS-FLOWTECH (SHIPFLOW)	Pot.Flow	BL	S	4.754	3.097	1.657	0.226	0.0279	7.572	–
			E%D	1.180	–	–	−4.340	−0.080	2.930	–
CNR/INSEAN (XNAVIS-HO)	No	BX	S	N/A	N/A	N/A	0.187	0.0268	–	N/A
			E%D	N/A	–	–	13.825	3.943	–	N/A
ECN/CNRS (ISISCFD/LRN-EASM-AD)	V	BX	S	4.625	3.108	1.517	0.214	0.0291	7.600	–
			E%D	3.870	–	–	1.240	−4.410	2.560	–
ECN/CNRS (ISISCFD/WF-EASM-AD)	V	BX	S	4.620	3.125	1.495	0.213	0.0291	7.620	–
			E%D	3.970	–	–	1.840	−4.190	2.310	–
HSVA (FreSCo+)	V	BV	S	4.652	–	–	0.206	0.0255	7.803	–
			E%D	−3.310	3.126	1.526	−5.020	−8.650	0.040	–
KRISO (WAVIS)	LS	BS	S	4.813	3.207	1.606	0.216	0.0280	7.863	–
			E%D	−0.042	–	–	0.599	0.000	−0.812	–
MIJAC (OpenFOAM)	No	BX	S	4.555	2.957	1.598	0.210	0.0270	–	16.784

(continued)

Table 2 (continued)

Organization (Code)	Free surface	Prop. M.		C_T × 1000	C_F × 1000	C_{PV} × 1000	K_T	K_Q	n(rps)	R_T-T(N)
			E%D	5.319	–	–	3.273	3.292	–	7.783
NMRI (NAGISA)	LS	BX	S	4.922	3.211	1.711	0.233	0.0293	7.560	–
			E%D	−2.310	–	–	−7.520	−5.090	3.130	–
SJTU (naoeFOAM)	V	BX	S	4.621	2.839	1.782	0.222	0.0285	7.691	–
			E%D	3.949	–	–	−2.304	−2.151	1.397	–
SOTON (OpenFOAM)	No	BP	S	5.190	3.636	1.554	0.180	0.0248	–	11.339
			E%D	−7.872		–	17.143	11.111	–	37.696
SOTON (STAR-CCM+)	No	BP	S	4.762	3.391	1.371	0.158	0.0226	–	11.601
			E%D	1.021	–	–	27.189	18.996	–	36.256
YNU (SURFv7)	No	BX	S	4.573	3.084	1.490	0.218	0.0281	7.450	–
			E%D	4.950	–	–	−0.460	−0.720	4.490	–

Of 14 submissions using an actual propeller model, 4 solutions are with the force balance approach and 10 are with the fixed revolution approach. Of 12 submissions using body force models, 8 are with the force balance and 4 are with the fixed revolution.

3.2 *With ESD (Case 1.6a)*

Table 3 (with an actual propeller model) and Table 4 (with body force models) show 25 solutions submitted for Case 1.6a. Again, 14 submissions are using actual propeller models and 11 submissions are with body force propeller models. Free surface effects are not considered in 7 submissions. The other 17 submissions include free surface effects via VOF model (12), level-set model (4) or potential flow model (1). Of 14 submissions using an actual propeller model, 4 solutions are with the force balance approach and 10 are with the fixed revolution approach. Of 12 submissions using body force models, 8 are with the force balance and 4 are with the fixed revolution. Although most of the submissions cover are both Cases 1.5a and 1.6a, some organizations submitted one case only.

4 Uncertainty Estimation

Participants are requested to perform uncertainty analysis and to report the results. Iterative uncertainty $U_I$of C_T, K_T and K_Q in the self-propulsion case is larger than that of C_T in the resistance case. The criterion of iterative convergence, $|U_I/\varepsilon_{12}| < 0.1$, is not satisfied in 38% of the submissions in case of C_T and K_Q and 50% in case of K_T.

Grid convergence for K_T and K_Q for Cases 1.5a and 1.6a is visualized in Figs. 1 and 2. Most of submissions seem to show convergent behaviors though divergence and oscillation are observed in some cases. The averaged values of the estimated U_G except outliers are 4.84%S_1 and 3.59%S_1 for K_T of Cases 1.5a and 1.6a, respectively and 6.42%S_1 and 4.07%S_1 for K_Q of Cases 1.5a and 1.6a, respectively as shown in Tables 5 and 6.

5 Propeller Parameters

Figure 3 shows the comparison errors of all data of the thrust coefficient K_T for Case 1.5a. All the submissions except the one with 67.7 M grid points employ the grid less than 25 M points. The comparison errors are generally in the range between −10%D and +5%D except for some outliers not depending on the number of grid points. Larger errors are observed for the cases using body force models with fixed

Table 3 Submissions with actual propeller model for Case 1.6a

Organization (Code)	Free surface	Prop. M.		C_T × 1000	C_F × 1000	C_{PV} × 1000	K_T	K_Q	n(rps)	R_T-T(N)
EFD (NMRI)	–	–	D	4.762	–	–	0.233	0.0295	7.5	18.1
ABS (OpenFOAM)	V	A	S	4.786	3.159	1.627	0.227	0.0297	7.673	–
			E%D	−0.497	–	–	2.746	−0.540	−2.303	–
CNR/INSEAN (XNAVIS)	No	A	S	N/A	N/A	N/A	0.242	0.0337	–	N/A
			E%D	N/A	–	–	−0.039	−0.142	–	N/A
CSSRC (FLUENT)	No	A	S	4.682	3.122	1.560	0.236	0.0302	–	17.257
			E%D	−0.016	–	–	0.014	0.024	–	−0.047
ECN/CNRS (ISISCFD/WF-EASM-RANSE)	V	A	S	4.572	3.133	1.439	0.227	0.0310	–	17.320
			E%D	3.990	–	–	2.780	−3.520	–	4.310
HHI (HiFoam)	V	A	S	4.884	3.254	1.629	0.217	0.0279	–	20.300
			E%D	−2.550	–	–	0.223	5.520	–	−12.280
HHI (STRA-CCM+)	V	A	S	4.758	–	–	0.236	0.0303	–	18.100
			E%D	0.080	–	–	−1.120	−2.710	–	0.280
IIHR (REX-DDES)	LS	A	S	4.731	3.190	1.540	0.237	0.0308	7.430	–

(continued)

Table 3 (continued)

Organization (Code)	Free surface	Prop. M.		C_T × 1000	C_F × 1000	C_{PV} × 1000	K_T	K_Q	n(rps)	R_T-T(N)
			E%D	0.650	–	–	−1.720	−4.410	0.930	–
IIHR (REX-DES)	LS	A	S	5.158	3.130	2.030	0.238	0.030	7.995	–
			E%D	−8.320	–	–	−2.150	−2.930	−6.600	–
IIHR (REX-RANS)	LS	A	S	4.697	3.180	2.330	0.231	0.031	7.481	–
			E%D	1.360	–	–	1.000	−5.080	0.250	–
MARIC (FINEMarine)	V	A	S	5.025	–	–	0.240	0.0286	7.500	18.100
			E%D	5.523	–	–	3.005	−3.051	–	
MARIN (ReFreSCo)	V	A	S	4.674	3.074	1.619	0.225	0.0291	–	18.392
			E%D	1.857	–	–	3.582	1.462	–	−1.611
PNU (FLUENTv15)	V	A	S	5.190	3.173	2.017	0.250	0.0300	7.829	18.202
			E%D	−8.988	–	–	−7.296	−1.695	−4.387	−0.564
SHIME (OpenFOAM)	No	A	S	4.841	3.069	1.772	0.219	0.029	–	20.300
			E%D	−1.669	–	–	6.087	2.240	–	−11.923
UDE (STAR-CCM+)	V	A	S	4.858	3.005	1.862	0.235	0.0280	–	18.900
			E%D	−2.008	–	–	−0.944	5.085	–	−4.420

Table 4 Submissions with body force propeller model for Case 1.6a

Organization (Code)	Free surface	Prop. M.		C_T × 1000	C_F × 1000	C_{PV} × 1000	K_T	K_Q	n(rps)	R_T-T(N)
EFD (NMRI)	–	–	D	4.762	–	–	0.233	0.0295	7.5	18.1
CHALMERS-FLOWTECH (SHIPFLOW)	Pot.Flow	BL	S	4.740	3.109	1.632	0.243	0.0292	7.306	–
			E%D	0.450	–	–	−4.480	0.980	2.590	–
CNR/INSEAN (XNAVIS-HO)	No	BX	S	N/A	N/A	N/A	0.222	0.0300	–	N/A
			E%D	N/A	–	–	0.047	−0.017	–	N/A
ECN/CNRS (ISISCFD/LRN-EASM-AD)	V	BX	S	4.660	3.103	1.563	0.239	0.0306	7.310	–
			E%D	2.140	–	–	−2.360	−3.660	2.530	–
ECN/CNRS (ISISCFD/WF-EASM-AD)	V	BX	S	4.617	3.116	1.501	0.233	0.0305	7.330	–
			E%D	3.040	–	–	0.130	−3.250	2.270	–
HSVA (FreSCo+)	V	BV	S	4.584	3.111	1.473	0.219	0.0272	7.522	–
			E%D	−3.730	–	–	−6.190	−7.640	0.290	–
KRISO (WAVIS)	LS	BP	S	4.760	3.204	1.556	0.231	0.0290	7.580	–
			E%D	0.042	–	–	0.858	1.017	−1.067	–
MIJAC (OpenFOAM)	No	BX	S	4.523	2.926	1.597	0.233	0.0287	–	16.146

(continued)

Table 4 (continued)

Organization (Code)	Free surface	Prop. M.		C_T × 1000	C_F × 1000	C_{PV} × 1000	K_T	K_Q	n(rps)	R_T-T(N)
			E%D	5.013	–	–	0.214	2.766	–	10.795
NMRI (NAGISA)	LS	BX	S	4.893	3.214	1.680	0.240	0.0300	7.440	–
			E%D	−2.760	–	–	−3.050	−1.760	0.760	–
SJTU (naoeFOAM)	V	BX	S	4.596	2.548	2.048	0.239	0.0299	7.402	–
			E%D	3.486	–	–	−2.575	−1.356	1.307	–
SOTON (OpenFOAM)	No	BP	S	4.468	3.196	1.272	0.180	0.0248	–	9.226
			E%D	6.172		–	22.833	15.932	–	49.029
YNU (SURFv7)	No	BX	S	4.552	3.082	1.470	0.233	0.0294	7.220	–
			E%D	4.410	–	–	0.000	0.340	3.730	–

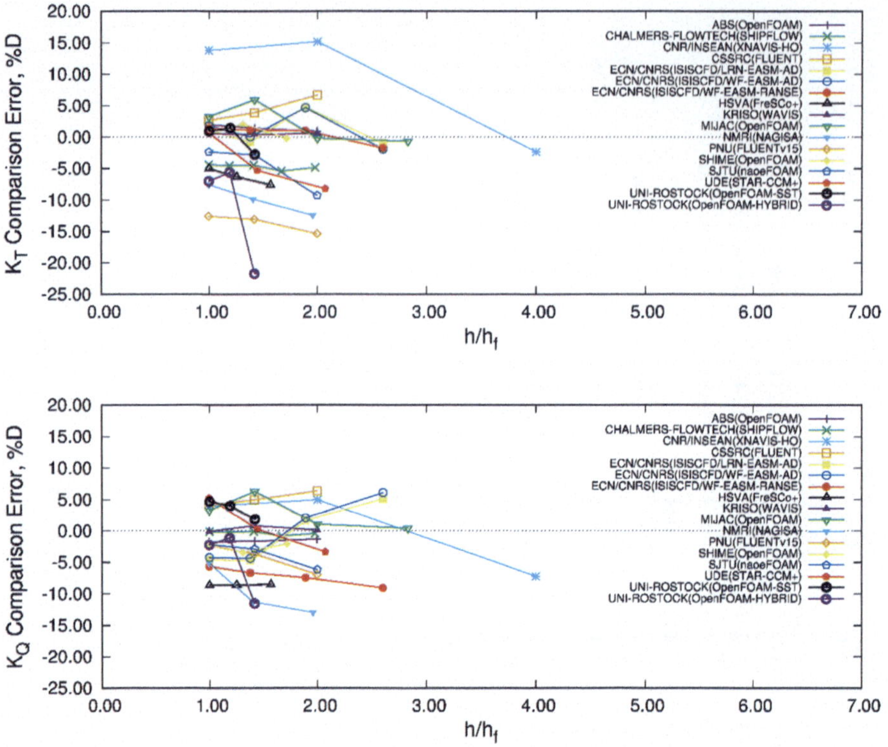

Fig. 1 Grid convergence of K_T and K_Q for Case 1.5a

revolution. Figure 4 is the same plot for the torque coefficient K_Q. The error range is from -10%D to $+5$%D, again except for outliers. Figure 5 is the errors of the towing force $R_{T(SP)} - T$ for 13 solutions using the fixed revolution approach and Fig. 6 is the errors of propeller revolution n for 11 solutions using the force balance approach. Scatter of the towing force is in the range of ±10%D and it does not depend on the grid size, whereas, the error of propeller revolution is from -1%D to 5%D.

Averages and standard deviations of the comparison errors of K_T and K_Q are listed in Table 5. Average of absolute errors are 3 to 4%D. Differences between actual propeller models and body force models are not large.

Figures 7 and 8 show the comparison errors of K_T and K_Q in Case 1.6a, respectively. Numbers of grid points in Case 1.6a are distributed in the range less than 25 M except the large grid cases with 70.9 M points. The grid points are, in general, slightly larger than those of Case 1.5a since the presence of the duct increases the complexity of the geometry. The ranges of errors in K_T and K_Q are from -10%D and 5%D which is almost the same as Case 1.5a shown in Figs. 5 and 6.

Figures 9 and 10 is the comparison errors of the towing force $R_{T(SP)} - T$ and the propeller revolution n. The errors of $R_{T(SP)} - T$ are in the range of ±10%D except

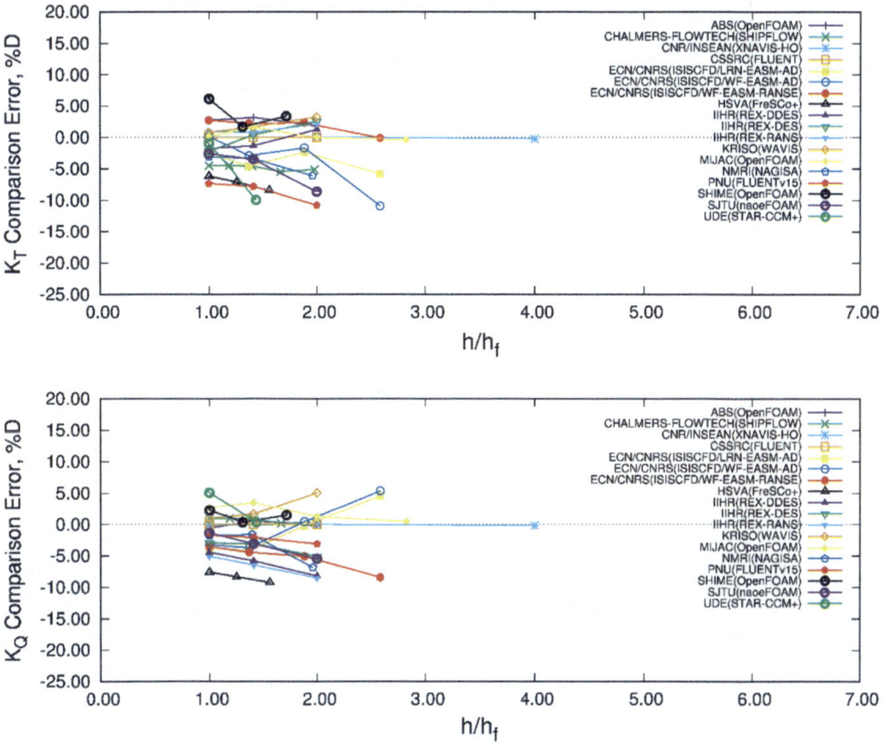

Fig. 2 Grid convergence of K_T and K_Q for Case 1.6a

Table 5 Errors of thrust and torque coefficients in Case 1.5a

Items	K_T				K_Q			
	E%D	|E|%D	σ_{SD}%D	UG%D	E%D	|E|%D	σ_{SD}%D	UG%D
Actual Prop (14/24)	−0.30	3.09	4.41		−1.37	3.42	3.94	
Modeled Prop (10/24)	0.11	4.04	5.60		−1.81	3.25	3.70	
Force balance (12/24)	−1.85	3.65	4.62		−2.82	2.82	2.48	
Fixed n (12/24)	1.59	3.32	4.64		−0.29	3.88	4.50	
All (24)	−0.13	3.49	4.94	4.84	−1.55	3.35	3.85	6.42

the outliers and this is in the same level as Case 1.5a. The scatter of the revolution rate n is also in the same range as Case 1.5a.

Averages and standard deviations of the comparison errors of K_T and K_Q are listed in Table 6. Average of the absolute errors $|E|$ of K_T and K_Q are about 2%D in Case 1.6a with EFD and lightly smaller than those in Case 1.5a without ESD which is 3% D. Average error of actual propeller models is slightly better than body force propeller models in this case.

Table 6 Errors of thrust and torque coefficients in Case 1.6a

Items	K_T				K_Q			
	E%D	|E|%D	σ_{SD}%D	UG%D	E%D	|E|%D	σ_{SD}%D	UG%D
Actual Prop (14/26)	0.44	2.34	3.11		−0.70	2.74	3.19	
Modeled Prop (12/26)	2.36	5.46	9.39		1.61	4.55	6.91	
Force balance (15/26)	−1.58	2.68	2.98		−2.36	2.69	2.36	
Fixed n (11/26)	4.71	5.06	8.35		3.55	4.61	6.11	
All (26)	−0.43	2.02	2.87	3.59	−0.86	2.35	2.96	4.07

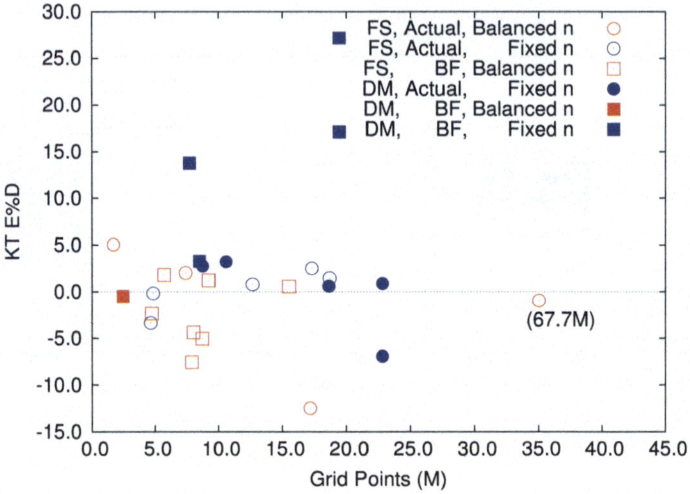

Fig. 3 Comparison errors of K_T in Case 1.5a

6 Self-prolusion Factors

One of the goals of self-propulsion simulations is to obtain self-propulsion factors, i.e., thrust deduction and the wake coefficients, $1 - t$ and $1 - w$, respectively and the relative rotative efficiency η_R for power estimation. To estimate the self-propulsion factors, the resistance data and the propeller open water characteristics are needed in addition to the self propulsion data. Although the resistance data is available from the corresponding submissions to Cases 1.1a and 1.2a, the open water characteristics data of individual propeller models are not included in the requested items. Therefore, the result of the experimental propeller open test is used for the present analysis. Figure 11 shows the propeller open water characteristics of the experiment for the model propeller MP.687 which was provided to participants.

The procedure to estimate self-propulsion factors is essentially same as that in the experiment as summarized below:

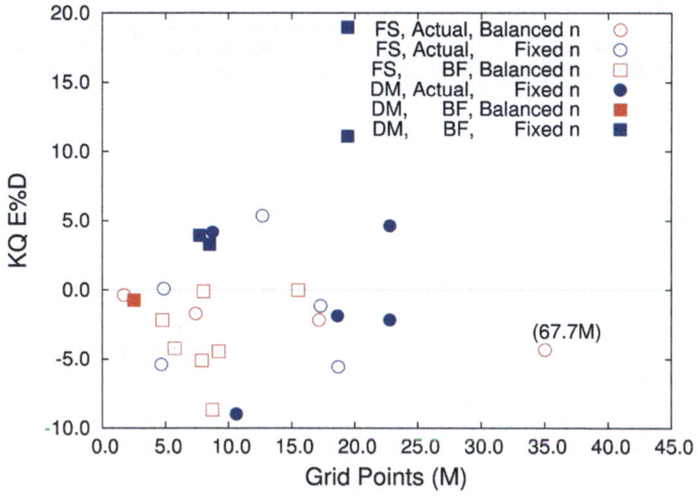

Fig. 4 Comparison errors of K_Q in Case 1.5a

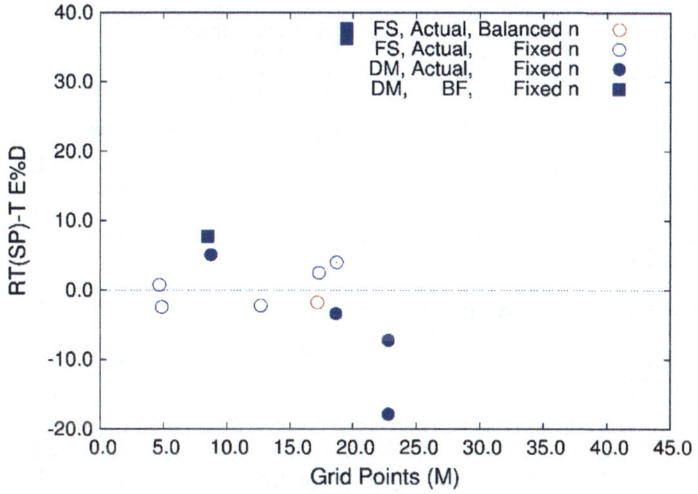

Fig. 5 Comparison errors of $R_{T(SP)} - T$ in Case 1.5a

First, from the computed K_T, K_Q and n, the wake coefficient $1 - w$ is obtained by using the thrust identity method with the measured propeller open water characteristics and the propeller efficiency behind hull η_B is computed using thrust T, torque Q and the ship speed U as,

$$\eta_B = \frac{T(1-w)U}{2\pi n Q}$$

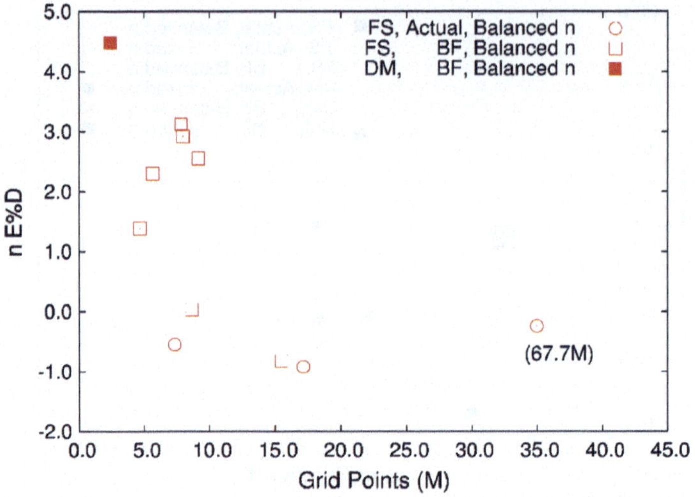

Fig. 6 Comparison errors of n in Case 1.5a

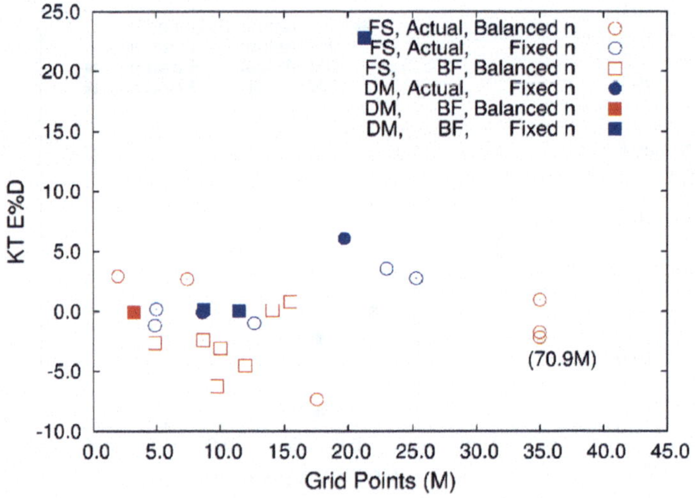

Fig. 7 Comparison errors of K_T in Case 1.6a

The relative rotative efficiency is $\eta_R = \eta_B/\eta_o$, where η_o is the propeller open water efficiency. From thrust T and total resistance in towing condition $R_{(tow)}$ from $Ct_{(tow)}$ in Case 1.1a or 1.2a with the towing force SFC for the ship point condition, thrust deduction coefficient $1 - t$ is obtained by

$$1 - t = \frac{R_{(tow)} - SFC}{T}$$

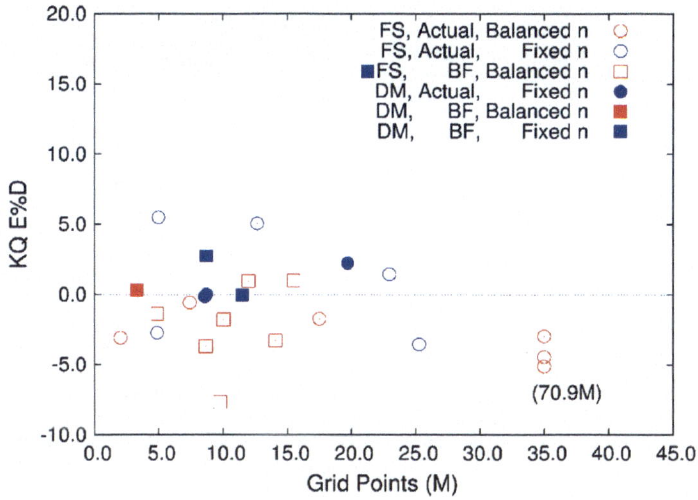

Fig. 8 Comparison errors of K_Q in Case 1.6a

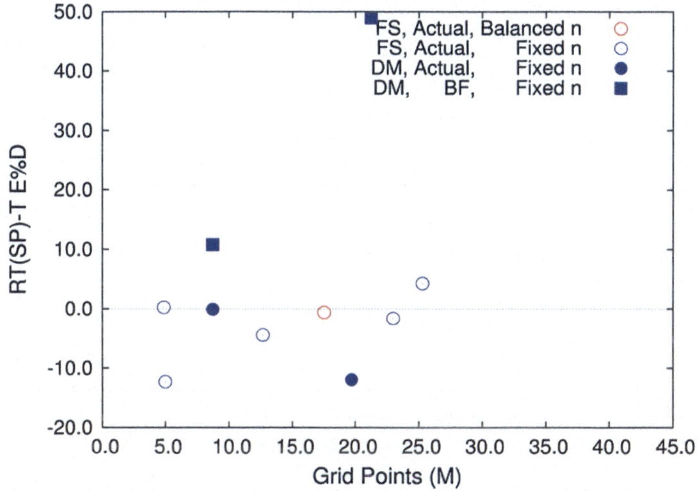

Fig. 9 Comparison errors of $R_{T(SP)} - T$ in Case 1.6a

Finally, the model scale Delivered Power (*DP*) is estimated by

$$DP = \frac{T(1 - w)U}{\eta_B}$$

Estimated self-propulsion factors and *DP*s in Case 1.5a are listed in Tables 7 for actual propeller models and Table 8 for body force propeller models. Comparison

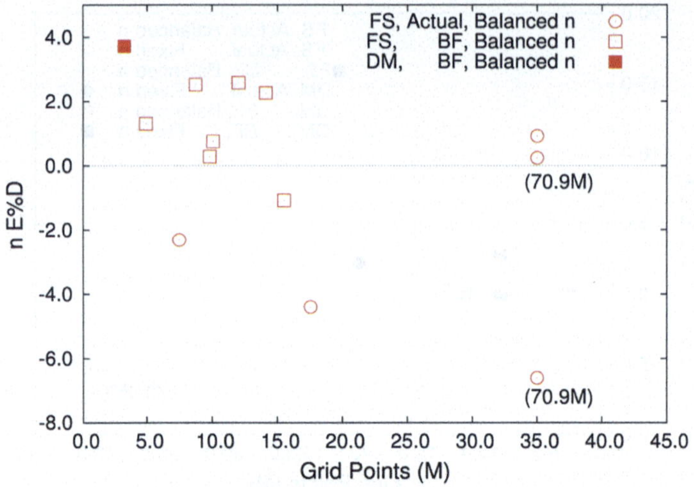

Fig. 10 Comparison errors of n in Case 1.6a

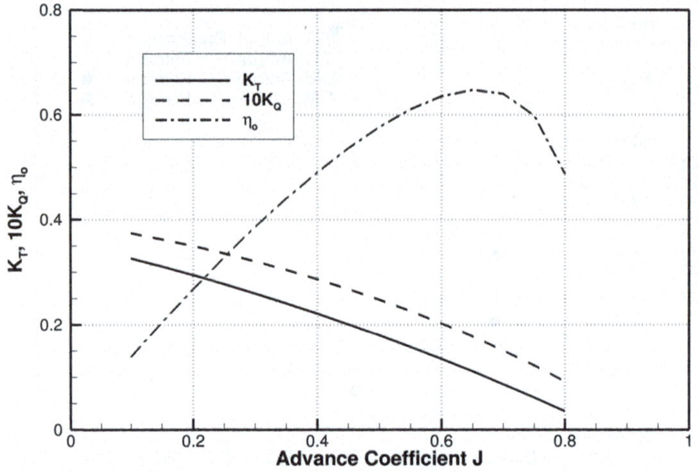

Fig. 11 Measured propeller open characteristics of model propeller MP. 687

errors of $1-t$ and $1-w$ are shown in Figs. 12 and 13, respectively. Error ranges are from -20%D to $+10\%$D for $1-t$ and from -10%D to 5%D for $1-w$ without outliers and the scatters are not dependent on the number of grids. Average of errors are listed in Table 9. Absolute errors for $1-t$ and $1-w$ are 11%D and 7%D, which are considerably larger than errors of K_T and K_Q which are around 3%D as shown in Table 5. The standard deviations values are also quite large (24%D and 11%D). From the definition, self-propulsion factors are the results of complicated hydrodynamic interaction between a ship hull and a propeller. Not only K_T, K_Q

Table 7 Self-propulsion factors with actual propeller models for Case 1.5a

Organization (Code)	Free surface	Prop. M.		$1-t$	$1-w$	η_R	η_O	DP(Kw)
EFD (NMRI)	–	–	D	0.804	0.554	1.015	0.501	0.0290
ABS (OpenFOAM)	V	A	S	0.808	0.568	0.984	0.511	0.0296
			E%D	−0.440	−2.481	3.834	−1.840	−2.026
CSSRC (FLUENT)	No	A	S	–	0.532	0.992	0.487	0.0299
			E%D	–	4.082	2.859	2.857	−2.870
ECN/CNRS (ISISCFD/WF-EASM-RANSE)	V	A	S	0.792	0.562	0.967	0.508	0.0298
			E%D	1.482	−1.336	5.969	−1.385	−2.517
HHI (HiFoam)	V	A	S	0.828	0.552	1.015	0.501	0.0287
			E%D	−2.919	0.421	0.000	0.000	1.258
HHI (STRA-CCM+)	V	A	S	0.804	0.530	0.985	0.485	0.0304
			E%D	−0.002	4.471	3.733	3.345	−4.652
IIHR (REX-DDES)	LS	A	S	–	0.546	0.979	0.497	0.0301
			E%D	–	1.435	4.419	0.939	−3.584
MARIC (FINEMarine)	V	A	S	0.790	0.515	1.054	0.475	0.0285
			E%D	1.747	7.136	−4.935	5.329	1.725

(continued)

Table 7 (continued)

Organization (Code)	Free surface	Prop. M.		$1-t$	$1-w$	η_R	η_O	*DP*(Kw)
MARIN (ReFreSCo)	V	A	S	0.803	0.568	0.987	0.513	0.0289
			E%D	0.177	−2.527	3.374	−2.291	0.311
PNU (FLUENTv15)	V	A	S	0.728	0.464	1.078	0.432	0.0300
			E%D	9.516	16.365	−7.930	13.768	−3.525
SHIME (OpenFOAM)	No	A	S	0.773	0.555	1.008	0.504	0.029
			E%D	3.897	−0.140	0.859	−0.465	1.018
UDE (STAR-CCM+)	V	A	S	–	0.559	1.065	0.506	0.0271
			E%D	–	−0.739	−6.303	−0.926	6.675
UNI-ROSTOCK (OpenFOAM-SST)	No	A	S	0.712	0.559	1.042	0.506	0.028
			E%D	11.513	−0.739	−3.359	−0.926	4.553
UNI-ROSTOCK (OpenFOAM-HYBRID)	No	A	S	0.637	0.501	1.023	0.464	0.030
			E%D	20.734	9.602	−1.055	7.364	−2.517

Table 8 Self-propulsion factors with body force propeller models for Case 1.5a

Organization (Code)	Free surface	Prop. M.		$1-t$	$1-w$	η_R	η_O	DP(Kw)
EFD (NMRI)	–	–	D	0.804	0.554	1.015	0.501	0.0290
CHALMERS-FLOWTECH (SHIPFLOW)	Pot.Flow	BL	S	0.813	0.506	1.044	0.480	0.0262
			E%D	−1.150	8.661	−3.659	4.331	9.771
ECN/CNRS (ISISCFD/LRN-EASM-AD)	V	BX	S	0.871	0.547	0.963	0.508	0.0276
			E%D	−8.306	1.263	6.382	−1.385	4.842
ECN/CNRS (ISISCFD/WF-EASM-AD)	V	BX	S	0.766	0.552	0.960	0.511	0.0278
			E%D	4.795	0.420	6.777	−1.840	4.089
HSVA (FreSCo+)	V	BV	S	0.770	0.588	1.070	0.526	0.0262
			E%D	4.230	−6.114	−6.909	−4.936	9.752
KRISO (WAVIS)	LS	BS	S	0.805	0.560	1.008	0.504	0.0294
			E%D	−0.147	−0.949	0.859	−0.465	−1.399
MIJAC (OpenFOAM)	No	BX	S	0.816	0.575	1.024	0.517	0.0277
			E%D	−1.418	−3.713	−1.230	−3.186	4.553
NMRI (NAGISA)	LS	BX	S	0.829	0.482	1.016	0.462	0.0274
			E%D	−3.046	12.984	−0.132	7.881	5.693

(continued)

Table 8 (continued)

Organization (Code)	Free surface	Prop. M.		$1-t$	$1-w$	η_R	η_O	DP(Kw)
SJTU (naoeFOAM)	V	BX	S	0.807	0.528	1.009	0.489	0.0280
			E%D	−0.331	4.824	0.661	2.373	3.416
SOTON (OpenFOAM)	No	BP	S	1.393	0.671	1.000	0.577	0.0254
			E%D	−73.253	−20.998	1.768	−15.047	12.331
SOTON (STAR-CCM+)	No	BP	S	1.471	0.738	1.002	0.610	0.0232
			E%D	−82.955	−33.059	1.595	−21.692	20.108
YNU (SURFv7)	No	BX	S	0.795	0.524	1.011	0.499	0.0251
			E%D	1.083	5.500	0.492	0.468	13.446

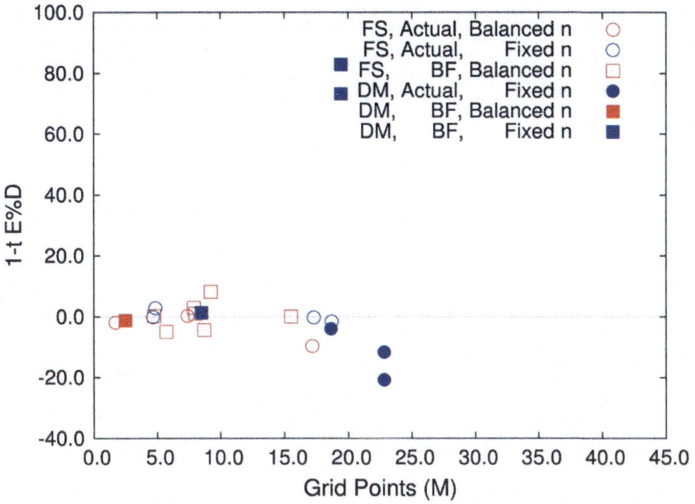

Fig. 12 Comparison errors of $1-t$ in Case 1.5a

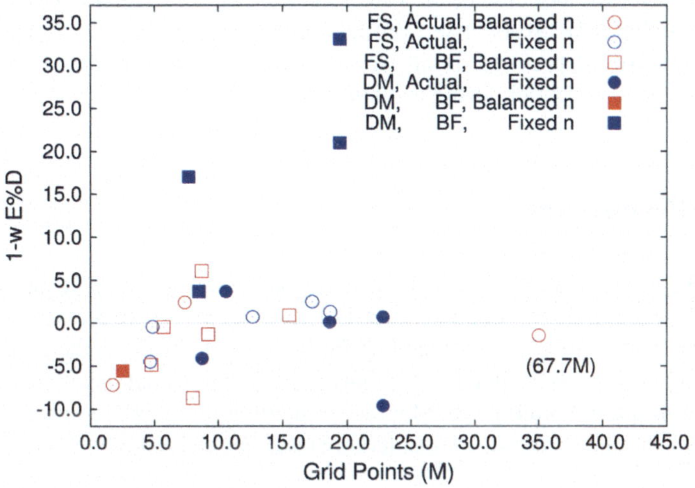

Fig. 13 Comparison errors of $1-w$ in Case 1.5a

and n from self-propulsion simulations but also C_T from resistance simulations are involved in the analysis. These points are attributed to the larger errors and scatters of self-propulsion factors.

For Case 1.6a, the estimated values of self-propulsion factors and *DP*s are listed in Tables 10 and 11 for actual propeller models and for body force models, respectively. Comparison errors of $1-t$ and $1-w$ are shown in Figs. 14 and 15, respectively. Errors of $1-t$ are from -10%D to 5%D and errors of $1-w$ are in the range of

Table 9 Errors of self-propulsion factors and delivered power in Case 1.5a

Items	$1-t$			$1-w$			DP		
	E%D	|E|%D	σ_{SD}%D	E%D	|E|%D	σ_{SD}%D	E%D	|E|%D	σ_{SD}%D
Actual Prop (14/26)	−4.57	5.24	6.88	−2.27	3.94	5.39	0.97	3.19	3.64
Modeled Prop (12/26)	14.59	16.43	30.20	4.02	9.63	12.85	−7.66	7.89	5.47
Force balance(12/26)	−0.72	3.16	6.22	−4.09	5.68	6.22	−3.52	5.27	5.34
Fixed n (14/26)	12.28	19.84	33.66	4.67	7.33	10.97	−2.58	5.44	6.96
All (26)	5.47	11.10	24.33	0.63	6.57	10.09	−3.01	5.36	6.28

± 10%D except outliers. Error ranges are slightly smaller than Case 1.5a. Average errors and the standard deviations are shown in Table 12. Absolute errors for $1-t$ and $1-w$ are 3%D and 5%D and smaller than Case 1.5a. Due to presence of ESD, stern flow fields are smoothened to some extent and this may improve the accuracy of self-propulsion factors.

Absolute errors of model scale *DP*s are 5%D for Case 1.5a and 4%D for Case1.6a, which are smaller than errors of self-propulsion factors in Case 1.5a. Since *DP*s are integrated quantity depending on all factors such as resistance, thrust and torque, some errors may cancel out during estimations.

7 ESD Performance

Based on the model scale Delivered Powers (*DP*s) derived above, ESD performance which is the reduction *DP* due to the ESD is estimated. *DP* reduction rate which is defined as $DP_{(ESD)}/DP_{(Bare)}$ is listed in Tables 10 and 11. From the experimental result, *DP* reduction is 0.94, that is, 6% power reduction in model scale. The estimated *DP* reductions are plotted Fig. 16. Most of simulations show *DP* reductions to some extent. However, the scatters are considerably large and reduction rates are from 0.88 to 1.04.

In view that a few percent power reduction is the typical ESD performance, the current status of ESD performance estimation is not sufficient. In order to achieve reliable prediction of ESD effects, the accuracy improvement in all the aspects, resistance, wake, propeller effect is required, particularly in the estimations of self-propulsion factors.

Table 10 Self-propulsion factors with actual propeller models for Case 1.6a

Organization (Code)	Free surface	Prop. M.		$1-t$	$1-w$	η_R	η_O	*DP*(Kw)	*DP*(Kw) bare	ESD/ Bare
EFD (NMRI)	–	–	D	0.81107	0.48087	1.00878	0.462	0.0273	0.0290	0.940
ABS (OpenFOAM)	V	A	S	0.807	0.510	0.984	0.477	0.0290	0.0296	0.980
			E%D	0.460	−6.026	3.080	−3.314	−6.312	−2.026	−4.200
CSSRC (FLUENT)	No	A	S	–	0.469	0.994	0.454	0.0275	0.0299	0.923
			E%D	–	2.534	1.789	1.703	−0.953	−2.870	1.864
ECN/CNRS (ISISCFD/WF-EASM-RANSE)	V	A	S	0.852	0.498	0.943	0.477	0.0283	0.0298	0.950
			E%D	−4.992	−3.636	8.167	−3.314	−3.627	−2.517	−1.083
HHI (HiFoam)	V	A	S	0.793	0.531	1.015	0.501	0.0254	0.0287	0.888
			E%D	2.246	−10.358	−0.718	−8.555	6.736	1.258	5.547
HHI (STRA-CCM+)	V	A	S	0.829	0.469	0.991	0.454	0.0276	0.0304	0.910
			E%D	−2.218	2.534	2.194	1.703	−1.287	−4.652	3.215
IIHR (REX-DDES)	LS	A	S	–	0.461	0.978	0.451	0.0273	0.0301	0.909
			E%D	–	4.129	3.821	2.277	−0.103	−3.584	3.361
IIHR (REX-DES)	LS	A	S	–	0.493	1.007	0.449	0.033	0.0285	1.162
			E%D	–	−2.421	0.241	2.855	−21.480	1.725	−23.613

(continued)

Table 10 (continued)

Organization (Code)	Free surface	Prop. M.		$1-t$	$1-w$	η_R	η_O	*DP*(Kw)	*DP*(Kw) bare	ESD/ Bare
IIHR (REX-RANS)	LS	A	S	–	0.484	0.954	0.467	0.028	0.0289	0.970
			E%D	–	−0.652	6.732	−1.118	−2.842	0.311	−3.163
MARIC (FINEMarine)	V	A	S	0.815	0.455	1.062	0.443	0.0261	0.0300	0.868
			E%D	−0.514	5.312	−6.599	4.020	4.396	−3.525	7.651
MARIN (ReFreSCo)	V	A	S	0.783	0.505	0.998	0.482	0.0265	0.029	0.924
			E%D	3.464	−4.991	1.347	−4.390	2.724	1.018	1.723
PNU (FLUENTv15)	V	A	S	0.718	0.440	1.042	0.415	0.0311	0.0271	1.149
			E%D	11.486	8.515	−4.081	10.052	−14.069	6.675	−22.227
SHIME (OpenFOAM)	No	A	S	0.762	0.524	0.982	0.497	0.026	0.028	0.955
			E%D	6.004	−9.025	3.248	−7.535	3.058	4.553	−1.566
UDE (STAR-CCM+)	V	A	S	–	0.472	1.069	0.457	0.0255	0.030	0.858
			E%D	–	1.842	−7.450	1.132	6.401	−2.517	8.699

Table 11 Self-propulsion factors with body force propeller models for Case 1.6a

Organization (Code)	Free surface	Prop. M.		$1-t$	$1-w$	η_R	η_O	*DP* (Kw)	*DP*(Kw) bare	ESD/ bare
EFD (NMRI)	–	–	D	0.81107	0.48087	1.00878	0.462	0.0273	0.0290	0.940
CHALMERS-FLOWTECH (SHIPFLOW)	Pot.Flow	BL	S	0.822	0.434	1.050	0.435	0.0246	0.0262	0.940
			E%D	−1.369	9.805	−5.023	5.793	9.770	9.771	0.000
ECN/CNRS (ISISCFD/LRN-EASM-AD)	V	BX	S	0.823	0.447	0.990	0.446	0.0258	0.0276	0.936
			E%D	−1.450	7.031	2.318	3.436	5.289	4.842	0.470
ECN/CNRS (ISISCFD/WF-EASM-AD)	V	BX	S	0.789	0.468	0.976	0.462	0.0260	0.0278	0.933
			E%D	2.736	2.721	4.078	0.000	4.821	4.089	0.764
HSVA (FreSCo+)	V	BV	S	0.774	0.526	1.047	0.497	0.0250	0.0262	0.956
			E%D	4.532	−9.345	−4.768	−7.535	8.273	9.752	−1.639
KRISO (WAVIS)	LS	BP	S	0.810	0.490	1.020	0.467	0.0273	0.0294	0.928
			E%D	0.142	−1.984	−1.382	−1.118	−0.077	−1.399	1.304
MIJAC (OpenFOAM)	No	BX	S	0.817	0.479	1.037	0.462	0.0262	0.0277	0.945
			E%D	−0.678	0.464	−3.467	0.000	4.061	4.553	−0.516
NMRI (NAGISA)	LS	BX	S	0.827	0.452	1.013	0.443	0.0267	0.0274	0.976
			E%D	−1.909	6.069	−0.486	4.020	2.103	5.693	−3.806

(continued)

Table 11 (continued)

Organization (Code)	Free surface	Prop. M.		$1-t$	$1-w$	η_R	η_O	*DP* (Kw)	*DP*(Kw) bare	ESD/ bare
SJTU	V	BX	S	0.811	0.453	1.013	0.446	0.0262	0.0280	0.935
(naoeFOAM)			E%D	−0.032	5.861	−0.539	3.436	3.917	3.416	0.519
SOTON	No	BP	S	1.575	0.645	1.000	0.577	0.0226	0.0254	0.889
(OpenFOAM)			E%D	−94.159	−34.148	1.036	−24.889	17.098	12.331	5.438
YNU	No	BX	S	0.802	0.461	1.012	0.462	0.0239	0.0251	0.952
(SURFv7)			E%D	1.177	4.180	−0.423	0.000	12.323	13.446	−1.298

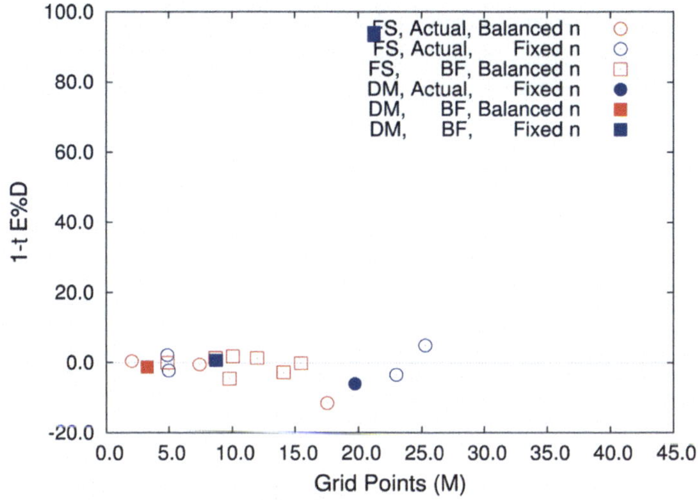

Fig. 14 Comparison errors of $1 - t$ in Case 1.6a

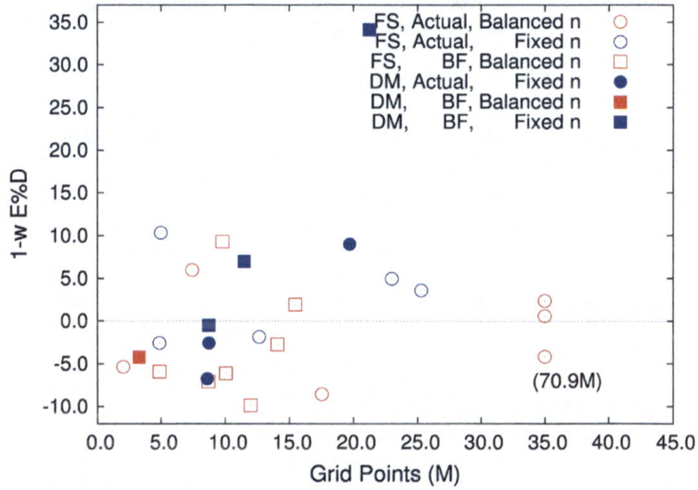

Fig. 15 Comparison errors of $1 - w$ in Case 1.6a

8 Conclusions

Conclusions from the analysis of submitted data for JBC self-propulsion cases are summarized as follows:

- Grid uncertainty U_G in general is about 4–5%S for K_T and K_Q.

Table 12 Errors of self-propulsion factors and delivered power in Case 1.6a

Items	$1-t$			$1-w$			DP		
	E%D	|E|%D	σ_{SD}%D	E%D	|E|%D	σ_{SD}%D	E%D	|E|%D	σ_{SD}%D
Actual Prop (8, 13, 8)	1.99	3.92	4.79	−0.94	4.77	5.49	−0.34	4.00	4.87
Modeled Prop (9, 9, 9)	0.35	1.56	2.01	2.76	5.27	5.42	0.89	2.82	3.71
Force balance (11, 14, 14)	1.39	2.35	3.67	2.37	5.29	5.44	1.63	3.50	4.11
Fixed n (6, 8, 8)	0.64	3.27	3.67	−2.58	4.42	4.87	−2.41	3.54	3.89
All (17, 22, 17)	1.12	2.67	3.69	0.57	4.97	5.76	0.16	3.51	4.48

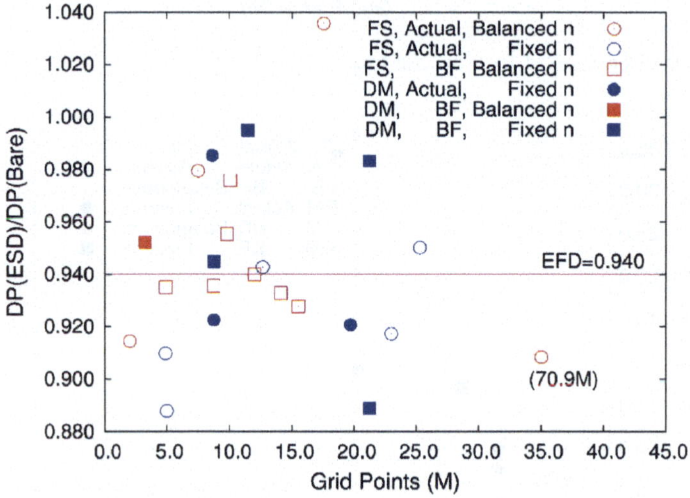

Fig. 16 Comparison of *DP* Reductions due to the Duct

- Averages of absolute comparison errors of K_T and K_Q are 3–4%D for a bare hull case and 2–3%D with ESD case. ESD case is slightly smaller than a bare hull case. Actual propeller models give slightly better results than body force models though the difference is smaller in case of a bare hull.
- Averages of absolute comparison errors of self-propulsion factors, $1-t$ and $1-w$ are 7–10%D in a bare hull case, which are considerably larger than the errors of thrust and torque. This is due to the increased number of quantities involved to estimate the factors. Errors become smaller with the presence of a duct, maybe due to the smoothening of stern flow fields by a duct.
- Average absolute errors of model scale delivered power (*DP*) is in the range of 5–6%, about the same level of other quantities. However, the estimated *DP* reduction

rates due to a duct scatter from 0.88 to 1.0, whereas the experimental value is 0.94. The current accuracy of CFD estimation of ESD performance seems to be reasonably well in terms of the magnitude, although this may not be sufficient for design/performance prediction of an energy saving duct which aims the reduction rate of a few percent.

Evaluation of Resistance, Sinkage, Trim and Self-propulsion Predictions for KCS

Jin Kim

Abstract This chapter discusses the results of resistance predictions including trim and sinkage (the workshop case 2.1) and the results of self-propulsion predictions (the workshop case 2.5) for KCS. The resistance predictions for 6 different Froude numbers in trim and sinkage free condition were requested. The comparison error at the design speed (Fr = 0.26) is −0.2% and the standard deviation is 1.5% of the data value. The mean comparison error for all 6 speeds is 0.43% and the mean standard deviation is 2.48%. The increased error and standard deviation is caused by the results for low Fr simulation submissions. The submitted trim and sinkage also shows larger comparison error in low Fr region (Fr < 0.2). Self-propulsion results are reported with both towing force (FD) fixed and propeller revolution (rps) fixed. The participants using a body force propeller model based on potential theory selected FD fixed condition for self-propulsion and the participants who adopted actual propeller rotating simulation preferred rps fixed condition. The mean comparison errors of KT and KQ are 0.5 and −3.5% respectively. The standard deviations are 2.7 and 2.4% respectively. Self-propulsion parameters are slightly better predicted by body force models (FD fixed). However, local flow characteristics are better predicted by actual propeller rotating simulation (rps fixed).

1 Resistance, Sinkage & Trim

1.1 Participants

The KCS resistance cases of the Workshop were the case 2.1 for the bare hull with a rudder, which was the same case in Gothenburg 2010 case 2.2b (Larsson et al. 2010, 2013). The KCS model ship is towed in a condition of free to heave and pitch motion with 6 different speeds in a calm water. This case received 13 submissions

J. Kim (✉)
Korea Research Institute of Ships & Ocean Engineering (KRISO), Daejeon, Korea
e-mail: jkim@kriso.re.kr

T. Hino et al. (eds.), *Numerical Ship Hydrodynamics*, Lecture Notes in Applied and Computational Mechanics 94, https://doi.org/10.1007/978-3-030-47572-7_8

of computational results from 11 different organizations. The 13 cases with reported computational scheme and grid information are listed in the Table 1.

The number of participants is increased compared to the previous Gothenburg 2010 workshop. Most participants used commercial and open source codes except of WAVIS, an in-house code of KRISO. The finite volume method with unstructured grid is used by most participants. The grid size varies from 1.1 to 6.6 M. All submissions except one use two equation turbulence models. For the ship motion of pitch and heave, all moving, deforming grid and overset schemes are used.

The Fig. 1 shows the total resistance (C_T) results of all participants for 6 different Froude numbers (Fr). The submitted results are still showing large deviation from the experimental values. The results are more scattered as Fr decreases. The Figs. 2 and 3 show sinkage and trim results of all participants for 6 different Fr's. The error bars show the standard deviation of CFD submissions except the outliers.

Table 1 Computational schemes and grid information of all participants for the case 2.1

Participants	Grid type	Free surface	Turbulence	Grid motion	Discretization	Grid size	Code
CTO	Unstr.	VOF	Two-eqn.	All moving	FV	–	Star-CCM
MARIC	–	VOF	Two-eqn.	Deforming grid	FV	–	FINEMarine
UNIZAG	Unstr.	VOF	Two-eqn.	All moving	FV	4.7 M	NavalFoam
HHI	Unstr.	VOF	RSM	All moving	FV	1.1 M	Star-CCM
UDE(C)	–	–	–	–	–	–	Comet
UDE(F)	Unstr.	VOF	Two-eqn.	Deforming grid	FV	1.1 M	FINEMarine
UDE(O)	Unstr.	VOF	Two-eqn.	Deforming grid	FV	1.1 M	OpenFoam
SJTU	Unstr.	VOF	Two-eqn.	Overset	FV	3.4 M	NaoeFoam
KRISO	Str.	Level-Set	Two-eqn.	All moving	FV	4.3 M	WAVIS
DAMEN	Unstr.	VOF	Two-eqn.	Deforming grid	FV	2.6 M	FINEMarine
NUMECA	Unstr.	VOF	Two-eqn.	Deforming grid	FV	6.2 M	ISIS
UM	Unstr.	VOF	Two-eqn.	All moving	FV	2.6 M	OpenFoam
PNU	Unstr.	VOF	Two-eqn.	All moving	FV	6.6 M	Fluent

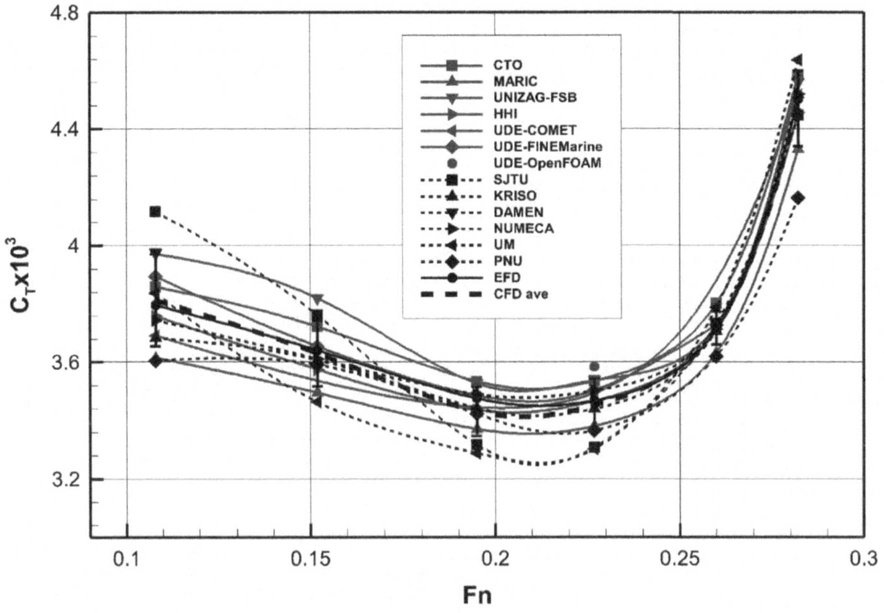

Fig. 1 Total resistance results for all participants

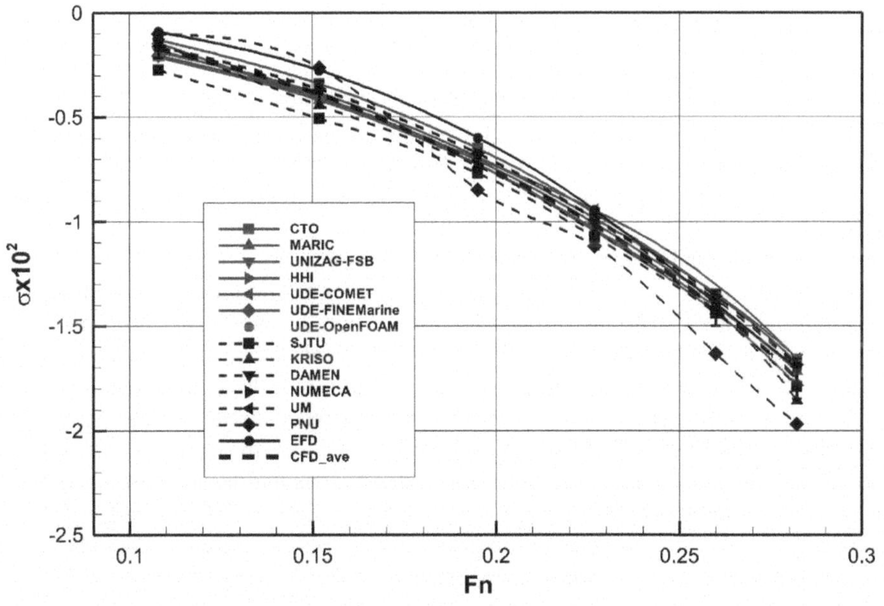

Fig. 2 Sinkage results for all participants

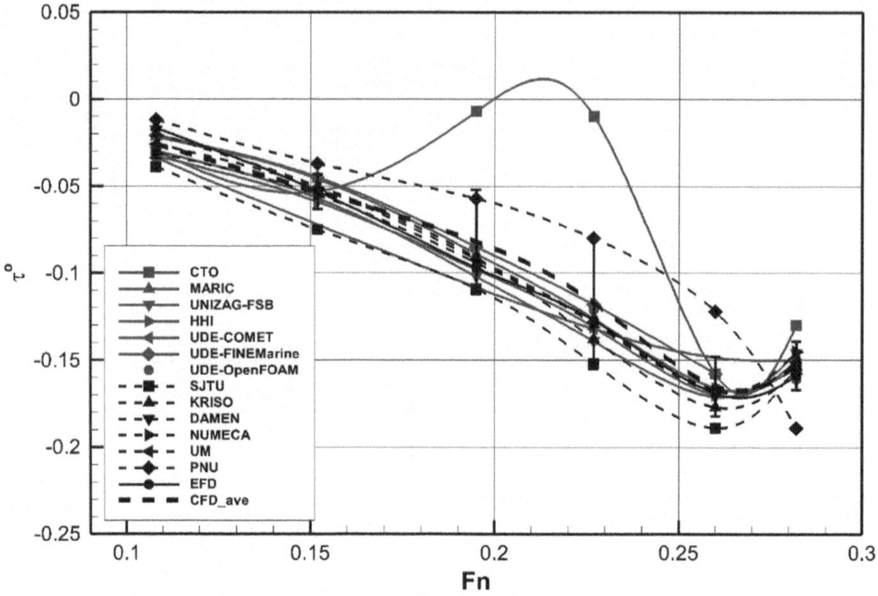

Fig. 3 Trim results for all participants

1.2 Statistics and Error Estimation

The case is exactly same as the case 2.1b of the previous Gothenburg 2010 workshop. We are interested in both how well resistance, sinkage and trim can be predicted and how much the results are improved compared to the previous workshop. Especially, resistance is the primary parameter for the evaluation of ship hydrodynamic performances.

All computational results are shown in the previous Sect. 1.1 and we can see that a couple of them are very different from other results. The results are considered as outliers when the absolute error, $|E|$ is larger than 2σ, where $E = D\text{-}S$ and σ is the standard deviation. The Tables 2, 3 and 4 show all submitted results for total resistance, sinkage and trim respectively and outlier ($|E| > 2\sigma$) is marked as X.

The analyzed statistics and error estimation with and without outliers are shown for each *Fr* in Tables 5, 6, 7, 8, 9 and 10, where T2015 means the statistics with outlier and T2015*, without outliers. The errors and standard deviations are compared with the previous Gothenburg 2010 workshop and the absolute error $|E|$ and standard deviation based on the absolute error is analyzed in the present analysis. The submitted frictional resistance (C_F) and pressure resistance (C_P) are separately shown in Figs. 4, 5, 6, 7, 8 and 9 compared with the experimental results (*EFD*) and averaged computational results (CFD_{ave}). In the group discussion for this case, Dr. Starke from *MARIN* pointed out that a couple of entries unrealistically had friction resistances well below a plate friction line (ATTC line). We saw the misuse of wall functions as a likely

Table 2 All submitted results for total resistance

Fn	0.108	0.152	0.195	0.227	0.260	0.282
CTO	O	O	O	O	O	O
MARIC	O	O	O	O	O	O
UNIZAG-FSB	O	O	O	O	O	O
HHI	O	O	O	O	O	O
UDE-Comet	O	N/A	N/A	O	N/A	O
UDE-FINEMarine	O	O	O	O	O	N/A
UDE-OpenFOAM	N/A	N/A	N/A	O	N/A	O
SJTU	O	O	O	O	O	O
KRISO	O	O	O	O	O	O
DAMEN	O	N/A	N/A	N/A	N/A	N/A
NUMECA	O	O	O	O	O	O
UM	O	O	O	O	O	O
PNU	O	O	O	O	O	X

Table 3 All submitted results for sinkage

Fn	0.108	0.152	0.195	0.227	0.260	0.282
CTO	O	O	O	O	O	O
MARIC	O	O	O	O	O	O
UNIZAG-FSB	O	O	O	O	O	O
HHI	O	O	O	O	O	O
UDE-Comet	O	N/A	N/A	O	N/A	O
UDE-FINEMarine	O	O	O	O	O	N/A
UDE-OpenFOAM	N/A	N/A	N/A	O	N/A	O
SJTU	X	O	O	O	O	O
KRISO	O	O	O	O	O	O
DAMEN	O	N/A	N/A	N/A	N/A	N/A
NUMECA	O	O	O	O	O	O
UM	O	O	O	O	O	O
PNU	O	O	X	X	X	X

candidate to blame: two of the participants with low Cf-values, also reported mean y + values around 15, i.e. in the buffer layer. The ITTC and ATTC lines based on a flat plate friction are also given in the Figs. 4–9.

Mean values of the total resistance, sinkage and trim for corresponding *Fr* are plotted in the Figs. 10, 11 and 12 respectively. The Table 11 shows statistics and error estimation based on the mean values of computational results. The error, *E* and absolute error, |*E*| are obtained as 0.43 and 2.0%*D* respectively. Both errors are

Table 4 All submitted results for trim

Fn	0.108	0.152	0.195	0.227	0.260	0.282
CTO	O	O	X	X	O	O
MARIC	O	O	O	O	O	O
UNIZAG-FSB	O	O	O	O	O	O
HHI	O	O	O	O	O	O
UDE-Comet	O	N/A	N/A	O	N/A	O
UDE-FINEMarine	O	O	O	O	O	N/A
UDE-OpenFOAM	N/A	N/A	N/A	O	N/A	O
SJTU	O	X	O	O	O	O
KRISO	O	O	O	O	O	O
DAMEN	O	N/A	N/A	N/A	N/A	N/A
NUMECA	O	O	O	O	O	O
UM	O	O	O	O	O	O
PNU	O	O	O	O	X	X

Table 5 Statistics and error estimation for $Fn = 0.108$

	G2010	T2015	T2015*
Participants	4	12	
C_T E_{mean}%D	+1.6	−0.4	0.3
C_T σ_{SD}%D	1.4	4.2	3.5
Sinkage E_{mean}%D	−67.1	−89.0	−78.7
Sinkage σ_{SD}%D	8.2	50.3	36.8
Trim E_{mean}%D	−60.3	−54.8	−48.0
Trim σ_{SD}%D	38.6	45.5	40.8
C_T $\lvert E\rvert_{mean}$%D	N/A	3.4	3.0
CT $\sigma_{SD}\lvert E\rvert$%D	N/A	2.2	1.6
Sinkage $\lvert E\rvert_{mean}$%D	N/A	89.0	78.7
Sinkage $\sigma_{SD}\lvert E\rvert$%D	N/A	50.3	36.8
Trim $\lvert E\rvert_{mean}$%D	N/A	59.7	53.4
Trim $\sigma_{SD}\lvert E\rvert$%D	N/A	38.1	32.7

slightly increased since $E = -0.3\%$D and $|E| = 1.64\%D$ in G2010. The standard deviation is also increased from 1.3 to $2.48\%D$.

Note that the number of participants is three times larger than at G2010. Presumably more unexperienced users participated in 2015.

For the predictions of resistance, sinkage and trim, the number of grid points are varied from 1.1 to 6.6 M. The submitted results do not show large dependency of grid points as shown in Fig. 17. It means that CFD users who have efficient grid making

Table 6 Statistics and error estimation for $Fn = 0.152$

	G2010	T2015	T2015*
Participants	3	10	
C_T E_{mean}%D	−0.1	0.3	0.7
C_T σ_{SD}%D	1.1	3.1	3.0
Sinkage E_{mean}%D	−44.1	−41.9	−37.2
Sinkage σ_{SD}%D	15.1	23.5	19.3
Trim E_{mean}%D	0.7	0.4	5.1
Trim σ_{SD}%D	4.9	19.0	12.7
C_T $\|E\|_{mean}$%D	N/A	2.3	2.4
CT $\sigma_{SD}\|E\|$%D	N/A	1.7	1.8
Sinkage $\|E\|_{mean}$%D	N/A	43.9	38.2
Sinkage $\sigma_{SD}\|E\|$%D	N/A	21.7	17.0
Trim $\|E\|_{mean}$%D	N/A	14.7	10.0
Trim $\sigma_{SD}\|E\|$%D	N/A	13.0	8.9

Table 7 Statistics and error estimation for $Fn = 0.195$

	G2010	T2015	T2015*
Participants	4	10	
C_T E_{mean}%D	−0.9	1.3	1.6
C_T σ_{SD}%D	1.7	2.4	2.5
Sinkage E_{mean}%D	−12.9	−20.2	−19.0
Sinkage σ_{SD}%D	6.6	9.5	5.5
Trim E_{mean}%D	1.6	15.0	2.0
Trim σ_{SD}%D	3.8	30.8	8.0
C_T $\|E\|_{mean}$%D	NA	2.0	2.2
CT $\sigma_{SD}\|E\|$%D	NA	1.7	1.9
Sinkage $\|E\|_{mean}$%D	NA	20.2	19.0
Sinkage $\sigma_{SD}\|E\|$%D	NA	9.5	5.5
Trim $\|E\|_{mean}$%D	NA	18.8	6.7
Trim $\sigma_{SD}\|E\|$%D	NA	28.4	4.2

techniques can have an acceptable simulation results for low block coefficient hull form such as KCS (Fig. 13).

Table 8 Statistics and error estimation for $Fn = 0.227$

	G2010	T2015	T2015*
Participants	6	12	
C_T E_{mean}%D	−1.0	0.5	0.5
C_T σ_{SD}%D	1.4	2.7	2.7
Sinkage E_{mean}%D	−4.2	−7.5	−7.0
Sinkage σ_{SD}%D	4.1	5.4	4.2
Trim E_{mean}%D	−3.2	7.9	−3.5
Trim σ_{SD}%D	4.6	29.8	7.7
C_T $\lvert E\rvert_{mean}$%D	NA	2.1	2.1
CT $\sigma_{SD}\lvert E\rvert$%D	NA	1.5	1.7
Sinkage $\lvert E\rvert_{mean}$%D	NA	7.5	7.0
Sinkage $\sigma_{SD}\lvert E\rvert$%D	NA	5.4	4.2
Trim $\lvert E\rvert_{mean}$%D	NA	15.7	5.9
Trim $\sigma_{SD}\lvert E\rvert$%D	NA	26.2	5.8

Table 9 Statistics and error estimation for $Fn = 0.260$

	G2010	T2015	T2015*
Participants	4	10	
C_T E_{mean}%D	−0.3	−0.2	−0.5
C_T σ_{SD}%D	1.2	1.5	1.3
Sinkage E_{mean}%D	1.5	−1.5	0.2
Sinkage σ_{SD}%D	4.8	6.0	2.7
Trim E_{mean}%D	−3.1	2.6	−0.2
Trim σ_{SD}%D	3.0	10.3	5.6
C_T $\lvert E\rvert_{mean}$%D	NA	1.1	1.0
CT $\sigma_{SD}\lvert E\rvert$%D	NA	1.0	1.0
Sinkage $\lvert E\rvert_{mean}$%D	NA	3.9	2.4
Sinkage $\sigma_{SD}\lvert E\rvert$%D	NA	4.8	0.9
Trim $\lvert E\rvert_{mean}$%D	NA	6.2	3.8
Trim $\sigma_{SD}\lvert E\rvert$%D	NA	8.4	3.8

2 Self-Propulsion

2.1 Participants

The KCS self-propulsion case of the Workshop was the case 2.5 for the bare hull with a rotating propeller and without a rudder. Both heave and pitch motions are fixed.

Table 10 Statistics and error estimation for $Fn = 0.282$

	G2010	T2015	T2015*
Participants	6	11	
C_T E_{mean}%D	−0.4	0.7	−0.0
C_T σ_{SD}%D	0.8	2.9	1.9
Sinkage E_{mean}%D	−2.1	−2.9	−1.6
Sinkage σ_{SD}%D	5.6	5.8	4.1
Trim E_{mean}%D	3.6	3.6	5.9
Trim σ_{SD}%D	4.2	9.0	5.2
C_T $\|E\|_{mean}$%D	NA	2.0	1.5
CT $\sigma_{SD}\|E\|$%D	NA	2.1	1.1
Sinkage $\|E\|_{mean}$%D	NA	4.5	3.3
Sinkage $\sigma_{SD}\|E\|$%D	NA	4.5	2.7
Trim $\|E\|_{mean}$%D	NA	7.3	6.1
Trim $\sigma_{SD}\|E\|$%D	NA	6.0	4.9

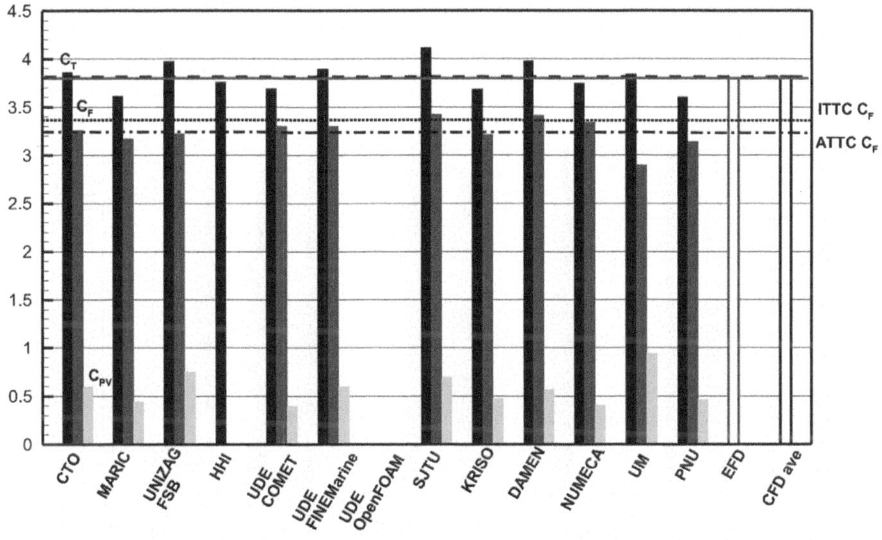

Fig. 4 Comparison of the resistance components and the experimental data for $Fn = 0.108$

The computations were requested for the model at the ship point that the model ship was towed to account for the larger skin friction at model scale compared to full scale. The towing force, the skin friction correction, SFC, was pre-computed and was the same as in the experiments. In the experiments the thrust T, was adjusted by varying the rps, n, such that $T = R_{T(SP)} - SFC$, where $R_{T(SP)}$ is the resistance with a rotating propeller. 3 participants followed this procedure, but the others computed

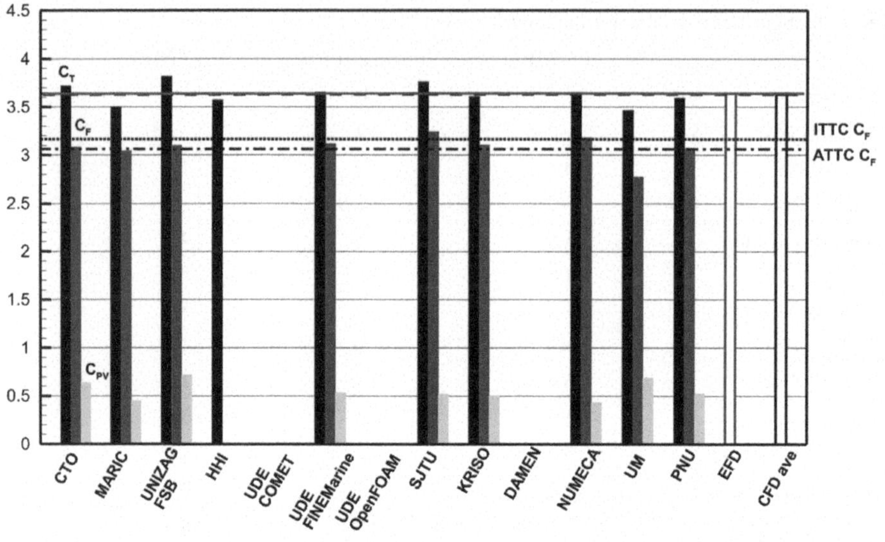

Fig. 5 Comparison of the resistance components and the experimental data for $Fn = 0.152$

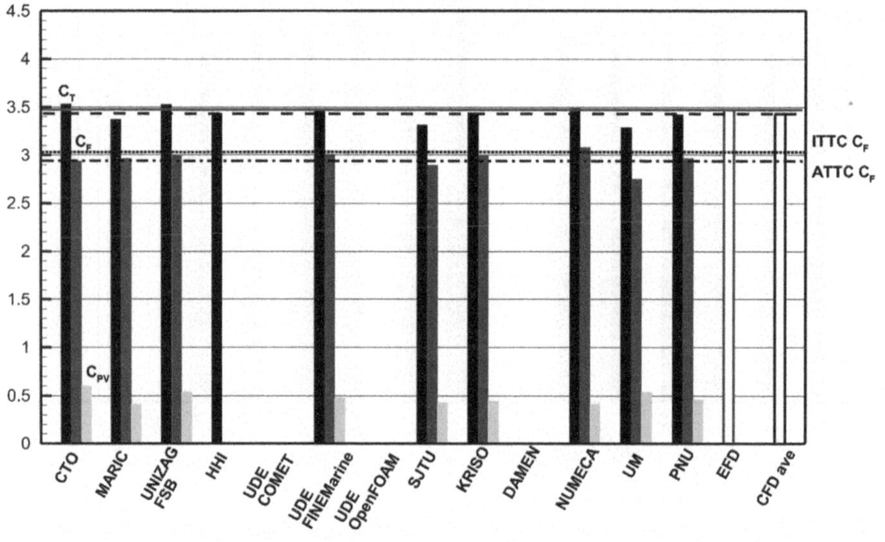

Fig. 6 Comparison of the resistance components and the experimental data for $Fn = 0.195$

the flow with the measured *rps* from the experiment. In the first case, the achieved *n* from numerical force balancing was requested, while in the second case the resulting towing force, $R_{T(SP)} - SFC$, was to be reported. The participants for the first case used a body force propeller with a Lifting Surface method, the others used an actual

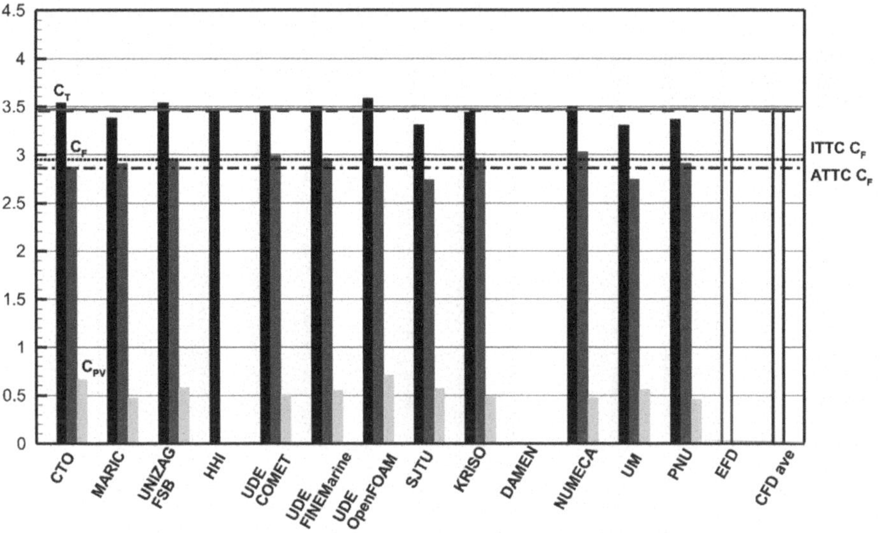

Fig. 7 Comparison of the resistance components and the experimental data for $Fn = 0.227$

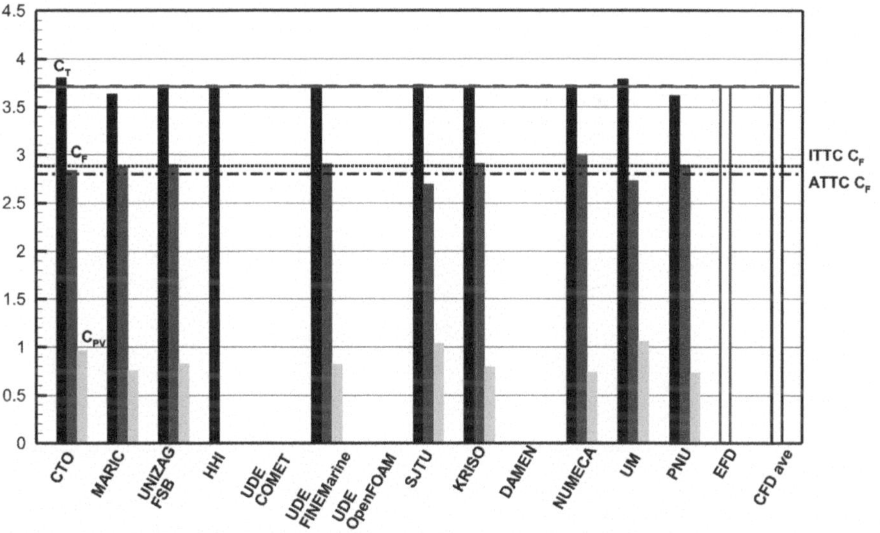

Fig. 8 Comparison of the resistance components and the experimental data for $Fn = 0.260$

rotating propeller. The computational schemes, grid information and the ship point are shown in the Table 12.

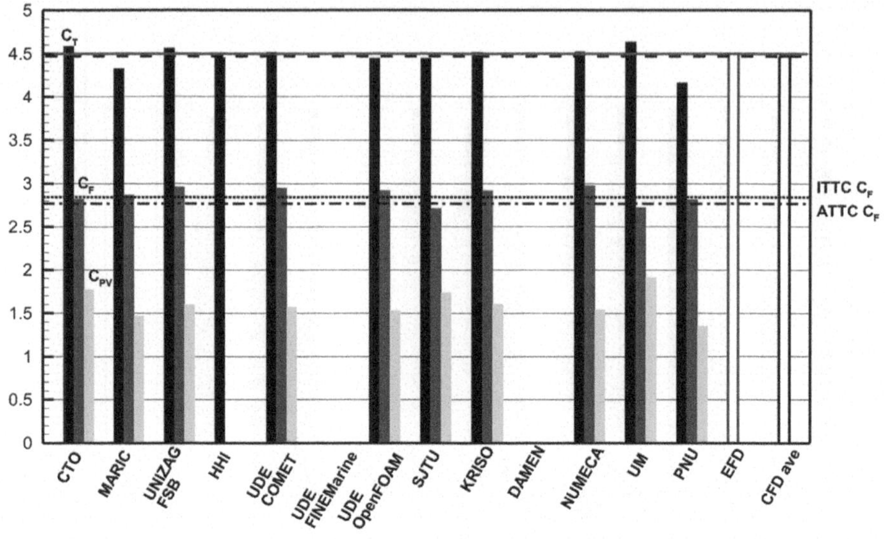

Fig. 9 Comparison of the resistance components and the experimental data for $Fn = 0.282$

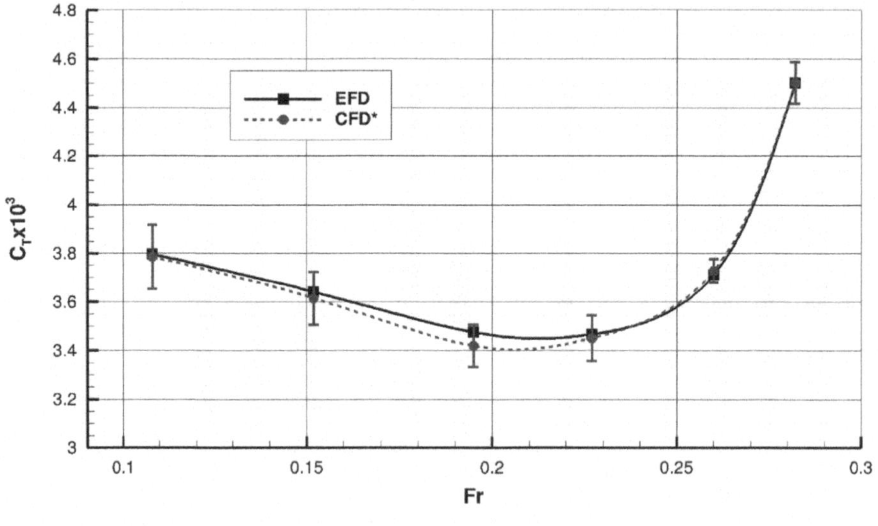

Fig. 10 Comparison with mean total resistance and experiment

2.2 Statistics and Error Estimation

All computational results of the resistance at self-propelled condition, $R_{T(SP)}$, are shown in the Fig. 14. For better statics and error estimation, two submissions are considered as outliers because the absolute error, $|E|$ is larger than 2σ, where $E =$

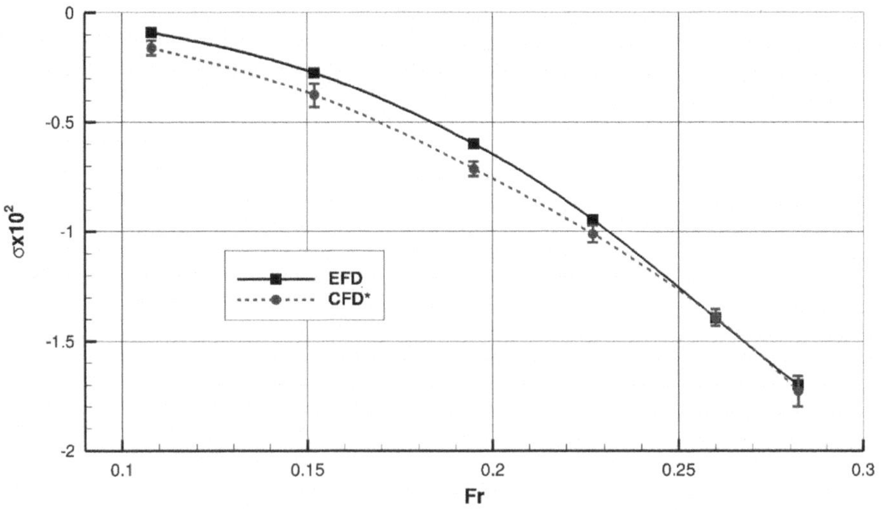

Fig. 11 Comparison with mean sinkage and experiment

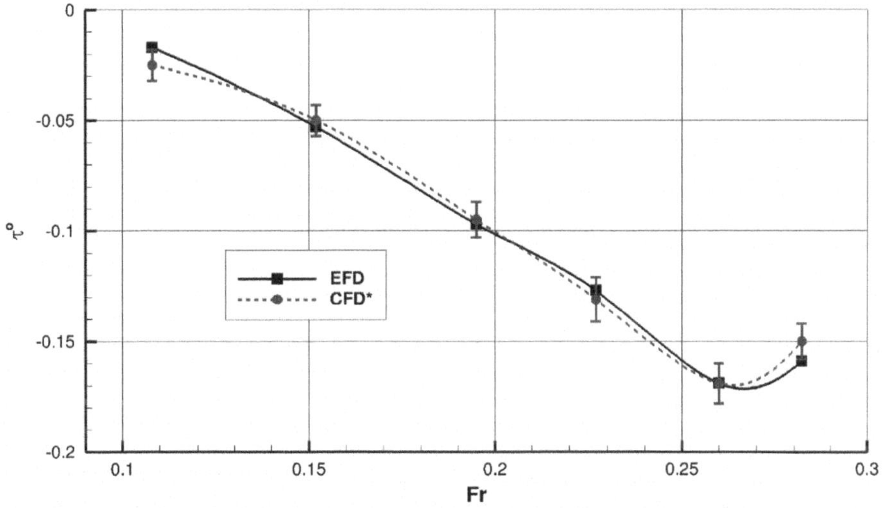

Fig. 12 Comparison with mean computed trim and experiment

D-S and σ is the standard deviation. The statics and errors are compared with the previous workshops in the Table 13. The errors of $R_{T(SP)}$ are 1%D which is slightly bigger than −0.3%D at G2010. Tables 14 and 15 show the error estimation for the FD fixed cases and rps fixed cases, respectively.

Table 11 Statistics and error estimation based on mean values for each *Fn*

	G2010	T2015*
Participants	4	13
C_T E_{mean}%D	−0.3	0.43
C_T σ_{SD}%D	1.3	2.48
Sinkage E_{mean}%D	−21.9	−23.87
Sinkage σ_{SD}%D	9.9	12.09
Trim E_{mean}%D	−9.6	−6.45
Trim σ_{SD}%D	11.2	13.34
C_T $\lvert E\rvert_{mean}$%D	1.64	2.00
Sinkage $\lvert E\rvert_{mean}$%D (Fr < 0.2)	55.6	45.26
Sinkage $\lvert E\rvert_{mean}$%D (Fr ≥ 0.2)	7.5	4.25
Trim $\lvert E\rvert_{mean}$%D (Fr < 0.2)	30.5	23.34
Trim $\lvert E\rvert_{mean}$%D (Fr ≥ 0.2)	3.62	5.29

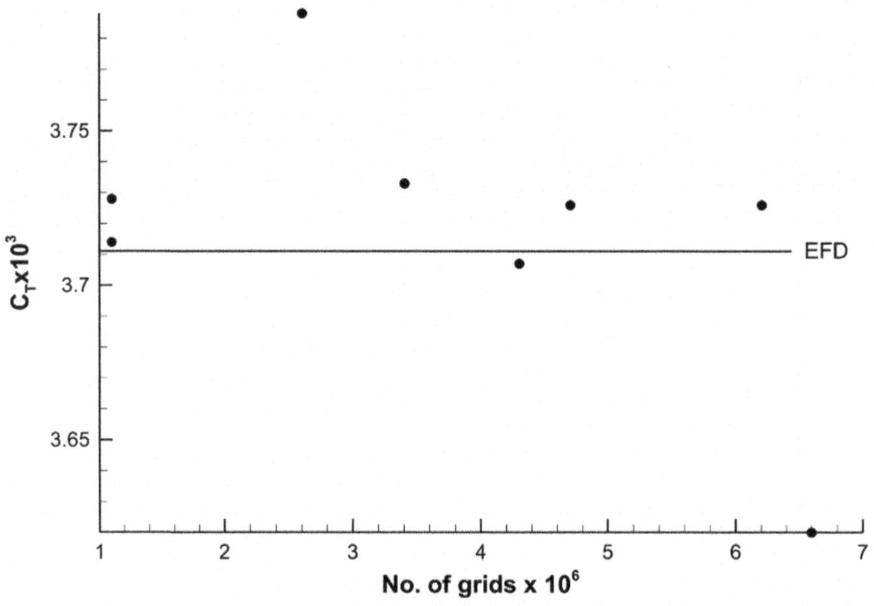

Fig. 13 CT comparison at $Fn = 0.26$ w.r.t. no. of grids

2.3 *Local Flow Characteristics*

Figure 15 shows the computed axial velocity contours and cross flow vectors at propeller plane and the velocity profiles downstream of propeller plane at z/Lpp = −0.03 are depicted in Fig. 16. It is confirmed that the methods using actual propeller

Table 12 Computational schemes and grid information of all participants for the case 2.5

Participants	Grid Type	Turbulence	Grid Size	Propeller model	Ship point	Code
CTO	Unstructured	Two-eqn.	–	–	FD fixed	Star-CCM
HHI	Unstructured	RSM	4.5 M	Actual rotating	RPS fixed	Star-CCM
MARIN	Structured	One-eqn.	12 M	Lifting Surface	FD fixed	PARNASSOS
MARIC	Unstructured	Two-eqn.	–	–	RPS fixed	FINEMarine
KRISO	Structured	Two-eqn.	8.6 M	Lifting Surface	FD fixed	WAVIS
UM	Unstructured	Two-eqn.	2.6 M	Actual rotating	RPS fixed	OpenFoam
PNU	Unstructured	Two-eqn.	15 M	Actual rotating	–	Fluent

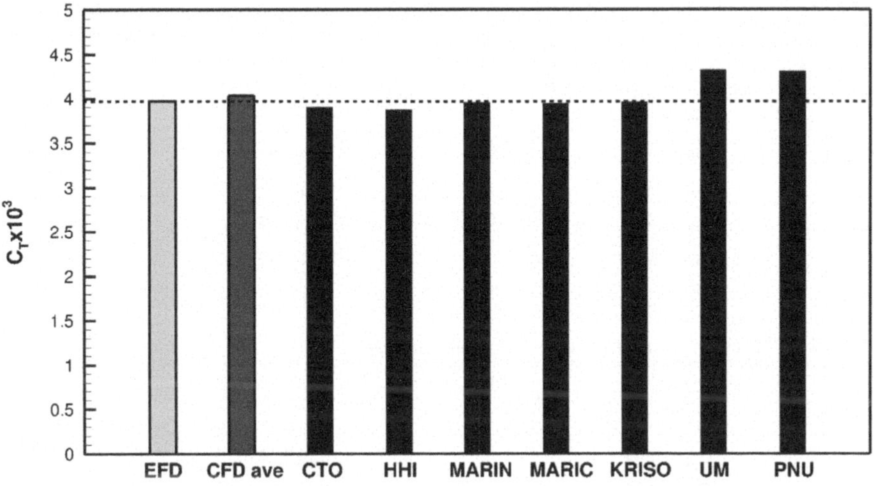

Fig. 14 The resistance at self-propelled condition

model better predict the crescent shape of high-speed region of the axial velocity contour. Figure 17 shows the comparison of hull surface pressure distributions.

Table 13 Statistics and error estimation for the case 2.5

		T2005	G2010	T2015
Participants		4	13	5/7
C_T	E_{mean}%D	−0.9	−0.3	1.0
	σ_{SD}%D	1.0	3.1	1.0
K_T	E_{mean}%D		−0.6	0.5
	σ_{SD}%D		7.2	2.7
K_Q	E_{mean}%D		−4.6	−3.5
	σ_{SD}%D		6.1	2.4
Rps	E_{mean}%D		0.6	−0.3
	σ_{SD}%D		2.8	1.8
R_T-T	E_{mean}%D		−7.8	3.9
	σ_{SD}%D		4.4	1.1

Table 14 Error estimation for the case 2.5 (fixed FD case)

		2015		2015
Participants		3		
C_T	E_{mean}%D	0.6	$\|E\|_{mean}$%D	0.6
K_T	E_{mean}%D	0.6	$\|E\|_{mean}$%D	2.9
K_Q	E_{mean}%D	−3.0	$\|E\|_{mean}$%D	3.4
Rps	E_{mean}%D	−0.3	$\|E\|_{mean}$%D	1.5
R_T-T	E_{mean}%D	–	$\|E\|_{mean}$%D	–

Table 15 Error estimation for the case 2.5 (fixed rps case)

		2015		2015
Participants		2		
C_T	E_{mean}%D	1.5	$\|E\|_{mean}$%D	1.5
K_T	E_{mean}%D	0.3	$\|E\|_{mean}$%D	0.9
K_Q	E_{mean}%D	−4.2	$\|E\|_{mean}$%D	4.2
Rps	E_{mean}%D	–	$\|E\|_{mean}$%D	–
R_T-T	E_{mean}%D	3.9	$\|E\|_{mean}$%D	3.9

3 Conclusions

- For the predictions of resistance, sinkage and trim, the number of grid points are varied from 1.1 M to 6.6 M. The submitted results do not show large dependency of grid points as shown in Fig. 17. It means that CFD users who have efficient

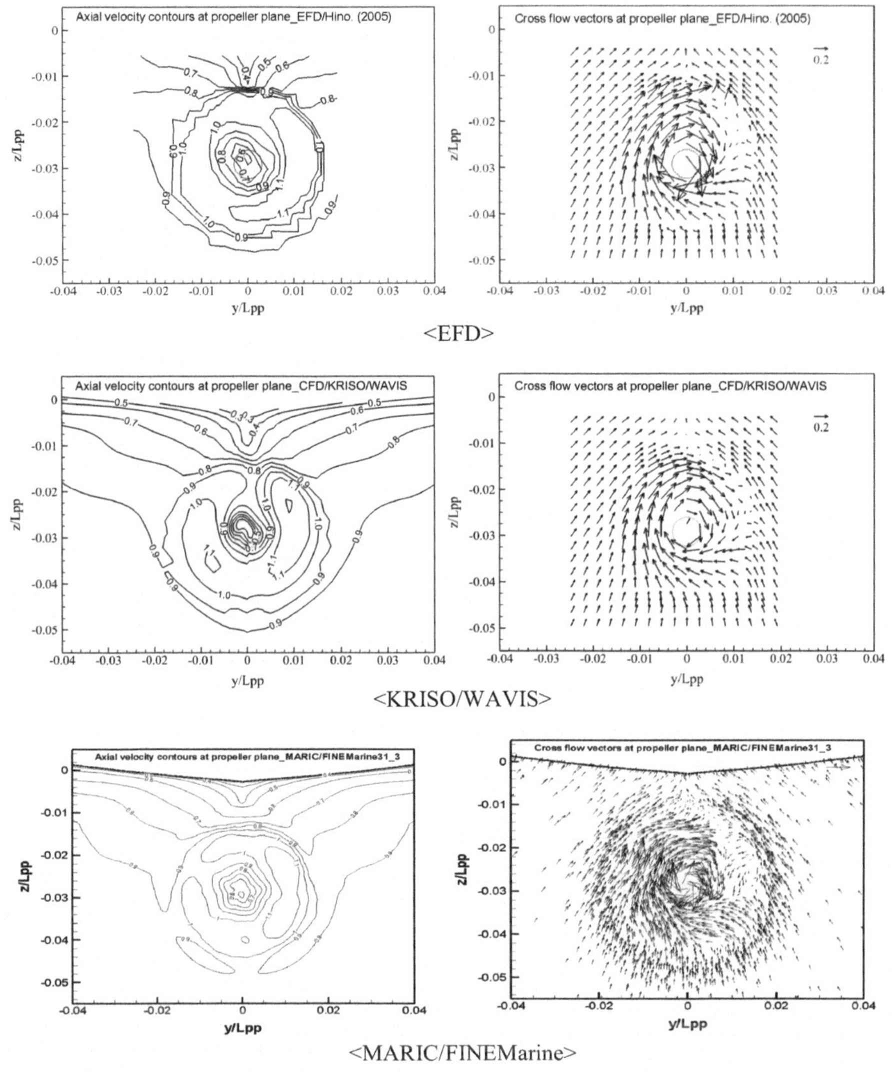

Fig. 15 Axial velocity contours and flow vectors

grid making techniques can have an acceptable simulation results for low block coefficient hull form such as KCS.

- Most participants use two equation turbulence model and only few participants use an advanced turbulence model. Based on these submissions, there is no visible improvement in accuracy of the resistance prediction for more advanced turbulence model than the two equation models.
- The comparison error for resistance prediction at the design speed ($Fr = 0.26$) is -0.2% of the data and the standard deviation is 1.5% of the data value. The mean

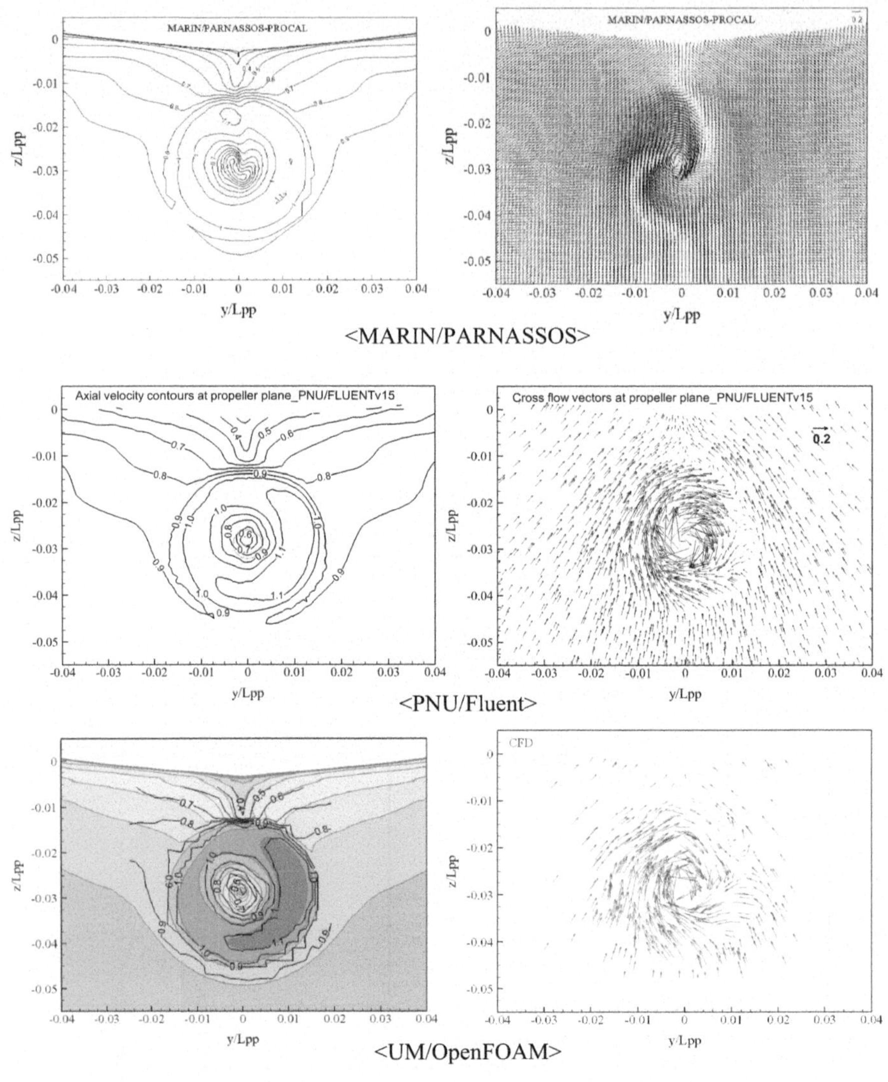

Fig. 15 (continued)

comparison error for all 6 speeds is 0.43% and the mean standard deviation is 2.48%. The increased error and standard deviation is caused by the results for low Fr simulation submissions. The mean absolute error |*E*| is 2%, which is slightly larger than 1.64% the results of Gothenburg 2010.

- The comparison errors and standard deviation of the sinkage and trim are much larger than for the resistance. The errors come from the results in a speed range below $Fr = 0.2$. Two possible reasons can be considered. This range generates very short waves and so it requires better grid distribution to resolve the waves. Another

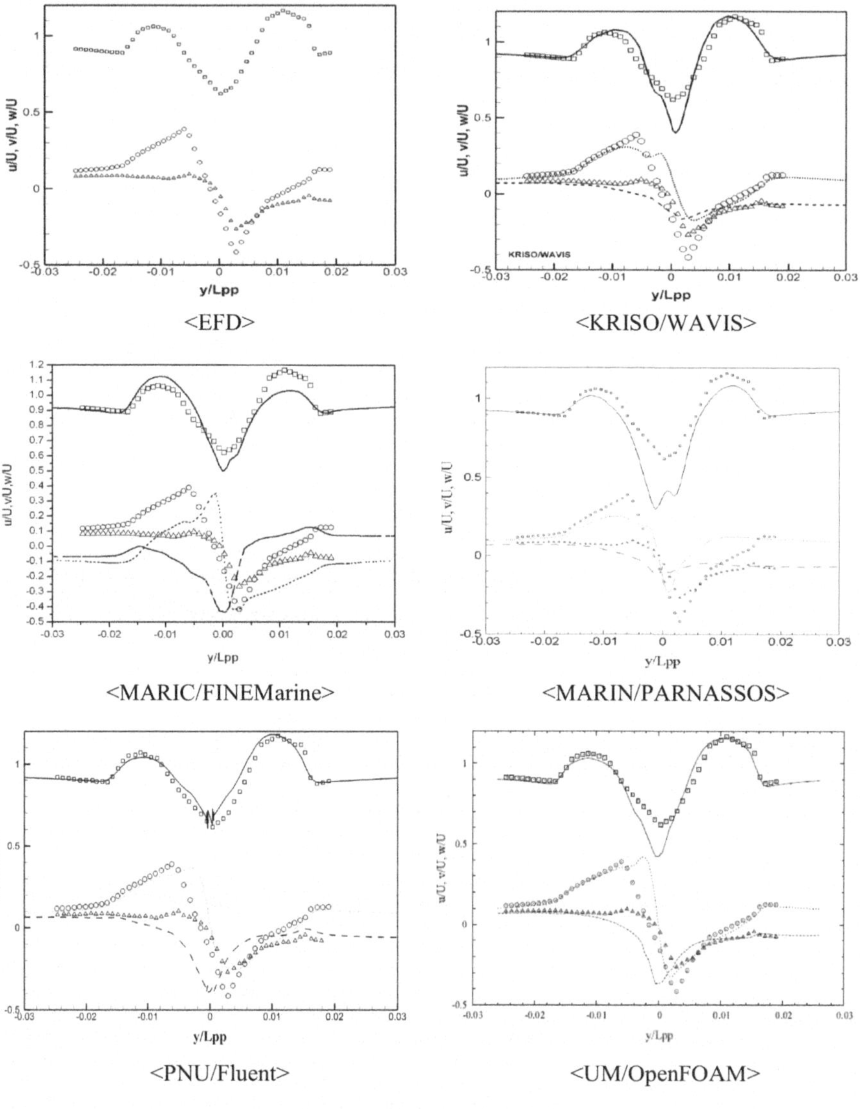

Fig. 16 Velocity downstream of propeller plane at $z/L_{pp} = -0.03$

problems are likely to be due to the difficulties of measuring two quantities at low speed in the experiment. The data might have larger uncertainties than for high speed.

- Self-propulsion submissions are slightly improved compared to the results of Gothenburg 2010 Workshop. The mean comparison errors of K_T and K_Q are 0.5 and −3.5% respectively (−0.6 and −4.6% for Gothenburg 2010 workshop). The standard deviations for of K_T and K_Q are 2.7 and 2.4% respectively, which are

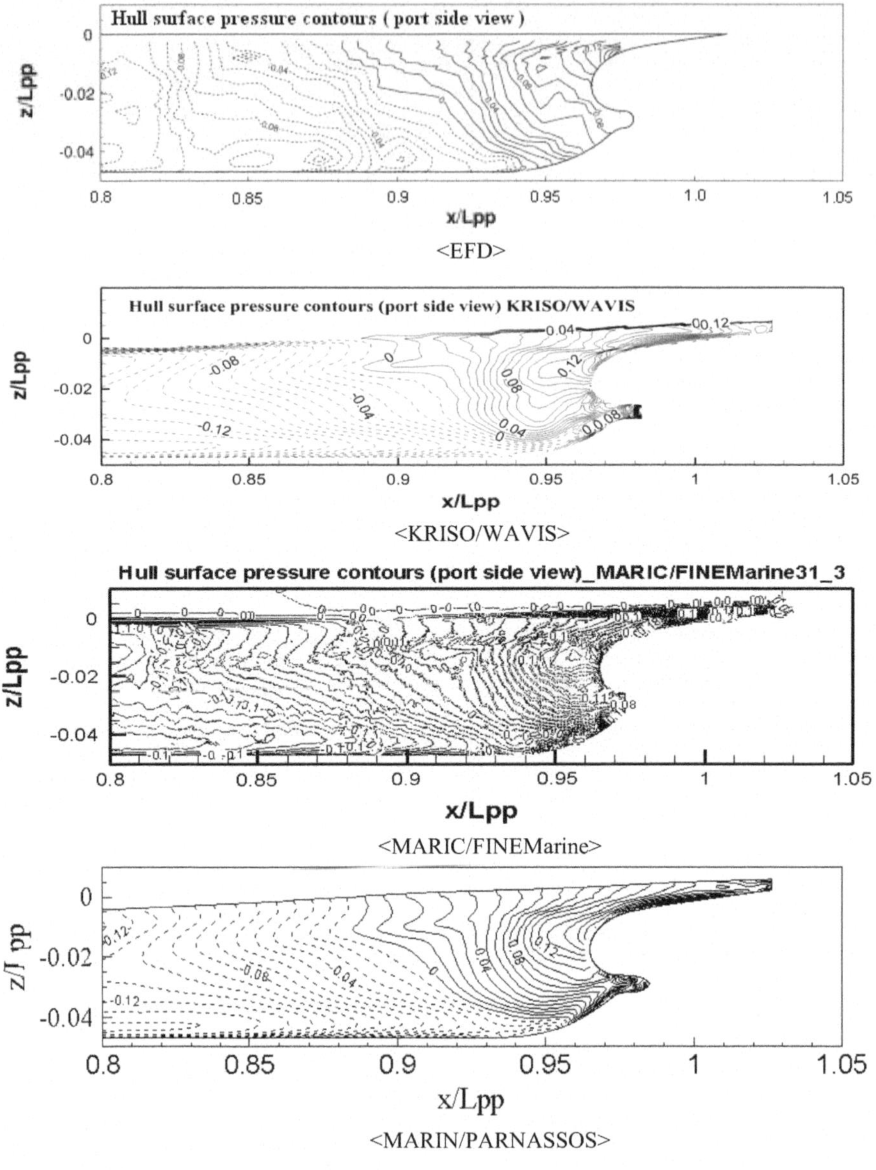

Fig. 17 Hull surface pressure contours (pressure coefficient)

smaller than the results of 7.2 and 6.1% respectively for Gothenburg 2010 workshop. However, prediction for K_Q is still less accurate than K_T. Self-propulsion parameters are slightly better predicted by body force models (FD fixed) but local flow characteristics are comparatively well predicted by actual propeller rotating simulation (rps fixed).

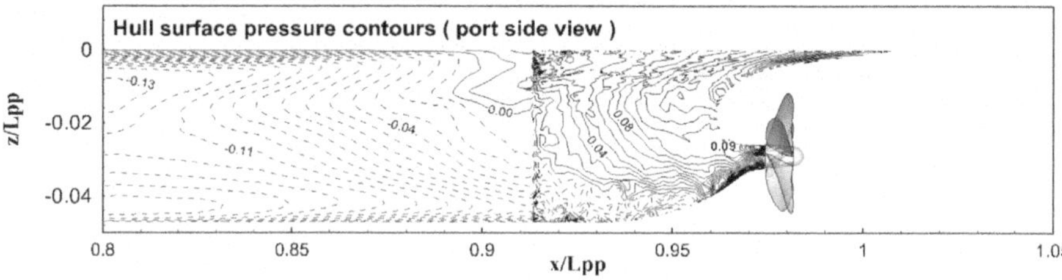

<PNU/Fluent>

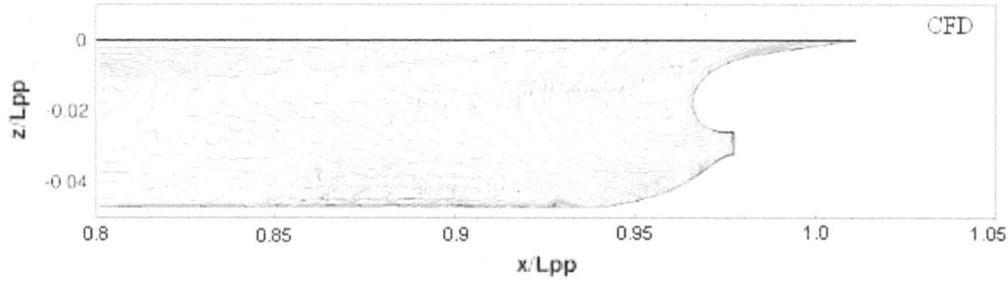

<UM/OpenFOAM>

Fig. 17 (continued)

- For the future workshops, new experimental data for KCS or other container hull would be recommended to simulate not only captive but free running 6DoF calm water self-propulsion condition.

References

Larsson, L., Stern, F., & Visonneau, M. (2010). Proceedings of Gothenburg 2010 A Workshop on Numerical Ship Hydrodynamics. Gothenburg, Sweden.

Larsson, L., Stern, F., Visonneau, M. (Eds.). (2013). Numerical Ship Hydrodynamics—An assessment of the Gothenburg 2010 Workshop, Springer.

Assessment of CFD for KCS Added Resistance and for ONRT Course Keeping/Speed Loss in Regular Head and Oblique Waves

Frederick Stern, Hamid Sadat-Hosseini, Timur Dogan, Matteo Diez, Dong Hwan Kim, Sungtek Park, and Yugo Sanada

Abstract CFD is assessed for added resistance for KCS (captive test cases 2.10 and 2.11) and course keeping/speed loss for ONRT (free running test cases 3.9/3.12/3.13) in head and oblique waves. The number of submissions were 10, 2, and 8 for test cases 2.10, 2.11, and 3.9/3.12/3.13, respectively. The assessment approach uses both solution and N-version validation. The former considers whether the absolute error $|E_i| = |D - S_i|$ is less, equal or greater than the validation uncertainty, which is the root sum square of the numerical and experimental uncertainties, i.e., $|E_i| \leq U_{V_i} = \sqrt{U_{SN_i}^2 + U_D^2}$. The latter considers whether the absolute error is less, equal or greater than the state-of-the-art SoA$_i$ uncertainty, i.e., $|E_i| \leq U_{SoA_i} = \sqrt{U_{V_i}^2 + P_{|E_i|}^2}$ where $P_{|E_i|} = k\sigma_{|E|}$ is the uncertainty due to the scatter in the solution absolute error. Errors and uncertainties are normalized using both the data value D and its dynamic range DR. The captive resistance C_T and free running self-propulsion propeller revolutions RPS $\overline{|E|} < 2\%$D with U_D less than but comparable $P_{|E_i|} < 3\%$D such that 3/1 solutions were validated but 8/4 codes/solutions were N-version validated for C_T/RPS. The head waves captive and free running heave and pitch $\overline{|E|}$ is less than 8%DR with U_D less than 5%DR and $P_{|E_i|}$ less than 13%DR such that about 5 for captive and 2 for free running solutions were validated and about 7 for captive and 5 for free running codes/solutions were N-version validated. The errors for added resistance and speed loss were less than 13%DR with U_D less than 7%DR and $P_{|E_i|} = 25$ and 13%DR such that about 4 for captive and 1 for free running solutions were validated and 7 for captive 4 for free

F. Stern (✉) · T. Dogan · S. Park · Y. Sanada
University of Iowa and IIHR-Hydroscience & Engineering, Iowa City, IA, USA
e-mail: frederick-stern@uiowa.edu

H. Sadat-Hosseini
University of North Texas, Denton, TX, USA

M. Diez
CNR-INM, National Research Council-Institute of Marine Engineering, Rome, Italy

D. H. Kim
Korea Research Institute of Ocean Science & Technology, Daejeon, South Korea

T. Hino et al. (eds.), *Numerical Ship Hydrodynamics*, Lecture Notes in Applied and Computational Mechanics 94, https://doi.org/10.1007/978-3-030-47572-7_9

running codes/solutions were N-version validated. For captive head waves, the errors and scatter are smaller than those for potential flow. The oblique waves captive and free running motion errors $\overline{|E|}$ are less than 10%DR except for roll with U_D large 23%DR for captive pitch and other wise small <2%DR. The errors for added resistance and speed loss were <8%DR with U_D < 9%DR. The largest errors were for roll, which had errors for captive of 12%DR and for free running of 20%DR. None of the KCS and ONRT test cases were able to achieve a programmatic requirement of 5%D and 5%DR for calm water and waves, respectively. Note that not all the experimental uncertainties are able to meet this requirement either such that both experiments and CFD need to reduce their uncertainties and in addition CFD its errors. Nonetheless, in view the comparable CFD capability for ONRT free running vs. KCS captive conditions, the prognosis for CFD capability is excellent. Future CFD assessment should extend the current test cases for consideration of added power in regular and irregular waves. Verification should be required, and more limited/most important validation variables should be confirmed and used for the V&V and SoA assessment.

Nomenclature

A	Desired wave amplitude (m)
AR	Added resistance (N)
B	Breadth (m)
B_{WL}	Maximum beam at the waterline (m)
C_B	Block coefficient
C_M	Midship Section Coefficient
CG	Center of gravity
C_T	Total resistance coefficient
C_{Ti}	I-th harmonic of total resistance coefficient
D	Measurement value
DDES	Delayed detached eddy simulation
E	Error
E_i	Error of i-th simulation
DR	Dynamic range
f_e	Wave encounter frequency (Hz)
f_{hz}	Hydrostatic estimates of natural heave frequency (Hz)
$f_{h\theta}$	Hydrostatic estimate of natural pitch frequency (Hz)
$f_{h\phi}$	Hydrostatic estimate of natural roll frequency (Hz)
Fr	Froude number
g	Gravitational acceleration (m/s^2)
H	Wave height (m)
IMO	International Maritime Organization
k	Wave number/Spring constant (N/m)
KCS	KRISO/MOERI Container Ship
KG	Vertical center of gravity (m)

K_Q	Torque coefficient
K_T	Thrust coefficient
k_{xx}	Radius of gyration in roll (m)
k_{yy}	Radius of gyration in pitch (m)
k_{zz}	Radius of gyration in yaw (m)
KVLCC2	KRISO Very Large Crude Carrier
L	Ship length (m)
LCB	Longitudinal center of buoyancy
L_{PP}	Length between perpendiculars (m)
LS	Level set method
L_{WL}	Length at the waterline (m)
n	Propeller revolutions (rps)
ONRT	ONR Tumblehome
P	The random/precision uncertainty
Re	Reynolds number
RSM	Reynolds stress turbulence model
S	Wetted surface area (m^2)/simulation value
Si	I-th simulation value
SoA	State of art
S_R	Wetted surface area of rudder (m^2)
S_W	Wetted surface area of hull (m^2)
t	Time (s)
T	Draft (m)
T_e	Wave encounter period (s)
u	Surge velocity (m/s)
u_i	I-th harmonic surge velocity amplitude (m/s)
U	Ship speed (m/s)
U_D	Experimental uncertainty
U_{SoA}	Mean code uncertainty
U_{SoAi}	Uncertainty of i-th code
v	Sway velocity (m/s)
v_i	I-th harmonic sway velocity amplitude (m/s)
V	Ship speed (m/s)
V_A	Desired approach speed (m/s)
VoF	Volume of fluid method
X	X force (N)/Coordinate (m)
X_C	X axis carriage coordinate (m)
X_i	Measurement value
X_H	Hydrodynamic force (N)
X_S	X axis ship coordinate (m)
Y_C	Y axis carriage coordinate (m)
Y_S	Y axis ship coordinate (m)
z	Heave (m)
Z_i	I-th harmonic heave amplitude (m)
Z_i	I-th harmonic heave amplitude (m)

δ	Rudder angle (deg)
Δ	Mass of ship (kg)
ζ_i	Wave elevation (m)
ζ_i	I-th harmonic wave amplitude (m)
θ	Pitch angle (rad)
θ_i	I-th harmonic pitch amplitude (rad)
λ	Wave length (m)
μ	Wave encounter angle (deg)
ν	Kinematic viscosity of water (m²/s)
ρ	Density of water (kg/m³)
σ	Standard deviation
σ/L	Sinkage
τ	Trim (deg)
ϕ	Roll angle (rad)
ϕ_i	I-th harmonic roll amplitude (rad)
ψ	Yaw angle (rad)
ψ_i	I-th harmonic yaw amplitude (rad)
ψ_c	Target yaw angle (deg)
∇	Displacement (m³)

1 Introduction

T2015 included test cases for KCS Fr = 0.26 and Re = $1.074 \times 10^7$and Re = 3.613×10^6, respectively, for added resistance in regular head (2.10) and oblique (2.11) waves; and for course keeping/speed loss for ONRT Fr = 0.2 and Re = 3.497×10^6 for free running in regular head (3.12) and oblique (3.13) waves. Test cases 2.10 and 2.11 included calm water resistance, sinkage and trim. ONRT calm water self-propulsion propeller revolutions, sinkage and trim was included as test case (3.9). Section 5 covers the test case 2.10; Sect. 6 covers test case 2.11; and Sect. 7 covers test cases 3.9, 3.12, and 3.13.

Section 2 summarizes the quantitative approach for the CFD assessment, which uses both solution and N-version validation. The former considers whether the absolute error $|E_i| = |D - S_i|$ is less, equal or greater than the validation uncertainty, which is the RSS of the numerical and experimental uncertainties, i.e., $|E_i| \leq U_{V_i} = \sqrt{U_{SN_i}^2 + U_D^2}$. The latter considers whether the absolute error is less, equal or greater than the state-of-the-art SoA$_i$ uncertainty, i.e., $|E_i| \leq U_{SoA_i} = \sqrt{U_{V_i}^2 + P_{|E_i|}^2}$ where $P_{|E_i|} = k\sigma_{|E|}$ is the uncertainty due to the scatter in the solution absolute error. The tables, figures and discussions use these variables; thus, the reader is encouraged to familiarize themselves with this terminology.

Section 3 summarizes the experimental N-order and M×N-order level testing and uncertainty analysis. The individual facility data value is Xi and when available its individual facility N-order level testing uncertainty estimate is U_{Xi}. The multiple facility M×N order level testing average value $\bar{X} = \frac{1}{M}\sum_{i=1}^{M} X_i$ and precision limit are $P_{Xi} = 2\sigma_{Xi}$, where $\sigma_{X_i}^2 = \frac{1}{M-1}\sum_{i=1}^{M}\left(X_i - \bar{X}\right)^2$ is the standard deviation SD such that the individual facility MxN-order level testing uncertainty estimate is $U_{Di} = \sqrt{U_{Xi}^2 + P_{Xi}^2}$. The discussions use both Xi and $\bar{X}$, which are often simply referred to as D and similarly for U_{Xi} and U_{Di} as U_D for convenience. Chapter 4 summarizes the available experimental data used for the KCS and ONRT test cases. Multiple facility data is available for KCS calm water resistance and head waves such that σ_{Xi} is evaluated, which was used for the KCS calm water resistance, sinkage and trim validation.

Section 4 summarizes the validation variables and experimental uncertainties used for the KCS and ONRT test cases. The validation variables for calm water KCS are resistance, sinkage and trim and for ONRT self-propulsion are propeller revolutions, sinkage and trim. Resistance and propeller revolutions are considered primary and motions secondary variables. For KCS head waves, added/0th harmonic resistance, and 1st harmonic heave and pitch are considered primary and 1st harmonic resistance as secondary variables; with additionally 1st harmonic roll as primary variable for oblique waves. Similarly, for ONRT free running in head and oblique waves primary variables are speed loss (0th harmonic axial velocity) and first harmonic motions which additionally include yaw are the primary variables and other 0th and 2nd harmonic variables as secondary. Most of the discussions focus on averages over wavelength or heading, including averages for roughly speaking first order (1st harmonic) and second order (0th and 2nd harmonic) terms. For KCS head waves test case 2.10, comparisons are made with the previous G2010 workshop, which had the same test case; and with results from 8 different potential flow codes provided by Bunnik et al. (2010).

2 Verification, Validation and SoA Approaches

The individual code solution verification and validation (V&V) follows Stern et al. (2001) overall procedures. For ship hydrodynamics, verification usually uses either the factor of safety (Xing and Stern 2010) or least square (Eça and Hoekstra 2014) methods. The numerical uncertainties U_{SN_i} use the root-sum-square of the iterative U_{I_i}, grid U_{G_i} and time step U_{T_i} uncertainties

$$U_{SN_i} = \sqrt{U_{G_i}^2 + U_{T_i}^2 + U_{I_i}^2}\%S_i \tag{1}$$

i indicates an individual code and S_i is the solution on the finest grid. ASME (2009) advocates adding U_{I_i} with $U_{G_i}^2$ and $U_{T_i}^2$. The comparison error E_i is

$$E_i\%D = \frac{(D - S_i)}{D} \times 100. \tag{2}$$

D is the experimental benchmark data. Validation compares the errors with the validation interval.

$$U_{V_i} = \sqrt{U_{SN_i}^2 + U_D^2}\%D \tag{3}$$

U_D is the experimental benchmark data uncertainty. The simulations are validated at an interval U_{V_i} if

$$|E_i|\%D \leq U_{V_i}\%D. \tag{4}$$

A shortcoming of this and other V&V approaches is that the justification for uncertainty estimates at the 95% confidence level is based on reasoning like that used for experimental bias uncertainties at the 0-order-replication level without additional statistical 1-order replication level precision uncertainties, which in combination provides N-order replication level and increased confidence for experimental uncertainty analysis.

Available solutions from multiple codes for the same validation variables enable evaluation of the mean and standard deviation, which is referred to as N-version testing (Hemsch 2004). Stern et al. (2006, 2007) extended N-version testing by also considering the individual code V&V, which was referred to as CFD certification: a process for assessing probabilistic confidence intervals for CFD codes for specific benchmark applications and certification variables. Certification of CFD codes is at the multiple codes or users, models, grid types, etc. level (code level). Multiple users are appropriate for many user codes, whereas multiple codes are appropriate for various few user codes (which is often the case for industrial applications). Differences between versions and implementations are due to a myriad of possibilities for modeling, numerical methods, and their implementation as CFD codes and simulations.

Stern et al. (2016, 2017) modified this approach to use the mean and standard deviation of the absolute value of the error, which provides a more conservative error estimate and smaller standard deviation. Here we follow this approach by considering the mean error $\bar{E}$, mean absolute error $\overline{|E|}$ and its standard deviation $\sigma_{|E|}$ and uncertainty U_{S_oA} in assessing the results. The use of the terminology "certification" was criticized as implies/requires an authorizing body; therefore, the new approach is called a statistical approach for CFD SoA assessment: N-version verification and validation.

At the multiple code level, the individual code/solution uncertainty includes both deterministic/bias and random/precision components

$$U_{S_i}^2 = B_{S_i}^2 + P_{S_i}^2 \tag{5}$$

$$B_{S_i} = U_{SN_i} \tag{6}$$

The mean code solution, bias and standard deviation are

$$\bar{S} = \frac{1}{N}\sum_{i=1}^{N} S_i \tag{7}$$

$$B_{\bar{S}}^2 = \frac{1}{N}\sum_{i=1}^{N} U_{SN_i}^2 \tag{8}$$

$$\sigma_S = \sqrt{\frac{1}{N-1}\sum_{i=1}^{N}\left(S_i - \bar{S}\right)^2} = \sqrt{\frac{1}{N-1}\sum_{i=1}^{N}\left(E_i - \bar{E}\right)^2} = \sigma_E \tag{9}$$

The individual code/solution and mean code random/precision numerical uncertainties are

$$P_{S_i} = 2\sigma_S \text{ and } P_{\bar{S}} = 2\sigma_S/\sqrt{N}. \tag{10}$$

Outliers can be identified and rejected similarly as with experimental data using, e.g., Chauvenet's criterion. Herein, for simplicity, a solution is rejected if its deviation from the mean is larger than $2\sigma_S$, i.e. $N \approx 10$. The mean code solution uncertainty is

$$U_{\bar{S}}^2 = B_{\bar{S}}^2 + P_{\bar{S}}^2. \tag{11}$$

The average error and average absolute error are

$$\bar{E} = \frac{1}{N}\sum_{i=1}^{N} E_i \tag{12}$$

$$\overline{|E|} = \frac{1}{N}\sum_{i=1}^{N} |E_i| \geq \left|\bar{E}\right| \tag{13}$$

The average absolute error is always greater than or equal to the absolute value of the signed average error. The average absolute error standard deviation is

$$\sigma_{|E|} = \sqrt{\frac{1}{N-1}\sum_{i=1}^{N}\left(|E_i| - \overline{|E|}\right)^2} \tag{14}$$

with absolute value statistical property

$$\sigma_{|E|}^2 = \sigma_E^2 + \bar{E}^2 - \overline{|E|}^2 \leq \sigma_E^2 = \sigma_S^2 \tag{15}$$

The random/precision uncertainty component for the individual code/solution and mean code are

$$P_{|E_i|} = 2\sigma_{|E|} \tag{16}$$

$$P_{\overline{|E|}} = \frac{2\sigma_{|E|}}{\sqrt{N}} \tag{17}$$

The uncertainty of the individual code/solution error E_i is

$$U_{SoA_i}^2 = U_D^2 + U_{SN_i}^2 + P_{|E_i|}^2 \tag{18}$$

The uncertainty of the mean code error $\overline{|E|}$ is

$$U_{SoA}^2 = U_D^2 + \bar{U}_{SN} + P_{\overline{|E|}}^2 \tag{19}$$

The SoA of the individual code/solution and mean code are defined by

$$|E_i| \leq U_{SoA_i} \tag{20}$$

$$\overline{|E|} \leq U_{SoA} \tag{21}$$

3 Experimental Uncertainties

The experimental uncertainty analysis considers both individual facility/model N-order testing U_{X_i} and M × N-order level testing U_{D_i} and $U_{\overline{X}}$ (Stern et al. 2005). Head waves U_{X_i} also includes post-workshop analysis of the statistical convergence uncertainty U_{SC_i}.

N-order level testing:

$$X_i = \frac{1}{N}\sum_{j=1}^{N} X_i^j \tag{22}$$

$$\sigma_{X_i^j}^2 = \frac{1}{N-1}\sum_{j=1}^{N}(X_i^j - X_i)^2 \tag{23}$$

$$U_{X_i^j} = \sqrt{B_{X_i}^2 + (2\sigma_{X_i^j})^2 + U_{SC_i^j}^2} \tag{24}$$

$$U_{X_i} = \sqrt{B_{X_i}^2 + (2\sigma_{X_i^j}/\sqrt{N})^2 + U_{SC_i}^2} \tag{25}$$

M × N-order level testing:

$$\overline{X} = \frac{1}{M}\sum_{i=1}^{M} X_i \tag{26}$$

$$B_{\overline{X}} = \frac{1}{M}\sum_{i=1}^{M} U_{X_i} \tag{27}$$

$$\sigma_{X_i}^2 = \frac{1}{M-1}\sum_{i=1}^{M} (X_i - \bar{X})^2 \tag{28}$$

$$P_{X_i} = 2\sigma_{X_i} \tag{29}$$

$$P_{\overline{X}} = \frac{2\sigma_{X_i}}{\sqrt{M}} \tag{30}$$

$$U_{\overline{X}} = \sqrt{B_{\overline{X}}^2 + P_{\overline{X}}^2} \tag{31}$$

$$U_{D_i} = \sqrt{U_{X_i}^2 + P_{X_i}^2} \tag{32}$$

The facility bias uncertainty is included as follows.

$$\Delta_i = \bar{X} - X_i \tag{33}$$

$$U_{\Delta_i} = \sqrt{U_{X_i}^2 + U_{\overline{X}}^2} \tag{34}$$

If the absolute value of Δ_i is less than U_{Δ_i}

$$|\Delta_i| \leq U_{\Delta_i} \tag{35}$$

the individual facility is certified at interval U_{Δ_i}. Whereas if the absolute value of Δ_i is greater than U_{Δ_i}

$$|\Delta_i| > U_{\Delta_i} \tag{36}$$

the facility bias and total uncertainty are defined as

$$U_{FB_i} = \sqrt{\Delta_i^2 - U_{\Delta_i}^2} \tag{37}$$

$$U_{T_i} = \sqrt{U_{X_i}^2 + U_{FB_i}^2}. \tag{38}$$

Lastly, U_{D_i} including facility bias is

$$U_{D_i} = \sqrt{U_{X_i}^2 + U_{FB_i}^2 + P_{X_i}^2} = \sqrt{U_{T_i}^2 + P_{X_i}^2} \quad (39)$$

M × N-order level testing U_{D_i} is only available for test case 2.10 and thus far only used for calm water resistance.

4 Validation Variables and Experimental Uncertainties

4.1 KCS and ONRT Particulars and Coordinate Systems

Tables 1 and 2 list the model particulars for the KCS and ONRT full and model scales. Figures 1 and 2 shows the carriage- and ship-fixed coordinate system used to report the validation variables for KCS and ONRT, respectively. For KCS, the

Table 1 Main particulars for KCS 2.10 and 2.11 (Fr = 0.26)

	Full scale	IIHR/FORCE L = 2.7 m	FORCE L = 6.0702 m
Main particulars			
L(= L_{PP}) (m)	230	2.7	6.0702
L_{WL} (m)	232.5	2.729	6.1357
B_{WL} (m)	32.2	0.378	0.8498
T (m)	10.8	0.1268	0.285
∇ (m^3)	52030	0.0842	0.9571
S_W (m^2)	9424	1.2987	6.6177
S_R (m^2)	115	0.0158	0.0801
C_B	0.6505	0.651	0.651
C_M	0.9849		
LCB (%L_{PP}), fwd+	−1.48	–	−1.48
KG/L	0.0622	0.0622	0.0622
k_{xx}/B	0.4	0.4	0.4
k_{yy}/L, k_{zz}/L	0.25	0.25	0.252
$f_{h\phi}$ (Hz)	0.025	0.23	0.15
f_{hz} (Hz)	0.11	1.09	0.73
$f_{h\theta}$ (Hz)	0.11	1.09	0.73
Ship speed			
U (m/s)	12.35	1.24	2.00
Re	2.84×10^9	3.613×10^6	1.074×10^7

Table 2 Main particulars for ONRT 3.9, 3.11, 3.12 (Fr = 0.2)

	Full scale	IIHR L = 3.147 m
Main particulars		
$L(=L_{WL})$ (m)	154.0	3.147
B_{WL} (m)	18.78	0.384
T (m)	5.494	0.112
Δ (kg)	8507000	72.6
S_W (m^2)	NA	1.5
S_R (m^2)	NA	0.012×2
C_B	0.535	0.535
LCB (%L_{PP}), fwd+	NA	1.625
KG/L	NA	0.156
k_{xx}/B	0.444	0.444
k_{yy}/L, k_{zz}/L	0.25	0.246
Propeller diameter	NA	0.1066
Propeller center, lateral location (from CL)	NA	0.9267
Propeller center, vert. location (below WL)	NA	0.02661
Propeller shaft angle (downward positive)	NA	0.03565
Propeller rotation direction (view from stern)	Inward	Inward
Shaft angle (deg)	5	5
Maximum rudder rate	NA	35
$f_{h\phi}$ (Hz)	NA	0.65
f_{hz} (Hz)	NA	1.6
$f_{h\theta}$ (Hz)	NA	1.6
Ship speed		
U (m/s)	7.77	1.11
Re	1.197×10^9	3.497×10^6

carriage-fixed coordinate system is located at the center of gravity, with x pointing toward the stern, y to starboard and z upward. Figure 1 shows the IIHR and Osaka University (OU) surge free mounts. FORCE used a fixed surge added resistance mount designed for large motions (same as used in waves) and not a conventional calm water resistance test setup. For ONRT, the ship-fixed coordinate system is also located at the center of gravity but with x pointing toward the bow, y to starboard and z downward. The evaluation of the motions and forces is at the center of gravity.

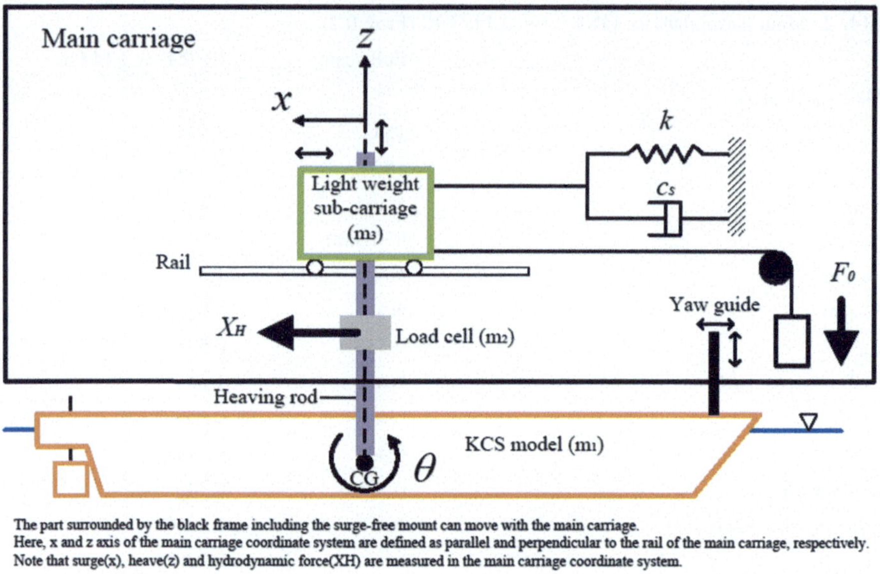

Fig. 1 Coordinate system for KCS 2.10 and 2.11 experiments and simulations

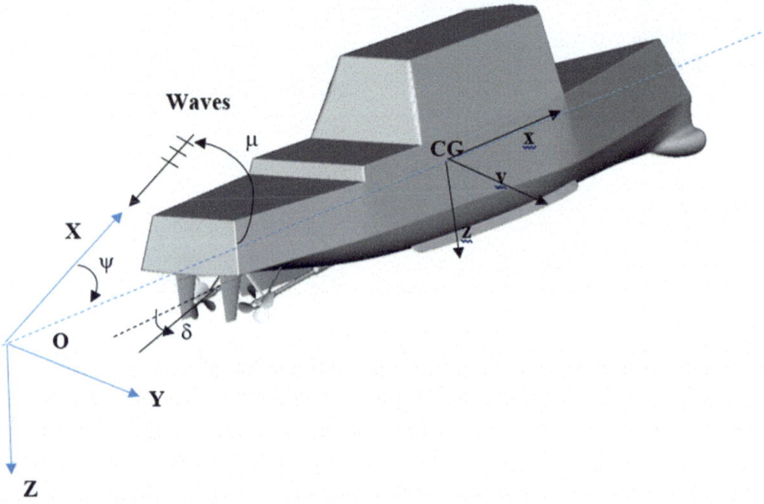

Fig. 2 Coordinate system for ONRT 3.9, 3.11, 3.12 experiments and simulations

4.2 *Wave Elevation*

For test cases 2.10, 2.11, 3.12 and 3.13 wave elevations are measured at FP, but the y/L values are different: for 2.10, y/L = 0.81; for 2.11, y/L = 0.037; and for 3.12

and 3.13, y/L = 0.032. The test case 2.11, 3.12 and 3.13 y/L locations are less than desirable to insure that they are not affected by the model diffraction/radiation waves but were restricted due to the experimental setup. For 2.10, the CFD did not report the location for the wave elevation submissions, whereas for 2.11, 3.12 and 3.13, the location was the same as the experiments.

For 2.10, the experimental 1st harmonic wave amplitude ζ_1 and wavelength λ were used as the CFD input, whereas for 2.11, 3.12 and 3.13, the ideal wave amplitude A and length λ values were used. For 2.11 and 3.12/3.13 the experimental differences with the ideal values was $\leq$3% based on stationary basin fixed wave gauge data, which meets IMO criteria.

For 2.10, the wavelength conditions are $\lambda/L = 0.65, 0.85, 1.15, 1.37$ and 1.95 with the wave steepness of $H/\lambda = 1/60$. For 2.11, $\lambda/L = 1$, $H/\lambda = 1/60$ and wave headings of $\mu = 0, 45, 90, 135$ and 180°. For 3.12 and 3.13, the conditions are the same as 2.11, except $H/\lambda = 1/50$.

The wave conditions are close to the usual assumptions for small amplitude ship motions response. The experimental waves were largely 1st harmonic dominant. For 2.10, the second harmonic of wave amplitude ζ_2 is $\leq 1\%$ of the first harmonic wave amplitude ζ_1 for short waves and $\zeta_2 \leq 10\%\zeta_1$ for long waves. The experimental wave uncertainty was not reported for 2.10. For 2.11 and 3.12/3.13, $\zeta_2 = 5\%$ and $3\%\zeta_1$, respectively. The experimental wave uncertainty was $\leq$1.5%D for 2.11 and $\leq$3%D for 3.12 and 3.13. Herein only qualitative assessment of the experimental and CFD wave elevations is conducted via comparisons of the time histories.

4.3 Motions and Forces

4.3.1 Test Case 2.10

FORCE conducted experiments for Test Case 2.10 using L = 6.1 m model. IIHR used the 2.7 m model for Test Case 2.11. Both models are appended with a horn type rudder. The hydrostatic estimates of heave, pitch and roll natural frequencies for IIHR/FORCE models are (f_{hz}, $f_{h\theta}$, $f_{h\phi}$) = (1.09/0.73, 1.09/0.73, 0.23/0.15 Hz).

The validation variables for calm water are resistance $C_T = \frac{X}{1/2\rho SV^2}$, sinkage $\frac{\sigma}{L}$ and trim τ [deg]. FORCE provided the experimental values X_i of the validation variables, as shown in Table 3. The data uncertainty U_{X_i} is only provided for C_T. P_{X_i} and U_{D_i} values are evaluated based on data from 5 facilities using 7.3 m (NMRI and

Table 3 Uncertainties for KCS resistance tests in calm water at Fr = 0.26 (Case 2.10)

Variable	X_i	$U_{X_i}\%D$	$P_{X_i}\%D$	$U_{D_i}\%D$
$C_T \times 10^3$	3.835	0.95	1.47	1.75
$\sigma/L \times 10^2$	−0.2074	NA	7.38	7.38
τ [deg]	−0.165	NA	14.25	14.25

KRISO), 4.4, 6.1, 2.7 m (FORCE) and 2.7 m (IIHR) and 3.2 m (OU) models. The P_{X_i} and U_{D_i} values in Table 3 exclude the data for the 2.7 and 3.2 m models. U_{D_i} is used for the CFD assessment, as provides a more robust U_D estimate than U_{X_i}.

The validation variables for head waves are wave elevation $\frac{\zeta}{A}$, resistance $C_T = \frac{X}{1/2\rho SV^2}$, added resistance $AR = \frac{X_{TO}-X_{TCalm}}{\rho g A^2 B^2/L} = \frac{SV^2}{2gA^2B^2/L}(C_{TO} - C_{TCalm})$, heave $\frac{z}{A}$ and pitch $\frac{\theta}{Ak}$ (where $A = \zeta_1$ for 2.10 and A for 2.11 and ONRT test cases). Resistance coefficient is based on wetted surface area $S/L^2 = 0.1818$ with rudder in calm water. The computational and experimental values were compared for time histories of wave elevation, resistance, heave and pitch. Fourier series 0th, 1st and 2nd harmonic amplitudes and phases for resistance, heave and pitch were also evaluated for both experiments and computations, as follows:

$$
\begin{aligned}
P(t) &= \frac{P_0}{2} + \sum_{n=1}^{N} P_n \cos(2\pi f_e t + P_{\varepsilon n}) \\
a_n &= \frac{2}{T}\int_0^T P(t)\cos(2\pi f_e t)dt \\
b_n &= \frac{2}{T}\int_0^T P(t)\sin(2\pi f_e t)dt \\
P_n &= \sqrt{a_n^2 + b_n^2} \\
P_{\varepsilon n} &= tan^{-1}\left(-\frac{b_n}{a_n}\right) - \gamma_I
\end{aligned}
\tag{40}
$$

where P is time series, P_n is the n-th harmonic amplitude and $P_{\varepsilon n}$ is the corresponding phase, $f_e = \omega_e/2\pi$, $\omega_e = \omega - \frac{\omega^2}{g}U\cos\mu$ is the wave encounter frequency, ω is the wave frequency, μ is the wave encounter angle ($\mu = 0$ for head waves) and γ_I is the incident wave phase at the bow at $t = 0$.

The experimental values X_i for the 0th, 1st and 2nd harmonic amplitudes and phases of the validation variables are provided by FORCE, as shown in Tables 4, 5, 6, 7 and 8. U_{X_i} is also provided for AR and 1st harmonic amplitude of heave and pitch. IIHR evaluated the statistical convergence uncertainty U_{SC_i} for the FORCE X_i using a running Fourier series analysis. For each wave length, two different window sizes were selected for the running Fourier series analysis; one was based on two encounter wave periods, and the other was based on the $\sqrt{N}$ encounter wave periods where N was the number of encounter wave periods for which the FORCE data was available. $\sqrt{N}$ was 8, 7, 6, 6, 5 for $\lambda/L = 0.65, 0.85, 1.15, 1.37$, and 1.95, respectively. U_{SC_i} values based on two encounter wave periods are reported in Tables 4, 5, 6, 7 and 8, which are slightly larger than those based on $\sqrt{N}$ encounter wave periods. Overall, the U_{SC_i} values are often large and are included in U_D estimate.

Table 4 0th harmonic amplitude uncertainties for KCS seakeeping in waves at Fr = 0.26 (Case 2.10)

Variable/Dynamic range	λ/L	$f_e/f_{hz/\theta}$	X_i	$U_{X_i}\%D$	$U_{SC_i}\%D$	$U_{X_i}\%D$	$U_{X_i}\%DR$
$C_T \times 10^3$ *DR = 5.927*	0.65	1.50	8.237		5.2	5.2	7.2
	0.85	1.24	9.269		5.6	5.6	8.7
	1.15	1.00	14.164		2.8	2.8	6.7
	1.37	0.89	13.948		2.9	2.9	6.7
	1.95	0.70	10.798		9.1	9.1	16.5
			Ave.		**5.1**	**5.1**	**9.2**
AR DR = 8.002	0.65	1.50	3.386	8.2	6.3	10.3	4.4
	0.85	1.24	6.102	7.9	4.2	8.9	6.8
	1.15	1.00	9.911	7.9	2.0	8.2	10.1
	1.37	0.89	6.512	7.9	3.1	8.5	6.9
	1.95	0.70	1.909	8.0	25.7	26.9	6.4
			Ave.	**8.0**	**8.3**	**12.6**	**6.9**
z/A DR = 0.610	0.65	1.50	−0.810		6.2	6.2	8.3
	0.85	1.24	−0.628		5.5	5.5	5.7
	1.15	1.00	−0.278		9.4	9.4	4.3
	1.37	0.89	−0.249		8.1	8.1	3.3
	1.95	0.70	−0.200		18.6	18.6	6.1
			Ave.		**9.6**	**9.6**	**5.5**
θ/Ak DR = 0.128	0.65	1.50	−0.108		9.1	9.1	7.7
	0.85	1.24	−0.130		4.6	4.6	4.6
	1.15	1.00	−0.002		701.3	701.3	11.8
	1.37	0.89	−0.007		212.3	212.3	12.4
	1.95	0.70	−0.057		57.9	57.9	25.8
			Ave.		**197.0**	**197.0**	**12.5**

4.3.2 Test Case 2.11

The test case 2.11 validation variables for calm water and waves are the same as 2.10, but also include roll $\frac{\phi}{Ak}$. IIHR provided the X_i and U_{X_i} for resistance, sinkage and trim in calm water, as shown in Table 9. IIHR provided the X_i and U_{X_i} for added resistance and 0th, 1st and 2nd harmonic amplitudes of wave elevation and motions in waves, as shown in Tables 10, 11 and 12. Only one facility/model is available and M × N analysis is not possible.

Table 5 1st harmonic amplitude uncertainties for KCS seakeeping in waves at Fr = 0.26 (Case 2.10)

Variable/Dynamic range	λ/L	$f_e/f_{hz/\theta}$	X_i	$U_{X_i}\%D$	$U_{SC_i}\%D$	$U_{X_i}\%D$	$U_{X_i}\%DR$
$\zeta/A\,DR = 0.011$	0.65	1.50	0.005		1.3	1.3	0.6
	0.85	1.24	0.006		0.8	0.8	0.5
	1.15	1.00	0.010		0.2	0.2	0.2
	1.37	0.89	0.012		0.3	0.3	0.3
	1.95	0.70	0.016		0.8	0.8	1.2
			Ave.		**0.7**	**0.7**	**0.6**
$C_T \times 10^3\,DR = 21.749$	0.65	1.50	3.348		20.4	20.4	3.1
	0.85	1.24	4.206		6.6	6.6	1.3
	1.15	1.00	3.929		5.5	5.5	1.0
	1.37	0.89	12.768		2.7	2.7	1.6
	1.95	0.70	25.097		0.7	0.7	0.8
			Ave.		**7.2**	**7.2**	**1.6**
$z/A\,DR = 0.803$	0.65	1.50	0.130	4.0	18.0	18.4	3.0
	0.85	1.24	0.244	4.0	13.2	13.8	4.2
	1.15	1.00	0.899	4.0	1.8	4.4	4.9
	1.37	0.89	0.874	4.0	1.9	4.4	4.8
	1.95	0.70	0.933	4.0	1.1	4.2	4.8
			Ave.	**4.0**	**7.2**	**9.0**	**4.3**
$\theta/Ak\,DR = 1.102$	0.65	1.50	0.017	3.8	45.4	45.5	0.7
	0.85	1.24	0.146	4.0	9.6	10.4	1.4
	1.15	1.00	0.748	4.0	1.7	4.3	3.0
	1.37	0.89	0.966	4.0	1.5	4.2	3.7
	1.95	0.70	1.119	4.0	0.5	4.0	4.1
			Ave.	**4.0**	**11.7**	**13.7**	**2.6**

4.3.3 Test Case 3.9, 3.12 and 3.13

The validation variables for calm water self-propulsion are propeller RPS n, sinkage $\frac{\sigma}{L}$ and trim τ [deg]. The validation variables for waves are wave elevation $\frac{\zeta}{A}$, heave $\frac{z}{A}$, roll $\frac{\phi}{Ak}$, pitch $\frac{\theta}{Ak}$, yaw $\frac{\psi}{Ak}$, surge velocity $\frac{u}{U}$ and sway velocity $\frac{v}{U}$. The computational and experimental values were compared for time histories of wave elevation, resistance, heave and pitch. Fourier series 0th, 1st and 2nd harmonic amplitudes and phases for resistance, heave and pitch were also evaluated for both experiments and computations, as per Eqs. (40).

IIHR provided X_i and U_{X_i} for propeller RPS, sinkage and trim for self-propulsion in calm water, as shown in Table 13. X_i and U_{X_i} were also provided by IIHR for 0th, 1st and 2nd harmonic amplitudes and phases of wave elevation and motions

Table 6 1st harmonic phase uncertainties for KCS seakeeping in waves at Fr = 0.26 (Case 2.10)

Variable	λ/L	X_i	U_{X_i} %2π	U_{SC_i} %2π	U_{X_i} %2π
ζ/A	0.65	0.525			
	0.85	−2.641			
	1.15	0.483			
	1.37	−1.123			
	1.95	−1.040			
	Ave.				
$C_T \times 10^3$	0.65	−0.712		3.0	3.0
	0.85	−1.286		3.6	3.6
	1.15	−1.546		0.7	0.7
	1.37	−1.333		0.4	0.4
	1.95	−0.610		0.2	0.2
	Ave.			**1.6**	**1.6**
z/A	0.65	1.907		2.4	2.4
	0.85	0.970		3.4	3.4
	1.15	−3.101		0.3	0.3
	1.37	−2.397		0.5	0.5
	1.95	−1.971		0.2	0.2
	Ave.			**1.4**	**1.4**
θ/Ak	0.65	1.098		7.4	7.4
	0.85	1.337		3.3	3.3
	1.15	−2.495		0.3	0.3
	1.37	−1.702		0.4	0.4
	1.95	−0.832		0.2	0.2
	Ave.			**2.3**	**2.3**

for course keeping/speed loss, as shown in Tables 14, 15, 16, 17 and 18. Only one facility/model is available and M × N analysis is not possible.

After T2015 in July 2016, Dr. Michel Visonneau pointed out inconsistent differences between the CFD and the measured surge velocity for ONRT. A sign error was discovered in the equation to get the ship coordinates (X_S, Y_S) in the earth fixed coordinate system from the carriage coordinates (Xc, Yc) in the earth fixed coordinate system and ship deviations (ΔX, ΔY) from the carriage, which led to incorrect values for Xs, Ys, u and v. The final proceedings use the corrected Xs, Ys, u and v values. The maximum differences between original and corrected values are 1.6%DR for u_0 at $\mu = 135°$ and 87%DR for v_0 at $\mu = 90°$. The differences for the first harmonic u and v amplitudes are 102%DR at $\mu = 45°$ and 193%DR at $\mu = 90°$, respectively. The differences for the second harmonics and phases were larger.

After the T2015 in September 2018, Dr. Michel Visonneau pointed out a time delay between the yaw and rudder of about 0.233 s, which is attributed to a latency

Table 7 2nd harmonic amplitude uncertainties for KCS seakeeping in waves at Fr = 0.26 (Case 2.10)

Variable/Dynamic range	λ/L	$f_e/f_{hz/\theta}$	X_i	$U_{X_i}\%D$	$U_{SC_i}\%D$	$U_{X_i}\%D$	$U_{X_i}\%DR$
$C_T \times 10^3$ $DR = 2.045$	0.65	1.50	0.652		88.6	88.6	28.3
	0.85	1.24	0.755		53.3	53.3	19.7
	1.15	1.00	2.697		11.0	11.0	14.5
	1.37	0.89	2.201		14.4	14.4	15.5
	1.95	0.70	0.675		55.5	55.5	18.3
			Ave.		**44.6**	**44.6**	**19.3**
z/A $DR = 0.015$	0.65	1.50	0.005		115.6	115.6	39.3
	0.85	1.24	0.008		100.2	100.2	54.2
	1.15	1.00	0.013		17.2	17.2	14.4
	1.37	0.89	0.007		52.2	52.2	23.4
	1.95	0.70	0.020		46.9	46.9	62.9
			Ave.		**66.4**	**66.4**	**38.8**
θ/Ak $DR = 0.036$	0.65	1.50	0.001		96.3	96.3	3.2
	0.85	1.24	0.002		94.0	94.0	5.8
	1.15	1.00	0.014		12.3	12.3	4.7
	1.37	0.89	0.012		26.4	26.4	8.7
	1.95	0.70	0.037		18.4	18.4	19.0
			Ave.		**49.5**	**49.5**	**8.3**

of the yaw angle data acquisition in the experiments. The yaw angle used for the rudder control in the course keeping/speed loss experiments at IIHR is obtained by the carriage tracking system. There is a latency due to internal processing and communication between the systems. This latency cannot be eliminated based on the current technology even if we use an onboard gyroscope. For future validation studies, we propose the following:

1. For rudder control based on yaw angle in the CFD simulations, the latency for the yaw angle data acquisition should be considered as the simulation condition in addition to the rudder speed. In the CFD codes, the yaw angle obtained from the previous time step which is corresponding to the latency needs to be used for the rudder control to mimic the latency of the yaw angle data acquisition, as per the following new rudder controller equation:

$$\begin{aligned} \Delta\psi(t) &= \psi_C - \psi(t - t_L) \\ \delta(t) &= K_{\delta P}\left(\Delta\psi(t) + K_{\delta I}\int_0^t \Delta\psi(\tau)d\tau + K_{\delta D}\frac{d\Delta\psi(t)}{dt}\right) \end{aligned} \tag{41}$$

Here, t_L is latency of yaw angle acquisition.

Table 8 2nd harmonic phase uncertainties for KCS seakeeping in waves at Fr = 0.26 (Case 2.10)

Variable	λ/L	X_i	U_{X_i} %2π	U_{SC_i} %2π	U_{X_i} %2π
$C_T \times 10^3$	0.65	0.068		24.8	24.8
	0.85	−1.122		11.4	11.4
	1.15	1.829		2.2	2.2
	1.37	2.903		16.6	16.6
	1.95	−0.446		11.3	11.3
	Ave.			**13.3**	**13.3**
z/A	0.65	−0.046		61.4	61.4
	0.85	0.810		64.8	64.8
	1.15	0.214		2.7	2.7
	1.37	0.580		9.3	9.3
	1.95	0.794		8.7	8.7
	Ave.			**29.4**	**29.4**
θ/Ak	0.65	0.605		78.8	78.8
	0.85	1.215		40.9	40.9
	1.15	0.201		2.1	2.1
	1.37	0.708		4.2	4.2
	1.95	1.646		3.0	3.0
	Ave.			**25.8**	**25.8**

Table 9 Uncertainties for KCS resistance test in calm water at Fr = 0.26 (Case 2.11)

Variable	X_i	U_{X_i} %D
$C_T \times 10^3$	4.888	2.5
$\sigma/L \times 10^2$	−0.1704	0.6
τ [deg]	−0.022	65.5

For new experiments, it is desirable to develop a new method to evaluate the latency for the yaw angle data acquisition in course keeping/speed loss experiments. It is necessary to evaluate the worst execution time for each process inside the system and to obtain the actual latency due to the communication and controlling using a global master clock to issue time stamps for each measurement variable. In addition, when sensor fusion is done to get the maneuvering data, it is also necessary to evaluate time jittering of the internal clock inside each device based on the global master clock.

Table 10 0th harmonic amplitude uncertainties for KCS seakeeping in waves (Case 2.11)

Variable/Dynamic range	μ [deg]	$f_e/f_{hz/\theta}$	X_i	$U_{X_i}\%D$	$U_{X_i}\%DR$
$C_T \times 10^3$ $DR = 3.696$	0	0.57	13.135	0.8	5.8
	45	0.50	12.572	0.7	4.7
	90	0.35	9.994	1.2	6.3
	135	0.19	9.683	0.8	4.1
	180	0.12	9.438	0.8	4.1
		Ave.		**0.9**	**5.0**
AR $DR = 8.343$	0	0.57	9.296	6.0	6.7
	45	0.50	8.027	6.0	5.8
	90	0.35	2.208	26.3	7.0
	135	0.19	1.504	27.4	4.9
	180	0.12	0.953	43.4	5.0
		Ave.		**21.8**	**5.9**
z/A $DR = 0.214$	0	0.57	−0.322	7.7	23.1
	45	0.50	−0.260	5.3	12.9
	90	0.35	−0.298	4.9	13.6
	135	0.19	−0.269	5.3	13.3
	180	0.12	−0.474	3.4	14.9
		Ave.		**5.3**	**15.6**
θ/Ak $DR = 0.135$	0	0.57	0.088	21.8	28.5
	45	0.50	−0.027	105.0	42.1
	90	0.35	−0.047	61.2	42.2
	135	0.19	−0.021	164.4	50.7
	180	0.12	0.031	524.0	240.6
		Ave.		**175.3**	**80.8**
ϕ/Ak $DR = 1.69$	0	0.57	0.647	5.8	4.5
	45	0.50	−0.264	4.3	1.4
	90	0.35	1.177	4.2	5.8
	135	0.19	1.026	9.6	11.6
	180	0.12	−0.514	5.8	3.5
		Ave.		**5.9**	**5.4**

4.4 Conclusions

More attention is needed to quantify the CFD input wave condition and uncertainty; and to specify the CFD wave elevation evaluation locations.

KCS calm water and head waves experimental data was available from several facilities such that resistance, sinkage and trim data provided relatively robust validation/SoA assessment. Nonetheless, more attention is needed for individual facility

Table 11 1st harmonic amplitude uncertainties for KCS seakeeping in waves (Case 2.11)

Variable/Dynamic range	μ [deg]	$f_e/f_{hz/\theta}$	X_i	$U_{X_i}\%D$	$U_{X_i}\%DR$
$\zeta/LDR =$	0	0.57	0.0083	1.4	
	45	0.50	0.0083	1.5	
	90	0.35	0.0083	1.4	
	135	0.19	0.0083	1.1	
	180	0.12	0.0083	1.2	
		Ave.		**1.3**	
$C_T \times 10^3$ *DR = 5.551*	0	0.57	2.647	2.5	1.2
	45	0.50	6.193	1.0	1.2
	90	0.35	1.424	5.7	1.5
	135	0.19	3.775	3.4	2.3
	180	0.12	0.642	10.2	1.2
		Ave.		**4.6**	**1.5**
z/A *DR = 1.101*	0	0.57	0.608	3.5	1.9
	45	0.50	1.161	1.2	1.3
	90	0.35	1.132	1.8	1.9
	135	0.19	0.404	4.4	1.6
	180	0.12	0.059	24.6	1.3
		Ave.		**7.1**	**1.6**
θ/Ak *DR = 0.637*	0	0.57	0.492	6.1	4.7
	45	0.50	0.685	4.4	4.8
	90	0.35	0.048	60.0	4.5
	135	0.19	0.365	8.4	4.8
	180	0.12	0.138	451.2	97.4
		Ave.		**106.0**	**23.2**
ϕ/Ak *DR = 2.777*	0	0.57	0.012	236.6	1.0
	45	0.50	0.558	5.3	1.1
	90	0.35	0.378	8.3	1.1
	135	0.19	2.789	1.0	1.0
	180	0.12	0.247	14.3	1.3
		Ave.		**53.1**	**1.1**

uncertainty analysis and for comparing data that is measured in conventional or dedicated calm water resistance tests setups in all facilities, since calm water resistance normally is measured that way.

KCS oblique waves and ONRT head and oblique wave data and uncertainties need more data from more facilities including uncertainties for more robust validation/SoA assessment.

Table 12 2nd harmonic amplitude uncertainties for KCS seakeeping in waves (Case 2.11)

Variable/Dynamic range	μ [deg]	$f_e/f_{hz/\theta}$	X_i	$U_{X_i}\%D$	$U_{X_i}\%DR$
$C_T \times 10^3$ $DR = 2.932$	0	0.57	0.402	16.3	1.0
	45	0.50	3.219	2.0	1.0
	90	0.35	0.908	7.2	1.0
	135	0.19	0.715	9.9	1.1
	180	0.12	0.287	23.7	1.1
		Ave.		**11.8**	**1.1**
z/A $DR = 0.033$	0	0.57	0.020	70.7	0.2
	45	0.50	0.008	161.1	0.2
	90	0.35	0.034	40.9	0.2
	135	0.19	0.001	1374.9	0.2
	180	0.12	0.024	67.3	0.3
		Ave.		**343.0**	**0.2**
θ/Ak $DR = 0.018$	0	0.57	0.013	213.2	0.4
	45	0.50	0.010	275.9	0.4
	90	0.35	0.020	146.3	0.5
	135	0.19	0.028	113.7	0.5
	180	0.12	0.019	172.1	0.5
		Ave.		**184.2**	**0.5**
ϕ/Ak $DR = 0.142$	0	0.57	0.006	5076.3	4.7
	45	0.50	0.148	19.4	0.5
	90	0.35	0.067	45.1	0.5
	135	0.19	0.094	32.9	0.5
	180	0.12	0.046	63.7	0.5
		Ave.		**1047.5**	**1.3**

Table 13 Uncertainties for ONRT self-propulsion tests in calm water at Fr = 0.2 (Case 3.9)

Variable	X_i	$U_{X_i}\%D$
n [RPS]	8.97	0.0834
$\sigma/L \times 10^2$	0.0718	20.0
τ [deg]	−0.0386	134.2

5 CFD Assessment Test Case 2.10

As shown in Table 19, 10 institutes from 8 countries using 9 different CFD codes submitted results for both calm water and waves. Two equation turbulence models were used with 8 using wall functions and 2 using near wall models. The solvers were mostly Finite Volume (FV) except one is Finite Difference (FD). All but 2 used unstructured grids. 4 used level set and 6 used VoF free surface models. Grid motion

Table 14 0th harmonic amplitude uncertainties for ONRT course keeping in waves (Case 3.12 & 3.13)

Variable/Dynamic range	μ [deg]	$f_e/f_{hz/\theta}$	X_i	$U_{X_i}\%D$	$U_{X_i}\%DR$
z/A $DR = 0.134$	0	0.62	0.057	27.0	11.5
	45	0.58	0.013	108.3	10.4
	90	0.44	−0.049	27.5	10.1
	135	0.29	0.008	159.3	9.7
	180	0.22	0.085	16.1	10.2
		Ave.		**67.7**	**10.4**
θ/Ak $DR = 0.025$	0	0.62	−0.026	30.1	31.9
	45	0.58	−0.030	15.4	18.8
	90	0.44	−0.022	30.2	27.5
	135	0.29	−0.023	12.7	12.0
	180	0.22	−0.005	200.6	45.1
		Ave.		**57.8**	**27.1**
ϕ/Ak $DR = 0.279$	0	0.62	0.019	68.7	4.7
	45	0.58	−0.139	10.9	5.5
	90	0.44	0.098	18.7	6.6
	135	0.29	−0.181	12.0	7.8
	180	0.22	0.003	494.1	5.1
		Ave.		**120.9**	**5.9**
ψ/Ak $DR = 0.612$	0	0.62	−0.078	82.4	10.5
	45	0.58	−0.690	4.3	4.8
	90	0.44	−0.139	6.4	1.5
	135	0.29	−0.245	17.4	7.0
	180	0.22	−0.091	14.9	2.2
		Ave.		**25.1**	**5.2**
u/U $DR = 0.180$	0	0.62	0.818	0.993	4.516
	45	0.58	0.912	2.885	14.639
	90	0.44	0.981	2.277	12.431
	135	0.29	0.981	1.389	7.579
	180	0.22	0.998	1.671	9.272
		Ave.		**1.84**	**9.69**
v/U $DR = 0.037$	0	0.62	0.004	0.000	0.000
	45	0.58	−0.030	44.079	35.896
	90	0.44	−0.001	5023.561	85.149
	135	0.29	−0.016	140.405	59.808
	180	0.22	0.007	510.349	94.739
		Ave.		**1143.68**	**55.12**

Table 15 1st harmonic amplitude uncertainties for ONRT course keeping in waves (Case 3.12 and 3.13)

Variable/Dynamic range	μ [deg]	$f_e/f_{hz/\theta}$	X_i	$U_{X_i}\%D$	$U_{X_i}\%DR$
ζ/L $DR = 0.001$	0	0.62	0.010	1.0	13.7
	45	0.58	0.010	1.2	15.5
	90	0.44	0.010	1.0	13.4
	135	0.29	0.010	3.0	39.3
	180	0.22	0.009	1.6	20.1
		Ave.		**1.6**	**20.4**
z/A $DR = 0.89$	0	0.62	0.597	2.2	1.4
	45	0.58	0.821	1.7	1.6
	90	0.44	1.044	1.3	1.5
	135	0.29	0.517	2.7	1.6
	180	0.22	0.154	8.4	1.5
		Ave.		**3.3**	**1.5**
θ/Ak $DR = 0.672$	0	0.62	0.690	0.6	0.6
	45	0.58	0.684	1.1	1.1
	90	0.44	0.018	10.5	0.3
	135	0.29	0.436	0.9	0.6
	180	0.22	0.248	1.6	0.6
		Ave.		**2.9**	**0.6**
ϕ/Ak $DR = 1.788$	0	0.62	0.001	1421.0	0.6
	45	0.58	0.386	3.0	0.7
	90	0.44	1.180	1.6	1.0
	135	0.29	1.789	1.0	1.0
	180	0.22	0.017	82.8	0.8
		Ave.		**301.9**	**0.8**
ψ/Ak $DR = 0.568$	0	0.62	0.003	7.1	0.0
	45	0.58	0.158	0.4	0.1
	90	0.44	0.033	2.8	0.2
	135	0.29	0.571	0.8	0.8
	180	0.22	0.011	22.4	0.4
		Ave.		**6.7**	**0.3**
u/U $DR = 0.071$	0	0.62	0.005	15.495	1.053
	45	0.58	0.018	109.131	27.647
	90	0.44	0.003	900.697	36.884
	135	0.29	0.074	27.517	28.644
	180	0.22	0.062	35.378	31.291

(continued)

Table 15 (continued)

Variable/Dynamic range	μ [deg]	$f_e/f_{hz/\theta}$	X_i	$U_{X_i}\%D$	$U_{X_i}\%DR$
		Ave.		**217.64**	**25.10**
v/U $DR = 0.074$	0	0.62	0.000		
	45	0.58	0.003		
	90	0.44	0.054	2.168	1.581
	135	0.29	0.074	27.776	27.946
	180	0.22	0.001		
		Ave.		**14.97**	**14.76**

used dynamic patched, dynamic overset and deforming grids. Grid sizes varied from 0.6–29 M with average 9.8 M. Average number CPU and time is 60 processors and 3484 CPU hours, respectively.

It is very unfortunate that participants did not submit individual code grid and time step verification and an important oversight of the workshop organizers. One submission did verification for head waves but did not submit results and others have done for different hulls and similar conditions but not for present test case. Larsson et al. (2014) Table 11 summarizes results from several studies. For current grid sizes/time steps, individual solution verification U_{SN_i} was estimated at about 4 and 8%S_1 for 1st and 2nd order variables, respectively, which would increase intervals of validation. Therefore, neglecting U_{SN_i} provides estimates of validation intervals with warning that intervals are optimistic. Therefore, future workshops need to require that all participants submit verification.

Table 20 provides the calm water V&V and SoA variables. Figure 3 shows the C_T, sinkage and trim absolute error running record and SoA metrics. In this case the experimental and computational uncertainties include multiple facilities/models and codes/solutions, respectively.

At the solution level, E_i shows 60% over, 100% under and 56% over predictions for C_T, sinkage and trim, respectively. 3, 5 and 9 solutions are validated at $U_{V_i} = U_D = 1.75, 7.38$ and 14.25%D for C_T, sinkage and trim, respectively.

At the code level, the mean code shows over, under and over predictions with $\overline{|E|} = 2, 6$ and 5%D for C_T, σ/L and τ, respectively. $\overline{|E|}$ values are larger than signed $\overline{E}$. The code level solution absolute error scatter $P_{|E_i|} = 3, 4$ and 7%D such that $U_{SoA_i} = 3, 8$ and 16%D for C_T, σ/L and τ, respectively. There is one outlier for sinkage. Thus, 8, 7 and 9 codes/solutions[1] are SoA at U_{SoA_i} intervals $\left(|E_i| \le U_{SoA_i}\right)$. The mean code is not N-version validated for resistance but is for sinkage and trim at $U_{SoA} = 2, 8$ and 14%D, respectively.

For calm water resistance, the experimental 2%D and absolute error scatter 3%D uncertainties are comparable to the mean absolute error 2%D; thus, all three must be

[1]Code solutions are N-version validated; however, to distinguish between solution and N-version levels when discussing the latter for simplicity the terminology code is used when referring to each code solution.

Table 16 1st harmonic phase uncertainties for ONRT course keeping in waves (Case 3.12 and 3.13)

Variable/Dynamic range	μ [deg]	$f_e/f_{hz/\theta}$	X_i	U_{X_i} %2π
z/A	0	0.62	0.225	4.1
	45	0.58	0.415	4.3
	90	0.44	2.279	0.6
	135	0.29	−0.951	4.0
	180	0.22	0.314	2.5
		Ave.		**3.1**
θ/Ak	0	0.62	−2.458	5.1
	45	0.58	−1.791	4.4
	90	0.44	−2.333	3.5
	135	0.29	0.695	62.8
	180	0.22	2.240	2.4
		Ave.		**15.6**
ϕ/Ak	0	0.62	−0.534	36.3
	45	0.58	−1.673	11.1
	90	0.44	1.563	10.2
	135	0.29	−0.238	11.3
	180	0.22	1.706	11.0
		Ave.		**16.0**
ψ/Ak	0	0.62	−1.334	3.6
	45	0.58	0.296	5.1
	90	0.44	1.734	2.0
	135	0.29	2.072	2.0
	180	0.22	−1.899	24.3
		Ave.		**7.4**
u/U	0	0.62	−0.700	4.080
	45	0.58	0.037	2.740
	90	0.44	3.139	7.050
	135	0.29	2.326	8.260
	180	0.22	−2.409	2.970
		Ave.		**5.02**
v/U	0	0.62	−1.083	4.820
	45	0.58	1.015	3.190
	90	0.44	2.143	0.600
	135	0.29	−0.987	4.010
	180	0.22	−0.949	3.030
		Ave.		**3.13**

Table 17 2nd harmonic amplitude uncertainties for ONRT course keeping in waves (Case 3.12 and 3.13)

Variable/Dynamic range	μ [deg]	$f_e/f_{hz/\theta}$	X_i	$U_{X_i}\%D$	$U_{X_i}\%DR$
z/A $DR = 0.076$	0	0.62	0.019	66.2	16.8
	45	0.58	0.012	102.9	16.9
	90	0.44	0.067	19.2	17.0
	135	0.29	0.088	14.6	17.0
	180	0.22	0.019	69.7	17.1
		Ave.		**54.5**	**17.0**
θ/Ak $DR = 0.047$	0	0.62	0.020	5.6	2.4
	45	0.58	0.005	25.8	2.7
	90	0.44	0.016	10.9	3.6
	135	0.29	0.052	11.9	13.1
	180	0.22	0.015	7.8	2.4
		Ave.		**12.4**	**4.8**
ϕ/Ak $DR = 0.046$	0	0.62	0.004	296.7	23.8
	45	0.58	0.016	67.7	24.3
	90	0.44	0.040	28.6	24.9
	135	0.29	0.050	27.3	29.5
	180	0.22	0.013	103.8	29.7
		Ave.		**104.8**	**26.4**
ψ/Ak $DR = 0.022$	0	0.62	0.001	45.7	1.8
	45	0.58	0.008	3.6	1.4
	90	0.44	0.004	28.2	5.3
	135	0.29	0.023	31.4	32.7
	180	0.22	0.003	44.2	5.5
		Ave.		**30.6**	**9.3**
u/U $DR = 0.004$	0	0.62	0.004	13.834	13.320
	45	0.58	0.005	0.000	0.000
	90	0.44	0.001	2292.002	448.644
	135	0.29	0.004	160.672	159.293
	180	0.22	0.001	2519.754	494.136
		Ave.		**997.25**	**223.08**
v/U $DR = 0.002$	0	0.62	0.000		
	45	0.58	0.000		
	90	0.44	0.001		
	135	0.29	0.002		
	180	0.22	0.001		

Table 18 2nd harmonic phase uncertainties for ONRT course keeping in waves (Case 3.12 and 3.13)

Variable/Dynamic range	μ [deg]	$f_e/f_{hz/\theta}$	X_i	U_{X_i} %2π
z/A	0	0.62	−2.796	8.8
	45	0.58	0.092	6.7
	90	0.44	−0.082	0.8
	135	0.29	2.077	11.1
	180	0.22	0.089	6.6
		Ave.		**6.8**
θ/Ak	0	0.62	1.327	7.8
	45	0.58	−2.080	58.2
	90	0.44	2.228	2.2
	135	0.29	0.242	10.3
	180	0.22	3.094	54.1
		Ave.		**26.5**
ϕ/Ak	0	0.62	1.342	12.3
	45	0.58	−3.007	58.0
	90	0.44	−1.004	10.4
	135	0.29	−1.207	13.5
	180	0.22	−1.799	10.8
		Ave.		**21.0**
ψ/Ak	0	0.62	−2.468	35.2
	45	0.58	−2.632	8.7
	90	0.44	1.868	8.0
	135	0.29	−1.318	12.6
	180	0.22	−1.765	11.1
		Ave.		**15.1**
u/U	0	0.62	1.672	6.350
	45	0.58	−2.466	5.690
	90	0.44	1.464	37.600
	135	0.29	0.884	8.430
	180	0.22	−2.467	5.820
		Ave.		**12.78**
v/U	0	0.62	1.754	21.940
	45	0.58	−0.574	54.480
	90	0.44	1.850	4.790
	135	0.29	−0.193	9.960
	180	0.22	−0.848	8.560
		Ave.		**19.95**

Table 19 Submissions for KCS resistance test and seakeeping in calm water and head waves at Fr = 0.26

No.	Institute	Country	Code	Turbulence	Free surface	Discretization method	Grid motion	Grid type	Grid size	No .CPU	CPU time [CPUh]
1	FORCE	Denmark	StarCCM+	k-ε/k-ω SST	VoF	FV	Dynamic Overset	Unstructured	18– 28 M	NA	NA
2	HSVA	Germany	FreSCo+	k-ε/k-ω	VoF	FV	Dynamic	Unstructured	NA	NA	NA
3	IIHR	United States	CFDShip-Iowa V4.5	k-ε/k-ω SST	LS Single Ph.	FD	Dynamic Overset	Structured	29.2 M	224	16128
4	KRISO	South Korea	WAVIS	k-ε/k-ω EARSM	LS	FV	Dynamic	Structured	10 M Half B.	38	170.3
5	MARIC	China	FINEMarine	k-ε/k-ω SST	VoF	FV	Deforming	Unstructured	1.18-6.9 M	NA	NA
6	NMRI	Japan	NAGISA	k-ε/k-ω EARSM	LS	FV	Deforming	Unstructured	4.4 M	2	250
7	NUMECA	Belgium	FINEMarine	k-ε/k-ω SST	VoF	FV	Deforming	Unstructured	2–6 M	NA	3116
8	UDE	Germany	OpenFOAM	k-ε/k-ω SST	VoF	FV	Deforming	Unstructured	1.06 M	8	104
9	UM	United States	OF23x	k-ε/k-ω SST	VoF	FV	Dynamic	Unstructured	2.6 M	48	3800
10	UNIZAG-FSB	Croatia	swenseFoam	k-ε/k-ω SST	LS	FV	Dynamic	Unstructured	0.61–1.6 M	40	817
Ave.									9.80 M	60	3484
G2010 (6 sub.)									2.62 M (0.55–7.6 M)		

Table 20 Comparison errors and uncertainties for KCS calm water resistance test submissions at Fr = 0.26

Variable	$C_{T0} \times 10^3$		σ/L		τ (deg)		Motion abs. average
D	3.835		−0.00207		−0.165		
$U_D\%\|D\|$	1.75		7.38		14.25		7.79
# submissions	10		9		9		9
# outliers	0		1		0		0
CFD submissions i	S_i	$E_i\%\|D\|$	S_i	$E_i\%\|D\|$	S_i	$E_i\%\|D\|$	$\|E_i\|\%\|D\|$
1	3.741	2.45	−0.00180	13.04	−0.1574	4.37	8.71
2	3.955	−3.13	−0.00198	4.35	−0.1605	2.49	3.42
3	3.845	−0.26	−0.00188	9.18	−0.1797	−9.17	9.18
4	3.738	2.53	−0.00198	4.35	−0.1725	−4.80	4.57
5	3.836	−0.03	−0.00199	3.86	−0.1560	5.22	4.54
6	3.983	−3.86	−0.00195	5.80	−0.1644	0.12	2.96
7	3.898	−1.64	−0.00195	5.80	−0.1698	−3.16	4.48
8	3.730	2.74					
9	3.758	2.01	−0.00190	8.21	−0.1820	−10.57	6.93
10	4.015	−4.69	−0.00191	7.73	−0.1754	−6.56	6.33
$\overline{S}$	3.850		−0.00194		−0.169		
$\overline{E}\%\|D\|$	−0.39 (−2.55)		6.16 (3.67)		−2.45 (−0.24)		4.30 (2.15)
$\sigma_E\%\|D\|$	2.83		2.00		5.82		3.91
$\overline{\|E\|}\%\|D\|$	2.33 (3.03)		6.16 (4.33)		5.16 (3.46)		5.66 (3.61)
$\sigma_{\|E\|}\%\|D\|$	1.45 (2.83)		2.00 (1.46)		3.25 (2.34)		2.62 (2.21)
$P_{\|E_i\|}\%\|D\|$	2.90		3.99		6.50		5.25
$P_{\overline{\|E\|}}\%\|D\|$	0.92		1.33		2.17		1.75
$U_{SoA_i}\%\|D\|$	3.39		8.39		15.66		12.03
$U_{SoA}\%\|D\|$	1.98		7.50		14.41		10.96
# validated	3		5		9		7
# N-version validated	8		7		9		8
Mean Code N-version validated	No		Yes		Yes		Yes

Note The gray-shaded numbers are outliers and values in parenthesis are obtained from G2010

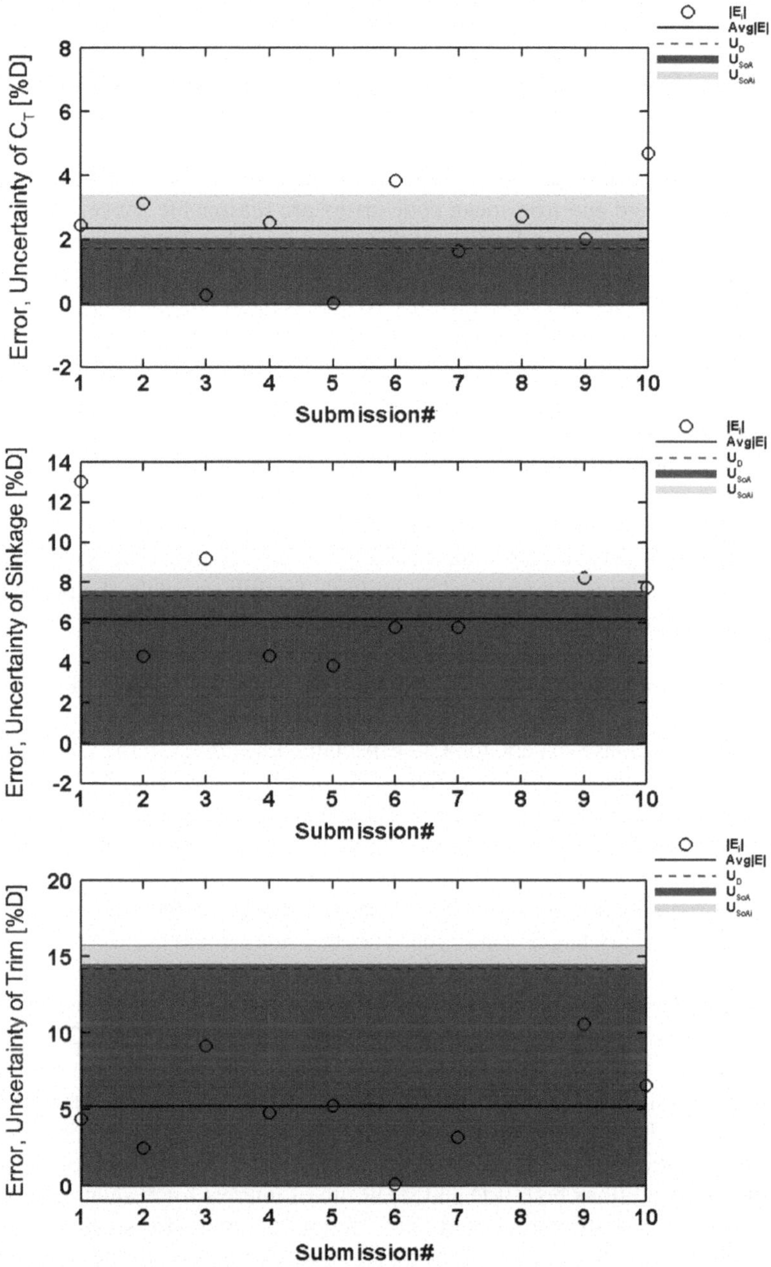

Fig. 3 EFD data and submission values for resistance and motions in calm water for KCS at Fr = 0.26

reduced to achieve mean code N-version validation. Nonetheless, 8 codes/solutions were N-version validated at $U_{SoAi} = 3\%D$. For sinkage/trim, the experimental uncertainties 7/14%D are larger than the absolute error scatter uncertainties 4/7%D and the mean absolute error 6/5%D; thus, reducing the experimental uncertainties is the most critical issue. Note that the present experimental uncertainties are based on the scatter in the KCS sinkage and trim Fr = 0.26 measurements from several facilities. Current sinkage and trim mean code errors are reasonable. Acceptance of the accuracy of E_i without consideration of the experimental and computational uncertainties excludes the possibility of establishing confidence levels/intervals; therefore, no assurance of the actual CFD capability. Whereas including uncertainties enables evaluation of both E_i and its confidence level/interval. Decomposition of the uncertainties into systematic and random components for both the experiments and simulations provides robust uncertainty estimates and identification of where to focus resources for the reduction in the intervals of validation/SoA.

Table 20 includes results from G2010 for C_T, σ/L and τ based on 6 CFD submissions for the same test case but using FORCE 4.4 m model validation data (Simonsen et al. 2013). T2015 results are improved for resistance and comparable for motions. Since G2010 is only based on 6 submissions vs. 10 for T2015 and the present experimental uncertainty estimate is based on multiple facilities/models, the present data is considered a more robust estimate of current CFD code capabilities.

Figures 4, 5,6, 7 and 8 show experimental time histories for a ten-period segment for wave elevation, resistance, heave and pitch for each wavelength condition: λ/L = 0.65, 0.85, 1.15, 1.37 and 1.95, respectively. Figures display mostly dominant first harmonic response except for C_T, especially for λ/L = 1.15, which is near heave/pitch resonance i.e. the frequency of encounter is close to 0.73 Hz based on the hydrostatic estimates as reported in Table 1. FORCE reported that the mount natural frequency was close to the heave and pitch resonance condition. Most clearly seen in C_T for λ/L equal to 0.85 and 1.15 in Figs. 5b and 6b. CFD shows similar trends as the experiments but several solutions do not display statistical convergence. CFD shows seemingly larger scatter for shorter vs. longer waves.

Tables 21 and 22 provide average results for each wave length and over all wave lengths of the CFD solutions for the 0th/AR and 1st and 2nd harmonic amplitude validation and SoA variables using %D and %DR, respectively. Table 23 is similar but provides the 1st and 2nd harmonic phase validation and SoA variables using %2π. The participants submitted their own harmonic analysis values for amplitudes and phases, which were used for reconstruction of the time histories. For validation at a single condition %D is usually used unless D nearly zero, whereas for validation for a range of conditions both %D and %DR are of interest for evaluation of trends with recognition that the former and latter weight the errors for small vs. large D values, respectively. Since large responses are mostly of interest %DR is often more meaningful. 0th and 2nd harmonics are considered 2nd order and 1st harmonic 1st order terms from a ship design point of view. The discussions mostly use values for λ/L averages and %DR.

Figures 9, 10, 11, 12 and 13 show the results for the 0th harmonic, 1st and 2nd harmonic amplitudes and 1st and 2nd harmonic phases, respectively, versus λ/L. The

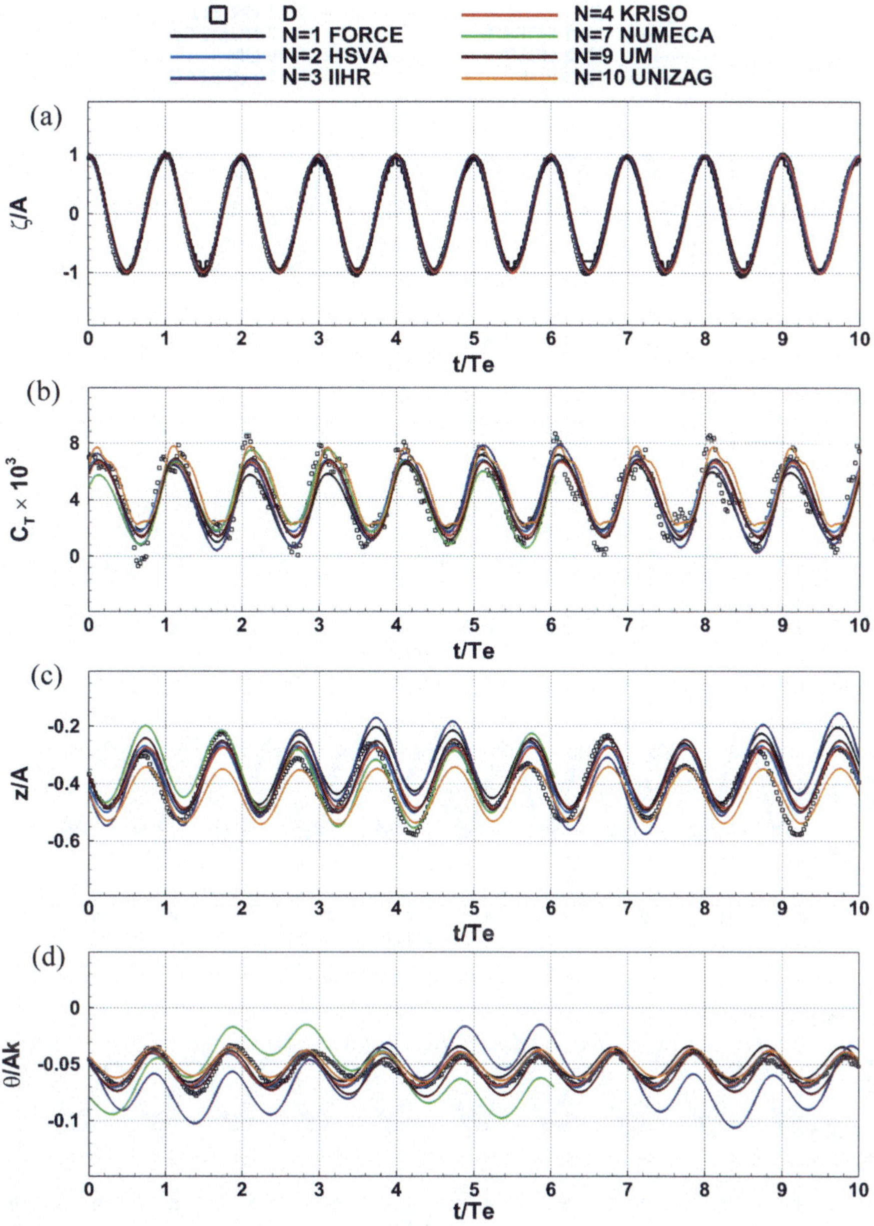

Fig. 4 Time series for head waves with $\lambda/L = 0.65$, $H/\lambda = 0.0157$ ($A/L = 0.00511$) and $f_e/f_n = 1.5$: **a** Wave; **b** Resistance coefficient; **c** Heave; and **d** Pitch

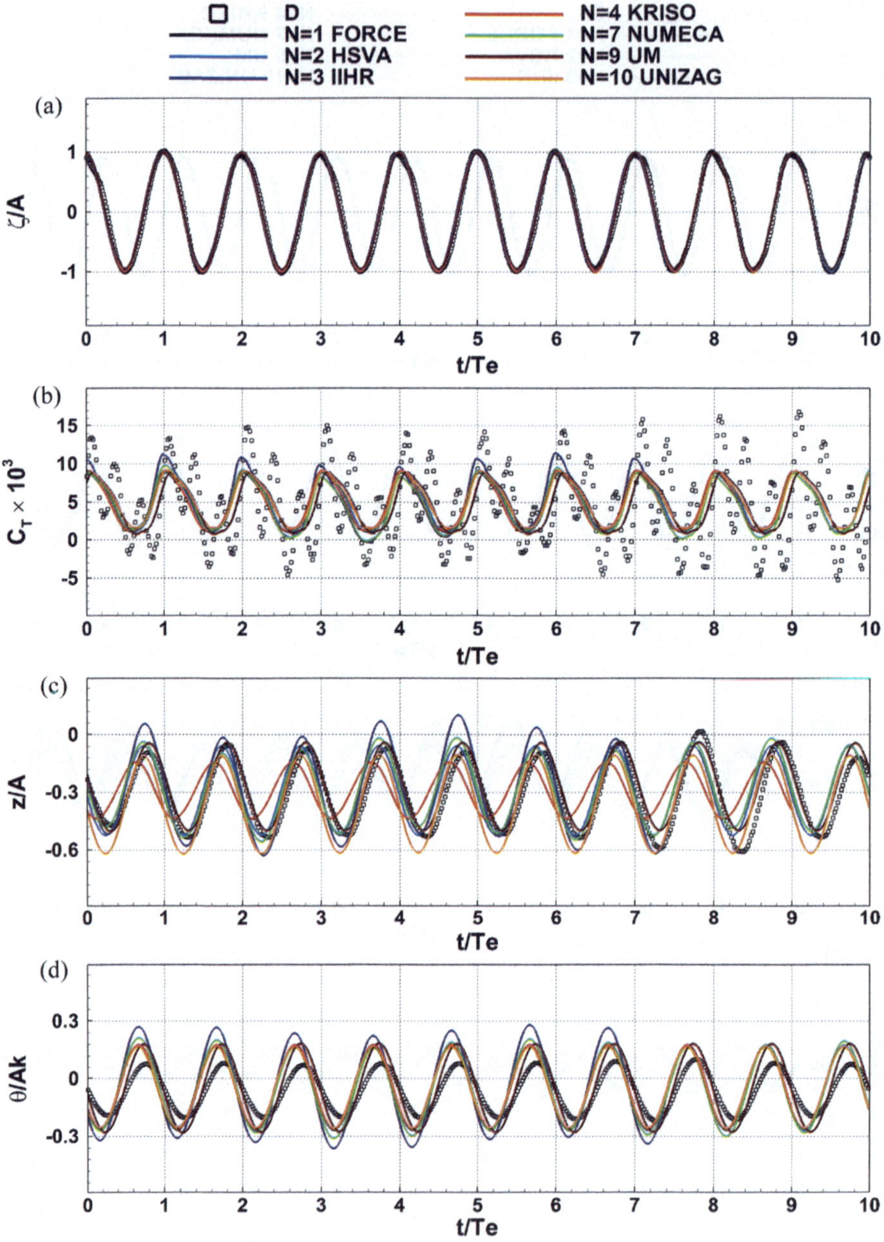

Fig. 5 Time series for head waves with $\lambda/L = 0.85$, $H/\lambda = 0.0151$ ($A/L = 0.00642$) and $f_e/f_n = 1.24$: **a** Wave; **b** Resistance coefficient; **c** Heave; and **d** Pitch

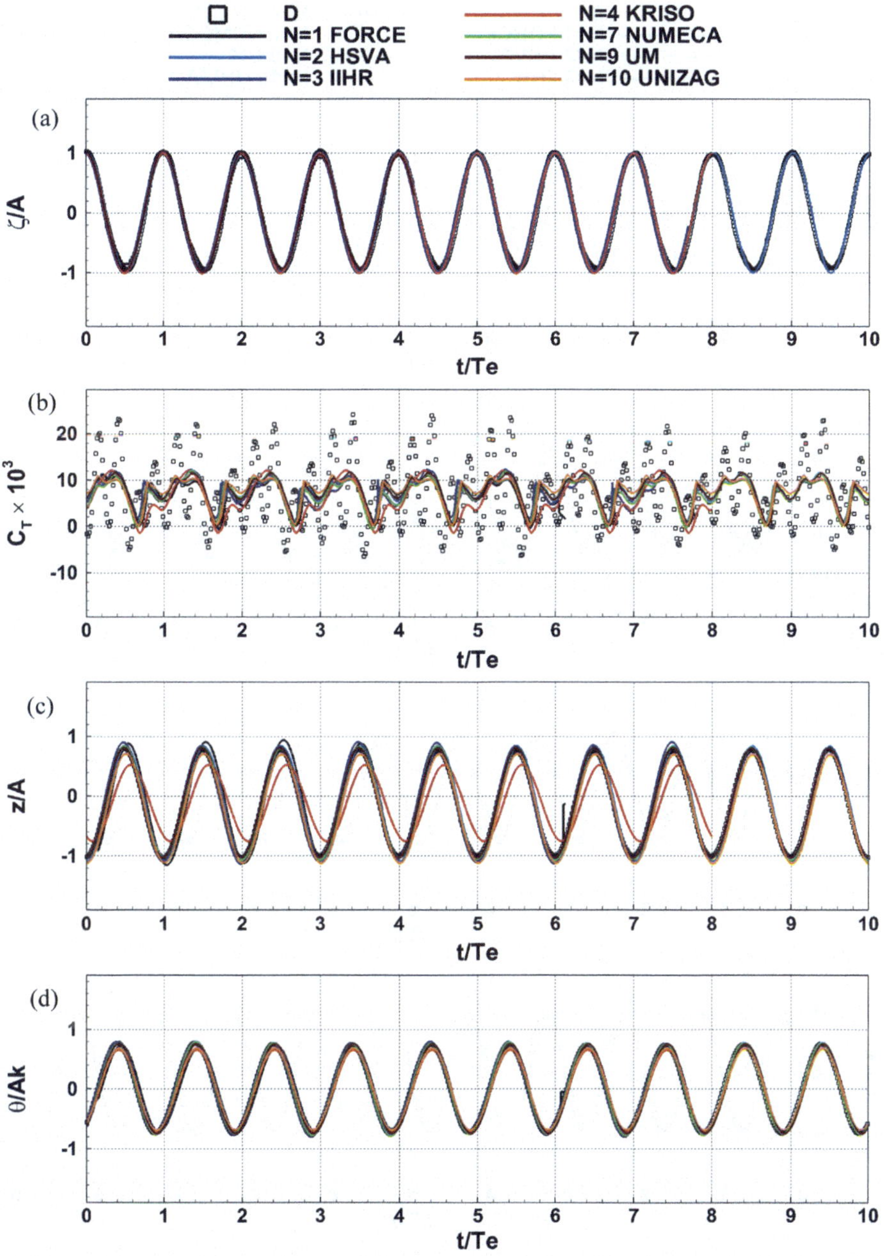

Fig. 6 Time series for head waves with $\lambda/L = 1.15$, $H/\lambda = 0.0176$ ($A/L = 0.0101$) and $f_e/f_n = 1.0$: **a** Wave; **b** Resistance coefficient; **c** Heave; and **d** Pitch

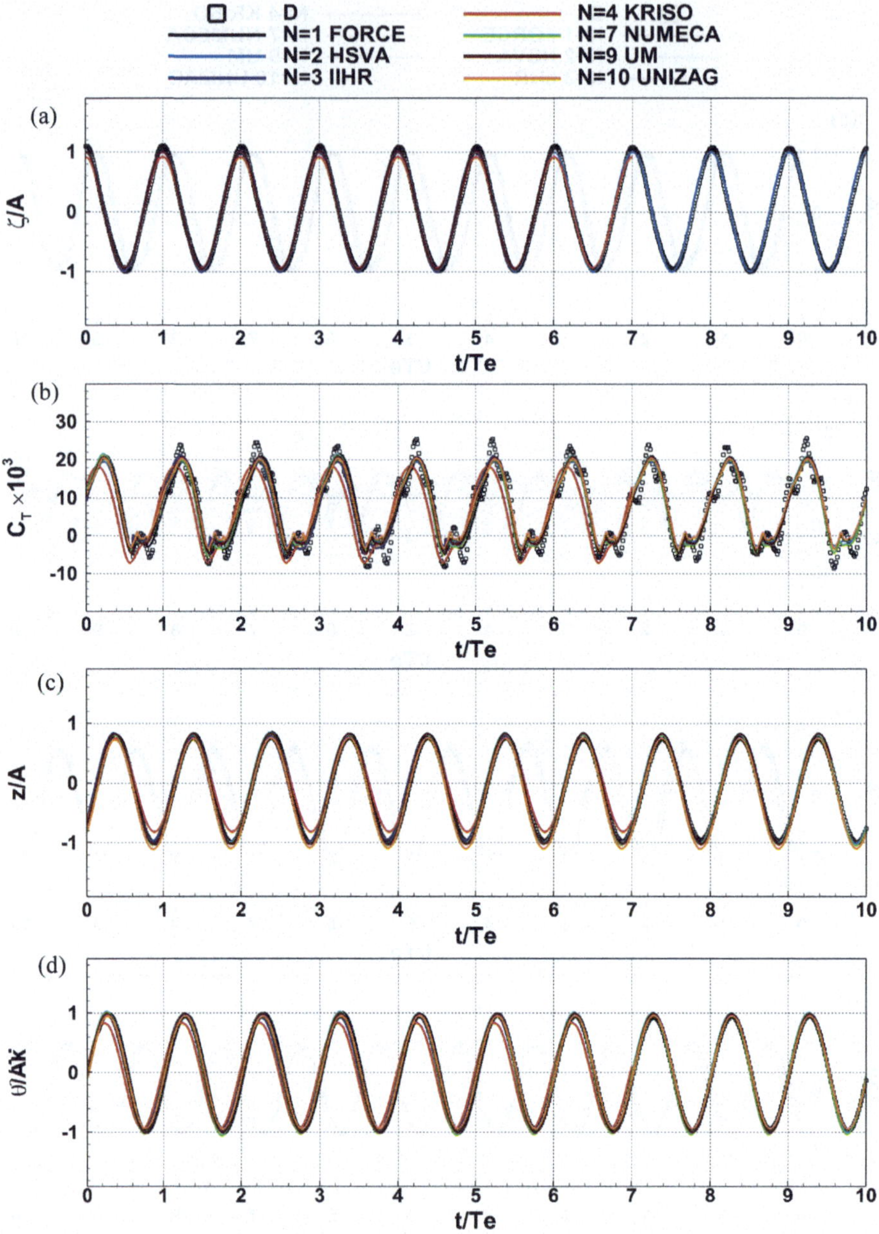

Fig. 7 Time series for head waves with $\lambda/L = 1.37$, $H/\lambda = 0.0179$ ($A/L = 0.0123$) and $f_e/f_n = 0.89$: **a** Wave; **b** Resistance coefficient; **c** Heave; and **d** Pitch

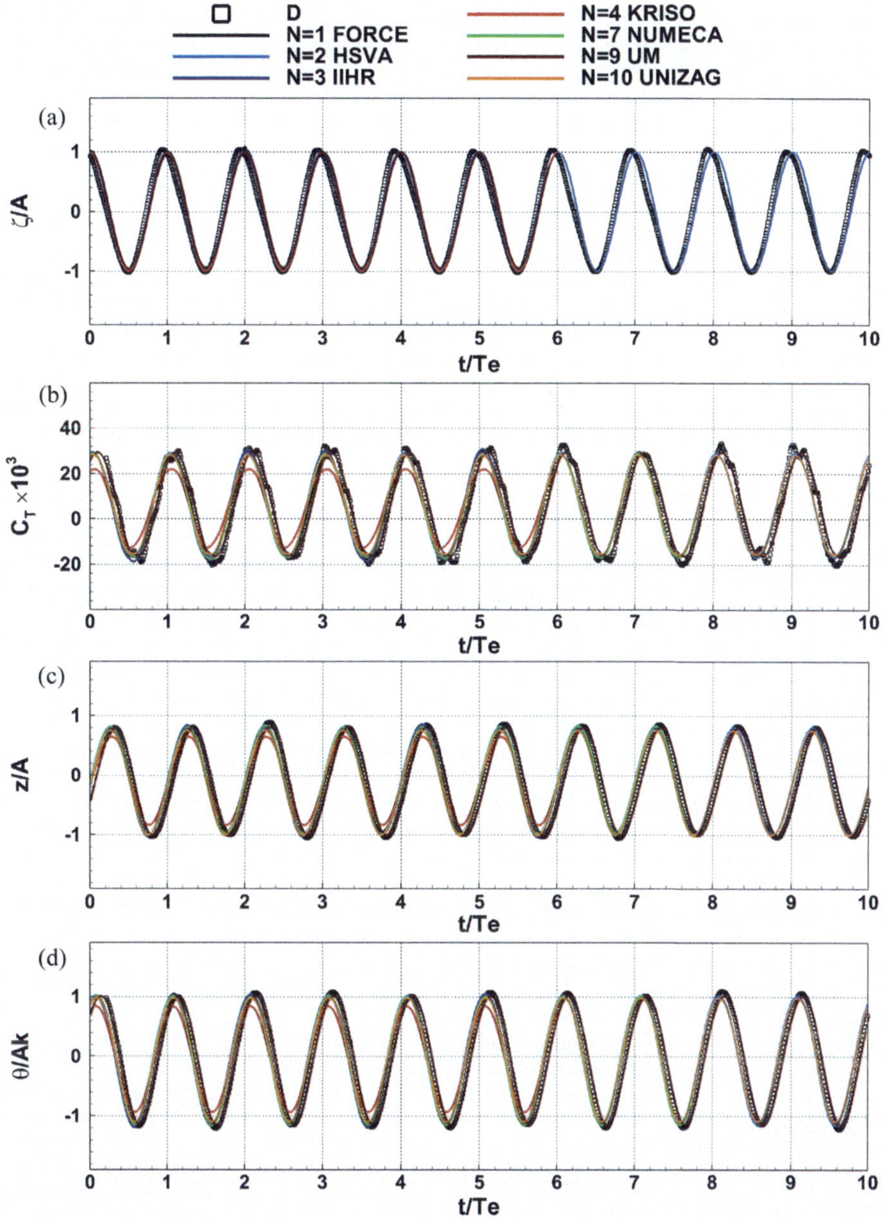

Fig. 8 Time series for head waves with λ/L = 1.95, H/λ = 0.0165 (A/L = 0.0161) and f_e/f_n = 0.70: **a** Wave; **b** Resistance coefficient; **c** Heave; and **d** Pitch

Table 21 (%D). Evaluations of individual and mean code values on harmonic amplitudes of resistance and motions for KCS seakeeping

λ/L	Variable	0th harmonic amplitude				1st harmonic amplitude				2nd harmonic amplitude			Ave	
		$C_{T0} \times 10^3$	AR	z_0/L	θ_0/Ak	ζ_1/L	$C_{T1} \times 10^3$	z_1/L	θ_1/Ak	$C_{T2} \times 10^3$	z_2/L	θ_2/Ak	1st	2nd
0.65	U_D%|D|	5.20	10.34	6.20	9.10	1.30	20.40	18.44	45.56	88.60	115.60	96.30	21.42	47.33
	# submissions	10	10	10	9	7	10	10	9	10	9	9	9.00	9.57
	# outliers	0	0	1	1	0	1	0	0	1	0	1	0.25	0.57
	$P_{\|Ei\|}$%|D|	11.33	122.23	7.11	9.49	4.28	13.98	19.69	38.27	27.22	61.32	112.45	19.06	50.16
	$P_{\|\bar{E}\|}$%|D|	3.58	38.65	2.37	3.35	1.62	4.66	6.23	12.76	9.07	20.44	39.76	6.32	16.75
	U_{SoAi}%|D|	12.47	122.67	9.43	13.14	4.48	24.73	26.98	59.50	92.69	130.86	148.05	28.92	75.62
	U_{SoA}%|D|	6.31	40.01	6.64	9.70	2.08	20.93	19.46	47.31	89.06	117.39	104.18	22.44	53.33
	$\|\bar{E}\|$%|D|	4.57	44.60	8.08	6.93	1.68	21.16	11.45	21.77	60.39	66.21	43.49	14.01	33.47
	# validated	8	3	2	6	5	4	7	8	9	9	7	6.00	6.29
	# N validated	8	8	7	7	6	7	10	8	9	9	7	7.75	7.86
	Mean N validated	Yes	No	No	Yes	Yes	No	Yes	Yes	Yes	Yes	Yes	–	–
0.85	U_D%|D|	5.60	8.95	5.50	4.60	0.80	6.60	13.79	10.40	53.30	100.20	94.00	7.90	38.88
	# submissions	9	9	9	9	7	9	9	9	9	8	8	8.50	8.71
	# outliers	0	0	1	1	0	0	0	1	0	1	1	0.25	0.57
	$P_{\|Ei\|}$%|D|	7.01	27.05	14.29	14.35	3.07	9.59	32.68	38.07	27.06	33.48	92.94	20.85	30.88
	$P_{\|\bar{E}\|}$%|D|	2.34	9.02	5.05	5.07	1.16	3.20	10.89	13.46	9.02	12.65	35.13	7.18	11.18
	U_{SoAi}%|D|	8.97	28.49	15.31	15.07	3.18	11.64	35.47	39.46	59.77	105.64	132.19	22.44	52.21
	U_{SoA}%|D|	6.07	12.70	7.47	6.85	1.41	7.33	17.58	17.01	54.06	101.00	100.35	10.83	41.21

(continued)

Table 21 (continued)

λ/L	Variable	0th harmonic amplitude				1st harmonic amplitude				2nd harmonic amplitude			Ave	
		$C_{T0} \times 10^3$	AR	z_0/L	θ_0/Ak	ζ_1/L	$C_{T1} \times 10^3$	z_1/L	θ_1/Ak	$C_{T2} \times 10^3$	z_2/L	θ_2/Ak	1st	2nd
	$\lvert\bar{E}\rvert\%\lvert D\rvert$	4.28	21.78	11.38	29.74	1.26	9.14	17.94	56.53	28.64	27.64	42.06	21.22	23.65
	# validated	7	2	1	0	4	2	5	0	9	7	6	2.75	4.57
	# N validated	7	5	7	0	5	6	7	1	9	7	7	4.75	6.00
	Mean N validated	Yes	No	No	No	Yes	No	No	No	Yes	Yes	Yes	–	–
1.15	$U_D\%\lvert D\rvert$	2.80	8.15	9.40	701.30	0.20	5.50	4.39	4.35	11.00	17.20	12.30	3.61	108.88
	# submissions	10	10	10	9	8	3	10	9	3	9	9	7.50	8.57
	# outliers	1	1	1	1	0	0	1	1	0	1	0	0.50	0.71
	$P_{\lvert Ei\rvert}\%\lvert D\rvert$	10.09	20.24	11.13	359.34	4.58	8.51	7.24	6.98	11.40	43.52	16.93	6.83	67.52
	$P_{\lvert\bar{E}\rvert}\%\lvert D\rvert$	3.36	6.75	3.71	127.05	1.62	4.91	2.41	2.47	6.58	15.38	5.64	2.85	24.07
	$U_{SoAi}\%\lvert D\rvert$	10.48	21.82	14.57	788.00	4.59	10.13	8.47	8.22	15.84	46.79	20.93	7.85	131.20
	$U_{SoA}\%\lvert D\rvert$	4.38	10.58	10.11	712.71	1.63	7.37	5.01	5.00	12.82	23.08	13.53	4.75	112.46
	$\lvert\bar{E}\rvert\%\lvert D\rvert$	4.73	8.31	14.07	238.14	1.46	19.81	6.40	4.44	23.85	21.35	13.32	8.03	46.25
	# validated	5	5	1	8	4	0	4	4	0	4	5	3.00	4.00
	# N validated	8	8	4	8	7	0	6	7	0	7	6	5.00	5.86
	Mean N validated	No	Yes	No	Yes	Yes	No	No	Yes	No	Yes	Yes	–	–
1.37	$U_D\%\lvert D\rvert$	2.90	8.49	8.10	212.30	0.30	2.70	4.43	4.27	14.40	52.20	26.40	2.93	46.40

(continued)

Table 21 (continued)

λ/L	Variable	0th harmonic amplitude				1st harmonic amplitude				2nd harmonic amplitude			Ave	
		$C_{T0} \times 10^3$	AR	z_0/L	θ_0/Ak	ζ_1/L	$C_{T1} \times 10^3$	z_1/L	θ_1/Ak	$C_{T2} \times 10^3$	z_2/L	θ_2/Ak	1st	2nd
	# submissions	9	9	9	9	8	9	9	9	9	9	9	8.75	9.00
	# outliers	0	0	1	1	0	1	0	0	0	1	1	0.25	0.57
	$P_{\|Ei\|}$%|D|	12.12	25.87	42.60	238.41	5.19	20.16	10.11	15.69	40.30	27.52	47.25	12.79	62.01
	$P_{\|\overline{E}\|}$%|D|	4.04	8.62	15.06	84.29	1.83	7.13	3.37	5.23	13.43	9.73	16.71	4.39	21.70
	U_{SoAi}%|D|	12.46	27.23	43.37	319.24	5.19	20.34	11.04	16.26	42.79	59.01	54.13	13.21	79.75
	U_{SoA}%|D|	4.97	12.10	17.10	228.42	1.86	7.62	5.56	6.75	19.69	53.10	31.24	5.45	52.38
	$\|\overline{E}\|$%|D|	8.56	18.53	26.07	153.10	1.99	10.31	7.28	8.61	21.81	19.02	21.26	7.05	38.33
	# validated	1	4	1	5	3	1	2	4	4	8	6	2.50	4.14
	# N validated	6	6	7	7	7	7	7	7	8	8	7	7.00	7.00
	Mean N validated	No	No	No	Yes	No	No	No	No	No	Yes	Yes	–	–
1.95	U_D%|D|	9.10	26.92	18.60	57.90	0.80	0.70	4.15	4.03	55.50	46.90	18.40	2.42	33.33
	# submissions	10	10	10	9	8	10	10	9	10	9	9	9.25	9.57
	# outliers	1	1	1	1	1	1	1	1	1	1	1	1.00	1.00
	$P_{\|Ei\|}$%|D|	5.06	23.11	13.55	23.24	0.77	6.94	4.28	3.42	46.93	9.88	19.68	3.85	20.21
	$P_{\|\overline{E}\|}$%|D|	1.69	7.70	4.52	8.22	0.29	2.31	1.43	1.21	15.64	3.49	6.96	1.31	6.89
	U_{SoAi}%|D|	10.41	35.48	23.01	62.39	1.11	6.98	5.96	5.29	72.68	47.93	26.94	4.83	39.83
	U_{SoA}%|D|	9.26	28.00	19.14	58.48	0.85	2.42	4.39	4.21	57.66	47.03	19.67	2.97	34.18

(continued)

Table 21 (continued)

λ/L	Variable	0th harmonic amplitude				1st harmonic amplitude				2nd harmonic amplitude			Ave	
		$C_{T0} \times 10^3$	AR	z_0/L	θ_0/Ak	ζ_1/L	$C_{T1} \times 10^3$	z_1/L	θ_1/Ak	$C_{T2} \times 10^3$	z_2/L	θ_2/Ak	1st	2nd
	$\lvert\bar{E}\rvert\%\lvert D\rvert$	3.44	10.44	11.53	12.53	0.49	9.77	3.10	4.17	20.35	7.41	8.26	4.38	10.56
	# validated	9	8	7	8	6	0	7	6	8	8	7	4.75	7.86
	# N validated	9	8	8	8	7	2	7	6	9	8	7	5.50	8.14
	Mean N Validated	Yes	Yes	Yes	Yes	Yes	No	Yes	Yes	Yes	Yes	Yes	–	–
Ave	$U_D\%\lvert D\rvert$	5.12	12.57	9.56	197.04	0.68	7.18	9.04	13.72	44.56	66.42	49.48	7.66	54.96
	# submissions	9.60	9.60	9.60	9.00	7.60	8.20	9.60	9.00	8.20	8.80	8.80	8.60	9.09
	# outliers	0.40	0.40	1.00	1.00	0.20	0.60	0.40	0.60	0.40	0.80	0.80	0.45	0.69
	$P_{\lvert Ei\rvert}\%\lvert D\rvert$	9.12	43.70	17.73	128.97	3.58	11.84	14.80	20.48	30.58	35.14	57.85	12.67	46.16
	$P_{\lvert\bar{E}\rvert}\%\lvert D\rvert$	3.00	14.15	6.14	45.60	1.30	4.44	4.87	7.02	10.75	12.34	20.84	4.41	16.12
	$U_{SoAi}\%\lvert D\rvert$	10.96	47.14	21.14	239.57	3.71	14.76	17.58	25.75	56.76	78.05	76.45	15.45	75.72
	$U_{SoA}\%\lvert D\rvert$	6.20	20.68	12.09	203.23	1.57	9.13	10.40	16.06	46.66	68.32	53.80	9.29	58.71
	$\lvert\bar{E}\rvert\%\lvert D\rvert$	5.12	20.73	14.22	88.09	1.38	14.04	9.23	19.11	31.01	28.33	25.68	10.94	30.45
	# validated	6.00	4.40	2.40	5.40	4.40	1.40	5.00	4.40	6.00	7.20	6.20	3.80	5.37
	# N validated	7.60	7.00	6.60	6.00	6.40	4.40	7.40	5.80	7.00	7.80	6.80	6.00	6.97

Table 22 (%DR). Evaluations of individual and mean code values on harmonic amplitudes of resistance and motions for KCS seakeeping

λ/L	Variable	0th harmonic amplitude				1st harmonic amplitude				2nd harmonic amplitude			Ave	
		$C_{T0} \times 10^3$	AR	z_0/L	θ_0/Ak	ζ_1/L	$C_{T1} \times 10^3$	z_1/L	θ_1/Ak	$C_{T2} \times 10^3$	z_2/L	θ_2/Ak	1st	2nd
0.65	U_D%|DR|	7.23	4.38	8.24	7.67	0.61	3.14	2.99	0.71	28.24	39.27	3.21	1.86	14.03
	# submissions	10	10	10	9	7	10	10	9	10	9	9	9.00	9.57
	# outliers	0	0	1	1	0	1	0	0	1	0	1	0.25	0.57
	$P_{\|Ei\|}$%|DR|	15.75	51.72	9.45	7.99	2.01	2.15	3.19	0.60	8.68	20.83	3.74	1.99	16.88
	$P_{\|\overline{E}\|}$%|DR|	4.98	16.36	3.15	2.83	0.76	0.72	1.01	0.20	2.89	6.94	1.32	0.67	5.50
	U_{SoAi}%|DR|	17.32	51.91	12.54	11.08	2.10	3.81	4.37	0.93	29.55	44.46	4.93	2.80	24.54
	U_{SoA}%|DR|	8.78	16.93	8.82	8.17	0.98	3.22	3.15	0.74	28.39	39.88	3.47	2.02	16.35
	$\|\overline{E}\|$%|DR|	6.35	18.87	10.74	5.84	0.79	3.26	1.86	0.34	19.25	22.50	1.45	1.56	12.14
	# validated	8	3	2	6	5	4	7	8	9	9	7	6.00	6.29
	# N validated	8	8	7	7	6	7	10	8	9	9	7	7.75	7.86
	Mean N validated	Yes	No	No	Yes	Yes	No	Yes	Yes	Yes	Yes	Yes	–	–
0.85	U_D%|DR|	8.76	6.82	5.67	4.68	0.47	1.28	4.19	1.38	19.69	54.16	5.76	1.83	15.08
	# submissions	9	9	9	9	7	9	9	9	9	8	8	8.50	8.71
	# outliers	0	0	1	1	0	0	0	1	0	1	1	0.25	0.57
	$P_{\|Ei\|}$%|DR|	10.96	20.62	14.73	14.59	1.81	1.85	9.92	5.05	9.99	18.09	5.69	4.66	13.53
	$P_{\|\overline{E}\|}$%|DR|	3.65	6.87	5.21	5.16	0.68	0.62	3.31	1.79	3.33	6.84	2.15	1.60	4.75
	U_{SoAi}%|DR|	14.03	21.72	15.78	15.32	1.87	2.25	10.77	5.23	22.08	57.10	8.10	5.03	22.02
	U_{SoA}%|DR|	9.49	9.69	7.70	6.96	0.83	1.42	5.34	2.26	19.97	54.59	6.15	2.46	16.36
	$\|\overline{E}\|$%|DR|	6.70	16.61	11.73	30.24	0.74	1.77	5.45	7.50	10.58	14.94	2.58	3.86	13.34
	# validated	7	2	1	0	4	2	5	0	9	7	6	2.75	4.57

(continued)

Table 22 (continued)

λ/L	Variable	0th harmonic amplitude				1st harmonic amplitude				2nd harmonic amplitude			Ave	
		$C_{T0} \times 10^3$	AR	z_0/L	θ_0/Ak	ζ_1/L	$C_{T1} \times 10^3$	z_1/L	θ_1/Ak	$C_{T2} \times 10^3$	z_2/L	θ_2/Ak	1st	2nd
	# N validated	7	5	7	0	5	6	7	1	9	7	7	4.75	6.00
	Mean N validated	Yes	No	No	No	Yes	No	No	No	Yes	Yes	Yes	–	–
1.15	$U_D\%\|DR\|$	6.69	10.09	4.29	11.80	0.19	0.99	4.91	2.95	14.51	14.41	4.74	2.26	9.50
	# submissions	10	10	10	9	8	3	10	9	3	9	9	7.50	8.57
	# outliers	1	1	1	1	0	0	1	1	0	1	0	0.50	0.71
	$P_{\|Ei\|}\%\|DR\|$	24.12	25.07	5.08	6.04	4.26	1.54	8.11	4.74	15.03	36.45	6.52	4.66	16.90
	$P_{\|\overline{E}\|}\%\|DR\|$	8.04	8.36	1.69	2.14	1.51	0.89	2.70	1.67	8.68	12.89	2.17	1.69	6.28
	$U_{SoAi}\%\|DR\|$	25.03	27.02	6.65	13.26	4.26	1.83	9.48	5.58	20.89	39.19	8.06	5.29	20.01
	$U_{SoA}\%\|DR\|$	10.46	13.10	4.61	11.99	1.52	1.33	5.60	3.39	16.90	19.33	5.21	2.96	11.66
	$\|\overline{E}\|\%\|DR\|$	11.32	10.29	6.42	4.01	1.36	3.58	7.17	3.02	31.45	17.88	5.13	3.78	12.36
	# validated	5	5	1	8	4	0	4	4	0	4	5	3.00	4.00
	# N validated	8	8	4	8	7	0	6	7	0	7	6	5.00	5.86
	Mean N validated	No	Yes	No	Yes	Yes	No	No	Yes	No	Yes	Yes	–	–
1.37	$U_D\%\|DR\|$	6.82	6.91	3.31	12.39	0.34	1.59	4.82	3.74	15.50	23.37	8.69	2.62	11.00
	# submissions	9	9	9	9	8	9	9	9	9	9	9	8.75	9.00
	# outliers	0	0	1	1	0	1	0	0	0	1	1	0.25	0.57
	$P_{\|Ei\|}\%\|DR\|$	28.52	21.05	17.42	13.92	5.85	11.84	11.01	13.74	43.38	12.32	15.56	10.61	21.74
	$P_{\|\overline{E}\|}\%\|DR\|$	9.51	7.02	6.16	4.92	2.07	4.19	3.67	4.58	14.46	4.36	5.50	3.63	7.42
	$U_{SoAi}\%\|DR\|$	29.32	22.16	17.73	18.64	5.86	11.94	12.02	14.24	46.07	26.42	17.82	11.01	25.45

(continued)

Table 22 (continued)

λ/L	Variable	0th harmonic amplitude				1st harmonic amplitude				2nd harmonic amplitude			Ave	
		$C_{T0} \times 10^3$	AR	z_0/L	θ_0/Ak	ζ_1/L	$C_{T1} \times 10^3$	z_1/L	θ_1/Ak	$C_{T2} \times 10^3$	z_2/L	θ_2/Ak	1st	2nd
	U_{SoA}%|DR|	11.70	9.85	6.99	13.34	2.09	4.48	6.06	5.91	21.20	23.77	10.29	4.64	13.88
	$\|\overline{E}\|$%|DR|	20.14	15.08	10.66	8.94	2.24	6.05	7.93	7.55	23.48	8.51	7.00	5.94	13.40
	# validated	1	4	1	5	3	1	2	4	4	8	6	2.50	4.14
	# N validated	6	6	7	7	7	7	7	7	8	8	7	7.00	7.00
	Mean N validated	No	No	No	Yes	No	No	No	No	No	Yes	Yes	–	–
1.95	U_D%|DR|	16.58	6.42	6.12	25.79	1.18	0.81	4.82	4.09	18.33	62.83	19.01	2.72	22.16
	# submissions	10	10	10	9	8	10	10	9	10	9	9	9.25	9.57
	# outliers	1	1	1	1	1	1	1	1	1	1	1	1.00	1.00
	$P_{\|Ei\|}$%|DR|	9.23	5.51	4.46	10.35	1.13	8.01	4.97	3.47	15.50	13.24	20.33	4.40	11.23
	$P_{\|\overline{E}\|}$%|DR|	3.08	1.84	1.49	3.66	0.43	2.67	1.66	1.23	5.17	4.68	7.19	1.50	3.87
	U_{SoAi}%|DR|	18.97	8.46	7.57	27.79	1.63	8.05	6.93	5.37	24.01	64.21	27.84	5.49	25.55
	U_{SoA}%|DR|	16.86	6.68	6.30	26.05	1.25	2.79	5.10	4.27	19.05	63.01	20.33	3.35	22.61
	$\|\overline{E}\|$%|DR|	6.26	2.49	3.79	5.58	0.72	11.28	3.60	4.24	6.72	9.92	8.53	4.96	6.19
	# validated	9	8	7	8	6	0	7	6	8	8	7	4.75	7.86
	# N validated	9	8	8	8	7	2	7	6	9	8	7	5.50	8.14
	Mean N validated	Yes	Yes	Yes	Yes	Yes	No	Yes	Yes	Yes	Yes	Yes	–	–
Ave	U_D%|DR|	9.22	6.92	5.53	12.47	0.56	1.56	4.35	2.58	19.25	38.81	8.28	2.26	14.35
	# submissions	9.60	9.60	9.60	9.00	7.60	8.20	9.60	9.00	8.20	8.80	8.80	8.60	9.09
	# outliers	0.40	0.40	1.00	1.00	0.20	0.60	0.40	0.60	0.40	0.80	0.80	0.45	0.69

(continued)

Table 22 (continued)

λ/L	Variable	0th harmonic amplitude				1st harmonic amplitude				2nd harmonic amplitude			Ave	
		$C_{T0} \times 10^3$	AR	z_0/L	θ_0/Ak	ζ_1/L	$C_{T1} \times 10^3$	z_1/L	θ_1/Ak	$C_{T2} \times 10^3$	z_2/L	θ_2/Ak	1st	2nd
	$P_{\lvert Ei \rvert}\%\lvert DR \rvert$	17.72	24.80	10.22	10.58	3.01	5.08	7.44	5.52	18.52	20.19	10.37	5.26	16.06
	$P_{\lvert \overline{E} \rvert}\%\lvert DR \rvert$	5.85	8.09	3.54	3.74	1.09	1.82	2.47	1.89	6.91	7.14	3.67	1.82	5.56
	$U_{SoAi}\%\lvert DR \rvert$	20.94	26.25	12.05	17.22	3.14	5.58	8.71	6.27	28.52	46.28	13.35	5.93	23.52
	$U_{SoA}\%\lvert DR \rvert$	11.46	11.25	6.88	13.30	1.33	2.65	5.05	3.32	21.10	40.12	9.09	3.09	16.17
	$\lvert \overline{E} \rvert\%\lvert DR \rvert$	10.15	12.67	8.67	10.92	1.17	5.19	5.20	4.53	18.30	14.75	4.94	4.02	11.49
	# validated	6.00	4.40	2.40	5.40	4.40	1.40	5.00	4.40	6.00	7.20	6.20	3.80	5.37
	# N validated	7.60	7.00	6.60	6.00	6.40	4.40	7.40	5.80	7.00	7.80	6.80	6.00	6.97

Table 23 (%2π). Evaluations of individual and mean code values on harmonic phase of resistance and motions for KCS seakeeping

λ/L	Variable	1st harmonic phase			2nd harmonic phase			Ave	
		$C_{T1} \times 10^3$	z_1/L	θ_1/Ak	$C_{T2} \times 10^3$	z_2/L	θ_2/Ak	1st	2nd
0.65	U_D%2π	3.00	2.40	7.40	24.80	61.40	78.80	4.27	55.00
	# submissions	9	9	8	9	8	8	8.67	8.33
	# outliers	0	1	1	1	0	1	0.67	0.67
	$P_{\|Ei\|}$%2π	1.90	5.76	4.80	5.35	12.16	4.84	4.15	7.45
	$P_{\|\bar{E}\|}$%2π	0.63	2.04	1.81	1.89	4.30	1.83	1.49	2.67
	U_{SoAi}%2π	3.55	6.24	8.82	25.37	62.59	78.95	6.20	55.64
	U_{SoA}%2π	3.07	3.15	7.62	24.87	61.55	78.82	4.61	55.08
	$\|\bar{E}\|$%2π	4.46	5.59	2.46	6.71	39.17	27.12	4.17	24.34
	# validated	1	0	7	8	8	7	2.67	7.67
	# N validated	1	6	7	8	8	7	4.67	7.67
	Mean N validated	No	No	Yes	Yes	Yes	Yes	–	–
0.85	U_D%2π	3.60	3.40	3.30	11.40	64.80	40.90	3.43	39.03
	# submissions	8	8	8	8	7	7	8.00	7.33
	# outliers	0	1	0	0	1	1	0.33	0.67
	$P_{\|Ei\|}$%2π	9.14	10.79	5.57	16.03	10.61	10.58	8.50	12.41
	$P_{\|\bar{E}\|}$%2π	3.23	4.08	1.97	5.67	4.33	4.32	3.09	4.77
	U_{SoAi}%2π	9.82	11.32	6.48	19.67	65.66	42.25	9.20	42.53
	U_{SoA}%2π	4.84	5.31	3.84	12.73	64.94	41.13	4.66	39.60
	$\|\bar{E}\|$%2π	8.39	10.92	10.99	19.80	42.13	37.68	10.10	33.20
	# validated	2	0	0	1	6	4	0.67	3.67
	# N validated	4	5	0	3	6	4	3.00	4.33
	Mean N validated	No	No	No	No	Yes	Yes	–	–
1.15	U_D%2π	0.70	0.30	0.30	2.20	2.70	2.10	0.43	2.33
	# submissions	3	9	8	3	8	8	6.67	6.33
	# outliers	0	1	0	0	1	1	0.33	0.67
	$P_{\|Ei\|}$%2π	6.59	5.16	3.21	69.59	5.65	12.91	4.99	29.38
	$P_{\|\bar{E}\|}$%2π	3.81	1.83	1.14	40.18	2.13	4.88	2.26	15.73
	U_{SoAi}%2π	6.63	5.17	3.23	69.62	6.26	13.08	5.01	29.65
	U_{SoA}%2π	3.87	1.85	1.17	40.24	3.44	5.31	2.30	16.33

(continued)

Table 23 (continued)

λ/L	Variable	1st harmonic phase			2nd harmonic phase			Ave	
		$C_{T1} \times 10^3$	z_1/L	θ_1/Ak	$C_{T2} \times 10^3$	z_2/L	θ_2/Ak	1st	2nd
	$\lvert\bar{E}\rvert\%2\Pi$	3.99	2.93	2.32	20.91	3.96	8.08	3.08	10.98
	# validated	1	1	0	2	4	1	0.67	2.33
	# N validated	3	7	7	3	5	5	5.67	4.33
	Mean N validated	No	No	No	Yes	No	No	–	–
1.37	$U_D\%2\Pi$	0.40	0.50	0.40	16.60	9.30	4.20	0.43	10.03
	# submissions	8	8	8	8	8	8	8.00	8.00
	# outliers	0	0	0	1	0	0	0.00	0.33
	$P_{\lvert Ei\rvert}\%2\Pi$	4.42	1.66	2.75	5.46	16.78	7.34	2.94	9.86
	$P_{\lvert\bar{E}\rvert}\%2\Pi$	1.56	0.59	0.97	2.06	5.93	2.60	1.04	3.53
	$U_{SoAi}\%2\Pi$	4.44	1.73	2.78	17.47	19.19	8.46	2.98	15.04
	$U_{SoA}\%2\Pi$	1.61	0.77	1.05	16.73	11.03	4.94	1.15	10.90
	$\lvert\bar{E}\rvert\%2\Pi$	2.13	1.74	2.25	3.20	8.23	4.64	2.04	5.36
	# validated	0	1	0	7	5	4	0.33	5.33
	# N validated	7	5	6	7	7	7	6.00	7.00
	Mean N validated	No	No	No	Yes	Yes	Yes	–	–
1.95	$U_D\%2\Pi$	0.20	0.20	0.20	11.30	8.70	3.00	0.20	7.67
	# submissions	9	9	8	9	8	8	8.67	8.33
	# outliers	1	1	1	1	1	0	1.00	0.67
	$P_{\lvert Ei\rvert}\%2\Pi$	1.92	1.47	1.43	15.64	8.49	19.58	1.61	14.57
	$P_{\lvert\bar{E}\rvert}\%2\Pi$	0.68	0.52	0.54	5.53	3.21	6.92	0.58	5.22
	$U_{SoAi}\%2\Pi$	1.93	1.48	1.44	19.30	12.16	19.81	1.62	17.09
	$U_{SoA}\%2\Pi$	0.71	0.56	0.57	12.58	9.27	7.54	0.61	9.80
	$\lvert\bar{E}\rvert\%2\Pi$	5.48	5.12	4.92	8.76	6.81	13.36	5.17	9.64
	# validated	0	0	0	6	4	0	0.00	3.33
	# N validated	0	0	0	7	7	7	0.00	7.00
	Mean N validated	No	No	No	Yes	Yes	No	–	–
Ave	$U_D\%2\Pi$	1.58	1.36	2.32	13.26	29.38	25.80	1.75	22.81

(continued)

Table 23 (continued)

λ/L	Variable	1st harmonic phase			2nd harmonic phase			Ave	
		$C_{T1} \times 10^3$	z_1/L	θ_1/Ak	$C_{T2} \times 10^3$	z_2/L	θ_2/Ak	1st	2nd
	# submissions	7.40	8.60	8.00	7.40	7.80	7.80	8.00	7.67
	# outliers	0.20	0.80	0.40	0.60	0.60	0.60	0.47	0.60
	$P_{\lvert Ei \rvert}\%2\pi$	4.80	4.97	3.55	22.41	10.74	11.05	4.44	14.73
	$P_{\lvert \overline{E} \rvert}\%2\pi$	1.98	1.81	1.29	11.07	3.98	4.11	1.69	6.39
	$U_{SoAi}\%2\pi$	5.28	5.19	4.55	30.29	33.17	32.51	5.00	31.99
	$U_{SoA}\%2\pi$	2.82	2.33	2.85	21.43	30.05	27.55	2.67	26.34
	$\lvert \overline{E} \rvert\%2\pi$	4.89	5.26	4.59	11.88	20.06	18.18	4.91	16.70
	# validated	0.80	0.40	1.40	4.80	5.40	3.20	0.87	4.47
	# N validated	3.00	4.60	4.00	5.60	6.60	6.00	3.87	6.07

left column shows D, Si (including outliers) and $\bar{S}$ without outliers. The middle and right columns show U_D, U_{SoA_i}, U_{SoA}, $|E_i|$ and $\overline{|E|}$ using %D and %DR, respectively. The solutions show large scatter especially for resistance but also motions.

The 0th harmonic experiments show that C_{T0}/AR peak at the resonance condition with small/medium values for short/long waves; C_{T0} is considerably larger than C_T in calm water except for short waves. Heave and pitch magnitudes are mostly larger than their calm water values, except for outliers. The mean solution $\bar{S}$ closely follows the experimental trends; however, the scatter in the solutions is very large. The errors are largest for C_{T0}/AR near resonance and for motions for wave lengths just before and after resonance. The solution scatter mostly correlates with large errors.

The 1st harmonic amplitude experiments show that heave and pitch follow the expected trends: for short/long waves heave/pitch are small/unity. Heave shows a maximum at the resonance condition. C_{T1} increases with λ/L except for the resonance condition where it shows a dramatic reduction. $\bar{S}$ closely follows the experimental trends with reduced scatter compared to the 0th harmonic. The error and scatter for C_{T1} is larger for long waves, whereas for the motions its mostly like that for the 0th harmonic.

The 1st harmonic phase experiments shows expected trends that pitch leads heave by nearly π/2 for large λ/L and for small λ/L the phase difference for heave and pitch reduces to nearly zero. C_{T1} phase is mostly small except near resonance where it is near π/2. $\bar{S}$ follows the experimental trends. The errors and scatter for both C_{T1} and motions are largest for the wavelength just before resonance and for the longest wave.

The 2nd harmonic amplitude experiments show that C_{T2} trends are similar C_{T0}, heave is nearly zero and pitch increases with λ/L with similar trends as for C_{T1}.

For AR, $U_D = 7\%$DR and $\overline{|E|} = 13\%$DR such that only 4 solutions are validated. The code level scatter $P_{|E_i|} = 25\%$DR is much larger than U_D. $U_{SoA_i} = 26\%$DR such that 7 codes/solutions are N-version validated. $U_{SoA} = 11\%$DR and the mean code is not N-version validated.

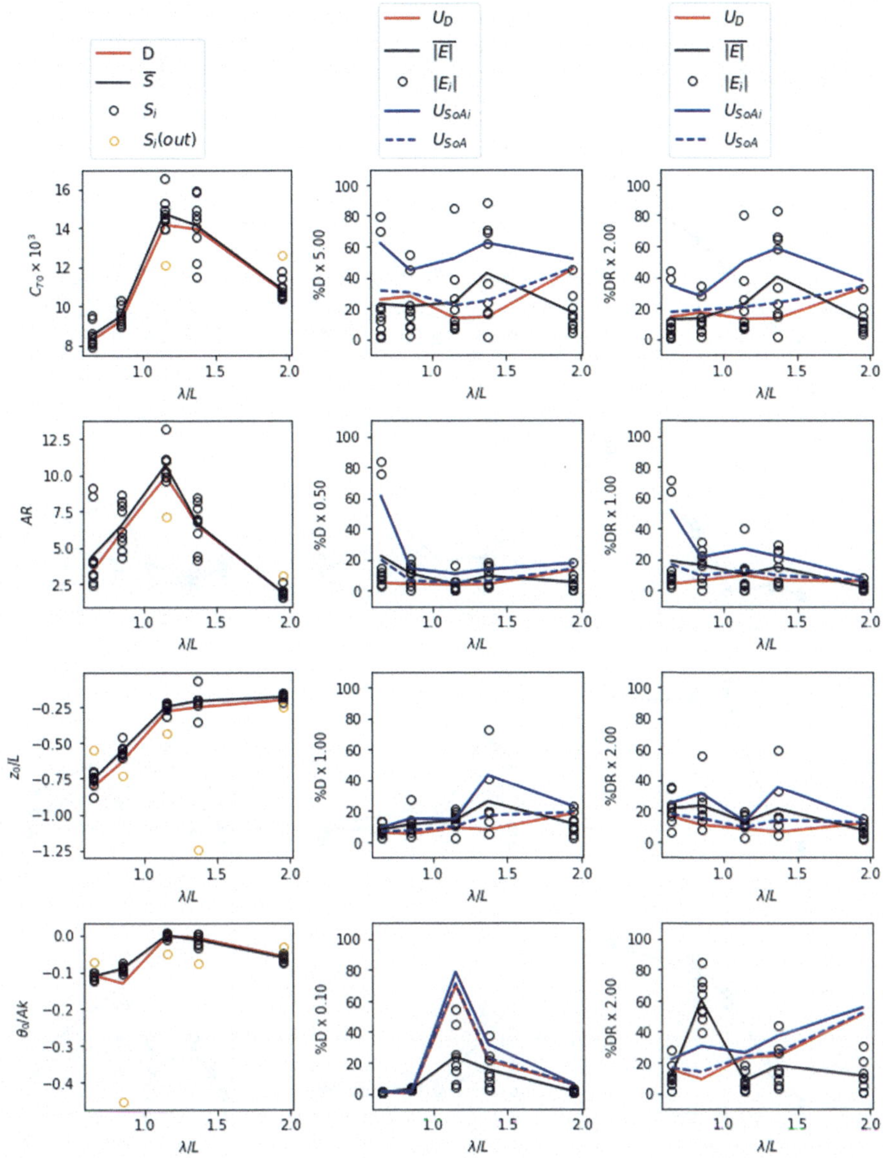

Fig. 9 EFD data and submission values for 0th harmonic amplitudes of resistance and motions for KCS seakeeping in regular waves at Fr = 0.26

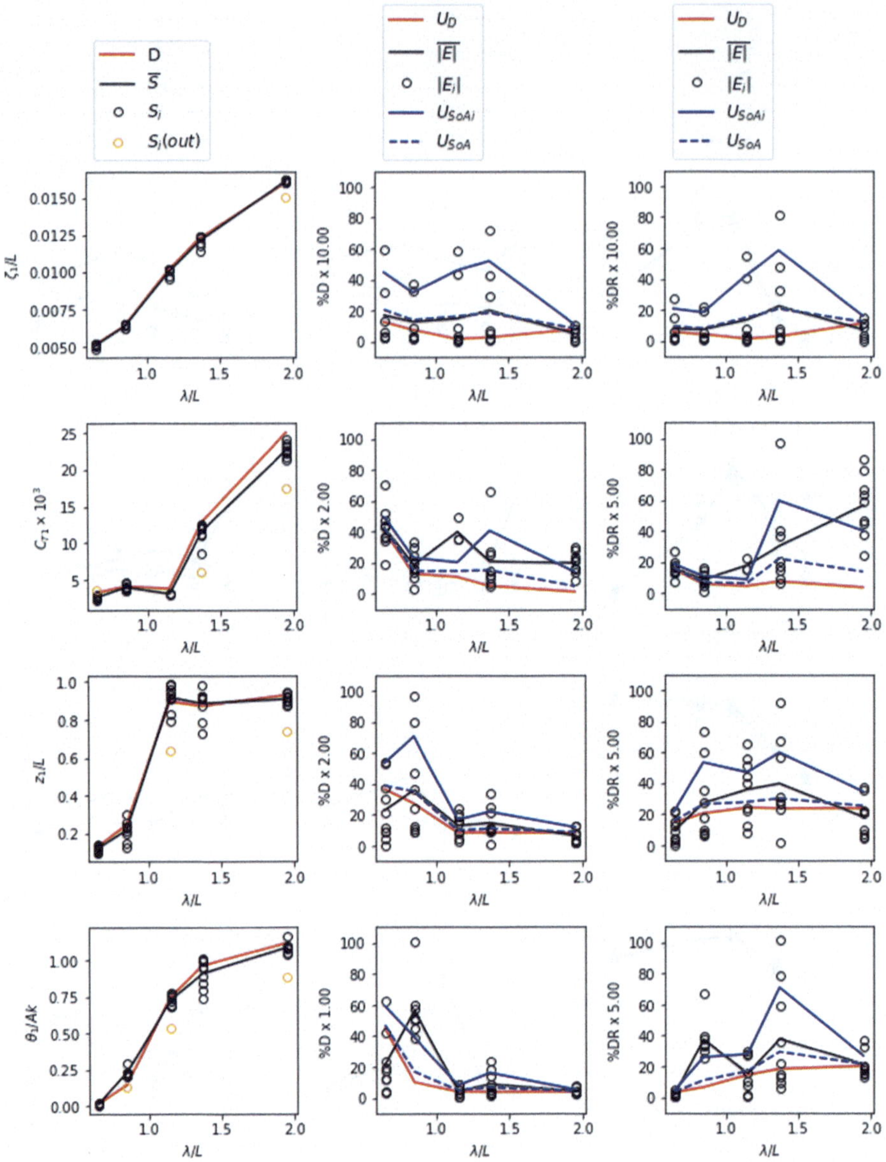

Fig. 10 EFD data and submission values for 1st harmonic amplitudes of resistance and motions for KCS seakeeping in regular waves at Fr = 0.26

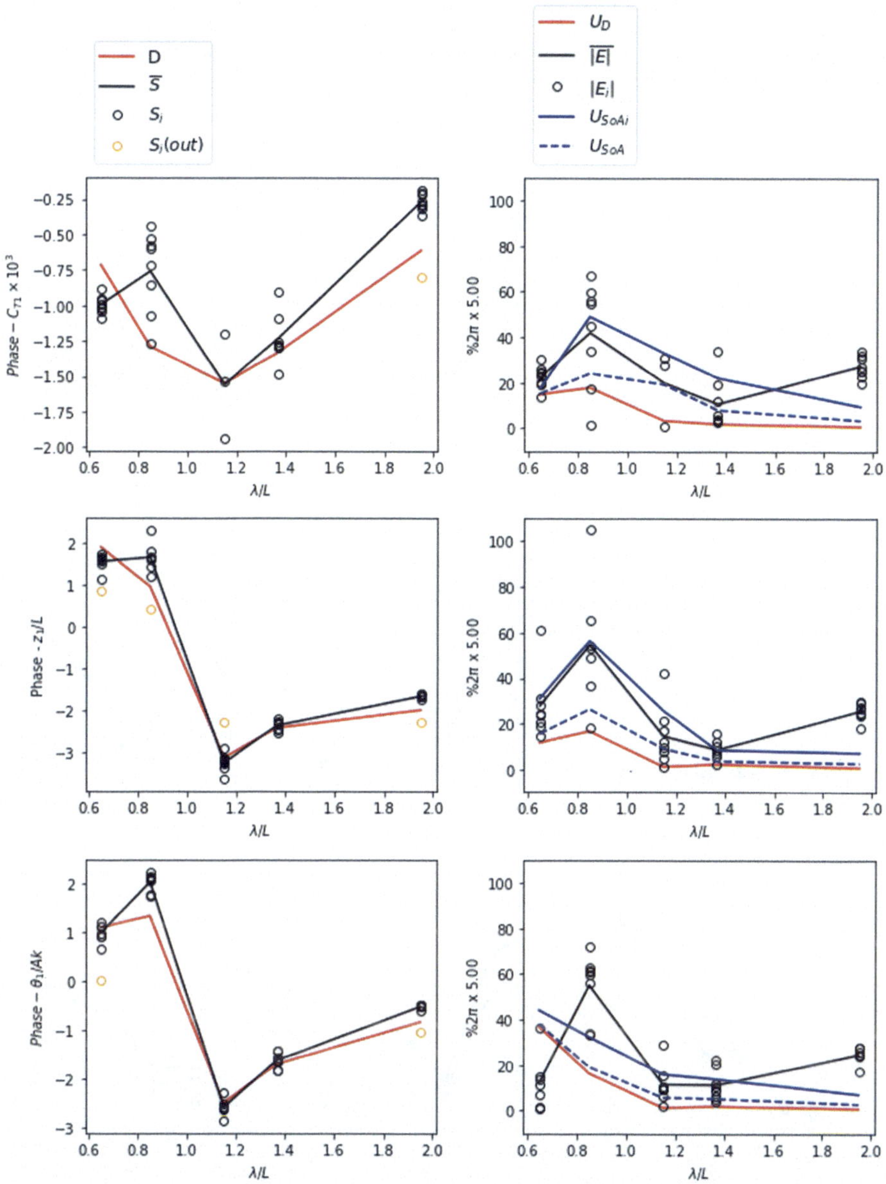

Fig. 11 EFD data and submission values for 1st harmonic phases of resistance and motions for KCS seakeeping in regular waves at Fr = 0.26

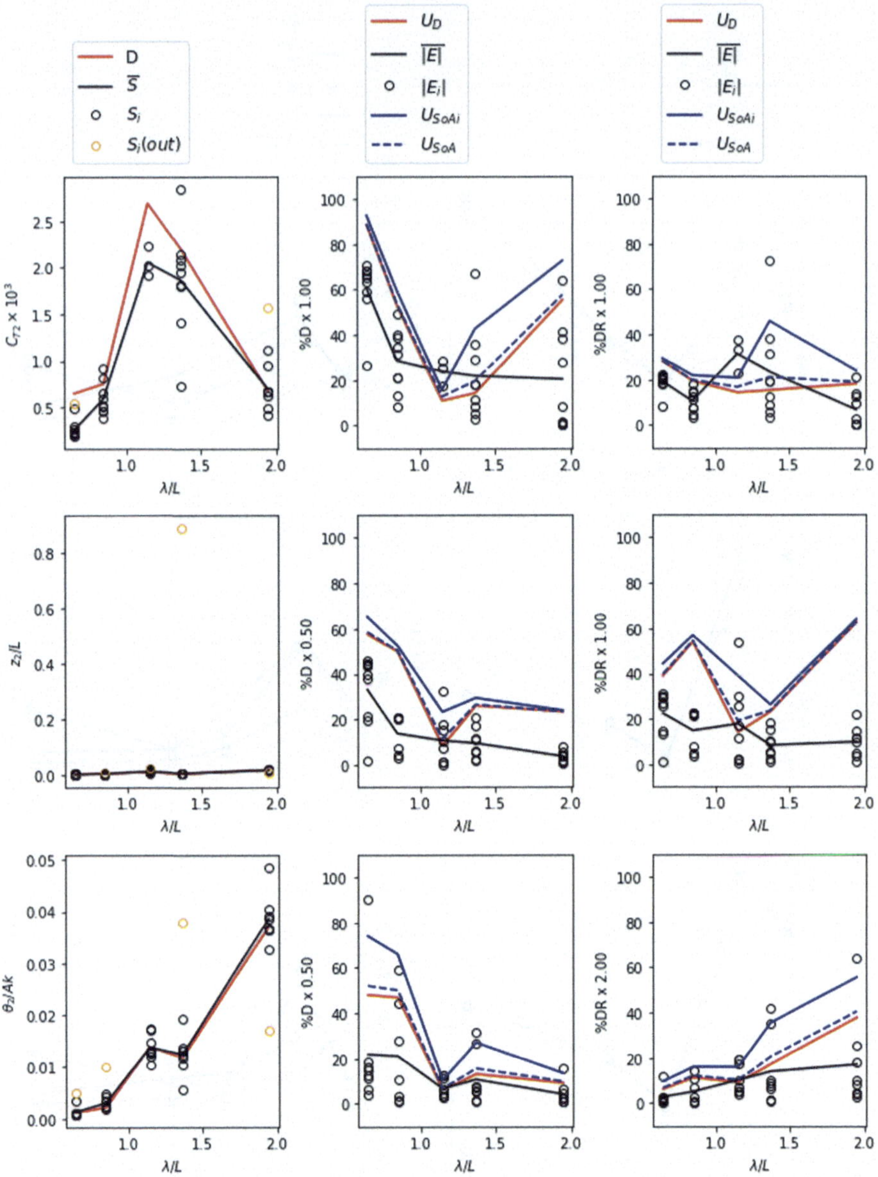

Fig. 12 EFD data and submission values for 2nd harmonic amplitudes of resistance and motions for KCS seakeeping in regular waves at Fr = 0.26

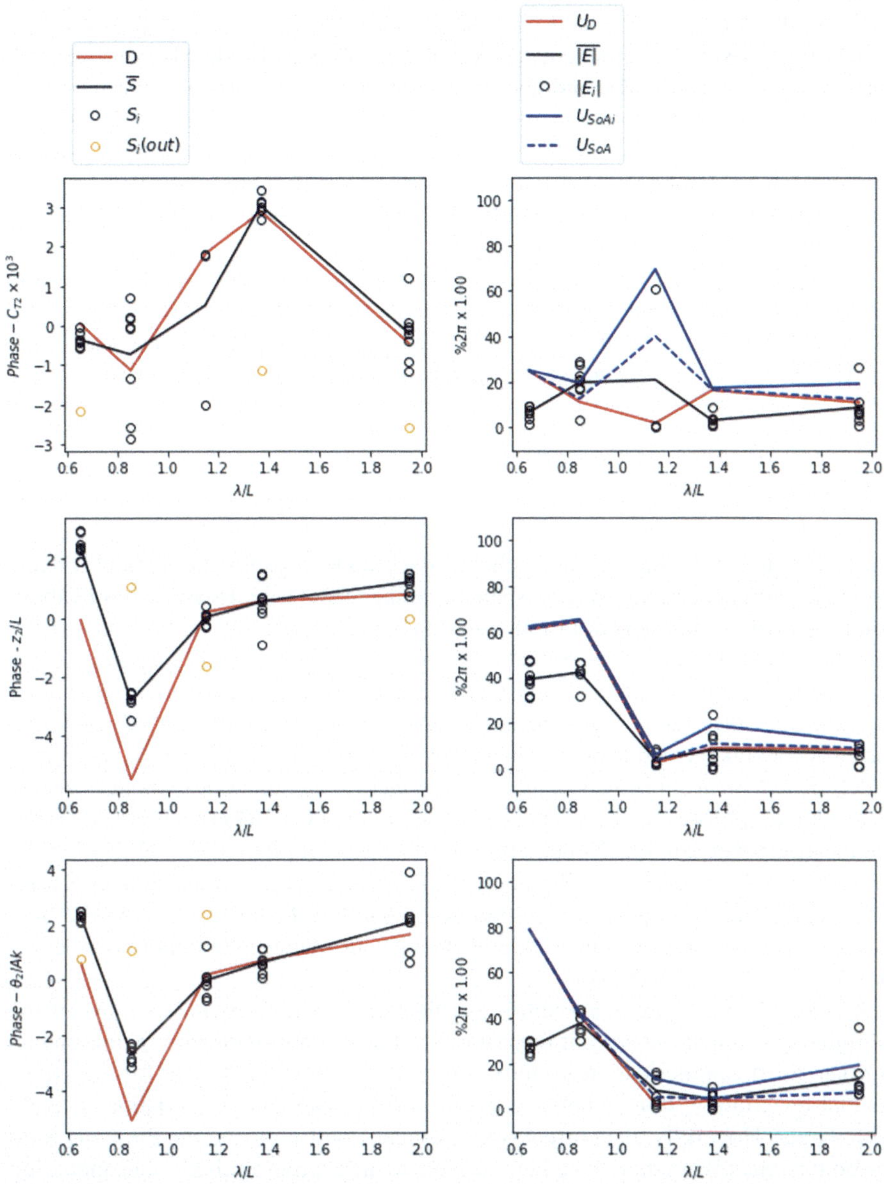

Fig. 13 EFD data and submission values for 2nd harmonic phases of resistance and motions for KCS seakeeping in regular waves at Fr = 0.26

For first harmonic heave and pitch amplitudes, U_D is about 4%DR and $\overline{|E|} = 5\%$DR, respectively, such that only about half the solutions validated. The code level scatter $P_{|E_i|}$ is about 7%DR and about twice as large as U_D. $U_{SoA_i} = 8\%$DR such that only about half the solutions are N-version validated. U_{SoA} is about 4%DR and the mean code is N-version validated for heave but not pitch. The trends for the first harmonic phases are like those for the first harmonic amplitudes with $\overline{|E|} = 5\%$DR. The trends for the 2nd harmonic amplitudes and phases for errors and uncertainties are similar but with much larger intervals.

The solution scatter uncertainty is larger than U_D especially for AR. The errors for AR are six times larger than for calm water resistance and for heave and pitch are comparable to those for sinkage and trim. The largest errors are for near resonance for AR, near resonance for heave, and just before/after resonance for pitch.

5.1 Conclusions

The calm water experimental data included assessment of facility biases/scale effects with reasonable uncertainties for resistance, sinkage and trim. However, more attention is needed for assessment of individual facility uncertainties and for comparing data that is measured in conventional or dedicated calm water resistance tests setups in all facilities, since calm water resistance normally is measured that way. Reducing experimental uncertainties is a deciding factor in reducing intervals of validation and SoA for sinkage and trim. The head wave experimental data showed significant effects of statistical convergence, mount resonance, and scatter due to model size and facility/mount. More attention is needed to document the CFD input wave condition and location for evaluation. Nonetheless, the data was useful for CFD validation/SoA with reasonable uncertainties. Reducing the experimental uncertainties is important for reducing the validation interval, but not a deciding factor in reducing the SoA interval since the order of magnitude of the current scatter in simulations is larger than U_D for 1st order variables.

It is very unfortunate that solution verification was not provided by any of the submitters such that the current intervals of validation are optimistic. A reasonable estimate for the numerical uncertainty is that it is comparable to that for the experimental uncertainty; thus, a better estimate for the intervals of validation is likely 30%. larger. Hopefully, the current SoA estimates will motivate the 8th CFD Ship Hydrodynamics Workshop to be held in 2021 to attract more submissions including verification, which will likely reduce the CFD scatter; and benchmark experiments from more facilities with reduced uncertainties. Ship hydrodynamics CFD workshops should follow the example of the AIAA CFD Drag Prediction Workshops by requiring the participants to submit verification and possibly use provided or specified grids. This will provide more robust uncertainty estimates and validation/SoA confidence levels/intervals.

The current sample size of 10 institutes from 8 countries using 9 different CFD codes submitting results for both calm water and waves is enough, especially as the codes differ in modeling and numerical methods. The grid size variation was large from 0.6–29 M cells.

For calm water, 3, 5 and 9 solutions are validated at $U_{V_i} = 1.75, 7.38$ and 14.25%D for C_T, sinkage and trim, respectively. At the code level, the mean code shows $\overline{|E|} = 2$, 6 and 5%D for C_T, σ/L and trim, respectively. The code level $U_{SoA_i} = 3$, 8 and 16%D for C_T, sinkage/L and trim, respectively. Thus, 8, 7 and 9 codes are SoA at the U_{SoA_i} intervals. The mean code is not N-version validated for resistance but is for sinkage and trim at $U_{SoA} = 2$, 8 and 14%DR, respectively. Since G2010 is only based on 6 submissions vs. 10 for T2015 and the present experimental uncertainty estimate considers multiple facilities/models, the present results are considered a more robust estimate of the current CFD code capabilities.

For head waves (averaged over all λ/L), the scatter in the solutions is very large, i.e., 5 and 16%DR for 1st order and 2nd order variables, respectively. Interestingly the mean code solution shows relatively close agreement with the experiments both for 1st and 2nd order variables. This suggests that the present estimates using standard deviations well represent the multiple code level random/precision simulation and error uncertainty components. 4/5 solutions are validated at 2/14%DR for 1st and 2nd order variables. 6/7 codes are N version validated SoA_i at 6/24%DR for 1st and 2nd order variables. The mean code is not SoA for 1st order variables with $\overline{|E|} = 4$%DR and $U_{SoA} = 3$%DR, whereas it is SoA for 2nd order variables at interval 16%DR with $\overline{|E|} = 12$%DR

Table 24 summarizes the present results for short, medium and long waves, including comparison with G2010 and potential flow and CFD results from Bunnik et al. (2010). Error/scatter from Bunnik et al. (2010) were extracted from the provided figures, as quantitative values were not provided; also, H/λ values are not known but presumably in the linear range.

The present error/scatter with 10 submissions are larger for short than long waves. The average errors/scatter are 8/7, 16/10 and 24/28 for heave, pitch and AR, respectively. The errors/scatter increase from heave to pitch to AR.

The G2010 error/scatter with 6 submissions for KCS do not include short waves and AR. The average errors/scatter are 10/9 and 4/4 for heave and pitch, respectively, which are comparable T2015. The G2010 error/scatter with 5 submissions for KVLCC2 are mostly larger for short waves for heave and pitch, but not AR. The average errors/scatter are 11/7, 10/4 and 38/1 for heave, pitch and AR, respectively. The errors are similar for heave and pitch but larger for AR and the scatter is largest for heave and smaller for pitch and AR. The errors/scatter are larger/similar compared to those for KCS. Thus, T2015 and G2010 CFD errors/scatter are comparable but T2015 errors/scatter are considered most reliable since there are more submissions in T2015 with more rigorous analysis.

The Bunnik et al. (2010) error/scatter with 8 PF submissions for a container ship are mostly larger for short than long waves for heave, pitch and AR. The average errors/scatter are 25/13, 31/11 and 39/14 for heave, pitch and AR, respectively. The errors/scatter mostly increase from heave to pitch to AR. The error/scatter with 3 CFD submissions for a container ship are mostly larger for short than long waves for heave, pitch and AR. The average errors/scatter are 35/15, 22/12 and 36/13 for

Table 24 CFD and PF comparison for motions and added resistance computations in head waves

λ/L	$\|\bar{E}\|\%\|D\|$			$\sigma_{\|\bar{E}\|}\%\|D\|$		
	z_1/A	θ_1/Ak	AR	z_1/A	θ_1/Ak	AR
Container ship: KCS (T2015; CFD; N = 10)						
Short (λ/L = 0.65)	11.5	21.8	44.6	9.8	19.1	61.1
Medium (λ/L = 0.85, 1.15, 1.37)	10.5	23.2	16.2	8.3	10.1	12.2
Long (λ/L = 1.95)	3.1	4.2	10.4	2.1	1.7	11.6
Average	8.4	16.4	23.7	6.8	10.3	28.3
Container ship: KCS (G2010; CFD; N = 6)						
Medium (λ/L = 1.15, 1.33)	8.9	3.9		5.2	3.1	
Long (λ/L = 2.0)	10.8	4.1		12.7	4.6	
Average	9.9	4		9	3.9	
Tanker: KVLCC2 (G2010; CFD; N = 5)						
Short (λ/L = 0.6, 0.64)	16.6	15	34.5	8.5	6.6	
Medium (λ/L = 0.92, 1.1)	10.3	5.9	2.9	9.1	2.8	1.9
Long (λ/L = 1.6)	6.7	9.6	76	3.6	3.9	0.6
Average	11.2	10	37.8	7.1	4.4	1.3
G2010–Average	10.6	7.0	37.8	8.1	4.2	1.3
Container ship (Bunnik et al. 2010; PF; N = 8)						
Short (λ/L = 0.3, 0.5, 0.6, 0.7)	36.5	58.8	53.8	9.5	17.4	23.8
Medium (λ/L = 0.9, 1.1)	29.4	23	23.3	23.8	7.4	11.2
Long (λ/L = 1.4)	10	12	39.6	5.7	9	7.8
Average	25.3	31	38.9	13	11	14.3
Container ship (Bunnik et al. 2010; CFD; N = 3)						
Short (λ/L = 0.3, 0.5, 0.6, 0.7)	48.1	50.2	48.1	12.6	27.2	4.3
Medium (λ/L = 0.9,1.1)	45.1	13.6	33.7	28.1	8.6	26.3
Long (λ/L = 1.4)	10.7	2.1	26.4	4.5	1	9.7
Average	34.6	22	36.1	15	12	13.4
Ferry (Bunnik et al. 2010; PF; N = 7)						
Medium (λ/L = 0.85, 1.0, 1.17)	29.2	27.3		18.1	14	
Long (λ/L = 1.4, 1.75, 2.23, 2.90)	23.6	9.4		15.1	6.2	
Average	26.4	18		17	10	
Bunnik et al. 2010–Average	28.8	23.7	37.5	15.0	11.0	13.9

heave, pitch and AR, respectively. The error/scatter are larger for heave and AR than pitch. The error/scatter with 7 PF submissions for ferry ship are larger for short than long waves for heave and pitch. The average errors/scatter are 26/17 and 18/10 for heave and pitch. Bunnik et al. (2010) concluded that the scatter is large, there is a large difference between methods due to differences in the restoring force, added mass and damping, and diffraction models but the best methods used more accurate forward speed models, added resistance and internal loads were under predicted, and CFD did not show benefit in consideration larger CPU requirement.

Current CFD results from T2015 shows that CFD can predict head waves heave, pitch and AR with better accuracy than potential flow for short, medium and long waves. CFD results also show smaller scatter than potential flow solutions for heave and pitch. Use of finer grids and improved user best practices will likely reduce CFD errors and scatter. In addition, improved benchmark experimental data from multiple facilities and solution verification will provide estimates that are more robust for intervals of validation and SoA.

6 CFD Assessment Test Case 2.11

As shown in Table 25, only 2 institutes from 2 countries using two different CFD codes submitted results for both calm water and all the wave heading conditions. Both institutes also submitted results for test case 2.10 using the same codes and one used among the coarsest and the other among the finest grids. Although just 2 submissions are available, the results provide an opportunity for the assessment of CFD for AR variable heading.

Table 26 provides calm water results. Both codes/solutions have larger errors than shown for test case 2.10 calm water resistance, sinkage and trim.

Figure 14 shows the non-dimensional frequency of encounter f_eL/U for each heading vs. λ/L, including horizontal lines marking the heave/pitch and roll natural frequencies and parametric roll frequency. Test case 2.11 is for $\lambda/L = 1$ and $\mu = 0$, 45, 90, 135 and 180°. Head and bow quartering conditions are close to the heave/pitch natural frequency. Beam condition is far from all resonance conditions. Note that $B/\lambda = 0.14$. Stern quartering and following wave conditions are close to the roll natural frequency and parametric roll frequency.

The experimental and CFD time histories are shown in Figs. 15, 16, 17, 18 and 19 for $\mu = 0$, 45, 90, 135 and 180°., respectively. The experiments only cover $5T_e$ for $\mu = 0$ and 45°, $3T_e$ for $\mu = 90$, 135 and 180° and show more scatter compared to the 10 T_e available for test case 2.10. There is mostly good agreement between the 2 codes and the data, but clearly larger errors than for test case 2.10. The largest differences are for roll, phase shifts and poorer agreement for stern waves.

Tables 27 and 28 provide average results for each heading and over all headings of the CFD solutions for the 0th/AR and 1st and 2nd harmonic amplitude validation and SoA variables using %D and %DR, respectively. Table 29 is similar but provides the 1st and 2nd harmonic phase validation and SoA variables using $\%2\pi$. Similarly, as for test case 2.10, the participants submitted their own harmonic analysis values for amplitudes and phases, which were used for reconstruction of the time histories. The discussions mostly use values for μ averages and %DR.

Figures 20, 21, 22, 23 and 24 show the results for the 0th harmonic, 1st and 2nd harmonic amplitudes and 1st and 2nd harmonic phases, respectively, vs. μ. The left

Table 25 Submissions for KCS resistance test and seakeeping in calm water and head waves at Fr = 0.26 (Case 2.11)

No	Institute	Code	Turbulence	Free surface	Discretization	Grid motion	Grid type	Grid size	No.CPU	CPU time
1	IIHR	CFDShip-Iowa V4.5	k-ε/k-ω SST	LS Single Ph.	FD	Dynamic Overset	Structured	29.2 M	224	16128 CPUh
2	UNIZAG-FSB	swenseFoam	k-ε/k-ω SST	LS	FV	Dynamic	Unstructured	0.61–1.6 M	40	817 CPUh
Ave.								15.15 M	132	8472.5

Table 26 Comparison errors and uncertainties for KCS calm water resistance test submissions at Fr = 0.26 (Case 2.11)

Variable	$C_{T0} \times 10^3$		σ/L		τ (deg)	
D	4.888		−0.001704		−0.022	
$U_D\%\|D\|$	2.5		0.6		65.5	
CFD submissions i	S_i	$E_i\%\|D\|$	S_i	$E_i\%\|D\|$	S_i	$E_i\%\|D\|$
1	4.512	7.69	−0.00187	−9.74	−0.1816	−725.45
2	4.528	7.36	−0.00176	−3.29	−0.1717	−680.45
$\overline{E}\%D$	7.53		−6.51		−702.95	
$\|\overline{E}\|\%D$	7.53		6.51		702.95	

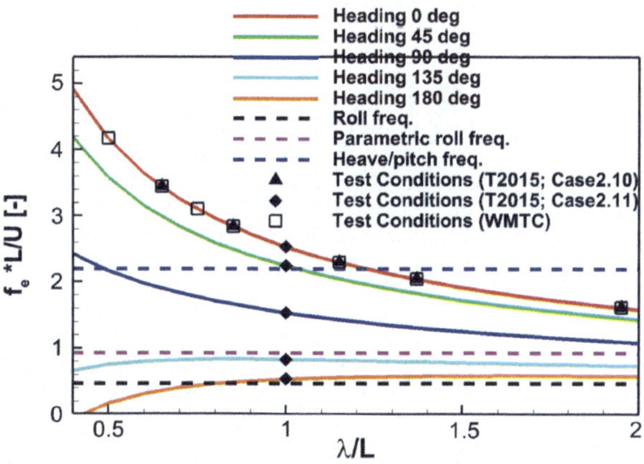

Fig. 14 Wave encounter frequencies compared to natural frequencies ($f_{z/\theta} = 2.20$; $f_\phi = 0.46$; $f_{PR} = 0.93$) for KCS test conditions in waves (WMTC: Sadat-Hosseini et al. 2015)

column shows D, Si and $\bar{S}$. The middle and right columns show U_D,$|E_i|$ and $|\overline{E}|$ using %D and %DR, respectively.

The 0th harmonic experiments show large C_{T0}, AR and θ_0/Ak for head waves (also following waves for θ_0/Ak). z_0/A is nearly constant except for following waves. ϕ_0/Ak is largest for $\mu = 90$ and 135°. Both CFD submissions mostly over predict C_{T0} and AR; show similar trends with each other and the data for z_0/A; show similar trends with each other but different from the data for θ_0/Ak; and mostly different with each other and the data for ϕ_0/Ak. The largest errors are for the motions.

C_{T1} and θ_1/Ak show similar oscillations vs. μ with peaks at 45 and 135 and trough at 90°; however, amplitudes are small/large for head/following waves, respectively. z_1/A is largest for bow and beam waves and smallest for stern waves. Since $B/\lambda = 0.14$ for beam waves, pitch is nearly zero and heave and wave amplitudes coincide, as the model simply moves up and down parallel with the wave. ϕ_1/Ak is largest for stern quartering waves. Both CFD solutions show overall good agreement with each other and the data for both amplitude and phase, except larger amplitude C_{T1} and some differences in phases for stern waves. The largest errors are for the C_{T1} amplitude for following waves.

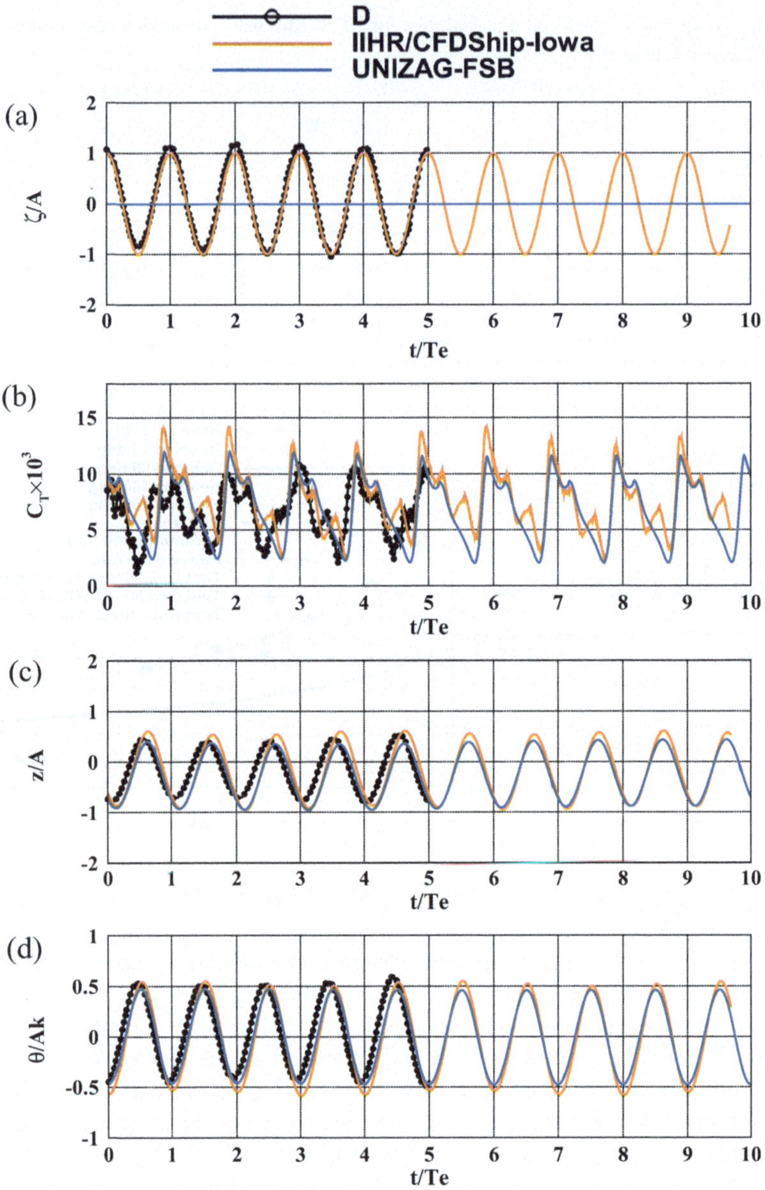

Fig. 15 Submissions with EFD for KCS course keeping in regular waves with $\mu = 0°$: **a** Wave; **b** Resistance coefficient; **c** Heave; **d** Pitch; **e** Roll; **f** x-velocity

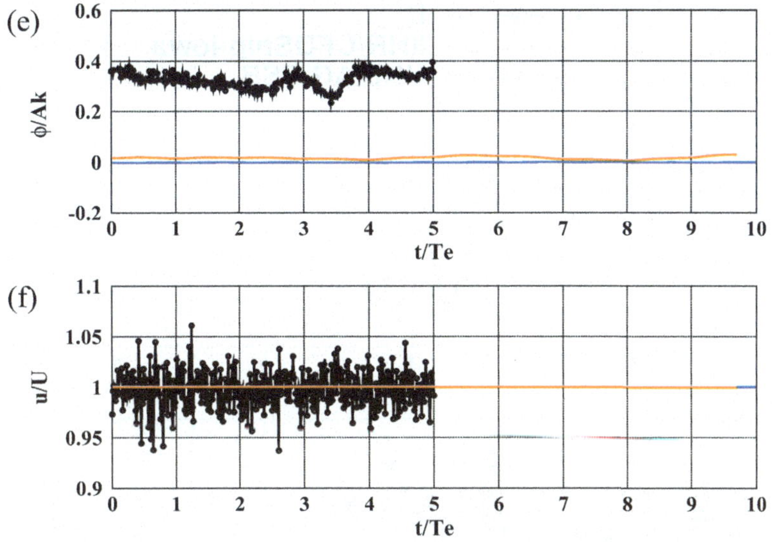

Fig. 15 (continued)

The 2nd harmonic experiments and CFD often show somewhat similar trends, but with larger CFD scatter than shown for the zeroth and first harmonic response.

The trends for $\overline{|E|}$ versus μ for %D vs. %DR show that.%DR provides a better assessment of the CFD capability since small D values bias the %D analysis. The primary variables (AR, z_1/A, θ_1/Ak and ϕ_1/Ak) $\overline{|E|}$ averaged over all headings are all less than about 12%DR, i.e., 7, 8, 6 and 12%DR, respectively. The overall average 1st/2nd order $\overline{|E|}$ and U_D are 18/40 and 7/43%DR for amplitudes and the overall 1st/2nd order $\overline{|E|}$ are 14/24%DR for phases. The largest errors are bow quartering for AR, beam waves for z_1/A, following waves for θ_1/Ak and bow and stern quartering waves for ϕ_1/Ak.

6.1 Conclusions

Improved benchmark experimental data from multiple facilities and more simulations from multiple institutes will provide estimates that are more robust for intervals of validation and SoA assessment. CFD can predict the effects of variable heading albeit with larger comparison errors than those for head waves. The agreement between the two codes solutions and the experimental data is remarkable.

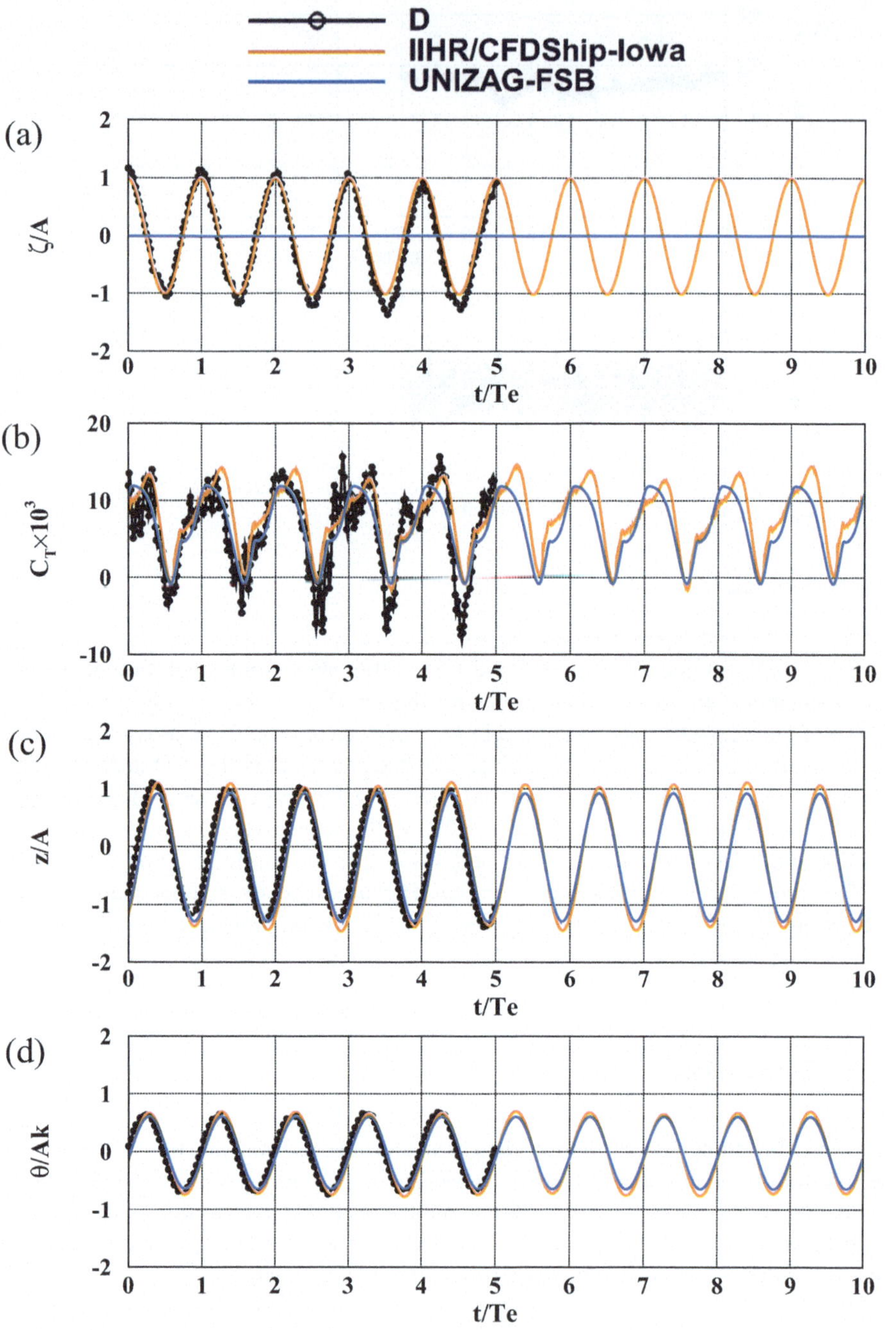

Fig. 16 Submissions with EFD for KCS course keeping in regular waves with $\mu = 45°$: **a** Wave; **b** Resistance coefficient; **c** Heave; **d** Pitch; **e** Roll; **f** x-velocity

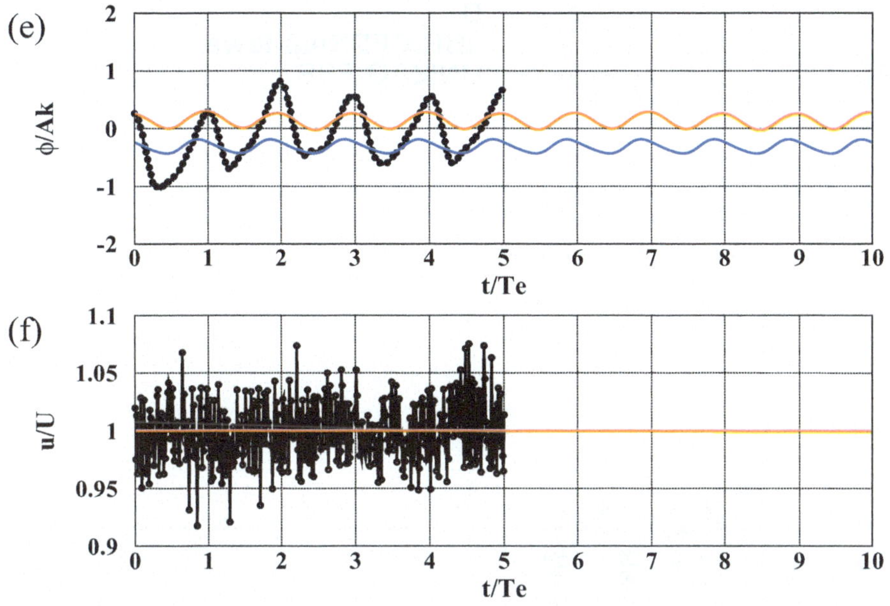

Fig. 16 (continued)

7 CFD Assessment Test Cases 3.9, 3.12, and 3.13

As shown in Table 30, 6 institutes from 4 countries using 5 different CFD codes submitted results for both calm water and waves. There were 8 submissions for calm water self-propulsion and (6,3,1,3,1) for μ = (0, 45, 90, 135, 180), respectively. Two equations, RSM and DDES turbulence models were used with 6 using wall functions and 2 using near wall models. Mostly discretized but also body force propeller modeling was used. It is not known which ones have used prescribed and which ones have applied interactive body forces. The solvers were mostly FV except one is FD. All but 1 used unstructured grids. 1 used single phase level set and 6 used VoF free-surface models. Grid motion used dynamic patched, dynamic overset and deforming grids. Most codes used reduced DOF and fixed rudder for test case 3.9 and in some cases for test cases 3.12 and 3.13. Grid sizes varied from 1–35 M with average 9.12 M. Average number CPU and CPU time is 87 processors and 100,000 h, respectively. Only IIHR also submitted to test cases 2.10 and 2.11.

7.1 Test Case 3.9

Table 31 provides the calm water self-propulsion V&V and SoA variables. Figure 25 graphically displays the sinkage, trim and propeller revolutions RPS validation. The

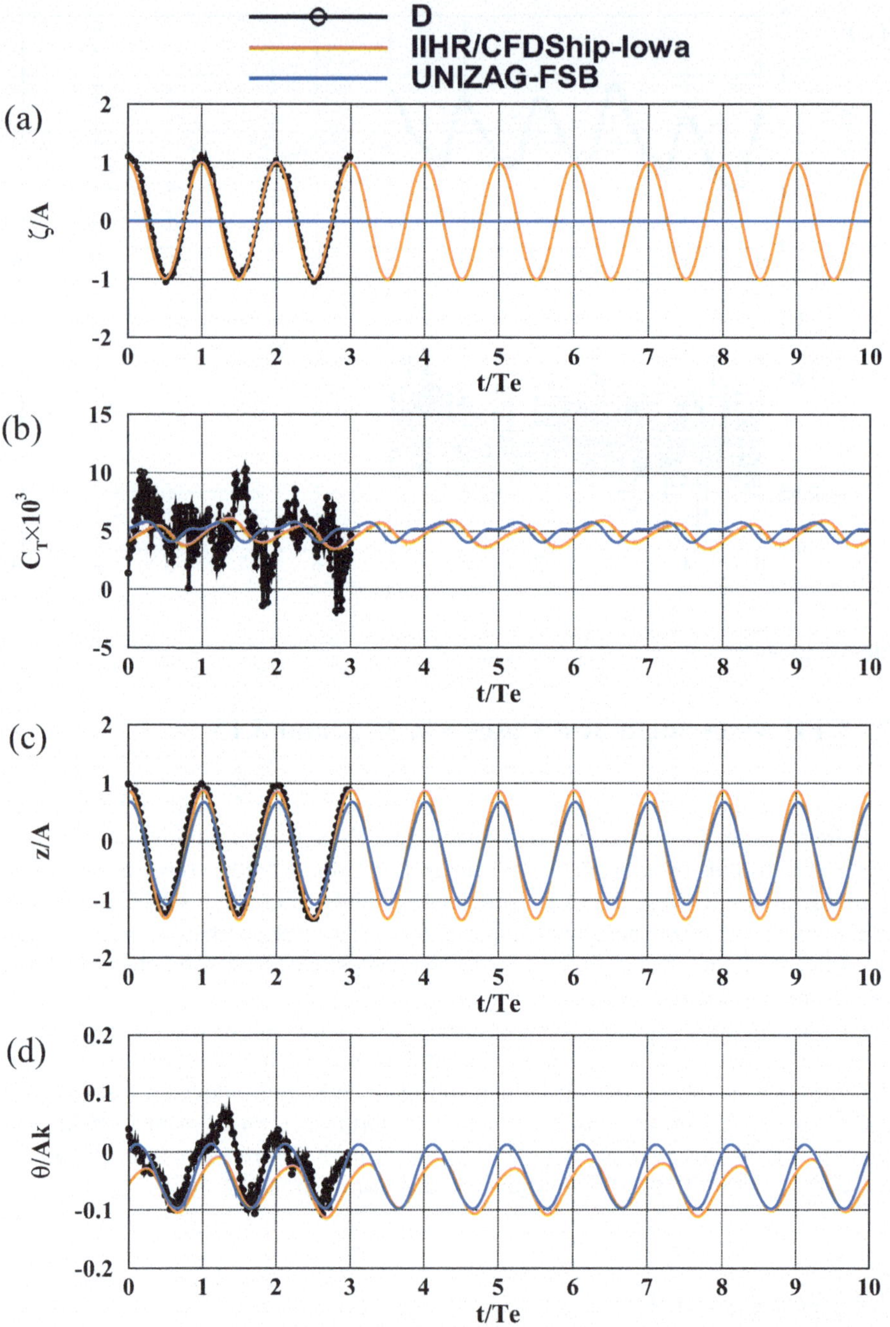

Fig. 17 Submissions with EFD for KCS course keeping in regular waves with $\mu = 90°$: **a** Wave; **b** Resistance coefficient; **c** Heave; **d** Pitch; **e** Roll; **f** x-velocity

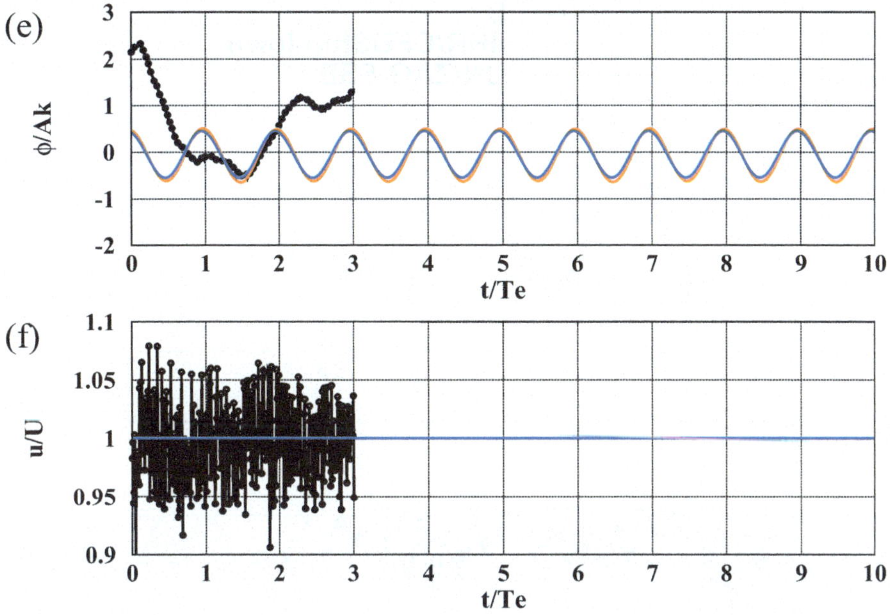

Fig. 17 (continued)

average errors are $\overline{|E|}$ = 1.5%D for RPS and 6 and 15%D for sinkage and trim, respectively. Most solutions under predict RPS with one outlier, whereas most solutions overpredict sinkage and trim. U_D is small for RPS such that only one solution is validated, whereas U_D is large for sinkage and trim such that all solutions are validated. The scatter in the solutions for RPS is small < 1%D, whereas it is large for sinkage and trim, i.e., 6 and 16%D, respectively. Since U_{SoA_i} is small for RPS and large for sinkage and trim 4 and 8 codes/solutions are N-version validated for RPS and sinkage/trim, respectively, and the mean code is not N-version validated for RPS but it is for sinkage and trim at U_{SoA} = <1, 20 and 134%D, respectively.

For calm water RPS, the mean absolute error 2%D and individual code uncertainty $U_{SoA_i} = 2\%D$ are acceptably small, which indicates free running self-propulsion can be predicted with simialr error and uncertainty as captive towed resistance. The sinkage and trim errors and uncertainties are mostly larger than those for captive towed resistance.

7.2 Test Cases 3.12 and 3.13

Test cases 3.12 and 3.13 are for λ/L = 1 and μ = 0 and 45, 90, 135 and 180°., respectively. Figure 26 shows head and bow quartering conditions are far from heave/pitch natural frequency. Beam condition is close to roll natural frequency. Note that B/λ

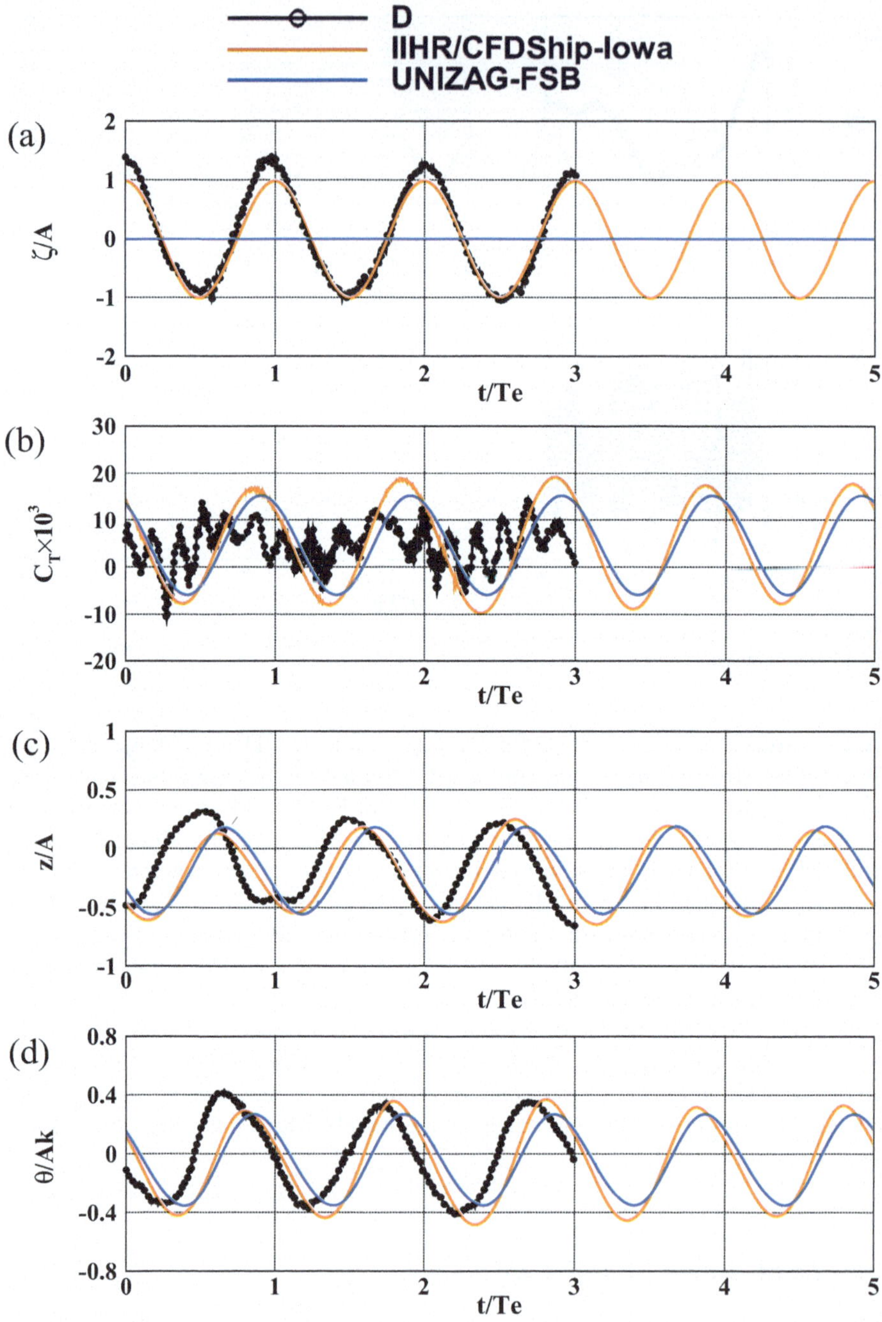

Fig. 18 Submissions with EFD for KCS course keeping in regular waves with $\mu = 135°$: **a** Wave; **b** Resistance coefficient; **c** Heave; **d** Pitch; **e** Roll; **f** x-velocity

(e)

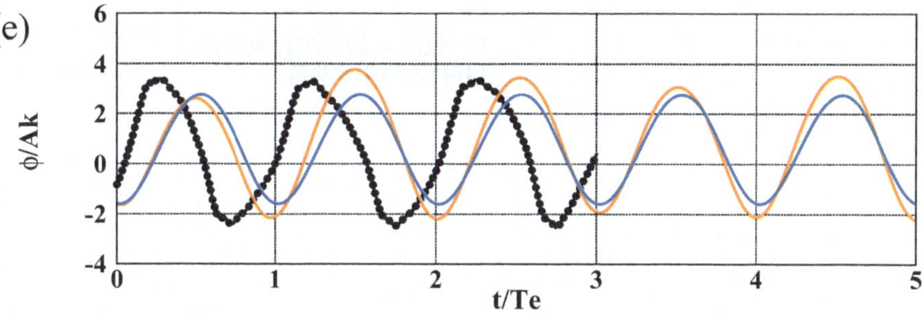

(f)

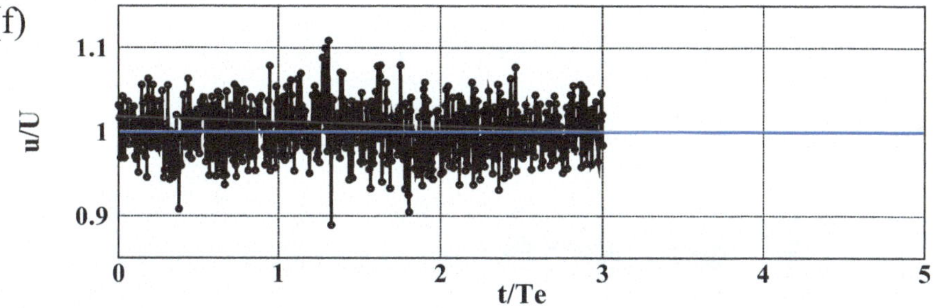

Fig. 18 (continued)

= 0.12. Stern quartering and following wave conditions are far from roll natural frequency. Parametric roll frequency is close to heave/pitch natural frequency and far from all heading conditions.

The experimental and CFD time histories are shown in Fig. 27, 28, 29, 30 and 31 for $\mu = 0$, 45, 90, 135 and 180°., respectively. The experiments and CFD cover 7, 6, 4, 6, 7 and 7, 7, 7, 6, 9 T_e, respectively, for $\mu = 0$, 45, 90, 135 and 180°. Different trends are observed for head/following, quartering and beam waves. For $\mu = 0$ and 180°. the errors are largest for ϕ, ψ, v and δ; and the experiments require much larger rudder control than the CFD. For $\mu = 45$ and 135, the errors are largest for ϕ, ψ, u, v and δ; and the experiments and CFD have similar heading control. For $\mu = 90$, the errors are largest for ϕ, ψ, v and δ; the experiments and CFD seem to approach the same steady state heading control, but with different initial transients. Thus, the largest differences are for horizontal motions and roll angle.

Tables 32 and 33 provide average results for each heading and over all headings of the CFD solutions for the 0th/AR and 1st and 2nd harmonic amplitude validation and SoA variables using %D and %DR, respectively. Table 24 is similar but provides the 1st and 2nd harmonic phase validation and SoA variables using $\%2\pi$. The participants submitted their own trajectories and time histories, which were harmonically analyzed using the last two cycles for all headings since the mean heading and encounter frequency were nearly constant. The discussions mostly use values for μ averages and %DR.

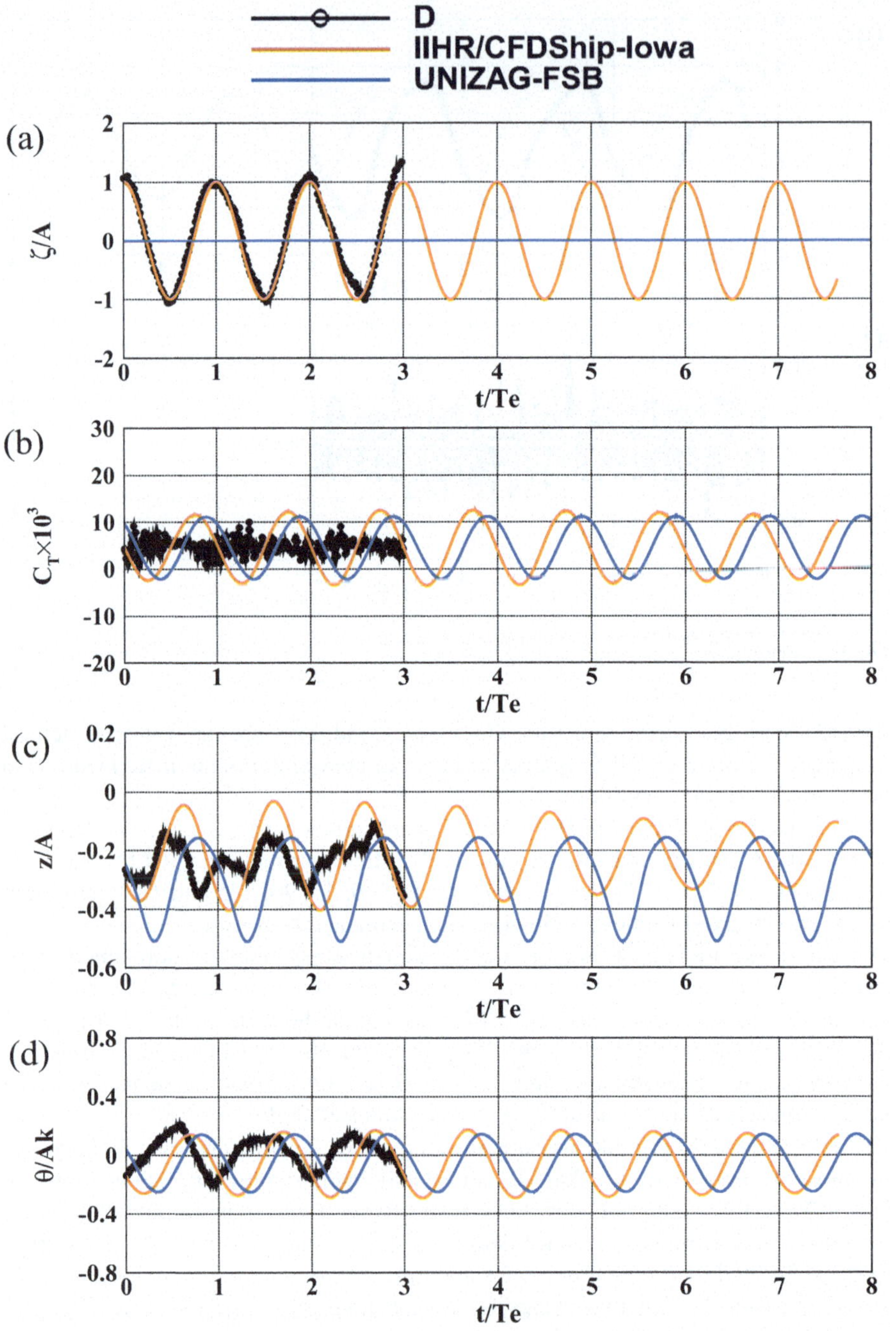

Fig. 19 Submissions with EFD for KCS course keeping in regular waves with $\mu = 180°$: **a** Wave; **b** Resistance coefficient; **c** Heave; **d** Pitch; **e** Roll; **f** x-velocity

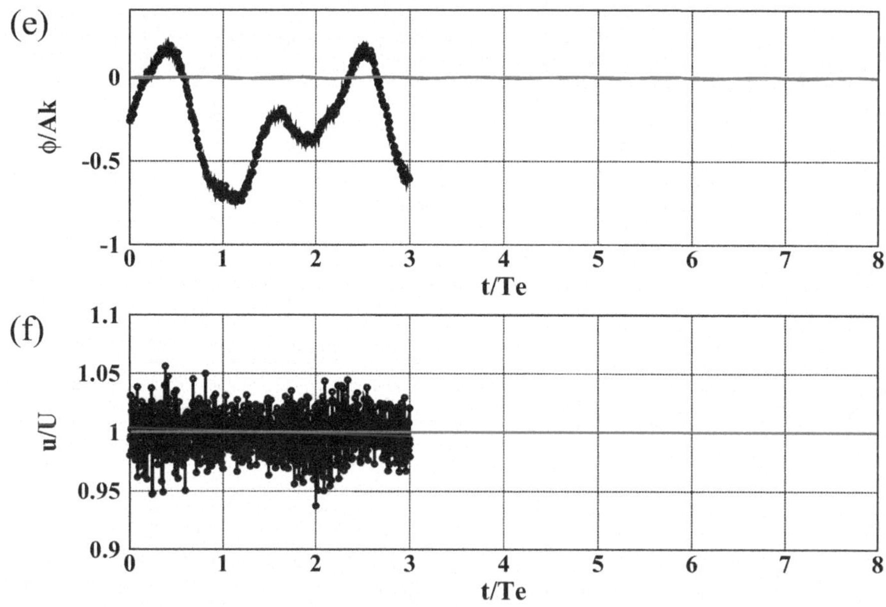

Fig. 19 (continued)

Figures 32, 33, 34, 35 and 36 show the results for the 0th harmonic, 1st and 2nd harmonic amplitudes and 1st and 2nd harmonic phases, respectively, versus μ. The left column shows D, Si and $\bar{S}$. The middle and right columns show U_D,$|E_i|$ and $\overline{|E|}$ using %D and %DR, respectively.

The 0th harmonic experiments show z_0/A peaks at $\mu = 0$ and 180, trough at 90, and near zero for 45 and 135°. θ_0/Ak is nearly constant. ϕ_0/Ak, ψ_0, and v_0 have peaks at $\mu = 0$, 90 and 180 and troughs at 45 and 135°. u_0 shows large speed loss for bow waves, which is attributed to large bow wave pitch motions. CFD shows large scatter for $\mu = 0$ and differences for other μ but fewer codes. The mean and code with results for all μ shows similar trends as the experiments, except for z_0/A. The errors for speed loss are relatively small, whereas the other variables show larger error with peaks for different headings.

z_1/A is maximum for beam waves and larger for bow vs. stern waves. Since B/λ = 0.12 for beam waves pitch is nearly zero and heave and wave amplitudes coincide, as the model simply moves up and down parallel with the wave. θ_1/Ak is maximum for bow, nearly zero for 90, and about half bow wave values for stern waves. ϕ_1/Ak increases from head to stern quartering and reduces for following waves. ψ_1/Ak is nearly zero for $\mu = 0$, 90 and 180 and large for $\mu = 45$ and especially 135°. u_1/U is small for bow and beam and larger for bow quartering and very largest for stern waves. v_1/U is small for bow and following and large for beam and bow and stern quartering waves. All variables show large response for stern quartering waves. The CFD trends for 1st harmonic amplitude and phase agree with the experiments but

Table 27 (%D). Evaluations of individual and mean code values on harmonic amplitudes of resistance and motions for KCS in regular wave (Case 2.11)

μ[deg]	Variable	0th harmonic amplitude					1st harmonic amplitude					2nd harmonic amplitude				Ave	
		$C_{T0} \times 10^3$	AR	z_0/L	θ_0/Ak	φ_0/Ak	ζ_1/L	$C_{T1} \times 10^3$	z_1/L	θ_1/Ak	φ_1/Ak	$C_{T2} \times 10^3$	z_2/L	θ_2/Ak	φ_2/Ak	1st	2nd
0	U_D%|D|	1.63	6.02	15.38	43.54	–	1.37	2.47	3.52	6.14	–	16.26	70.70	213.17	–	3.37	52.39
	# submissions	2	2	2	2	–	1	2	2	2	–	2	2	2	–	1.40	1.56
	$\|\bar{E}\|$%|D|	3.14	9.44	25.10	133.87	–	0.36	19.36	15.44	7.43	–	266.14	35.92	18.11	–	10.65	70.25
	# validated	0	0	1	0	–	1	0	0	1	–	0	2	2	–	0.40	0.56
45	U_D%|D|	1.39	5.98	10.68	209.94	8.67	1.51	1.04	1.24	4.45	5.29	2.02	161.11	275.85	19.36	2.71	77.22
	# submissions	2	2	2	2	2	1	2	2	2	2	2	2	2	2	1.80	2.00
	$\|\bar{E}\|$%|D|	4.77	16.21	33.63	31.70	271.90	0.36	14.39	6.04	6.13	76.67	68.76	22.97	9.99	92.61	20.72	61.39
	# validated	0	0	0	2	0	1	0	0	1	0	0	2	2	0	0.40	0.67
90	U_D%|D|	2.31	26.29	9.80	122.44	8.40	1.44	5.73	1.81	60.00	8.29	7.17	40.93	146.33	45.11	15.46	45.42
	# submissions	2	2	2	2	2	1	2	2	2	2	2	2	2	2	1.80	2.00
	$\|\bar{E}\|$%|D|	2.80	27.81	47.81	114.68	109.48	0.36	69.85	12.79	14.23	40.14	32.73	51.25	79.63	93.30	27.47	62.16
	# validated	1	1	0	1	0	1	0	0	2	0	1	0	2	0	0.60	0.67
135	U_D%|D|	1.58	27.39	10.63	328.77	19.18	0.94	3.41	4.41	8.41	0.98	9.94	1374.87	113.73	32.89	3.63	213.22
	# submissions	2	2	2	2	2	1	2	2	2	2	2	2	2	2	1.80	2.00
	$\|\bar{E}\|$%|D|	0.63	5.51	39.78	394.51	143.91	0.36	214.08	6.08	11.30	11.33	62.16	1472.34	23.99	56.29	48.63	244.35
	# validated	2	2	0	1	0	1	0	1	1	0	0	1	2	1	0.60	1.00
180	U_D%|D|	1.59	43.44	6.75	1048.03	–	1.22	10.19	24.61	451.21	–	23.74	67.34	172.14	–	121.81	194.72

(continued)

Table 27 (continued)

μ[deg]	Variable	0th harmonic amplitude					1st harmonic amplitude					2nd harmonic amplitude				Ave	
		$C_{T0} \times 10^3$	AR	z_0/L	θ_0/Ak	φ_0/Ak	ζ_1/L	$C_{T1} \times 10^3$	z_1/L	θ_1/Ak	φ_1/Ak	$C_{T2} \times 10^3$	z_2/L	θ_2/Ak	φ_2/Ak	1st	2nd
	# submissions	2	2	2	2	–	1	2	2	2	–	2	2	2	–	1.40	1.56
	$\lvert\bar{E}\rvert\%\lvert D\rvert$	0.69	21.02	18.98	451.26	–	0.36	1014.07	137.43	48.40	–	39.12	57.25	66.89	–	300.06	93.60
	# validated	2	2	0	2	–	1	0	0	2	–	0	1	2	–	0.60	1.00
Ave	$U_D\%\lvert D\rvert$	1.70	21.82	10.65	350.55	12.08	1.30	4.57	7.12	106.04	4.85	11.82	342.99	184.24	32.45	29.39	116.59
	# submissions	2	2	2	2	1.2	1	2	2	2	1.2	2	2	2	1.2	1.64	1.82
	$\lvert\bar{E}\rvert\%\lvert D\rvert$	2.41	16.00	33.06	225.21	175.09	0.36	266.35	35.56	17.50	42.71	93.78	327.95	39.72	80.73	81.51	106.35
	# validated	1	1	0.2	1.2	0	1	0	0.2	1.4	0	0.2	1.2	2	0.2	0.52	0.78

Table 28 (%DR). Evaluations of individual and mean code values on harmonic amplitudes of resistance and motions for KCS in regular wave (Case 2.11)

μ[deg]	Variable	0th harmonic amplitude					1st harmonic amplitude					2nd harmonic amplitude				Ave	
		$C_{T0} \times 10^3$	AR	z_0/L	θ_0/Ak	φ_0/Ak	ζ_1/L	$C_{T1} \times 10^3$	z_1/L	θ_1/Ak	φ_1/Ak	$C_{T2} \times 10^3$	z_2/L	θ_2/Ak	φ_2/Ak	1st	2nd
0	U_D%|DR|	5.80	6.71	23.08	28.52	–	1.90	1.18	1.94	4.74	–	2.23	42.18	151.51	–	2.44	37.15
	# submissions	2	2	2	2	–	1	2	2	2	–	2	2	2	–	1.40	1.56
	|$\bar{E}$|%|DR|	11.17	10.52	37.66	87.70	–	0.50	9.23	8.53	5.73	–	36.45	21.43	12.87	–	6.00	31.11
	# validated	0	0	1	0	–	1	0	0	1	–	0	2	2	–	0.40	0.56
45	U_D%|DR|	4.72	5.75	12.93	42.12	1.59	2.10	1.17	1.30	4.78	1.22	2.21	38.07	146.33	35.48	2.11	32.13
	# submissions	2	2	2	2	2	1	2	2	2	2	2	2	2	2	1.80	2.00
	|$\bar{E}$|%|DR|	16.24	15.59	40.73	6.36	49.82	0.50	16.05	6.37	6.59	17.74	75.49	5.43	5.30	169.71	9.45	42.74
	# validated	0	0	0	2	0	1	0	0	1	0	0	2	2	0	0.40	0.67
90	U_D%|DR|	6.26	6.96	13.60	42.23	6.86	2.00	1.47	1.86	4.48	1.30	2.22	42.19	156.43	37.55	2.22	34.92
	# submissions	2	2	2	2	2	1	2	2	2	2	2	2	2	2	1.80	2.00
	|$\bar{E}$|%|DR|	7.57	7.36	66.37	39.56	89.42	0.50	17.92	13.15	1.06	6.30	10.14	52.84	85.12	77.68	7.79	48.45
	# validated	1	1	0	1	0	1	0	0	2	0	1	0	2	0	0.60	0.67
135	U_D%|DR|	4.13	4.94	13.32	50.68	13.65	1.30	2.32	1.62	4.81	1.13	2.42	42.44	174.06	38.21	2.24	38.21
	# submissions	2	2	2	2	2	1	2	2	2	2	2	2	2	2	1.80	2.00
	|$\bar{E}$|%|DR|	1.64	0.99	49.83	60.81	102.48	0.50	145.58	2.23	6.46	13.11	15.16	45.45	36.72	65.39	33.58	42.05
	# validated	2	2	0	1	0	1	0	1	1	0	0	1	2	1	0.60	1.00
180	U_D%|DR|	4.07	4.96	14.92	240.55	–	1.70	1.18	1.33	97.44	–	2.32	47.81	179.41	–	25.41	70.58

(continued)

Table 28 (continued)

μ[deg]	Variable	0th harmonic amplitude					1st harmonic amplitude					2nd harmonic amplitude				Ave	
		$C_{T0} \times 10^3$	AR	z_0/L	θ_0/Ak	φ_0/Ak	ζ_1/L	$C_{T1} \times 10^3$	z_1/L	θ_1/Ak	φ_1/Ak	$C_{T2} \times 10^3$	z_2/L	θ_2/Ak	φ_2/Ak	1st	2nd
	# submissions	2	2	2	2	–	1	2	2	2	–	2	2	2	–	1.40	1.56
	$\lvert\bar{E}\rvert\%\lvert DR\rvert$	1.76	2.40	41.97	103.58	–	0.50	117.21	7.42	10.45	–	3.83	40.64	69.72	–	33.89	37.70
	# validated	2	2	0	2	–	1	0	0	2	–	0	1	2	–	0.60	1.00
Ave	$U_D\%\lvert DR\rvert$	5.00	5.86	15.57	80.82	7.37	1.80	1.46	1.61	23.25	1.22	2.28	42.54	161.55	37.08	6.88	42.60
	# submissions	2	2	2	2	1.2	1	2	2	2	1.2	2	2	2	1.2	1.64	1.82
	$\lvert\bar{E}\rvert\%\lvert DR\rvert$	7.67	7.37	47.31	59.60	80.57	0.50	61.20	7.54	6.06	12.38	28.21	33.16	41.94	104.26	18.14	40.41
	# validated	1	1	0.2	1.2	0	1	0	0.2	1.4	0	0.2	1.2	2	0.2	0.52	0.78

*Dynamic range was obtained from Mark Stocker's experimental data for 1st Amp ζ_1/L^* ($\zeta_1/L_{DR} = 0.006$)

Table 29 (%2π). Evaluations of individual and mean code values on harmonic phase of resistance and motions for KCS in regular wave (Case 2.11)

μ[deg]	Variable	1st harmonic phase				2nd harmonic phase				Ave	
		$C_{T1} \times 10^3$	z_1/L	θ_1/Ak	φ_1/Ak	$C_{T2} \times 10^3$	z_2/L	θ_2/Ak	φ_2/Ak	1st	2nd
0	# submissions	2	2	2	–	2	2	2	–	1.5	1.5
	$\lvert\bar{E}\rvert$%2п	7.65	8.94	6.75	–	21.27	17.82	17.29	–	7.78	18.79
45	# submissions	2	2	2	2	2	2	2	2	2	2
	$\lvert\bar{E}\rvert$%2п	4.54	4.54	4.43	4.08	30.40	25.82	9.43	47.88	4.40	28.38
90	# submissions	2	2	2	2	2	2	2	2	2	2
	$\lvert\bar{E}\rvert$%2п	16.16	1.47	2.69	22.23	19.98	6.73	12.45	57.48	10.64	24.16
135	# submissions	2	2	2	2	2	2	2	2	2	2
	$\lvert\bar{E}\rvert$%2п	16.52	15.28	14.48	25.25	30.79	9.56	32.02	30.60	17.89	25.74
180	# submissions	2	2	2	–	2	2	2	–	1.5	1.5
	$\lvert\bar{E}\rvert$%2п	38.24	17.82	26.20	–	18.34	22.16	21.53	–	27.42	20.68
Ave	# submissions	2.00	2.00	2.00	1.20	2.00	2.00	2.00	1.20	1.80	1.80
	$\lvert\bar{E}\rvert$%2п	16.62	9.61	10.91	17.19	24.15	16.42	18.55	45.32	13.62	23.55

with large differences between the solutions. The errors for z_1/A and θ_1/Ak are small compared to those for ϕ_1/Ak, ψ_1/Ak, u_1/U and v_1/U. Roll and heading have largest errors for stern quartering waves, u_1/U has largest errors for bow and stern waves, and v_1/U has largest errors for beam waves.

The 2nd harmonic experiments and CFD often show somewhat similar trends, but with larger CFD scatter than shown for the zeroth and first harmonic response.

The trends for $\overline{|E|}$ versus μ for %D versus %DR show that.%DR provides a better assessment of the CFD capability since small D values bias the %D analysis. For head waves primary variables (u_0, z_1/A, and θ_1/Ak), the code level scatter $P_{|E_i|}$ is about 12%DR and much larger than $U_D \approx 5\%DR$ for u_0 and 1%DR for motions. $\overline{|E|} = 11,\ 5$ and 8%DR, respectively, for amplitudes and <3%DR for phases such that only 1, 2 or 0 solutions are validated for amplitudes and 6 and 0 for heave and pitch phases. U_{SoA_i} is about 12%DR such that about 5 codes/solutions are N-version validated. U_{SoA} is about 5%DR and the mean code is only N-version validated for heave.

For oblique waves primary variables (u_0, z_1/A, θ_1/Ak, ϕ_1/Ak and ψ_1/Ak), the U_D values are small <2%DR except u_0 for which $U_D = 10\%DR$. $\overline{|E|} = 9$, 2, 3, 20, and 6%DR, respectively, for amplitudes and 12%DR for phases such that only about 1or 0 solutions are validated. The largest errors are bow and stern quartering waves for

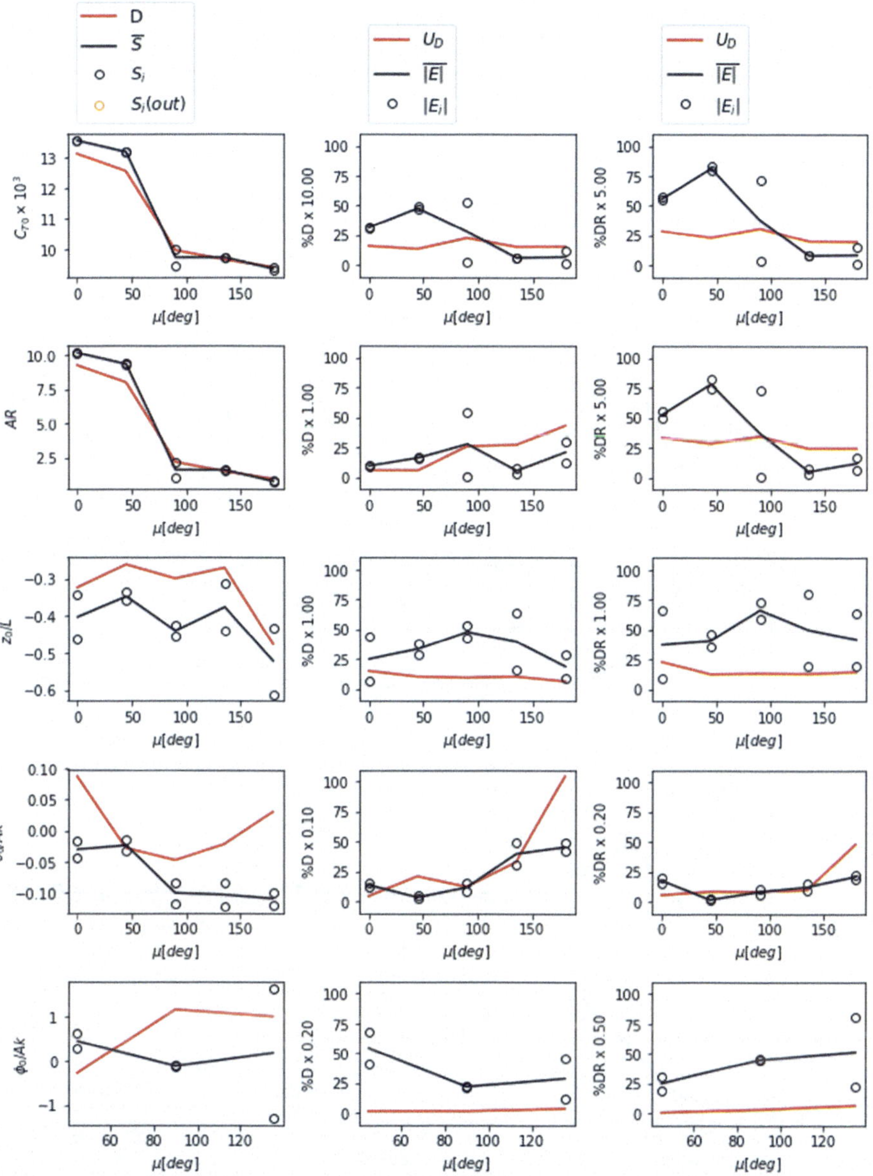

Fig. 20 Validations for 0th harmonic amplitudes of resistance and motions for KCS seakeeping in regular waves at various heading with Fn = 0.26

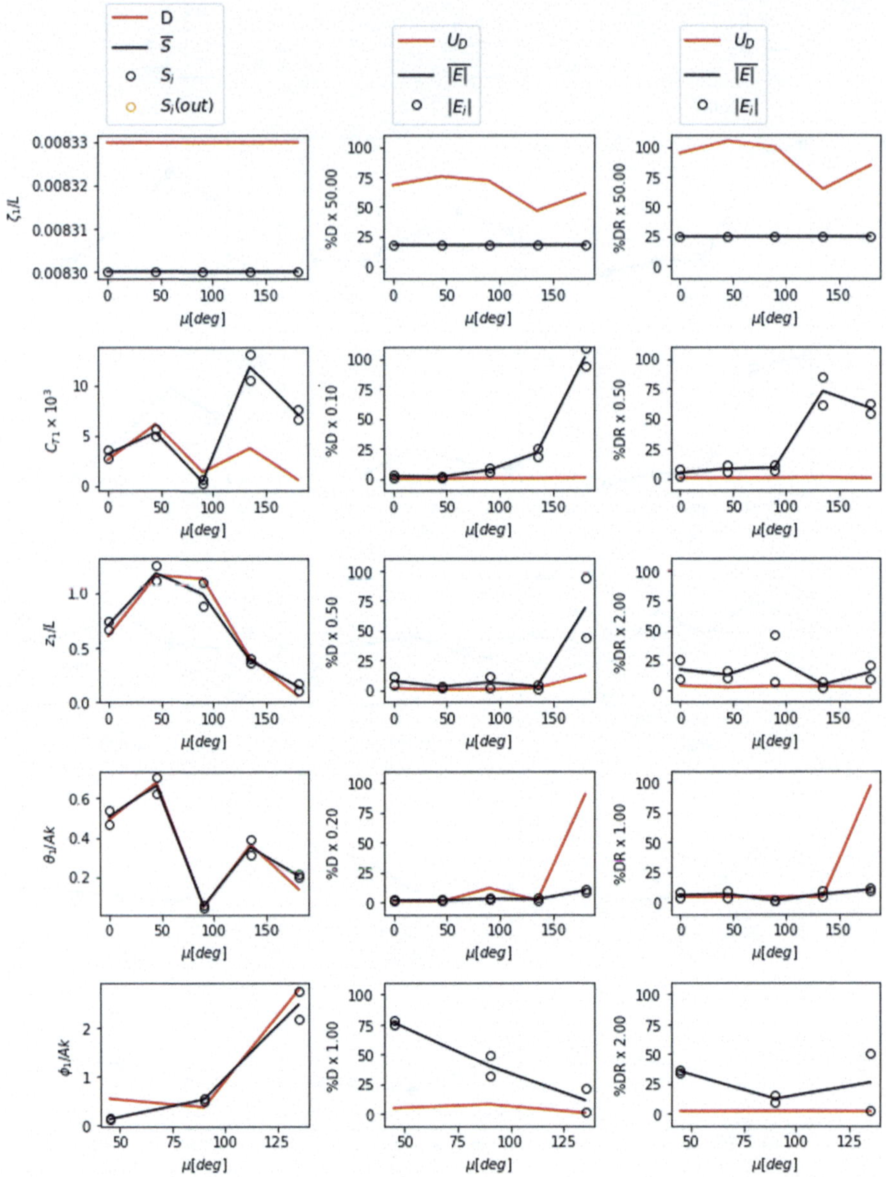

Fig. 21 Validations for 1st harmonic amplitudes of resistance and motions for KCS seakeeping in regular waves at various heading with Fr = 0.26

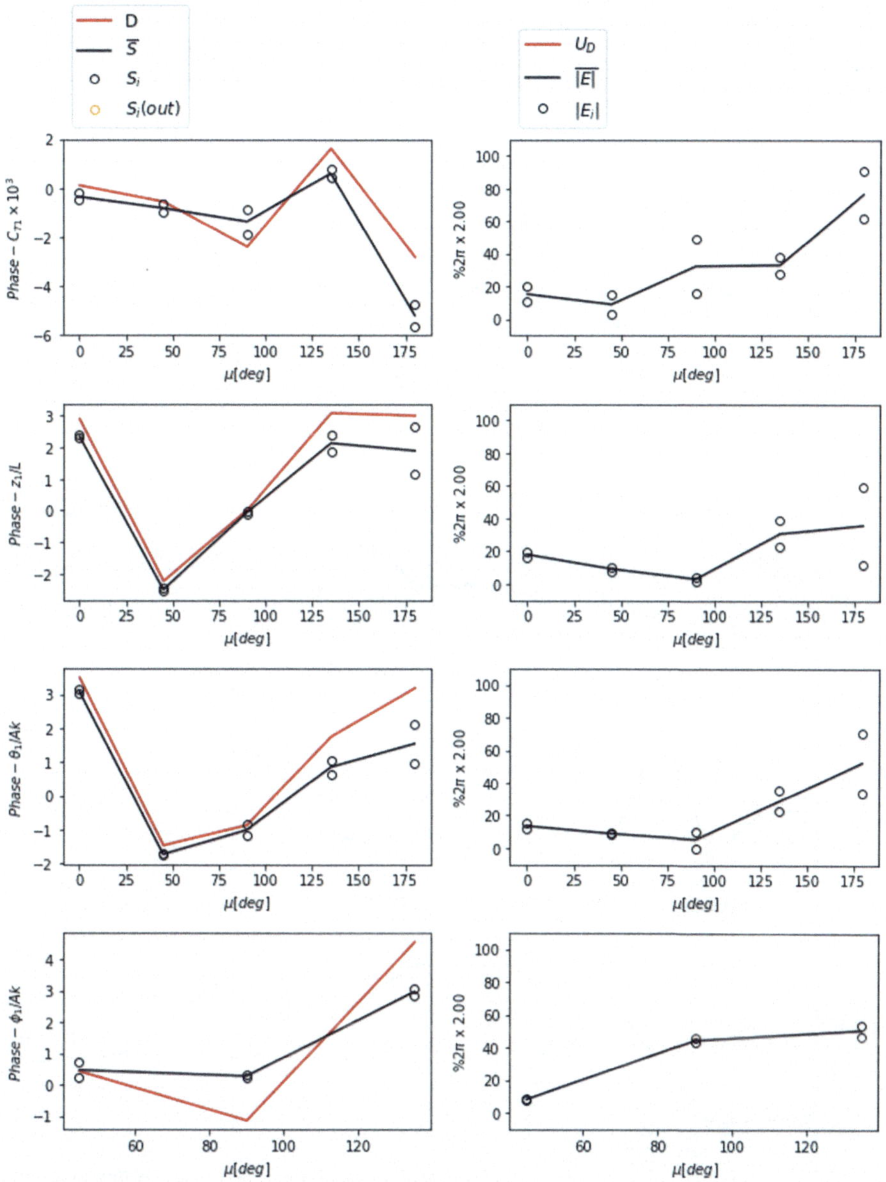

Fig. 22 Validations for 1st harmonic phases of resistance and motions for KCS seakeeping in regular waves at various heading with Fr = 0.26

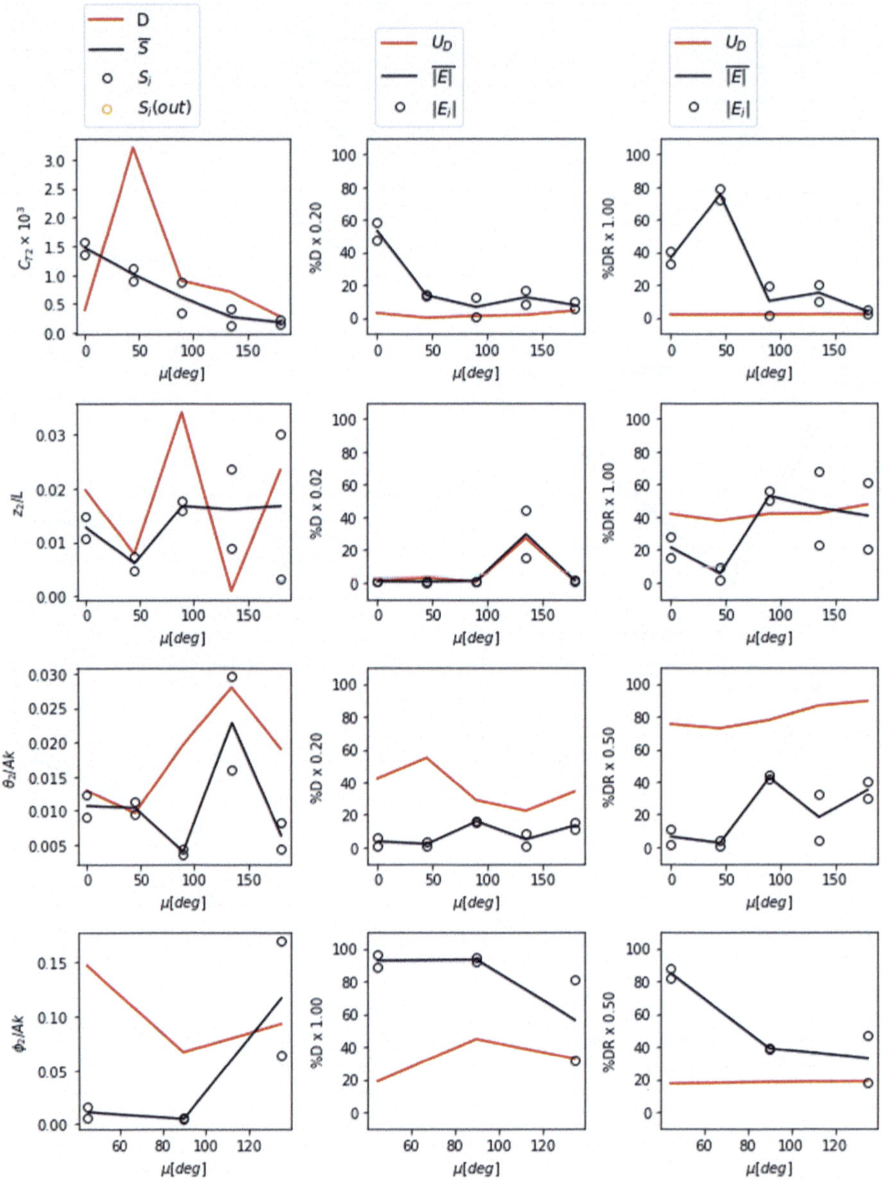

Fig. 23 Validations for 2nd harmonic amplitudes of resistance and motions for KCS seakeeping in regular waves at various heading with Fr = 0.26

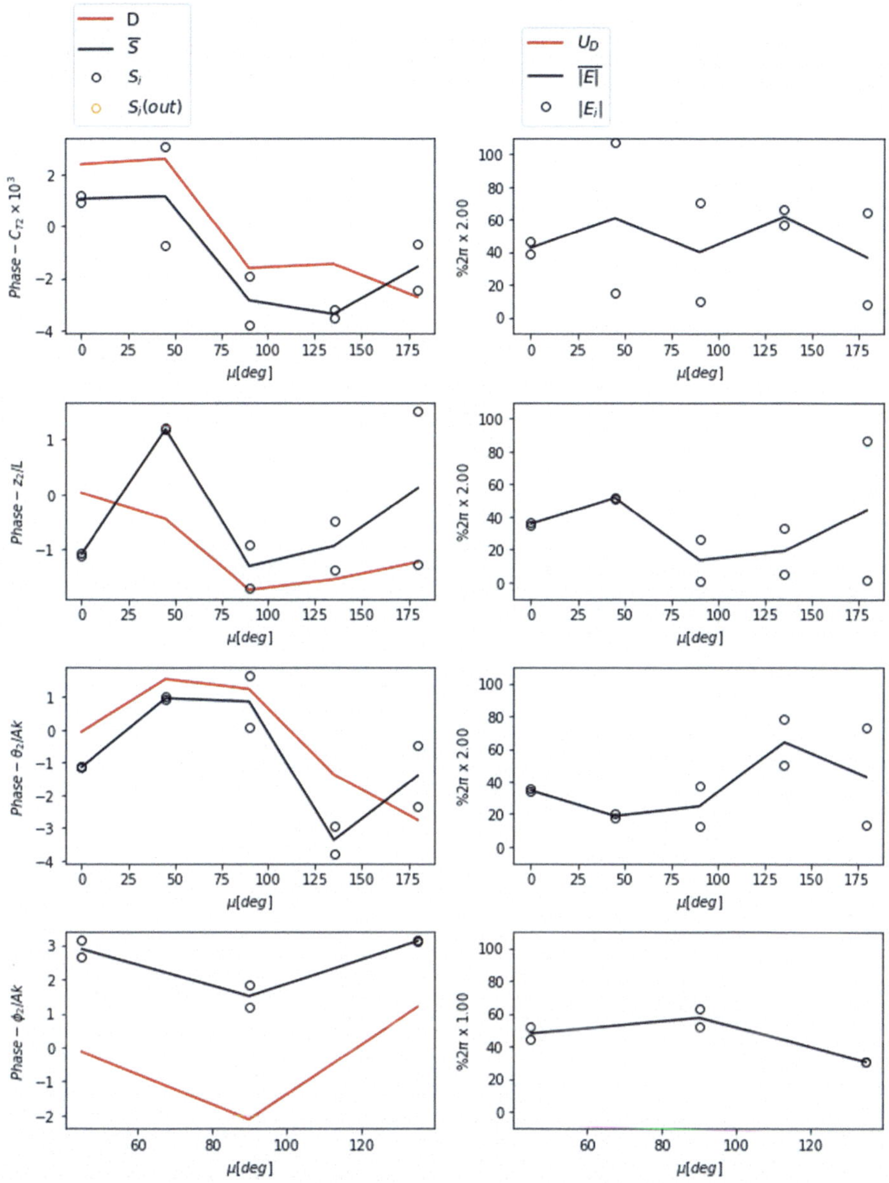

Fig. 24 Validations for 2nd harmonic phases of resistance and motions for KCS seakeeping in regular waves at various heading with Fr = 0.26

Table 30 Submissions for ONRT course keeping in head and variable heading waves at Fr = 0.2

No.	Institute	Code	Cases participated			Turbulence	Free surface	Propeller	Motions	Rudders	Discretization Method	Grid motion	Grid type	Grid size	No.CPU	CPU Time (CPUh)
1	ABS	OpenFOAM	3.9	3.12	3.13 (135)	Two-eq.	VoF	Discretized	6DOF	Active	FV	Overset	Unstructured	6.3–6.5 M	16	38301
																36002
2	IIHR	CFDShip-Iowa V4.5	3.9	3.12	3.13 (45, 90, 135, 180)	Two-eq.	LS	Body Force	4DOF (x, z, ϕ, θ)[1]	Fixed[1]	FD	Overset	Structured	23.2 M	224	215001
							Single Ph.		6DOF[2]	Active[2]						325002
3	CSSRC	FLUENT12	3.9	3.12		Two-eq.	VoF	Body Force and	2DOF (z, θ)	Fixed	FV	Dynamic patched	Unstructured	1.15–2.7 M	16	NA
								Discretized[1]								
								Body Force[2]								
4	ECN CNRS	ISISCFD-AD	3.9	3.12	3.13 (45)	Two-eq.	VoF	Body Force	3DOF (x, z, θ)[1]	Fixed[1]	FV	Deforming Grid/Overset	Unstructured	6.5 M	24–64	3201
									6DOF[2]	Active[2]						145002
5	IIHR	REX	3.9	3.12	3.13 (45, 135)	Two-eq. DDES	LS	Discretized	4DOF (x, z, ϕ, θ)[1]	Fixed[1]	FD	Overset	Structured	35 M	270	70000
							Single Ph.		6DOF[2]	Active[2]						
6	HHI	STAR-CCM+	3.9	3.12		RSM	VoF	Discretized	6DOF	Fixed	FV	Dynamic patched	Unstructured	10.7 M	32	NA
7	ECN CNRS	ISISCFD-FullRANSE	3.9			Two-eq.	VoF	Discretized	3DOF	Fixed	FV	Deforming Grid/Overset	Unstructured	9.3 M	72	20000
									(x, z, θ)					(Half Domain)		

(continued)

Table 30 (continued)

No.	Institute	Code	Cases participated			Turbulence	Free surface	Propeller	Motions	Rudders	Discretization Method	Grid motion	Grid type	Grid size	No.CPU	CPU Time (CPUh)
8	SJTU	OpenFOAM	3.9			Two-eq.	VoF	Discretized	6DOF	Active	FV	Overset	Unstructured	6.46 M	20	4100

Number of submissions for ONRT course keeping of Tokyo 2015 Workshop

μ[deg]	Heave	Pitch	Roll	Yaw	u2/U	v2/U
0	6	6	5	5	6	5
45	3	3	3	3	3	3
90	1	1	1	1	1	1
135	3	3	3	3	3	3
180	1	1	1	1	1	1

[1] Applied for Case 3.9
[2] Applied for Case 3.12 & 3.13
Case 3.9: Calm water condition
Case 3.12: Head wave condition (0°)
Case 3.13: Oblique wave condition (45–180°)

Table 31 Comparison errors and uncertainties for ONRT self-propulsion in calm water submissions at Fr = 0.2

Variable	σ/L		τ		N		Motion abs. average
D	0.072		−0.0386		8.970		
$U_D\%\|D\|$	20		134.2		0.083		77.1
# submissions	8		8		8		8
# outliers	0		0		1		0
CFD submissions i	S_i	$E_i\%\ \|D\|$	S_i	$E_i\%\ \|D\|$	S_i	$E_i\%\ \|D\|$	$\|E_i\|\%\|D\|$
1	0.074	−3.10	−0.0411	−6.48	8.800	1.90	4.79
2	0.070	2.21	−0.0387	−0.26	9.449	-5.34	1.24
3	0.084	−16.37	−0.0482	−24.87	8.970	0.00	20.62
4	0.078	−8.41	−0.0343	11.14	8.834	1.52	9.77
5	0.073	−1.77	−0.0521	−34.97	8.850	1.34	18.37
6	0.070	2.65	−0.0361	6.48	8.850	1.34	4.57
7	0.074	−3.36	−0.0437	−13.21	8.734	2.63	8.29
8	0.077	−6.64	−0.0464	−20.21	8.819	1.68	13.42
$\overline{S}$	0.075		−0.0426		8.837		
$\overline{E}\%\|D\|$	−4.35		−10.30		1.49		7.33
$\sigma_E\%\|D\|$	6.18		15.97		0.79		11.08
$\overline{\|E\|}\%\|D\|$	5.56		14.70		1.49		10.13
$\sigma_{\|E\|}\%\|D\|$	4.94		11.36		0.79		8.15
$P_{\|E_i\|}\%\|D\|$	9.89		22.72		1.59		16.31
$P_{\overline{\|E\|}}\%\|D\|$	3.50		8.03		0.56		5.77
$U_{SoA_i}\%\|D\|$	22.31		136.11		1.59		79.21
$U_{SoA}\%\|D\|nn$	20.30		134.44		0.57		77.37
# validated	8		8		1		8
# N-version validated	8		8		4		8
Mean code N-version validated	Yes		Yes		No		Yes

Note The gray-shaded numbers are outliers
**Propeller RPS is adjusted to the EFD value

speed loss, head and stern quartering waves for heave and pitch and stern quartering waves for roll and heading.

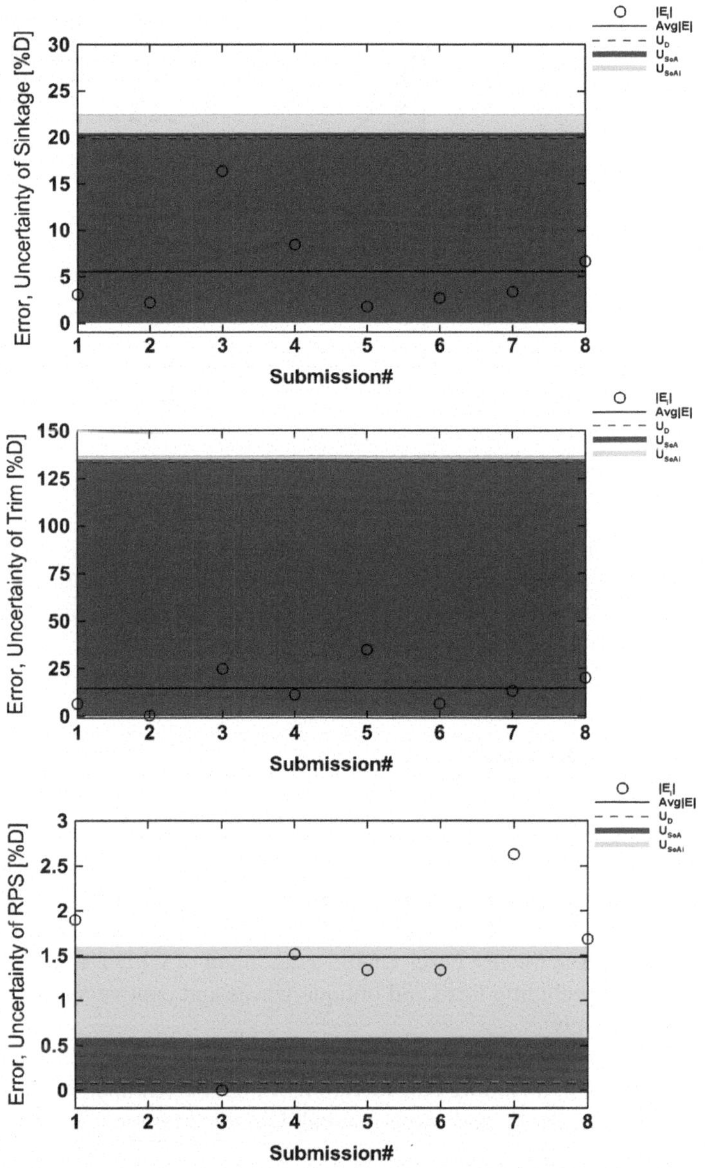

Fig. 25 EFD data and submission values for motions and propeller revolution for ONRT self-propulsion in calm water at Fr = 0.2 (**Propeller RPS is adjusted to the EFD value)

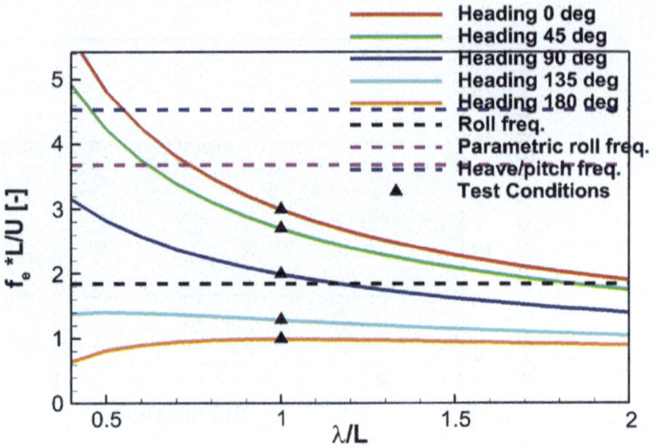

Fig. 26 Wave encounter frequencies compared to natural frequencies ($f_{z/\theta} = 4.53$; $f_{\varphi} = 1.84$; $f_{PR} = 3.68$) for ONR Tumblehome test conditions in waves

7.3 Conclusions

ONRT tests cases are the most challenging of all T2015 test cases. Nonetheless, 6 institutes using 5 codes participated. The errors and scatter were larger than test cases 2.10 and 2.11, however, clearly many institutes and codes displayed promising capability for free running course keeping simulations.

8 Overall Conclusions

T2015 test cases 2.10 and 2.11 for KCS and 3.9/3.12/3.13 for ONRT provide assessment of CFD capability for captive and free running model conditions in head and oblique waves. The number of submissions were relatively large for test cases 2.10 and 3.9/3.12/3.13, i.e., 10 and 8, respectively, but small for test case 2.11, i.e., 2. Nonetheless as a whole the test cases enable assessment of CFD for AR and course keeping/speed loss including head and oblique waves and captive and free running conditions, respectively.

Assessment approach includes not only evaluation of errors E using experimental data D, but also including consideration of both deterministic and stochastic numerical (U_{SN} and $P_{|E_i|}$) and experimental U_D uncertainties, which provides confidence levels/intervals for validation $U_V = \sqrt{U_{SN}^2 + U_D^2}$ and SoA (U_{SoA_i}) assessment. However, more attention is needed by individual facilities/codes on experimental/numerical uncertainty analysis to provide more robust confidence levels/intervals; and more attention is needed to quantify the CFD input wave condition and uncertainty and specify the CFD conditions and evaluation locations. KCS calm water and head waves experimental data was available from several facilities such that resistance, sinkage and trim data provided relatively robust validation/SoA assessment. Nonetheless, more attention is needed for individual facility uncertainty

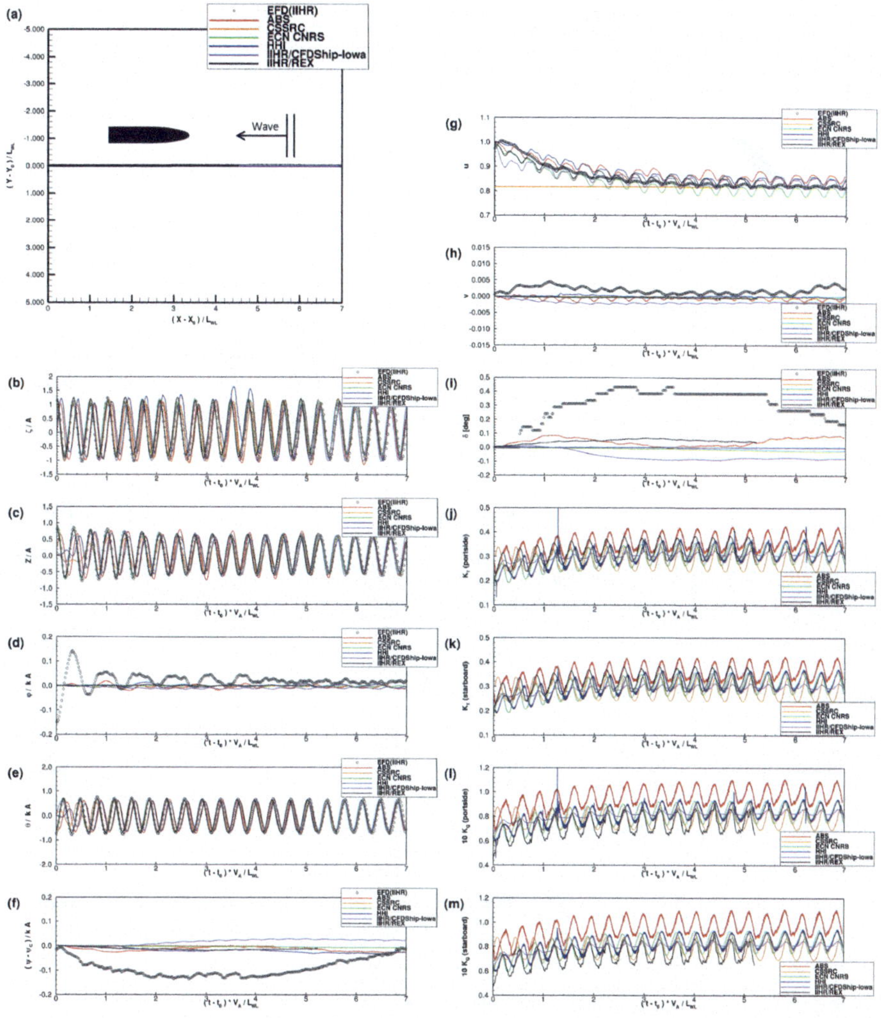

Fig. 27 Submissions with EFD for ONRT course keeping in waves with $\mu = 0°$: **a** Trajectory; **b** Wave; **c** Heave; **d** Roll; **e** Pitch; **f** Yaw; **g** x-vel.; **h** y-vel.; **i** Rudder angle; **j** Thrust coeff. (portside); **k** Thrust coeff. (starboard); **l** Torque coeff. (portside); **m** Torque coeff. (starboard) * **j** ~ **m** don't include EFD

analysis. KCS oblique waves and ONRT head and oblique wave data and uncertainties need more data from more facilities including uncertainties for more robust validation/SoA assessment.

Tables 35, 36 and 37 provide an overall summary of the CFD SoA assessment for calm water 2.10 and 3.9, head waves 2.10 and 3.12 and oblique waves 2.11 and 3.13 test cases, respectively. (Table 34)

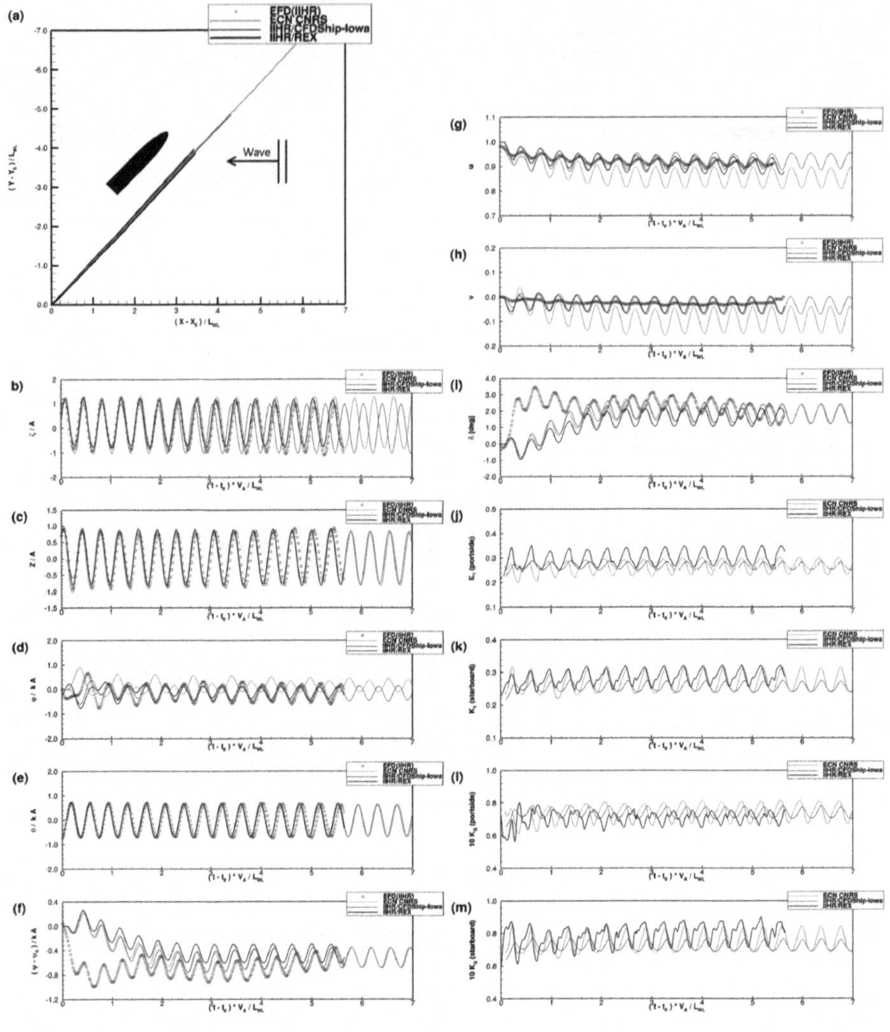

Fig. 28 Submissions with EFD for ONRT course keeping in waves with $\mu = 45°$: **a** Trajectory; **b** Wave; **c** Heave; **d** Roll; **e** Pitch; **f** Yaw; **g** x-vel.; **h** y-vel.; **i** Rudder angle; **j** Thrust coeff. (portside); **k** Thrust coeff. (starboard); **l** Torque coeff. (portside); **m** Torque coeff. (starboard) * **j** ~ **m** don't include EFD

Note that the Table 35 U_D%D values include the scatter in the experimental data due to scale effects/facility biases, i.e., $P_{X_i} = 2\sigma_{X_i}$, which is comparable to U_{Xi} for calm water resistance and used for U_D for sinkage and trim since U_{Xi} is not available, as per Table 3. For Table 36, U_D only includes U_{Xi} and U_{SCi} as per Tables 4, 5, 6, 7 and 8, but not P_{X_i} since σ_{X_i} was not available at the time of the SoA assessment. Recently, as included in Chapter 4, Tables 4, 5, 6, 7 and 8 $P_{X_i} = 28$, 6 and 24%DR for AR, heave and pitch, respectively, based on experiments at FORCE (test case 2.10),

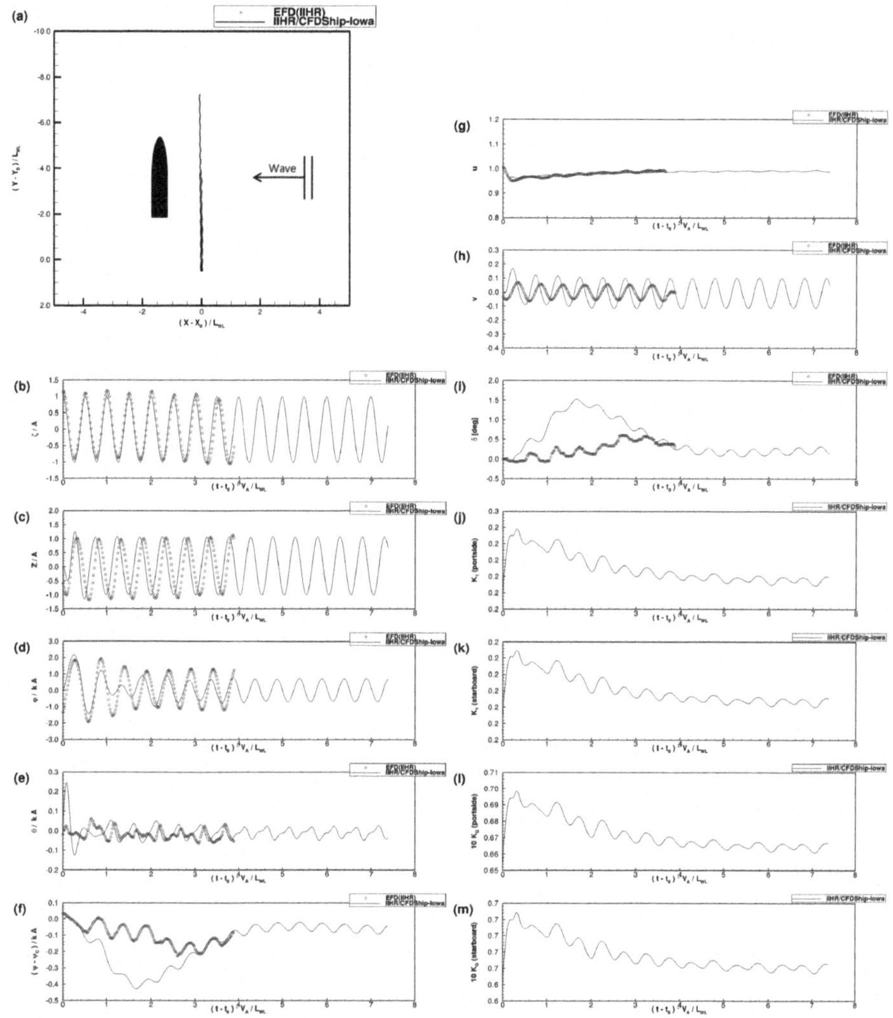

Fig. 29 Submissions with EFD for ONRT course keeping in waves with $\mu = 90°$: **a** Trajectory; **b** Wave; **c** Heave; **d** Roll; **e** Pitch; **f** Yaw; **g** x-vel.; **h** y-vel.; **i** Rudder angle; **j** Thrust coeff. (portside); **k** Thrust coeff. (starboard); **l** Torque coeff. (portside); **m** Torque coeff. (starboard) * **j** ~ **m** don't include EFD

IIHR (test case 2.11 head waves) and OU (head waves). Including these values would increase U_D considerably, i.e., = 29, 7 and 24%DR, respectively, such that the scatter in the experimental data is comparable to the scatter in the CFD for AR and heave and dominant for pitch.

For KCS calm water resistance (10 codes/solutions), the solution scatter $P_{|E_i|}$ is larger than the experimental uncertainty U_D for resistance but not for sinkage and trim. 3, 5 and 9 solutions are validated at U_D = 2, 7 and 14%D; 8, 7 and 9

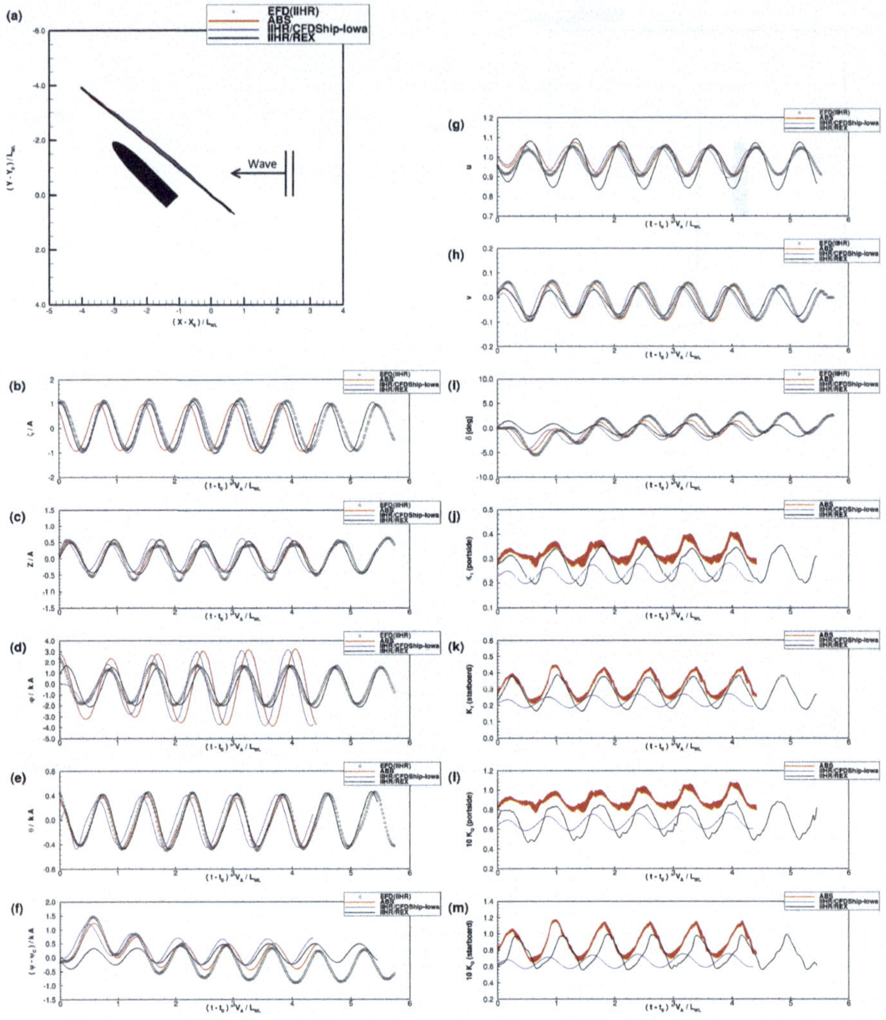

Fig. 30 Submissions with EFD for ONRT course keeping in waves with $\mu = 135°$: **a** Trajectory; **b** Wave; **c** Heave; **d** Roll; **e** Pitch; **f** Yaw; **g** x-vel.; **h** y-vel.; **i** Rudder angle; **j** Thrust coeff. (portside); **k** Thrust coeff. (starboard); **l** Torque coeff. (portside); **m** Torque coeff. (starboard) * **j** ~ **m** don't include EFD

codes/solutions are N-version validated at $U_{SoA_i} = 3,\ 8$ and 16%D; and $\overline{|E|} = 2,\ 6$ and 5%D, respectively, for resistance, sinkage and trim. The mean code is N-version validated for sinkage and trim, but not for resistance. Priorities for improvements are: for resistance, reduction in code/solution scatter; and for sinkage and trim, reduction in code/solution scatter and errors and experimental uncertainties.

Interestingly, scatter in the CFD codes/simulations for resistance is similar as the most recent AIAA CFD Drag Prediction Workshop (Vassberg, 2016) wherein

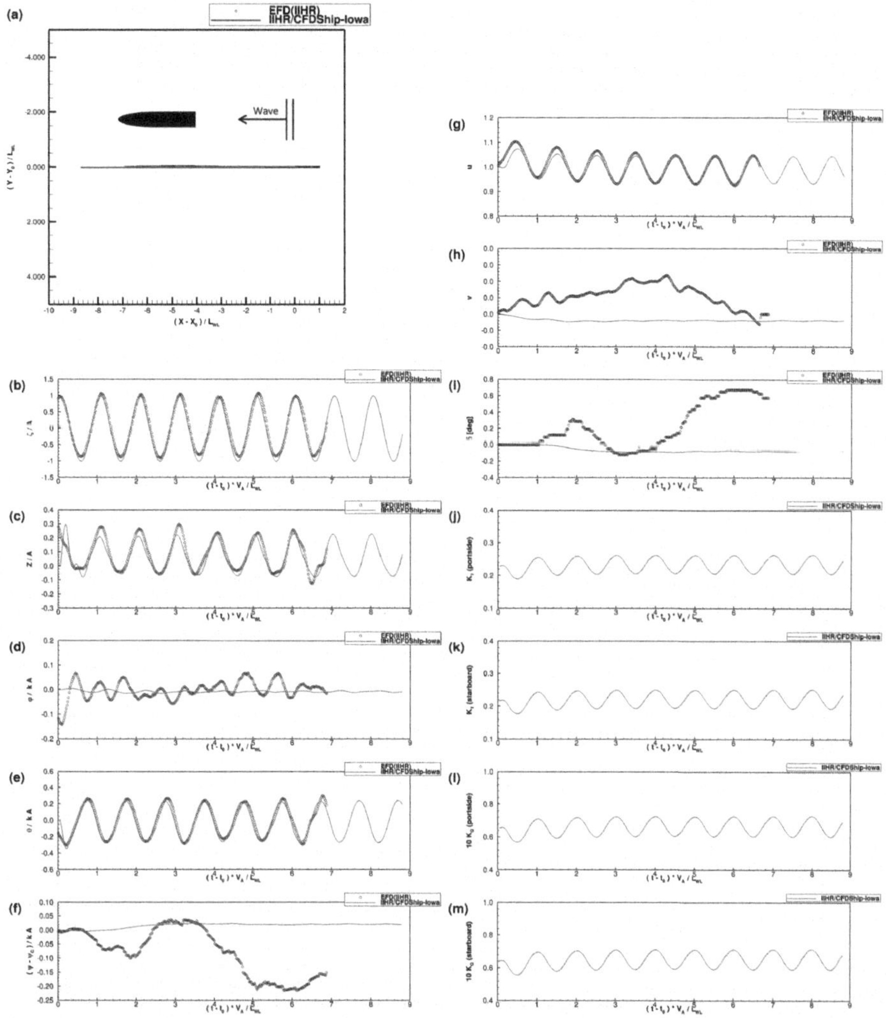

Fig. 31 Submissions with EFD for ONRT course keeping in waves with $\mu = 180°$: **a** Trajectory; **b** Wave; **c** Heave; **d** Roll; **e** Pitch; **f** Yaw; **g** x-vel.; **h** y-vel.; **i** Rudder angle; **j** Thrust coeff. (portside); **k** Thrust coeff. (starboard); **l** Torque coeff. (portside); **m** Torque coeff. (starboard) * **j** ~ **m** don't include EFD

the standard deviation is 2.1% of the mean value based on 44 CFD submissions, whereas the comparable value for ship hydrodynamics is 2.8% based on ten CFD submissions. Note that the AIAA test case is relatively less complex than that for ship hydrodynamics, i.e., low-Mach number compressible flow, fixed slender body.

For ONRT calm water free running self-propulsion (8 codes/solutions), the CFD SoA for propeller RPS is comparable to that for resistance. For sinkage and trim both U_D and $P_{|E_i|}$ are much larger than for captive tests especially for trim (partially due

Table 32 (%D). Evaluations of individual and mean code values on harmonic amplitudes of wave and motions for ONRT course keeping

μ [deg]	Variable	0th harmonic amplitude						1st harmonic amplitude						2nd harmonic amplitude						Ave	
		z_0/L	θ_0/Ak	φ_0/Ak	Ψ_0/Ak	u_0/U	v_0/U	z_1/L	θ_1/Ak	φ_1/Ak	Ψ_1/Ak	u_1/U	v_1/U	z_2/L	θ_2/Ak	φ_2/Ak	Ψ_2/Ak	u_2/U	v_2/U	1st	2nd
0	U_D%|D|	27.05	30.13	68.74	82.36	0.99	0.00	2.15	0.61	1421.00	7.07	15.50	0.00	66.22	5.64	296.71	45.68	13.83	0.00	241.05	53.11
	# submissions	6	6	5	5	6	5	6	6	5	5	6	5	6	6	5	5	6	5	5.50	5.50
	# outliers	0	0	0	0	0	0	0	0	0	0	1	0	0	0	0	0	1	0	0.17	0.08
	$P_{\|Ei\|}$%|D|	89.19	289.02	45.78	64.74	2.76	35.74	19.06	8.89	709.83	36.39	52.61	60.70	11.72	63.67	60.78	54.99	8.06	78.12	147.91	67.05
	$P_{\|\bar{E}\|}$%|D|	36.41	117.99	20.47	28.95	1.13	15.98	7.78	3.63	317.45	16.28	23.53	27.14	4.78	25.99	27.18	24.59	3.61	34.94	65.97	28.50
	U_{SoAi}%|D|	93.20	290.58	82.59	104.76	2.93	35.74	19.18	8.91	1588.43	37.07	54.84	60.70	67.25	63.92	302.87	71.48	16.01	78.12	294.85	100.79
	U_{SoA}%|D|	45.36	121.78	71.73	87.30	1.50	15.98	8.07	3.68	1456.03	17.74	28.17	27.14	66.40	26.60	297.95	51.88	14.30	34.94	256.81	69.64
	$\|\bar{E}\|$%|D|	41.83	103.78	106.02	85.41	2.37	113.90	7.42	7.77	294.32	70.56	271.53	75.67	11.07	28.28	80.26	68.03	7.62	68.28	121.21	59.74
	# validated	3	3	0	2	1	0	2	0	5	0	0	0	6	3	5	1	5	0	1.17	2.42
	# N validated	5	5	1	4	4	0	5	4	5	0	0	1	6	5	5	2	5	2	2.50	3.67
	Mean N validated	Yes	yes	No	yes	No	No	Yes	No	Yes	No	No	No	Yes	No	Yes	No	Yes	No	–	–
45	U_D%|D|	108.30	15.39	10.94	4.28	2.89	44.08	1.71	1.07	3.02	0.43	109.13	0.00	102.93	25.77	67.66	3.63	0.00	0.00	19.23	32.16
	# submissions	3	3	3	3	3	3	3	3	3	3	3	3	3	3	3	3	3	3	3.00	3.00
	$\|\bar{E}\|$%|D|	231.37	50.89	72.95	28.06	2.80	71.80	1.66	3.24	29.56	4.18	102.69	1392.70	16.29	73.50	11.60	23.62	60.47	1368.20	255.67	167.63
	# validated	0	0	2	0	2	2	2	0	0	0	2	0	3	0	3	0	0	0	0.67	1.00
90	U_D%|D|	27.55	30.15	18.71	6.40	2.28	5023.56	1.30	10.45	1.56	2.82	900.70	2.17	19.19	10.87	28.59	28.17	2292.00	0.00	153.17	623.96
	# submissions	1	1	1	1	1	1	1	1	1	1	1	1	1	1	1	1	1	1	1.00	1.00
	$\|\bar{E}\|$%|D|	192.47	56.34	73.16	62.42	1.04	764.45	0.43	49.71	40.40	18.94	29.06	99.79	97.99	23.46	69.76	65.70	67.67	55.31	39.72	127.48
	# validated	0	0	0	0	1	1	1	0	0	0	1	0	0	0	0	0	1	0	0.33	0.25
135	U_D%|D|	159.27	12.74	11.99	17.35	1.39	140.40	2.67	0.90	1.01	0.75	27.52	27.78	14.64	11.88	27.34	31.42	160.67	0.00	10.10	49.09

(continued)

Table 32 (continued)

μ [deg]	Variable	0th harmonic amplitude						1st harmonic amplitude						2nd harmonic amplitude						Ave	
		z_0/L	θ_0/Ak	φ_0/Ak	Ψ_0/Ak	u_0/U	v_0/U	z_1/L	θ_1/Ak	φ_1/Ak	Ψ_1/Ak	u_1/U	v_1/U	z_2/L	θ_2/Ak	φ_2/Ak	Ψ_2/Ak	u_2/U	v_2/U	1st	2nd
	# submissions	3	3	3	3	3	3	3	3	3	3	3	3	3	3	3	3	3	3	3.00	3.00
	$\lvert\bar{E}\rvert\%\lvert D\rvert$	514.22	8.32	101.81	155.28	1.33	44.73	4.86	6.24	65.99	26.05	24.38	29.64	49.81	43.55	65.46	68.16	44.61	194.01	26.19	107.61
	# validated	1	2	0	0	2	3	2	0	0	0	2	2	0	0	1	0	3	0	1.00	1.00
180	$U_D\%\lvert D\rvert$	16.11	200.57	494.08	14.94	1.67	510.35	8.44	1.64	82.75	22.43	35.38	0.00	69.70	7.81	103.77	44.24	2519.75	0.00	25.11	331.92
	# submissions	1	1	1	1	1	1	1	1	1	1	1	1	1	1	1	1	1	1	1.00	1.00
	$\lvert\bar{E}\rvert\%\lvert D\rvert$	23.41	161.85	365.28	123.71	0.86	127.09	4.75	0.91	84.30	89.61	13.09	87.41	3.81	80.28	89.43	91.37	38.96	97.93	46.68	100.33
	# validated	0	1	1	0	1	1	1	1	0	0	1	0	1	0	1	0	1	0	0.50	0.58
Ave	$U_D\%\lvert D\rvert$	67.66	57.80	120.89	25.07	1.84	1143.68	3.25	2.94	301.87	6.70	217.64	5.99	54.54	12.39	104.81	30.63	997.25	0.00	89.73	218.05
	# submissions	2.80	2.80	2.60	2.60	2.80	2.60	2.80	2.80	2.60	2.60	2.80	2.60	2.80	2.80	2.60	2.60	2.80	2.60	2.70	2.70
	$\lvert\bar{E}\rvert\%\lvert D\rvert$	200.66	76.23	143.84	90.98	1.68	224.40	3.82	13.57	102.92	41.87	88.15	337.04	35.79	49.81	63.30	63.38	43.86	356.75	97.89	112.56
	# validated	0.80	1.20	0.60	0.40	1.40	1.40	1.60	0.20	1.00	0.00	1.20	0.40	2.00	0.60	2.00	0.20	2.00	0.00	0.73	1.05

Table 33 (%DR). Evaluations of individual and mean code values on harmonic amplitudes of wave and motions for ONRT course keeping

μ[deg]	Variable	0th harmonic amplitude						1st harmonic amplitude						2nd harmonic amplitude						Ave	
		z_0/L	θ_0/Ak	φ_0/Ak	Ψ_0/Ak	u_0/U	v_0/U	z_1/L	θ_1/Ak	φ_1/Ak	Ψ_1/Ak	u_1/U	v_1/U	z_2/L	θ_2/Ak	φ_2/Ak	Ψ_2/Ak	u_2/U	v_2/U	1st	2nd
0	U_D%|DR|	11.46	31.89	4.72	10.46	4.52	0.00	1.44	0.63	0.64	0.04	1.05	0.00	16.77	2.42	23.83	1.80	13.32	0.00	0.63	10.10
	# submissions	6	6	5	5	6	5	6	6	5	5	6	5	6	6	5	5	6	5	5.50	5.50
	# outliers	0	0	0	0	0	0	0	0	0	0	1	0	0	0	0	0	1	0	0.17	0.08
	$P_{\|Ei\|}$%|DR|	37.79	305.81	3.14	8.23	12.56	3.83	12.79	9.12	0.32	0.20	3.58	0.37	2.97	27.28	4.88	2.16	7.76	3.43	4.40	34.99
	$P_{\|\bar{E}\|}$%|DR|	15.43	124.85	1.40	3.68	5.13	1.71	5.22	3.72	0.14	0.09	1.60	0.17	1.21	11.14	2.18	0.97	3.47	1.53	1.82	14.39
	U_{SoAi}%|DR|	39.49	307.47	5.67	13.31	13.35	3.83	12.87	9.14	0.72	0.20	3.73	0.37	17.03	27.39	24.33	2.81	15.42	3.43	4.50	39.46
	U_{SoA}%|DR|	19.22	128.85	4.92	11.09	6.83	1.71	5.42	3.78	0.66	0.10	1.92	0.17	16.82	11.40	23.93	2.04	13.77	1.53	2.00	20.18
	$\|\bar{E}\|$%|DR|	17.72	109.81	7.27	10.85	10.80	12.22	4.98	7.97	0.13	0.39	18.45	0.46	2.80	12.12	6.45	2.68	7.34	3.00	5.40	16.92
	# validated	3	3	0	2	1	0	2	0	5	0	0	0	6	3	5	1	5	0	1.17	2.42
	# N validated	5	5	1	4	4	0	5	4	5	0	0	1	6	5	5	2	5	2	2.50	3.67
	Mean N Validated	Yes	Yes	No	yes	No	No	Yes	No	Yes	No	No	No	Yes	No	Yes	No	Yes	No	–	–
45	U_D%|DR|	10.36	18.85	5.45	4.83	14.64	35.90	1.58	1.08	0.65	0.12	27.65	0.00	16.88	2.66	24.26	1.39	0.00	0.00	5.18	11.27
	# submissions	3	3	3	3	3	3	3	3	3	3	3	3	3	3	3	3	3	3	3.00	3.00
	$\|\bar{E}\|$%|DR|	22.14	62.32	36.36	31.63	14.19	58.47	1.53	3.29	6.39	1.16	26.01	51.71	2.67	7.60	4.16	9.02	72.30	163.16	15.01	40.33
	# validated	0	0	2	0	2	2	2	0	0	0	2	0	3	0	3	0	0	0	0.67	1.00
90	U_D%|DR|	10.08	27.51	6.58	1.45	12.43	85.15	1.53	0.27	1.03	0.16	36.88	1.58	17.02	3.61	24.93	5.31	448.64	0.00	6.91	53.56
	# submissions	1	1	1	1	1	1	1	1	1	1	1	1	1	1	1	1	1	1	1.00	1.00

(continued)

Table 33 (continued)

μ[deg]	Variable	0th harmonic amplitude						1st harmonic amplitude						2nd harmonic amplitude						Ave	
		z_0/L	θ_0/Ak	φ_0/Ak	Ψ_0/Ak	u_0/U	v_0/U	z_1/L	θ_1/Ak	φ_1/Ak	Ψ_1/Ak	u_1/U	v_1/U	z_2/L	θ_2/Ak	φ_2/Ak	Ψ_2/Ak	u_2/U	v_2/U	1st	2nd
	$\|\bar{E}\|\%\|DR\|$	70.41	51.41	25.73	14.17	5.67	12.96	0.50	1.30	26.66	1.09	1.19	72.76	86.88	7.80	60.83	12.39	13.25	38.58	17.25	33.34
	# validated	0	0	0	0	1	1	1	0	0	0	1	0	0	0	0	0	1	0	0.33	0.25
135	$U_D\%\|DR\|$	9.73	11.97	7.78	6.95	7.58	59.81	1.55	0.59	1.01	0.76	28.64	27.95	17.04	13.11	29.53	32.66	159.29	0.00	10.08	29.62
	# submissions	3	3	3	3	3	3	3	3	3	3	3	3	3	3	3	3	3	3	3.00	3.00
	$\|\bar{E}\|\%\|DR\|$	31.43	7.82	66.01	62.21	7.24	19.05	2.82	4.05	66.02	26.20	25.38	29.82	57.98	48.05	70.72	70.84	44.23	202.54	25.71	57.34
	# validated	1	2	0	0	2	3	2	0	0	0	2	2	0	0	1	0	3	0	1.00	1.00
180	$U_D\%\|DR\|$	10.22	45.06	5.15	2.22	9.27	94.74	1.46	0.61	0.80	0.43	31.29	0.00	17.15	2.41	29.67	5.46	494.14	0.00	5.76	59.62
	# submissions	1	1	1	1	1	1	1	1	1	1	1	1	1	1	1	1	1	1	1.00	1.00
	$\|\bar{E}\|\%\|DR\|$	14.85	36.36	3.81	18.35	4.76	23.59	0.82	0.34	0.82	1.70	11.58	0.95	0.94	24.78	25.57	11.28	7.64	36.67	2.70	17.38
	# validated	0	1	1	0	1	1	1	1	0	0	1	0	1	0	1	0	1	0	0.50	0.58
Ave	$U_D\%\|DR\|$	10.37	27.06	5.93	5.18	9.69	55.12	1.51	0.64	0.83	0.30	25.10	5.91	16.97	4.84	26.44	9.32	223.08	0.00	5.71	32.83
	# submissions	2.80	2.80	2.60	2.60	2.80	2.60	2.80	2.80	2.60	2.60	2.80	2.60	2.80	2.80	2.60	2.60	2.80	2.60	2.70	2.70
	$\|\bar{E}\|\%\|DR\|$	31.31	53.54	27.83	27.44	8.53	25.26	2.13	3.39	20.00	6.11	16.53	31.14	30.25	20.07	33.54	21.24	28.95	88.79	13.22	33.06
	# validated	0.80	1.20	0.60	0.40	1.40	1.40	1.60	0.20	1.00	0.00	1.20	0.40	2.00	0.60	2.00	0.20	2.00	0.00	0.73	1.05

Table 34 (%2π. Evaluations of individual and mean code values on harmonic phase of wave and motions for ONRT course keeping

μ[deg]	Variable	1st harmonic phase						2nd harmonic phase						Ave	
		z_1/L	θ_1/Ak	φ_1/Ak	Ψ_1/Ak	u_1/U	v_1/U	z_2/L	θ_2/Ak	φ_2/Ak	Ψ_2/Ak	u_2/U	v_2/U	1st	2nd
0	$U_D\%2\pi$	4.09	0.09	0.63	0.06	4.08	4.82	8.80	0.14	0.22	0.61	6.35	21.94	2.30	6.34
	# submissions	6	6	5	5	6	5	6	6	5	5	6	5	5.50	5.50
	# outliers	0	0	0	0	0	0	0	0	0	0	0	0	0.00	0.00
	$P_{\lvert Ei \rvert}\%2\pi$	2.22	5.51	10.92	42.29	13.91	15.05	19.18	28.51	44.72	36.51	14.90	24.61	14.98	28.07
	$P_{\lvert \bar{E} \rvert}\%2\pi$	0.91	2.25	4.88	18.91	5.68	6.73	7.83	11.64	20.00	16.33	6.08	11.01	6.56	12.15
	$U_{SoAi}\%2\pi$	4.65	5.51	10.94	42.29	14.50	15.81	21.10	28.51	44.72	36.52	16.20	32.97	15.62	30.00
	$U_{SoA}\%2\pi$	4.19	2.25	4.92	18.91	6.99	8.28	11.78	11.64	20.00	16.34	8.79	24.55	7.59	15.52
	$\lvert \bar{E} \rvert\%2\pi$	1.31	2.47	30.48	26.21	6.26	20.22	17.21	11.99	37.67	43.35	14.76	23.70	14.49	24.78
	# validated	6	0	0	0	3	0	2	0	0	0	0	3	1.50	0.83
	# N validated	6	5	0	3	5	1	4	5	3	2	4	4	3.33	3.67
	Mean N Validated	Yes	No	No	No	Yes	No	No	No	No	No	No	Yes	–	–
45	$U_D\%2\pi$	4.32	0.08	0.19	0.09	2.74	3.19	6.74	1.02	1.01	0.15	5.69	54.48	1.77	11.51
	# submissions	3	3	3	3	3	3	3	3	3	3	3	3	3.00	3.00
	$\lvert \bar{E} \rvert\%2\pi$	8.47	8.57	17.70	13.32	11.41	8.71	13.88	22.66	26.06	26.41	7.59	65.47	11.36	27.01
	# validated	0	0	0	0	0	2	0	0	0	0	2	1	0.33	0.50
90	$U_D\%2\pi$	0.62	0.06	0.18	0.04	7.05	0.60	0.77	0.04	0.18	0.14	37.60	4.79	1.42	7.25
	# submissions	1	1	1	1	1	1	1	1	1	1	1	1	1.00	1.00
	$\lvert \bar{E} \rvert\%2\pi$	9.57	9.97	4.12	12.05	27.62	11.76	1.93	4.28	23.34	40.69	55.63	38.25	12.52	27.35
	# validated	0	0	0	0	0	0	0	0	0	0	0	0	0.00	0.00
135	$U_D\%2\pi$	3.95	1.10	0.20	0.04	8.26	4.01	11.13	0.18	0.24	0.22	8.43	9.96	2.92	5.03

(continued)

Table 34 (continued)

μ[deg]	Variable	1st harmonic phase						2nd harmonic phase						Ave	
		z_1/L	θ_1/Ak	φ_1/Ak	Ψ_1/Ak	u_1/U	v_1/U	z_2/L	θ_2/Ak	φ_2/Ak	Ψ_2/Ak	u_2/U	v_2/U	1st	2nd
	# submissions	3	3	3	3	3	3	3	3	3	3	3	3	3.00	3.00
	$\lvert\bar{E}\rvert\%2\pi$	11.11	10.01	15.76	8.92	10.70	8.45	23.38	19.98	14.55	37.14	17.02	19.90	10.83	22.00
	# validated	0	0	0	0	2	1	1	0	0	0	0	1	0.50	0.33
180	$U_D\%2\pi$	2.51	0.04	0.19	0.42	2.97	3.03	6.60	0.94	0.19	0.19	5.82	8.56	1.53	3.72
	# submissions	1	1	1	1	1	1	1	1	1	1	1	1	1.00	1.00
	$\lvert\bar{E}\rvert\%2\pi$	1.31	0.98	36.94	3.91	3.28	2.14	8.37	11.65	19.22	37.96	63.38	26.80	8.09	27.90
	# validated	1	0	0	0	0	1	0	0	0	0	0	0	0.33	0.00
Ave	$U_D\%2\pi$	3.10	0.27	0.28	0.13	5.02	3.13	6.81	0.46	0.37	0.26	12.78	19.95	1.99	6.77
	# submissions	2.8	2.8	2.6	2.6	2.8	2.6	2.8	2.8	2.6	2.6	2.8	2.6	2.70	2.70
	$\lvert\bar{E}\rvert\%2\pi$	6.35	6.40	21.00	12.88	11.86	10.26	12.95	14.11	24.17	37.11	31.68	34.82	11.46	25.81
	# validated	1.4	0	0	0	1	0.8	0.6	0	0	0	0.4	1	0.53	0.33

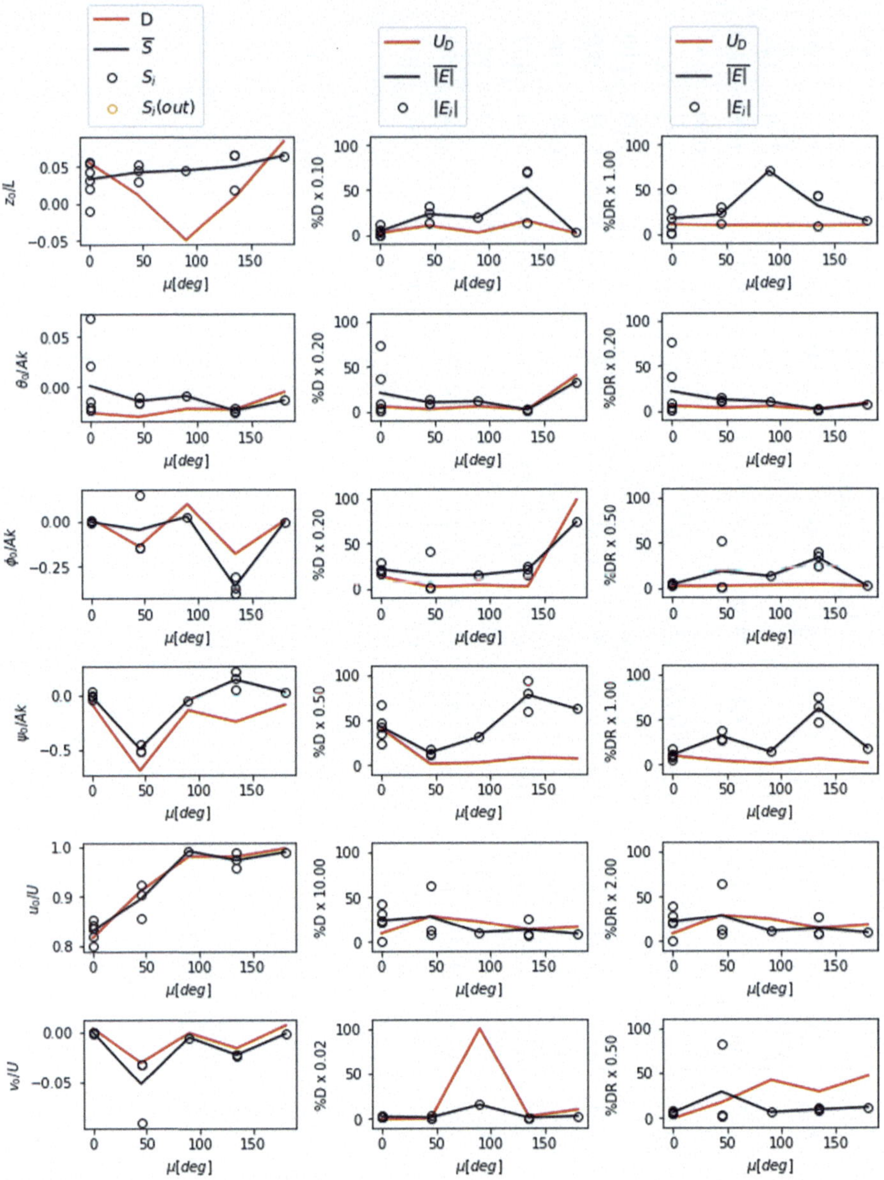

Fig. 32 Validations for 0th harmonic amplitudes of motions for ONRT seakeeping in regular waves at Fn = 0.2

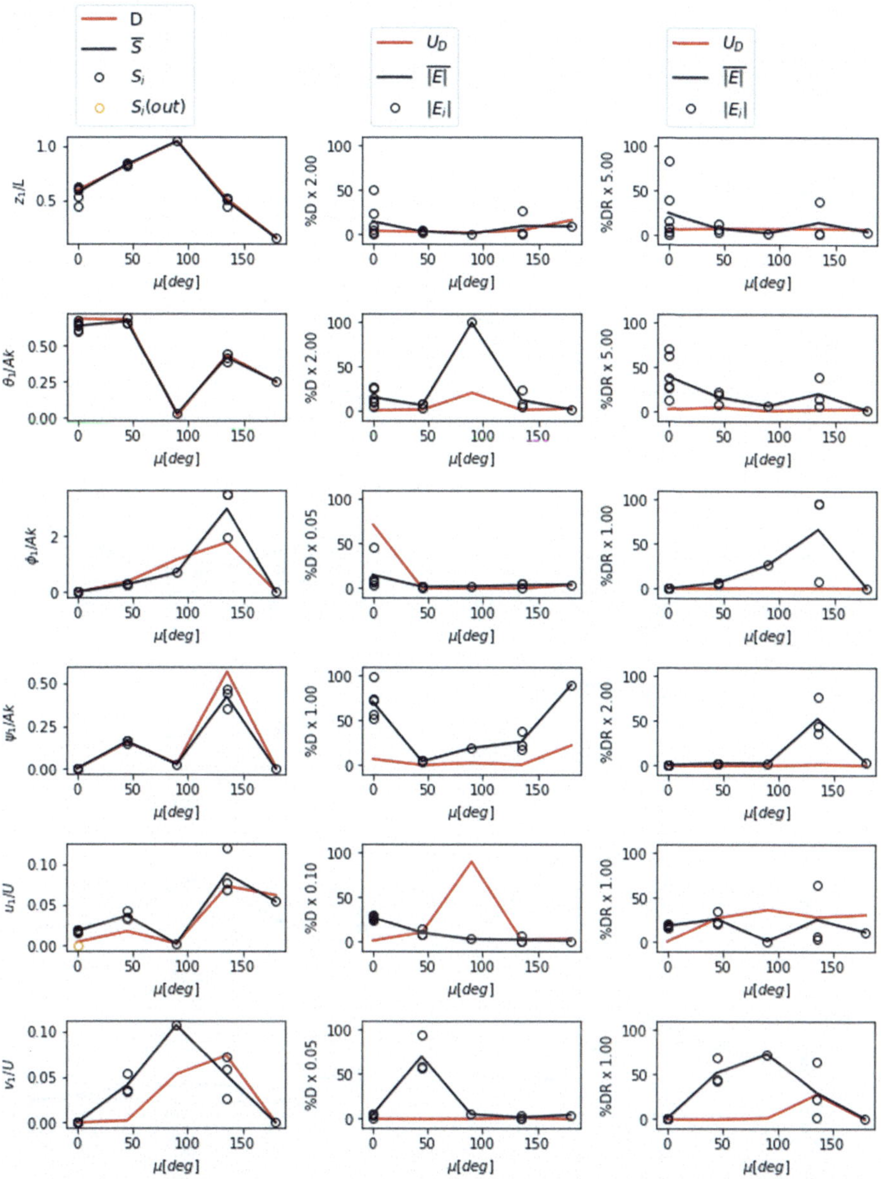

Fig. 33 Validations for 1st harmonic amplitudes of wave and motions for ONRT seakeeping in regular waves at Fr = 0.2

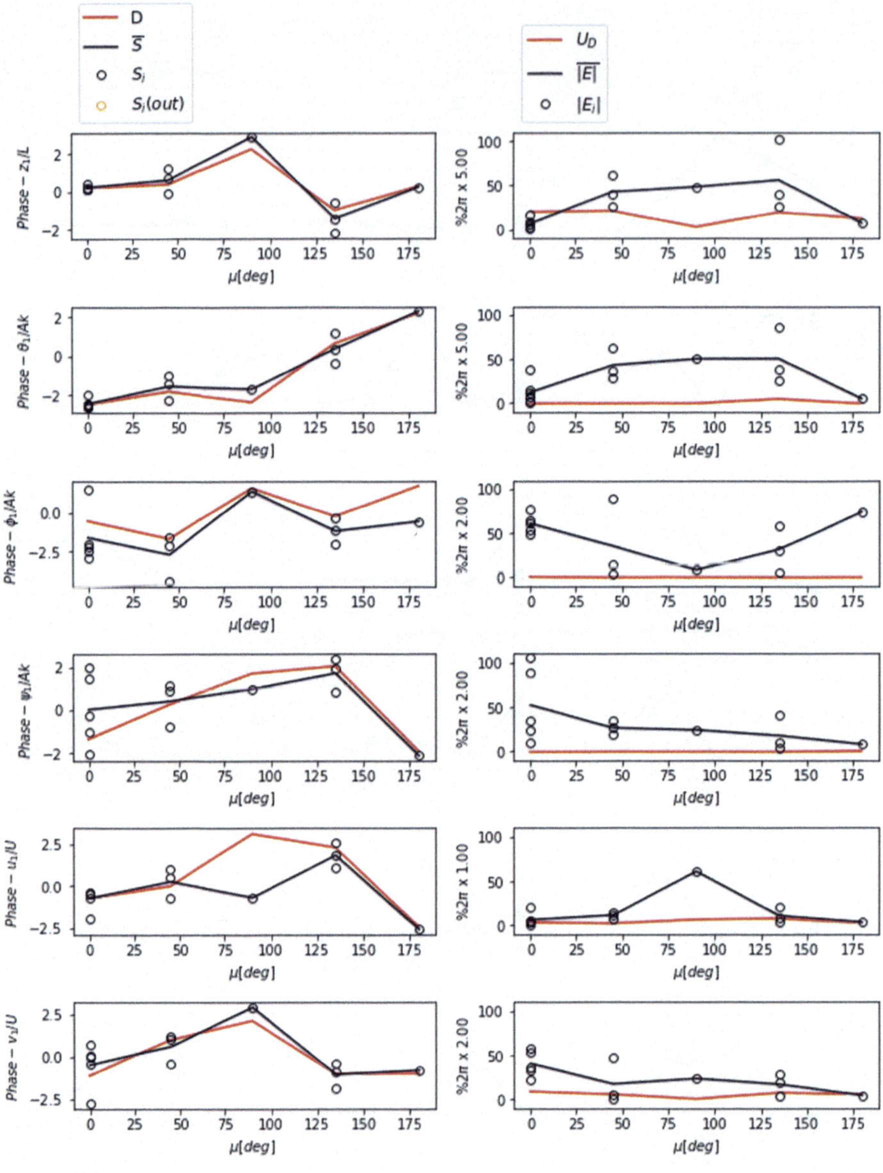

Fig. 34 Validations for 1st harmonic phases of motions for ONRT seakeeping in regular waves at Fr = 0.2

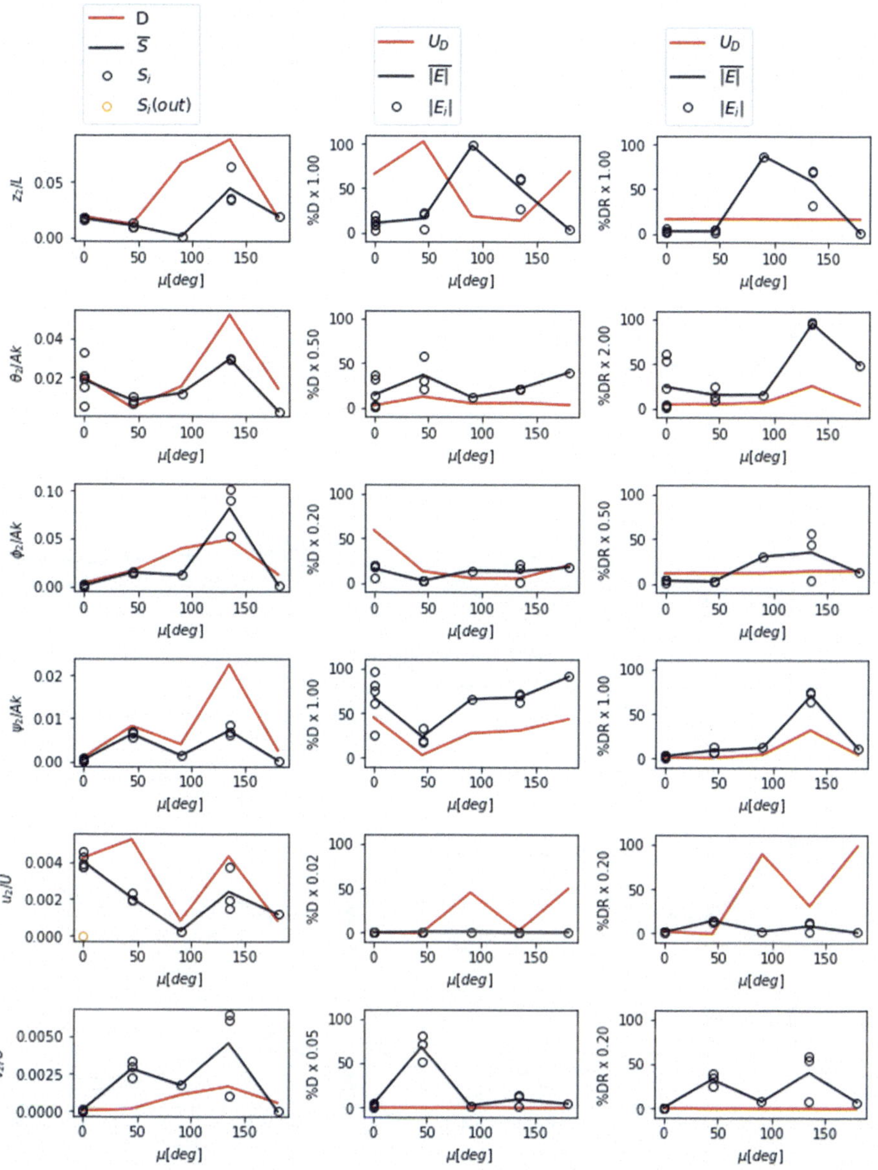

Fig. 35 Validations for 2nd harmonic amplitudes of motions for ONRT seakeeping in regular waves at Fr = 0.2

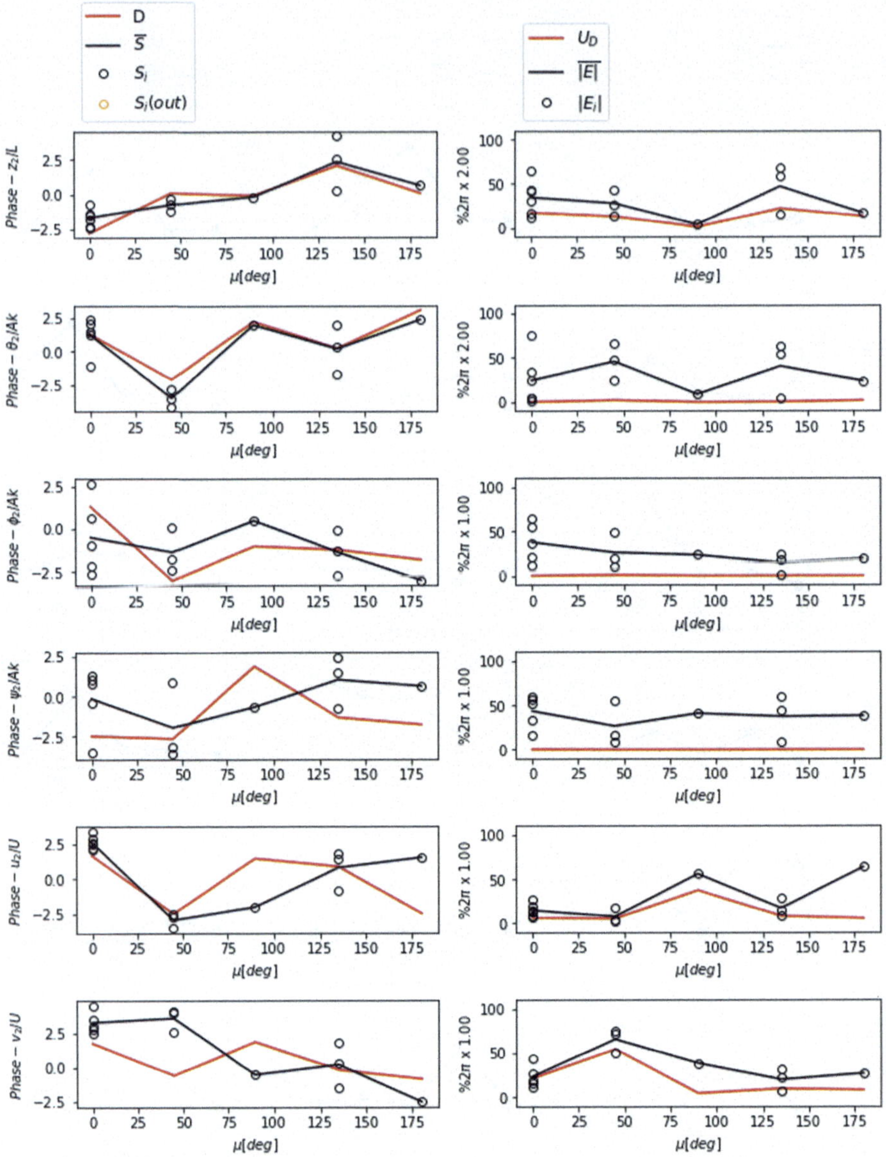

Fig. 36 Validations for 2nd harmonic phases of motions for ONRT seakeeping in regular waves at Fr = 0.2

Table 35 Summary of calm water cases

Variable	KCS 2.10 calm water			ONRT 3.9 calm water		
	$C_T \times 10^3$	σ/L	ς	N	σ/L	ς
$U_D\%\|D\|$	1.75	7.38	14.25	0.083	20	134.2
$P_{\|Ei\|}\%\|D\|$	2.90	3.99	6.50	1.59	9.89	22.72
$U_{SoAi}\%\|D\|$	3.39	8.39	15.66	1.59	22.31	136.11
$U_{SoA}\%\|D\|$	1.98	7.50	14.41	0.57	20.3	134.44
$\|\bar{E}\|\%\|D\|$	2.33	6.16	5.16	1.49	5.56	14.7

to the mount, as designed primarily for ship motions testing). For RPS, sinkage and trim, respectively: 1, 8 and 8 solutions are validated at $U_D = < 1$, 20 and 134%D; 4, 8 and 8 codes/solutions are N-version validated at $U_{SoA_i} = 2,\ 22$ and 136%D; and $\overline{|E|} = 2,\ 6$ and 15%D. The mean code is N-version validated for sinkage and trim, but not for RPS.

For KCS captive head waves averaged over λ/L (10 codes/solutions), the 1st (first harmonic) and 2nd (AR, 0th and 2nd harmonic) order variables solution scatter $P_{|E_i|}$ is larger than the experimental uncertainty U_D. For AR, heave and pitch, respectively: 4, 5 and 4 solutions are validated at $U_D = 7$, 4 and 3%DR; 7, 7 and 6 codes/solutions are N-version validated at $U_{SoA_i} = 26,\ 9$ and 6%DR; and $\overline{|E|} = 13,\ 5$ and 5%DR. The mean code is not N-version validated. The solution scatter uncertainty is larger than U_D especially for AR. The errors for AR are six times larger than for calm water resistance and for heave and pitch are comparable to those for sinkage and trim. The largest errors are near resonance for AR, near resonance for heave, and just before/after resonance for pitch.

The current CFD results from T2015 shows that CFD can predict head waves heave, pitch and AR with better accuracy than potential flow for short, medium and long waves. CFD results also show smaller scatter than potential flow for heave and pitch. G2010 CFD errors/scatter are comparable to T2015, but T2015 errors/scatter are considered most reliable since there are more submissions in T2015 with more rigorous analysis.

Use of finer grids and improved user best practices will likely reduce CFD errors and scatter. In addition, improved benchmark experimental data from multiple facilities and solution verification will provide estimates that are more robust for intervals of validation and certification.

For ONRT free running head waves with λ/L = 1 (6 codes/solutions), $P_{|E_i|}$ is much larger than U_D. For u_0, heave and pitch, respectively: 1, 2 and 0 solutions are validated at $U_D = 5$, 1 and 1%DR; 4, 5 and 4 codes/solutions are N-version validated at $U_{SoA_i} = 13,\ 13$ and 9%DR; and $\overline{|E|} = 11,\ 5$ and 8%DR.

For KCS captive oblique waves with λ/L = 1 (2 codes/solutions), $U_D = 7$ and 43%DR and $\overline{|E|} = 18$ and 40%DR, respectively, for 1st and 2nd order terms. The largest errors are bow quartering for AR, beam waves for Z_1/A, following waves for θ_1/Ak and bow and stern quartering waves for ϕ_1/Ak.

Table 36 Summary of head wave cases

Variable	KCS 2.10 head waves λ/L average							ONRT 3.12 head waves λ/L = 1						
	Amplitudes					Phases		Amplitudes					Phases	
	AR	z_1/L	θ_1/Ak	1st Ave	2nd Ave	1st Ave	2nd Ave	u_0/U	z_1/L	θ_1/Ak	1st Ave	2nd Ave	1st Ave	2nd Ave
$U_D\%\lvert DR\rvert/\%2\pi$	6.92	4.35	2.58	2.26	14.35	1.75	22.81	4.52	1.44	0.63	0.63	10.10	2.30	6.34
$P_{\lvert Ei\rvert}\%\lvert DR\rvert/\%2\pi$	24.80	7.44	5.52	5.26	16.06	4.44	14.73	12.56	12.79	9.12	4.40	34.99	14.98	28.07
$U_{SoAi}\%\lvert DR\rvert/\%2\pi$	26.25	8.71	6.27	5.93	23.52	5.00	31.99	13.35	12.87	9.14	4.50	39.46	15.62	30.00
$U_{SoA}\%\lvert DR\rvert/\%2\pi$	11.25	5.05	3.32	3.09	16.17	2.67	26.34	6.83	5.42	3.78	2.00	20.18	7.59	15.52
$\lvert\bar{E}\rvert\%\lvert DR\rvert/\%2\pi$	12.67	5.20	4.53	4.02	11.49	4.91	16.70	10.80	4.98	7.97	5.40	16.92	14.49	24.78

Table 37 Summary of oblique wave cases

Variable	KCS 2.11 oblique waves μ average								ONRT 3.13 oblique waves μ average								
	Amplitudes						Phases		Amplitudes							Phases	
	AR	z_1/L	θ_1/Ak	φ_1/Ak	1st Ave	2nd Ave	1st Ave	2nd Ave	u_0/U	z_1/L	θ_1/Ak	φ_1/Ak	Ψ_1/Ak	1st Ave	2nd Ave	1st Ave	2nd Ave
$U_D\%\|DR\|/\%2\pi$	5.86	1.61	23.25	1.22	6.88	42.60	–	–	9.69	1.51	0.64	0.83	0.30	5.71	32.83	1.99	6.77
$\|\bar{E}\|\%\|DR\|/\%2\pi$	7.37	7.54	6.06	12.38	18.14	40.41	13.62	23.55	8.53	2.13	3.39	20.00	6.11	13.22	33.06	11.46	25.81

For ONRT free running oblique waves (3 or 1 code/solution depending heading), $U_D = 6$ and 32%DR and $\overline{|E|} = 13$ and 33%DR, respectively, for 1st and 2nd order terms. The largest errors are bow and stern quartering waves for speed loss, head and stern quartering waves for heave and pitch and stern quartering waves for roll and heading.

Figure 37 compares the experimental and mean code 0th and 1st harmonic amplitudes for the three test cases ($\lambda/L = 1.0$ for ONRT and KCS $L = 2.7$ m model and $\lambda/L = 1.15$ for KCS 6.1 m model). The 0th harmonic values are small and trends not discernable. However, 1st harmonic values show clear and consistent trends, which correlate with wave heading. Heave/pitch values are large for bow waves, maximum/minimum for beam waves and small for stern waves. B/λ is small for both models (0.14 and 0.12 for KCS and ONRT, respectively) and they simply oscillate with the wave with minimum relative motion for beam waves. Roll is small for bow waves, increases for beam waves, maximum for stern quartering and reduced for following waves. Larger heave/pitch for bow quartering waves and roll for stern quartering waves is due to resonance conditions.

Figure 38 is like 37 but only includes experimental data; since, this condition was not used as a test case at T2015. Heave and pitch values for bow waves and roll for beam and stern quartering waves are smaller since the encounter frequency is not close to the resonance condition at those wave headings.

These trends seem reasonable for course keeping and likely typical for displacement ships with similar Fr, λ/L and H/λ. The agreement between the experiments and CFD is outstanding and surely provides motivation for increased use of CFD for industrial ship hydrodynamics applications.

As discussed, Stern et al. (2017), programmatic requirements/tolerances U_{req} can be considered similarly as for solution validation, but with U_V replaced by U_{SoA_i}. Since U_{SoA_i} is $\geq U_V$ it will always be a more conservative assessment. There are six possible combinations of $|E_i|$, U_{SoA_i} and U_{req} assuming none are equal:

$$\begin{array}{l} (1)\ |E_i| < U_{SoA_i} < U_{req} \\ (2)\ |E_i| < U_{req} < U_{SoA_i} \\ (3)\ U_{req} < |E_i| < U_{SoA_i} \\ (4)\ U_{SoA_i} < |E_i| < U_{req} \\ (5)\ U_{SoA_i} < U_{req} < |E_i| \\ (6)\ U_{req} < U_{SoA_i} < |E_i| \end{array} \tag{41}$$

In cases 1, 2 and 3 N-version validation is achieved at the U_{SoA_i} interval, i.e., the comparison error is below the noise level. From an uncertainty perspective, modeling errors cannot be isolated. In cases 4, 5 and 6, the comparison error is larger than the noise level, i.e.,$U_{SoA_i} < |E_i|$ such that from an uncertainty perspective the sign and magnitude of E can be used to estimate the modeling errors δ_{SM}. If $U_{SoA_i} \ll |E_i|$, E $= \delta_{SM}$. Only cases 1 and 4 meet the programmatic requirements.

Assuming $U_{req} = 5\%D$ for calm water resistance and self-propulsion, resistance and RPS are condition 6, whereas sinkage and trim are condition 3. Assuming U_{req}

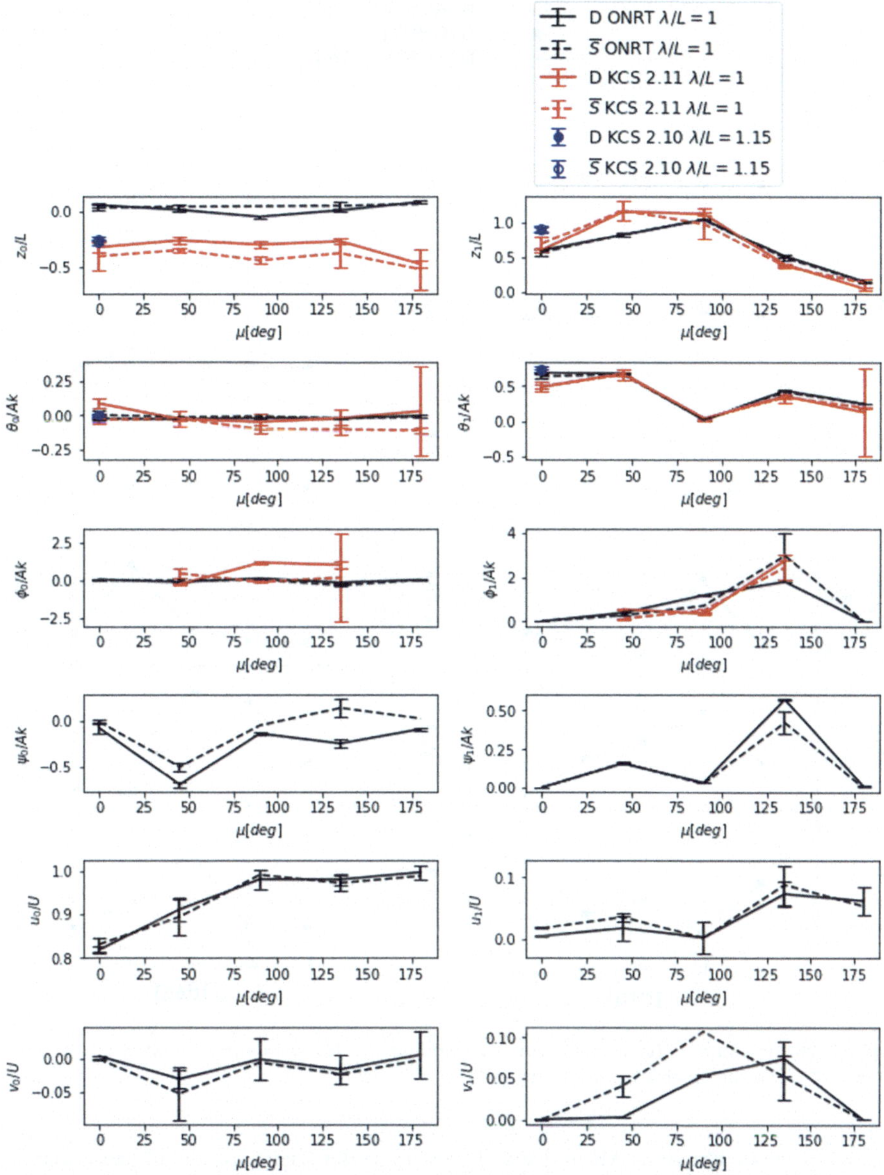

Fig. 37 0th harmonic (left) and1st harmonic amplitude (right) of motions for KCS L = 6.1 m for λ/L = 1.15 and KCS L = 2.7 m and ONRT for λ/L = 1.0

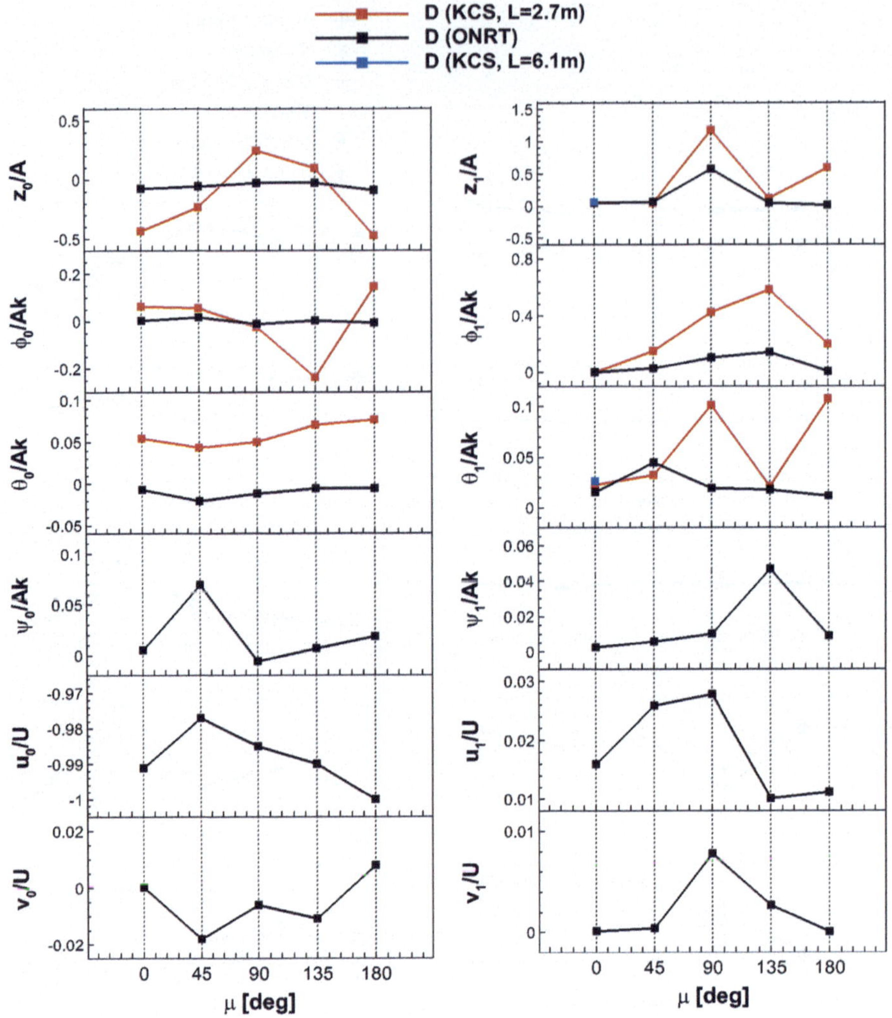

Fig. 38 0th harmonic (left) and1st harmonic amplitude (right) of motions for KCS L = 6.1 m, KCS L = 2.7 m, and ONRT for λ/L = 0.5

= 5%DR for head waves AR and speed loss, 1st order terms are conditions 6 and 5, respectively, and 2nd order terms are condition 3; and for oblique waves all errors and uncertainties are greater than U_{req}. Thus, none of the T2015 test cases were able to achieve a programmatic requirement of 5%D and 5%DR for calm water and waves, respectively. Note that not all the experimental uncertainties are able to meet this requirement either such that both experiments and CFD need to reduce their uncertainties and in addition CFD its errors. Nonetheless, in view the comparable CFD capability for ONRT free running vs. KCS captive conditions, the prognosis for CFD capability is excellent.

Future CFD assessment should extend the current test cases for consideration of added power in regular and irregular waves. Verification should be required, and more limited/most important validation variables should be confirmed and used for the V&V and SoA assessment.

Acknowledgments The Office of Naval Research supports the IIHR research under grants administered by Drs. Thomas Fu, Woei-Min Lin, Ki-Han Kim, and Robert Brizzolara.

References

ASME. (2009). Standard for verification and validation in computational fluid dynamics and heat transfer. *American Society of Mechanical Engineers* (New York).

Bunnik, T., van Daalen, E., Kapsenberg, G., Shin, Y., Huijsmans, R., Deng, G., Delhommeau, G., Kashiwagi, M., & Beck, B. (2010). A comparative study on state-of-the-art prediction tools for Seakeeping. In: *28th Symposium on Naval Hydrodynamics*. Pasadena, California, pp. 12–17.

Eça, L., & Hoekstra, M. (2014). A procedure for the estimation of the numerical uncertainty of CFD calculations based on grid refinement studies. *Journal of Computational Physics, 262,* 104–130.

Hemsch, M. (2004). Statistical analysis of computational fluid dynamics solutions from the drag prediction workshop. *Journal of Aircraft, 41*(1), 95–103.

Larsson, L., Stern, F., & Visonneau, M. (Eds.). (2014). Numerical ship hydrodynamics: An assessment of the gothenburg 2010 workshop. Springer.

Sadat-Hosseini, H., Toxopeus, S., Kim, D., Castiglione, T., Sanada, Y., Stocker, M., Simonsen, C. D., Otzen, J. F., Toda, Y., & Stern, F. (2015). Experiments and computations for KCS added resistance for variable heading, 5th world maritime technology conference november 3–7, Rhode Island Convention & Omni Hotel Providence, Rhode Island, USA.

Simonsen, C. D., Otzen, J. F., Joncquez, S., & Stern, F. (2013). EFD and CFD for KCS heaving and pitching in regular head waves. *Journal of Marine Science and Technology, 18,* 435–459.

Stern, F., Wilson, R. V., Coleman, H. W., & Paterson, E. G. (2001). Comprehensive approach to verification and validation of CFD simulations—Part 1: Methodology and procedures. *Journal of Fluids Engineering—Transactions of the ASME, 123*(4), 793–802.

Stern, F., Olivieri, A., Shao, J., Longo, J., & Ratcliffe, T. (2005). Statistical approach for estimating intervals of certification or biases of facilities or measurement systems including uncertainties. *Journal of Fluids Engineering—Transactions of the ASME, 127*(3), 604–610.

Stern, F., Wilson, R., & Shao, J. (2006). Quantitative approach to V&V of CFD simulations and certification of CFD codes with examples. *International Journal for Numerical Methods in Fluids, 50*(11), 1335–1355.

Stern, F. (2007) Quantitative V&V of CFD solutions and certification of CFD codes with examples for ship hydrodynamics, symposium on computational uncertainty, AVT-147. Athens, Greece.

Stern, F., Diez, M., & Sadat-Hosseini, H. (2016). *Improved statistical approach for certification of CFD codes with examples, verification and validation symposium* (pp. 16–20). Las Vegas, Nevada: The Westin Las Vegas Hotel.

Stern, F., Diez, M., Sadat-Hosseini, H., Yoon, H., & Quadvlieg, F. (2017). Statistical approach for CFD State-of-the-Art assessment: N-Version verification and validation. *ASME Journal of Verification, Validation and Uncertainty Quantification, 2*(3).

Vassberg, J. (2016). Challenges and accomplishments of the AIAA CFD drag prediction workshop series. In: AVT-246 specialists' meeting on progress and challenges in validation testing for computational fluid dynamics. Avila, Spain.

Xing, T., & Stern, F. (2010). Factors of safety for Richardson extrapolation. *Journal of Fluids Engineering-Transactions of the ASME, 132*(6), 061403.

The manufacturer's authorised representative in the EU is Springer
Nature Customer Service Centre GmbH, Europaplatz 3, 69115 Heidelberg,
Germany. If you have any concerns regarding our products, please
contact ProductSafety@springernature.com

Printed and bound by CPI Group (UK) Ltd, Croydon, CR0 4YY
30/09/2026
02225503-0016